CLOSE BINARY STARS: OBSERVATIONS AND INTERPRETATION

INTERNATIONAL ASTRONOMICAL UNION
UNION ASTRONOMIQUE INTERNATIONALE

SYMPOSIUM No. 88

HELD IN TORONTO, CANADA, AUGUST 7–10, 1979

CLOSE BINARY STARS: OBSERVATIONS AND INTERPRETATION

EDITED BY

MIREK J. PLAVEC, DANIEL M. POPPER, and ROGER K. ULRICH
University of California, Department of Astronomy
Los Angeles, California, U.S.A.

D. REIDEL PUBLISHING COMPANY
DORDRECHT : HOLLAND / BOSTON : U.S.A. / LONDON : ENGLAND

Library of Congress Cataloging in Publication Data
Main entry under title:

Close Binary Stars.

(Symposium – International Astronomical Union; No. 88)
Sponsored by the International Astronomical Union.
Includes index.
1. Stars, Double–Congresses. I. Plavec, Miroslav. II. Popper, D. M. III. International Astronomical Union. IV. Series: International Astronomical Union. Symposium; No. 88.
QB821.C59 523.8′41 80-16083
ISBN 90-277-1116-X
ISBN 90-277-1117-8 pbk.

Published on behalf of
the International Astronomical Union
by
D. Reidel Publishing Company, P.O. Box 17, 3300 AA Dordrecht, Holland

Sold and distributed in the U.S.A. and Canada
by Kluwer Boston Inc., Lincoln Building,
160 Old Derby Street, Hingham, MA 02043, U.S.A.

In all other countries, sold and distributed
by Kluwer Academic Publishers Group,
P.O. Box 322, 3300 AH Dordrecht, Holland

D. Reidel Publishing Company is a member of the Kluwer Group.

Printed in The Netherlands

TABLE OF CONTENTS

Preface and Acknowledgements xiii
Organizing Committees xv
List of Participants xvii

I. GENERAL REVIEWS AND SURVEYS

M. J. Plavec — Introductory Address — 3

L. B. Lucy — The Formation of Close Binaries — 7

A. V. Tutukov, L. R. Yungelson — Statistical Investigation of Spectroscopic Binary Stars — 15

L. Mantegazza, P. Paolicchi, P. Farinella, F. Luzny — The Period Distribution of Close Binary Systems — 23

R. Rajamohan, P. Venkatakrishnan — Rotation in Close Binaries — 27

D. M. Gibson — Binarity as a Factor in Stellar Radio Emission — 31

A. K. Dupree, L. Hartmann, J. C. Raymond — Ultraviolet Spectroscopy of Binary Systems — 39

F. Fekel, C. H. Lacy, J. Tomkin — Observations of Spectroscopic Binaries with a Solid-State Detector — 53

J. D. Dorren, M. J. Siah, E. F. Guinan, G. P. McCook — Hydrogen-Line Photometry of Close Binaries — 59

I. S. McLean — Polarimetry of Close Binary Systems: Observations and Interpretations — 65

C. Aspin, J. C. Brown, J. F. L. Simmons — The Effect of Orbital Eccentricity on Polarimetric Binary Diagnostics — 71

S. A. H. Smith, A. C. Theokas — A New Method of Light Curve Analysis: Application to V 444 Cygni — 73

II. DETACHED SYSTEMS

J. Andersen, J. V. Clausen, B. Nordström	Determination of Absolute Dimensions of Main Sequence Binaries	81
T. J. Herczeg	Observed Period Changes in Detached Binary Systems	89
M. Kitamura	Surface Distribution of Chemical Anomalies of Ap Components in Detached Close Binaries	95

III. MASS TRANSFER AND MASS LOSS - THEORY

J. P. De Grève, D. Vanbeveren	Close Binaries Before and After Mass Exchange: A Comparison of Observations through Theoretical Computations	103
Th. J. Van der Linden	Evolution of Intermediate Mass Close Binaries	109
D. Vanbeveren, C. de Loore	The Evolution of Massive Close Binary Systems: Estimates of the Parameters Governing Mass- and Angular Momentum Loss	115
J. L. Modisette, Y. Kondo	Mass Transfer between Binary Stars	123
R. F. Webbink	On the Onset of Contact in Early-Type Binaries	127
D. Sugimoto, S. Miyaji	Evolution of Early-Type Contact Binaries	133
K. Nomoto, K. Nariai, D. Sugimoto	Formation of Luminous Contact Binaries by Rapid Accretion onto White Dwarfs	139
F. Meyer, E. Meyer-Hofmeister	Do Blue Stragglers Form by Common Envelope Evolution?	145
J. K. West, K.-Y. Chen	The Evolution of Pre-Main Sequence Close Binaries	149
G. T. Bath	Cataclysmic Variables, Hubble-Sandage Variables, and η Carinae	155

IV. MASSIVE BINARY SYSTEMS

C. D. Garmany, P. S. Conti	Anomalous Mass Ratios in O - Star Binaries	163
D. Vanbeveren	Observational and Evolutionary Aspects of Wolf-Rayet Stars	169
V. S. Niemela	Observations of Wolf-Rayet Binaries	177
A. F. J. Moffat, W. Seggewiss	Wolf-Rayet Binaries	181
P. Massey	HD 192641 and HD 193077: Two Single, Hydrogen-Rich Wolf-Rayets?	187
A. M. Cherepaschuk, J. A. Eaton, Kh. F. Khaliullin	The Interpretation of Light Curves λλ 2460 - 35 000 A of the Eclipsing Wolf-Rayet Binary V 444 Cygni	193
H. Drechsel, Y. Kondo, J. Rahe, G. E. McCluskey	Mass Flow Analysis of the Ultraviolet Spectrum of UW Canis Majoris	195
R. Poeckert	The Be Star φ Persei	199

V. ALGOL AND ALGOLS

D. M. Popper	Spectroscopic Orbits of Algols: A Progress Report	203
G. Giuricin, F. Mardirossian, M. Mezzetti	Mass Loss and Mass Transfer in Algols	207
W. Packet	Accretion and Chemical Structure of Algols	211
S. Söderhjelm	What Do We Know About Algol?	217
M. Kondo, A. Okazaki	Abundance of Sodium in the Atmospheres of Algol A and B	221
S. M. Rucinski	Rotation of the Primary Component of Algol	223
C. T. Bolton, D. J. Zubrod	Hα Emission Line Profiles in the Spectrum of Algol	225
D. R. Florkowski	On the Location of Radio and X-Ray Emission in the Algol System	229

R. H. Kaitchuk, R. K. Honeycutt, S. L. Mufson	Time-Resolved Spectroscopy of RW Tauri and X Persei	233
Y. Kondo, G. E. McCluskey, R. E. Stencel	IUE Ultraviolet Spectra of the Interacting Binary U Cephei	237
E. C. Olson	Transient Disks and Mass Transfer in U Cephei	243
V. Piirola	Polarization Observations of U Cephei	249
M. J. Plavec	IUE Observations of Long Period Eclipsing Binaries: A Study of Accretion onto Non-Degenerate Stars	251
R. E. Wilson, L. W. Twigg	On the Existence of Double-Contact Binaries	263
S. Kríz	Structure of the Gaseous Disk in RX Cas	269
M. Hack, U. Flora, P. Santin	IUE Observations of the Peculiar Binary Systems β Lyrae and υ Sagittarii	271
G. J. Peters	Hα Observations of Algol-Type Interacting Binary Systems	287
R. S. Polidan, G. J. Peters	Ultraviolet Observations of Close Binary Stars	293
E. Budding	Narrow Band Photometry and Observable Effects in Algol-Type Systems	299
K. Walter	On the Location of Gas Streams in Algol Systems	305

VI. X-RAY BINARIES

D. Crampton	X-Ray Binary Systems	313
D. Sugimoto, S. Miyaji	Non-Conservative Mass Exchange and Origin of X-Ray Close Binaries	323
E. Ergma, T. Tutukov	Thermonuclear Burning in the Envelope of an Accreting Neutron Star	329
J. Ziolkowski	Are Most of X-Ray Binaries Recurrent Transients?	335

J. F. L. Simmons, C. Aspin, J. C. Brown	A Critique of the Polarimetric Evidence on the Nature of CBS	343
G. D. Nicolson	Radio Flares in the Circinus X-1 System	347
R. Kippenhahn, H. Ritter, H. U. Schmidt, H.-C. Thomas	Optical and Ultraviolet Observations of HZ Herculis	349
C. T. Bolton, D. R. Gies	Re-Evaluation of the Orbital Elements of HDE 226868 = Cygnus X-1	355
M. Pakull	Optical Observations of X-Ray Sources in the LMC	357
E. J. Zuiderwijk	Variations in the Profiles of Hβ and He II 4686 A in Vela X-1	359
J. D. Dorren, E. F. Guinan	Evidence for a 6 Year Periodicity in X Persei	361
T. S. Galkina	Periodical Variations in the Spectrum of X Persei. Discovery of a 22-Hour Period in the Variations of the Radial Velocities of Hα	367
G. W. Collins, G. H. Newsom, R. N. Boyd	SS 433 as a Magnetic X-Ray Binary	375

VII. THE RS CANUM VENATICORUM STARS

D. S. Hall, J. M. Kreiner, S. N. Shore	Period Changes and Mass Loss Rates in 34 RS CVn-Type Binaries	383
S. N. Shore, D. S. Hall	A Starspot Model for the RS CVn Stars	389
F. M. Walter, P. A. Charles, S. Bowyer	The Coronae of RS CVn Systems	397
P. A. Feldman	Radio Flare Activity of RS CVn Stars	403
S. Catalano, A. Frisina, M. Rodonò	Out-of-Eclipse Light Variation of RS CVn Stars	405

D. A. Fraquelli — Spectrophotometry of the Hα Region in HR 5110 — 413

S. M. Jakate — The Star Spot Model and the Observations of SZ Piscium — 415

E. F. Milone, S. A. Naftilan — Remarks on the Complicated Close Binary Systems RT Lacertae, CG Cygni, and RW Coronae Borealis — 419

E. H. Geyer — Observational Evidence for Large-Scale Spot Activity on the Primary Component of XY Ursae Majoris — 423

VIII. CATACLYSMIC VARIABLES AND POLARS

I. G. Mitrofanov — The Nature of the Difference Between Novae, Dwarf Novae, and Novalikes — 431

G. T. Bath, C. C. Brunt, P. P. Eggleton, A. C. Fabian, R. Harkness, S. K. Mayo, J. Papaloizou, J. E. Pringle, M. Rayne, G. S. Savonije, J. A. J. Whelan, C. A. Whyte, D. T. Wickramasinghe — Current Work on Cataclysmic Variables at Oxford and Cambridge — 439

J. Smak — Circumstellar Matter in Cataclysmic Binaries — 443

A. V. Holm, C.-C. Wu, W. M. Sparks, F. H. Schiffer, A. Boggess — IUE Observations of WZ Sagittae in Outburst — 447

I. G. Mitrofanov — The Cyclotron Lines in the Optical Spectra of the Stars of AM Herculis Group: Observations and Interpretation — 453

J. C. Raymond, G. Branduardi, A. K. Dupree, G. Fabbiano, L. Hartmann — Ultraviolet Observations of AM Herculis — 467

S. Tapia, G. V. Coyne — Upper Limits to the Luminosity of a dM Companion in AM Herculis Objects — 471

A. M. Cherepaschuk, E. S. Dmitrienko — Observational Structure of Disk-like Envelope in the System DQ Her — 473

IX. CONTACT BINARIES

F. H. Shu	Theories of Contact Binary Stars	477
L. Anderson, M. Raff, F. H. Shu	Rotation Broadening and the Shapes of W Ursae Majoris Stars	485
O. Vilhu, T. Rahunen	W UMa Stars and Angular Momentum Loss	491
D. H. Smith, R. C. Smith, J. A. Robertson	Energy Transfer by Circulation in W Ursae Majoris Systems	495
J. A. Eaton, C.-C. Wu, S. M. Rucinski	Low Color Temperature Variations over the Common Envelope of W UMa and the Cause of Its W-Type Light Curve	501
A. P. Linnell	The Physical Status of VW Cephei	505
K. Nariai	Mass Loss from Contact Binary Systems and their Dynamical Evolution	511
F. van't Veer	Solar Type Contact Binaries - The Initial Mass Ratio Distribution Function	517
O. Havnes	On Sudden Period Changes in Contact Binaries	521
K.-C. Leung	Current Observational Development in Early-Type Contact Binary Systems	527

X. ... AND THE REST (SYMBIOTICS, SUPERGIANTS, PLANETARIES, POPULATION II SYSTEMS)

C. D. Keyes, M. J. Plavec	The Symbiotic Binary System AG Pegasi	535
M. Friedjung	IUE Observations of Z Andromedae	543
D. Chochol, J. Grygar, L. Hric	A Binary Model for the Symbiotic Star V 1329 Cygni	547
R. Faraggiana	The UV Spectrum of VV Cep	549
R. E. Stencel, Y. Kondo, A. P. Bernat, G. E. McCluskey	Ultraviolet Observations of 31 and 32 Cygni	555

R. F. Webbink	Close Binary Systems in Globular Clusters: A Preliminary Report	561
R. H. Méndez	A Search for Spectroscopic Binaries Among Planetary Nuclei	567
J. Saltzman, M. Livio, G. Shaviv	The Formation of Planetary Nebulae with Close Binary Nuclei	571
C. T. Bolton, J. R. Thomson	HD 137569, a Population II Remnant of Mass Exchange Evolution	575
J. Smak	Concluding Remarks	577
R. K. Ulrich	Postscript: Some Thoughts on Interacting Binary Systems	581

Subject index 585

Index of Stars and other objects 589

PREFACE

Symposium No. 88 of the International Astronomical Union was devoted to a comprehensive review of all types of close binary stars. The nine sessions were organized according to the type of the objects to be discussed. We have preserved this system, but assembled twelve papers of a more general character (reviews and surveys) into a special chapter placed at the beginning of this book.

We would like to remind the reader that the Symposium was preceded by the IAU Colloquium No. 53 on White Dwarfs and Variable Degenerate Stars, and that the sessions on Cataclysmic Variables and related topics at the two meetings supplemented each other.

The discussion in Toronto was carefully recorded by Mr. Robert Gauthier. We also wish to thank Mr. Robert O'Daniel, Ms. Joan Kaufmann, and Ms. Linda Reimers for assisting us with the editorial work.

M. J. Plavec
D. M. Popper
R. K. Ulrich

ACKNOWLEDGEMENTS

The International Astronomical Union and the University of Toronto contributed substantial funds enabling a number of participants to receive travel grants.

The Symposium was sponsored by the following Commissions of the International Astronomical Union: 29 (Stellar Spectra), 30 (Radial Velocities), 35 (Stellar Structure), 42 (Close Binary Stars), and 44 (Astronomy from Space).

Our special thanks go to the members of the Scientific Organizing Committee.

The Symposium was attended by 170 participants from 26 countries. All the local organization was on the shoulders of the Local Organizing Committee, chaired by Dr. C. T. Bolton. Their work was splendid, and our sincere thanks are offered to all of them.

Last but no least, let me thank Mrs. N. M. Pols v.d. Heijden and the rest of the staff of the D. Reidel Publishing Company for their cooperation.

M. J. Plavec

M. J. Plavec, D. M. Popper and R. K. Ulrich (eds.), Close Binary Stars: Observations and Interpretation, xiii.

ORGANIZING COMMITTEES

SCIENTIFIC ORGANIZING COMMITTEE

M. J.	Plavec (Chairman)
A. H.	Batten
C. T.	Bolton
A. M.	Cherepaschuk
R.	Giacconi
M.	Hack
M.	Kitamura
Y.	Kondo
G.	Larsson-Leander
B.	Paczynski
F. H.	Shu
B.	Warner
J. A. J.	Whelan

LOCAL ORGANIZING COMMITTEE

C. T.	Bolton (Chairman)
J. R.	Percy
Gail	Archer
Dennis	Crabtree
Robert	Gauthier
Doug	Gies
Gerry	Grieve
Ron	Lyons
Pamela	Sullivan
Linda	Twitchin
Joan	Tryggve
Donna	Zubrod-Grieve

SESSION CHAIRMEN

G.	Larsson-Leander:	Tuesday morning
F. H.	Shu	Tuesday afternoon
T. J.	Herczeg	Wednesday morning
M.	Hack	Wednesday afternoon
C. T.	Bolton	Wednesday evening
Y.	Kondo	Thursday morning
D. M.	Popper	Thursday afternoon
J. A. J.	Whelan	Friday morning
M. J.	Plavec	Friday afternoon

OFFICIAL BANQUET SPEAKERS

J. D.	Fernie
F. B.	Wood

LIST OF PARTICIPANTS

C. AIKMAN, Dominion Astrophysical Observatory, Victoria, Canada
H.M.K. AL-NAIMY, Astronomical Observatory, Baghdad, Iraq
J. ANDERSEN, Copenhagen University Observatory, Tølløse, Denmark
L.S. ANDERSON, University of Toledo, Toledo, USA
C. ASPIN, University of Glasgow, Glasgow, Scotland
P. BOJETCHKO, University of Toronto, Toronto, Canada
C.T. BOLTON, David Dunlap Observatory, Richmond Hill, Canada
B.B. BOOKMYER, Clemson University, Clemson, USA
P. BROGLIA, Osservatorio Milano-Merate, Merate, Italy
J.C. BROWN, University of Glasgow, Glasgow, Scotland
C.C. BRUNT, Institute of Astronomy, Cambridge, England
E. BUDDING, University of Manchester, Manchester, England
R. CANAL, Universidad de Barcelona, Barcelona, Spain
S. CATALANO, Catania Astrophysical Observatory, Catania, Italy
C.R. CHAMBLISS, Kutztown State College, Kutztown, USA
W.Y. CHAU, Queen's University, Kingston, Canada
K.-Y. CHEN, University of Florida, Gainesville, USA
K.J. CHEON, Korean University, Tokyo, Japan
A.M. CHEREPASHCHUK, Sternberg Astronomical Institute, Moscow, USSR
T.T. CHIA, St. Mary's University, Halifax, Canada
D. CHOCHOL, Astronomical Institute, Slovak Academy of Sciences, Tatranska Lomnica, Czechoslovakia
J.V. CLAUSEN, Copenhagen University Observatory, Tølløse, Denmark
C. CLEMENT, University of Toronto, Toronto, Canada
M. CLEMENT, University of Toronto, Toronto, Canada
G.W. COLLINS, II, Ohio State University, Columbus, USA
D. CRABTREE, University of Toronto, Toronto, Canada
R. CRAWFORD, University of Illinois, Urbana, USA
J.-P. DeGREVE, Vrije Univ. Brussel, Brussels, Belgium
M. de GROOT, Armagh Observatory, Armagh, Northern Ireland
C.W.H. de LOORE, Vrije Univ. Brussel, Brussels, Belgium
J. DOMMANGET, Observatoire Royal de Belgique, Brussels, Belgium
J.D. DORREN, Biruni Observatory, Shiraz, Iran
M.J. ECCLES, Royal Observatory of Edinburgh, Edinburgh, Scotland
R. FARAGGIANA, Osservatorio Astron. di Trieste, Trieste, Italy
P. FARINELLA, Osservatorio Milano-Merate, Merate, Italy
F. FEKEL, University of Texas, Austin, USA
P. FELDMAN, Herzberg Institute of Astrophysics, Ottawa, Canada
J.D. FERNIE, David Dunlap Observatory, Richmond Hill, Canada
A. FINZI, Israel Institute of Technology, Haifa, Israel
D.R. FLORKOWSKI, University of Florida, Gainesville, USA
M. FRACASSINI, University of Milan, Milan, Italy
D.A. FRAQUELLI, David Dunlap Observatory, Richmond Hill, Canada
W. FREEMAN, University of Toronto, Toronto, Canada
M. FRIEDJUNG, Institut d'Astrophysique, Paris, France
M. FUJIMOTO, Niigata University, Niigata, Japan
T.S. GALKINA, Crimean Astrophysical Observatory, Nauchnyi, USSR
K. GARMANY, JILA, Boulder, USA
R.F. GARRISON, David Dunlap Observatory, Richmond Hill, Canada

R. GAUTHIER, David Dunlap Observatory, Richmond Hill, Canada
E.H. GEYER, Obs. Hoher List, Daun, Federal Republic of Germany
D. GIES, David Dunlap Observatory, Richmond Hill, Canada
A. GIMENEZ, Instituto de Astrofisica de Andalucia, Granada, Spain
G. GIURICIN, Osservatorio Astron. di Trieste, Trieste, Italy
P.C. GREGORY, University of British Columbia, Vancouver, Canada
G. GRIEVE, David Dunlap Observatory, Richmond Hill, Canada
E.F. GUINAN, Villanova University, Villanova, USA
A.F. GULLIVER, University of Alberta, Edmonton, Canada
M. HACK, Osservatorio Astron. di Trieste, Trieste, Italy
D.S. HALL, Vanderbilt University, Nashville, USA
O. HAVNES, University of Tromsø, Tromsø, Norway
T.J. HERCZEG, University of Oklahoma, Norman, USA
A. HOLM, Goddard Space Flight Center, Greenbelt, MD
Y. HOSOKAWA, Yamagata University, Yanagata-shi, Japan
D.P. HUBE, University of Alberta, Edmonton, Canada
H. HUBERT, Astronomical Institute, Amsterdam, The Netherlands
D.H.P. JONES, Royal Greenwich Observatory, Hailsham, England
R. KAITCHUK, Indiana University, Bloomington, USA
S. KAWABATA, Kyoto Gakuen University, Kyoto, Japan
J.C. KEMP, University of Oregon, Eugene, USA
C.D. KEYES, University of California, Los Angeles, USA
M. KITAMURA, Tokyo Astronomical Observatory, Tokyo, Japan
Y. KONDO, Goddard Space Flight Center, Greenbelt, USA
J. KRAUTTER, Landessternwarte Heidelberg, Koenigstuhl, Heidelberg, Federal Republic of Germany
J.M. KREINER, Jagiellonian University, Krakow, Poland
S. KRIZ, Charles University, Prague, Czechoslovakia
W. KRZEMINSKI, N. Copernicus Astronomical Center, Warszawa, Poland
K.K. KWEE, Sterrewacht Leiden, Leiden, The Netherlands
C. LACY, University of Texas, Austin, USA
M. LANE, David Dunlap Observatory, Richmond Hill, Canada
E.L. GOMAR, University of South Florida, Tampa, USA
G. LARSSON-LEANDER, Lund Observatory, Lund, Sweden
K.-C. LEUNG, University of Nebraska, Lincoln, USA
A.P. LINNELL, Michigan State University, East Lansing, USA
L.B. LUCY, Columbia, University, New York, USA
J. MacDONALD, University of Sussex, Falmer, England
J. MADEJ, Warsaw University Observatory, Warszawa, Poland
L. MANTEGAZZA, Osservatorio Milano-Merate, Merate, Italy
P. MASSEY, University of Colorado, Boulder, USA
S. MAYO, Royal Greenwich Observatory, Hailsham, England
T. MAZEH, Tel Aviv University, Ramat Aviv, Israel
I.S. McLEAN, University of Arizona, Tucson, USA
R.H. MENDEZ, Instituto de Astronomica y Fisica del Espacio, Buenos Aires, Argentina
J.E. MERRILL, University of Florida, Gainesville, USA
F. MEYER, MPIfA, Munich, Federal Republic of Germany
E. MEYER-HOFMEISTER, MPIfA, Munich, Federal Republic of Germany

E.F. MILONE, University of Calgary, Calgary, Canada
R. MITALAS, University of Western Ontario, London, Canada
I.G. MITROFANOV, USSR Academy of Sciences, Leningrad, USSR
A.F.J. MOFFAT, University of Montreal, Montreal, Canada
D. MORTON, Anglo-Australian Observatory, Epping, Australia
K. NARIAI, Tokyo Astronomical Observatory, Tokyo, Japan
G.D. NICOLSON, NITR, Johannesburg, South Africa
V. NIEMELA, Instituto de Astronomica y Fisica del Espacio, Buenos Aires, Argentina
K. NOMOTO, Ibaraki University, Mito, Japan
E.C. OLSON, University of Illinois, Urbana, USA
W. PACKET, Vrije Univ. Brussel, Brussels, Belgium
M.W. PAKULL, ESO, Geneva, Switzerland
P. PAOLICCHI, Osservatorio Milano-Merate, Merate, Italy
J. POPALOIZOU, Institute of Astronomy, Cambridge, England
L.E. PASINETTI, Osservatorio Milano-Merate, Merate, Italy
J.R. PERCY, University of Toronto, Toronto, Canada
G.J. PETERS, University of Southern California, Los Angeles, USA
V. PIIROLA, University of Helsinki, Helsinki, Finland
M.J. PLAVEC, University of California, Los Angeles, USA
R. POECKERT, Dominion Astrophysical Observatory, Victoria, Canada
R.S. POLIDAN, Princeton University, Princeton, USA
D.M. POPPER, University of California, Los Angeles, USA
J.E. PRINGLE, Institute of Astronomy, Cambridge, England
J. RAHE, Remeis-Observatory, Bamberg, Federal Republic of Germany
T. RAHUNEN, University of Helsinki, Helsinki, Finland
R. RAJOMOHAN, Indian Institute of Astrophysics, Bangalore, India
H. RITTER, MPIfA, Munich, Federal Republic of Germany
J.A. ROBERTS, CSIRO, Epping, Australia
J.A. ROBERTSON, University of Sussex, Falmer, England
M. RODONO, Catania Astrophysical Observatory, Catania, Italy
S. RUCINSKI, Warsaw University Observatory, Warszawa, Poland
A.R. SADIK, Astronomical Observatory, Baghdad, Iraq
H.B. SANWAL, Osmania University, Hyderabad, India
C.D. SCARFE, University of Victoria, Victoria, Canada
H.J. SCHOBER, Institut fuer Astronomie, Graz, Austria
E.R. SEAQUIST, University of Toronto, Toronto, Canada
W. SEGGEWISS, Obs. Hoher List, Daun, Federal Republic of Germany
M. SHARA, University of Montreal, Montreal, Canada
F.H. SHU, University of California, Berkeley, USA
J. SIMMONS, University of Glasgow, Glasgow, Scotland
J. SMAK, N. Copernicus Astronomical Center, Warszawa, Poland
D.H. SMITH, University of Sussex, Falmer, England
R.C. SMITH, University of Sussex, Falmer, England
S. SOEDERHJELM, Lund Observatory, Lund, Sweden
J.-E. SOLHEIM, University of Tromsø, Tromsø, Norway
R.E. Stencel, Goddard Space Flight Center, Greenbelt, USA
D. SUGIMOTO, University of Tokyo, Tokyo, Japan
S. TAPIA, University of Arizona, Tucson, USA

A.R. TAYLOR, University of British Columbia, Vancouver, Canada
A.C. THEOKAS, University of Manchester, Manchester, England
D. TURNER, David Dunlap Observatory, Richmond Hill, Canada
A.V. TUTUKOV, USSR Academy of Sciences, Moscow, USSR
D. VanBEVEREN, Vrije Univ. Brussel, Brussels, Belgium
T.J. van der LINDEN, Astronomical Institute, Amsterdam, The Netherlands
W. Van HAMME, Rijksuniv., Gent, Belgium
J. van PARADIJS, Massachusetts Institute of Technology, Cambridge, USA
F. VAN'T VEER, Institut d'Astrophysique, Paris, France
O. VILHU, University of Helsinki, Helsinki, Finland
F.M. WALTER, University of California, Berkeley, USA
K. WALTER, University of Tuebingen, Tuebingen, Federal Republic of Germany
B. WARNER, University of Cape Town, Rondebosch, South Africa
R.F. WEBBINK, University of Illinois, Urbana, USA
J.K. WEST, University of Florida, Gainesville, USA
J.A.J. WHELAN, Institute of Astronomy, Cambridge, England
N.E. WHITE, Goddard Space Flight Center, Greenbelt, USA
C.A. WHYTE, Institute of Astronomy, Cambridge, England
D.T. WICKRAMASINGHE, Royal Observatory of Edinburgh, Edinburgh, Scotland
R.M. WILLIAMON, Fernback Science Center, Atlanta, USA
R.E. WILSON, University of Florida, Gainesville, USA
G.W. WOLF, Southwest Missouri State University, Springfield, USA
F.B. WOOD, University of Florida, Gainesville, USA
J. ZIOLKOWSKI, N. Copernicus Astronomical Center, Warsaw, Poland
D. ZUBROD GRIEVE, David Dunlap Observatory, Richmond Hill, Canada
E.J. ZUIDERWIJK, ESO, Geneva, Switzerland

I. GENERAL REVIEWS AND SURVEYS

INTRODUCTORY ADDRESS

Mirek J. Plavec
Department of Astronomy, University of California
Los Angeles

Four years ago, in 1975, we met in Cambridge, England for our first IAU Symposium devoted entirely to binary stars. Most of the talks given there were theoretical, and at the end some of us felt that it would be appropriate to organize yet another Symposium soon, this time oriented more towards observations and their immediate interpretation. This is why we have come together here in Toronto. Four years is not a long interval of time, and few disciplines of astronomy repeat their Symposia on such a short time scale. While the need for another Symposium was clear to the binary stars investigators, it is not so obvious to other colleagues. We must therefore be very grateful to the Executive Committee of the International Astronomical Union for their great understanding of our needs. This understanding went far beyond approving and sponsoring this meeting: We have received a substantial financial assistance, which all went to the support of the travel expenses of some of the participants. Many more travel grants were possible thanks to a most generous support by the University of Toronto. I would like to thank both institutions.

Four years ago I said in Cambridge that the choice of the place had been most appropriate for a meeting dealing with the structure and evolution of stars: there certainly exists in Cambridge a genius loci maintaining the tradition of such studies. It is again not a mere politeness when I say now that the choice of Canada, and specifically of Toronto, is equally fortunate and symbolical for a meeting intended to deal mainly with observations. The tradition of Canadian astronomy is rich with work on binary stars and related subjects, and Toronto certainly shares with Victoria a prominent place in this tradition. Let us remember at least a few names of those who not long ago would have been among us here: Robert M. Petrie, John F. Heard, Ruth Northcott, and Carlyle S. Beals. It is so very sad to miss them here... Fortunately, Canadian astronomy continues to flourish in many disciplines, including binary star astronomy, and this makes our stay here doubly pleasant.

I am not sure that Dr. Beals -- who to our sorrow passed away just

M. J. Plavec, D. M. Popper and R. K. Ulrich (eds.), Close Binary Stars: Observations and Interpretation, 3–5.

a few weeks ago -- ever considered himself an investigator of binary stars. Probably not. Yet his important work on early-type stars with emission lines has become a vital part of our effort to understand interacting binary stars. I think we can say that the discipline of binary star studies has, in the past decade or two, considerably overflown its original boundaries. In a recent textbook on Astrophysics and Stellar Astronomy, the discipline of binary stars is introduced as one of the "service" areas of astronomy, i.e. "areas whose main importance lies in the gathering of data for other areas". There is no doubt that studies of simple, detached binary systems provide most of the fundamental data on stellar characteristics, and it would be hard to overestimate the value of this service to astronomy. In general, however, I am of the opinion that we supply astrophysics more with problems and less with data. Perhaps our discipline is like the various service areas I had to deal with, for example with my car, where they fix a few troubles and at the same time create a few new ones.

I think that it is the most attractive feature of binary systems that they create the most puzzling, exciting and interesting objects in astrophysics. It appears that without interacting binaries, we would hardly enjoy the spectacular phenomena of novae, dwarf novae, symbiotic variables, of most if not all Wolf-Rayet stars, and of course of all the beautifully regular, interestingly distorted, or outright weird eclipses! Most of the complex phenomena we observe are somewhat related to mass transfer or mass loss in interacting systems. It is hard to believe that it has been only 12 years since the topic of mass transfer was first discussed at a scientific meeting (at Uccle in 1967). (I do not claim that the discipline was born there; there were some early pioneers, for example Donald Morton and Joe Smak, who are with us here, who made some first calculations several years before that, and many observers and theoreticians well before them again who pointed the way). Nevertheless, over the past twelve years the concept of mass transfer and mass loss expanded so much and enveloped so may different objects that we can say with confidence that this concept has become of the most fruitful ideas in astrophysics ever conceived.

Combined with this victorious campaign of theoretical explorations is the enormous progress of observational methods, including X-ray studies, far ultraviolet spectroscopy, radio observations, high time-resolution photometry, and great advances in the spectroscopy of the optical spectral region, and the infrared. This brought so many participants to this Symposium, and created an unusually heavy demand on time. We had to ask most of the contributing speakers not to exceed ten minutes, which certainly is a very restrictive rule. I think I must abide by it myself, and conclude this Introduction as fast as possible.

Permit me to make only two brief comments in the hope that they may be useful in the subsequent talks and discussions, since they are meant to clarify a bit of terminology. There is an endless confusion in the use of the terms <u>primary</u> and <u>secondary</u> for the respective components of binary systems. The primary star may be either the one of greater surface brightness, or greater visual magnitude, or greater bolometric magnitude, or larger current mass, or larger initial mass, etc. Since in most cases we will be dealing with mass transfer and mass loss, I propose that we label the components simply the loser and the gainer, respectively, thereby often evading the confusion. You will hear papers here claiming that most mass leaves the system rather than being captured by the gainer. This does not affect my proposed scheme. It simply means that we might often have the Las Vegas type of mass transfer, where the losers lose a lot while the gainers gain very little.

My second comment deals with the classification of eclipsing binaries. The old scheme distinguishing the Algols, β Lyrae stars, and W UMa stars is clearly obsolete. It was based on the outward appearance of the light curve in the visual or photographic domains. Long ago Kopal (1955) and myself (Plavec, 1964) argued that this scheme is misleading, since it lumps together systems of a very different character and evolutionary stage. Kopal showed that the constant light between eclipses, typical for the Algol class, disappears in actual Algols if we go sufficiently far to longer wavelengths to make the light of the distorted loser perceptible. Then these stars, including Algol itself, become β Lyrae objects. Perhaps for the sake of cosmic justice, it appears now that -- from the evolutionary point of view -- β Lyrae is probably an Algol-type system!

I think that the two classes, the Algols and the W UMa systems, have survived the test of time and should be preserved but partly redefined. The term Algols should describe systems which underwent the first phase of mass transfer, are now semi-detached, and the less massive component is of later spectral type. W UMa stars are contact systems of short period and with spectral types F or later (the spectral boundary may be disputable here). The β Lyrae class is in my opinion useless and produces only confusion. The old classification criterion of a marked ellipticity effect between eclipses was done away with by Kopal. β Lyrae, though not unique (see my talk at this Symposium), is sufficiently exotic and cannot be invoked as a prototype of a large class of systems. Our criteria should in the first place be based on structure and evolution.

It is time to start the regular session. Let me welcome most cordially all the participants, and express our thanks to our hosts for organizing this meeting.

REFERENCES

Kopal, Z.: 1955, Ann. Astrophys. 18, 379.
Plavec, M.: 1964, Bull. Astron. Czechoslovakia 15, 156.
Swihart, T.L.: 1968, "Astrophysics and Stellar Astronomy", p. viii.

THE FORMATION OF CLOSE BINARIES

L. B. Lucy
Department of Astronomy
Columbia University
New York, New York 10027, U.S.A.

INTRODUCTION

Although the formation of close binaries has been the subject of numerous investigations in the last 100 years (for a recent review, see Tassoul 1978), we cannot yet claim decisive observational support for any theory, a state of affairs probably reflecting insufficient development of the various theories, rather than a failure to propose the correct mechanism(s). If this is indeed so, we can look forward to significant progress soon, since numerical techniques now allow us - albeit crudely - to follow through to completion binary formation by the two most likely mechanisms, fission and fragmentation. (Herein a binary will be described as having formed by <u>fission</u> if it results from the bifurcation of a rotating protostar as it contracts <u>quasi-statically</u> towards the main sequence and by <u>fragmentation</u> if it results from a rotating protostar's break up into two or more components during, or immediately following, a phase of <u>dynamical</u> collapse).

In investigating these theories numerically, one is tempted to claim qualitative agreement with observation if a solution of the relevant equations has a model binary as its outcome. The consequences of the necessarily limited spatial resolution of three-dimensional gas dynamic calculations suggest, however, that we be more demanding. The effective addition of unintended physical effects as a result of the non-negligible errors of difference approximations is one consequence; another is the suppression of genuine, and perhaps crucial, effects that are implicit in the original equations but which involve short length scales. Thus, even if we adopt the right equations, impose the correct boundary conditions, and provide appropriate initial conditions, the final outcome of our calculations might not correspond closely to reality: a binary might form when it shouldn't and not form when it should.

Without discussing the merits of other observational tests, the view taken here will be that a critical test of the value of numerical simulations is that they provide an understanding of the formation of

M. J. Plavec, D. M. Popper and R. K. Ulrich (eds.), Close Binary Stars: Observations and Interpretation, 7–14.

close binaries with identical components (Lucy and Ricco 1979). This then is the context in which recent 3-D calculations will be discussed in the remainder of this paper.

FISSION

Calculations of binary formation by fission have been carried out by the present author (Lucy 1977) using a somewhat novel technique - the finite-size particle scheme - which has the particular merit of behaving well at low spatial resolution. Although numerical considerations dictated some modifications to the ideal problem, these calculations seem to indicate that binaries formed by fission will have small mass ratios ($q \sim 0.2$), a result not inconsistent with the classical investigations, according to which a binary is the evolutionary end-point of a sequence of pear-shaped figures. On this basis, Lucy and Ricco (1979) conclude that fission can only be responsible for the $q=1$ binaries if a post-formation dynamical mass-exchange instability brings them to $q=1$.

Gingold and Monaghan (1978, 1979), also using the finite-size particle scheme, have studied the fission of damped rotating polytropes. Their starting models rotate so rapidly that they have positive total energy; but the ensuing radial motions are damped artificially, so that the later configurations are gravitationally bound. By this device, models are brought to the onset of fission without first evolving along a quasi-static contraction sequence. In one such calculation, Gingold and Monaghan (1979) obtain a binary with $q=0.28$ and call attention to their solution's similarity to that of Lucy (1977), which did include the quasi-static phase.

FRAGMENTATION

Having argued against fission as the formation mechanism for $q=1$ binaries, Lucy and Ricco (1979) cited the investigations of Norman and Wilson (1978) and Cook and Harlow (1978) as grounds for suspecting that such systems result from the fragmentation of the toroidal configurations that form as rotating protostars collapse dynamically (Larson 1972). When the collapse is due to $\bar{\Gamma} < 4/3$, this mechanism can produce a close binary (Larson 1972; Bodenheimer 1978). Moreover, since prior to this collapse the protostar was pressure supported, it should then have been rather accurately axisymmetric; fragmentation at this phase therefore proceeds from rather perfect - i.e., noise-free - initial conditions, as one might conjecture to be necessary for the $q=1$ mechanism.

Although the above-mentioned investigators do report calculations ending with $q=1$ binaries, these all start with finite $m=2$ perturbations and so cannot be accepted as definitive demonstrations of the formation of such systems. Accordingly, the fragmentation of toroidal protostars has been re-investigated using the finite-size particle scheme

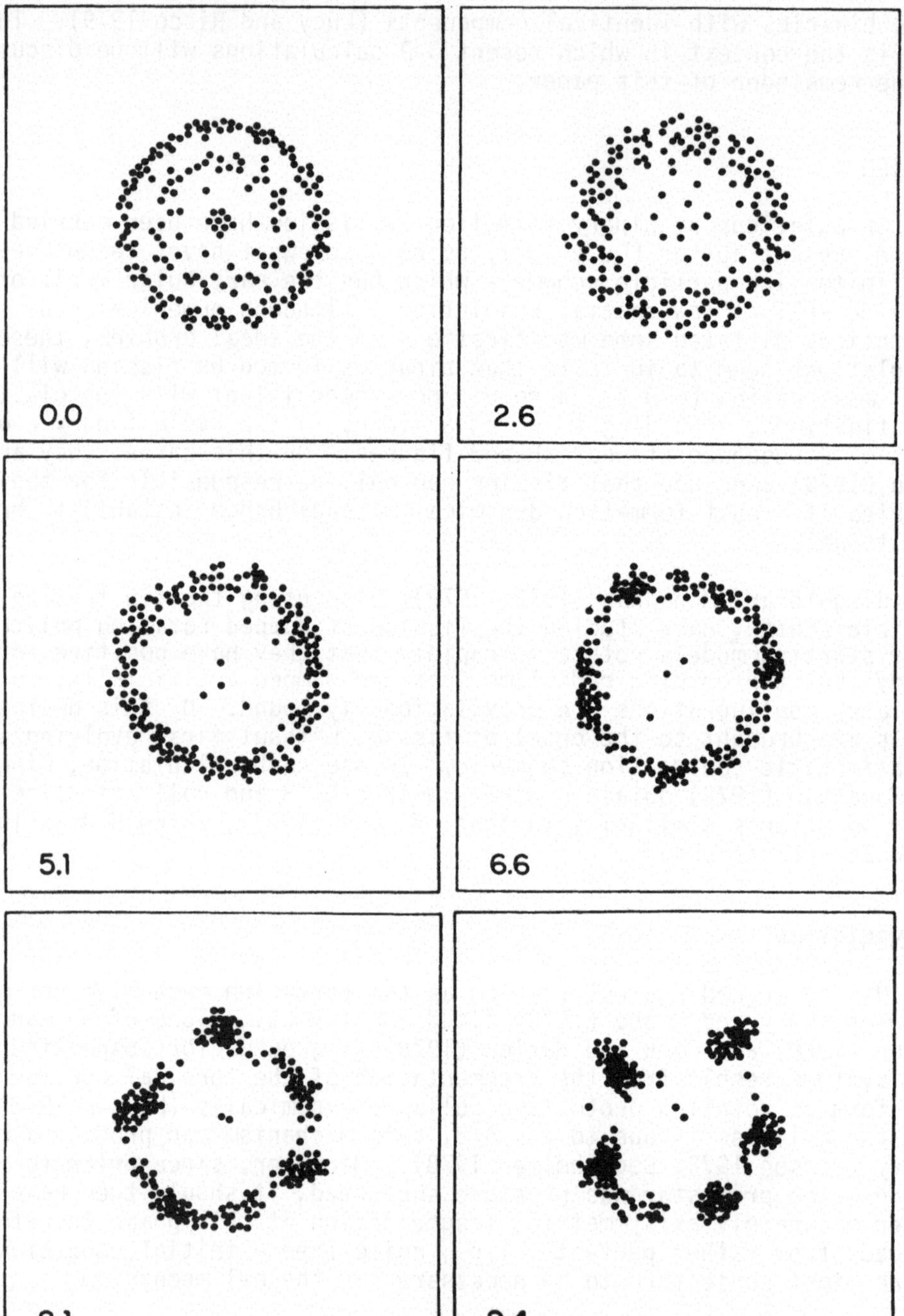

Fig. 1: Fragmentation of toroidal protostar having t_T=0.10 initially.

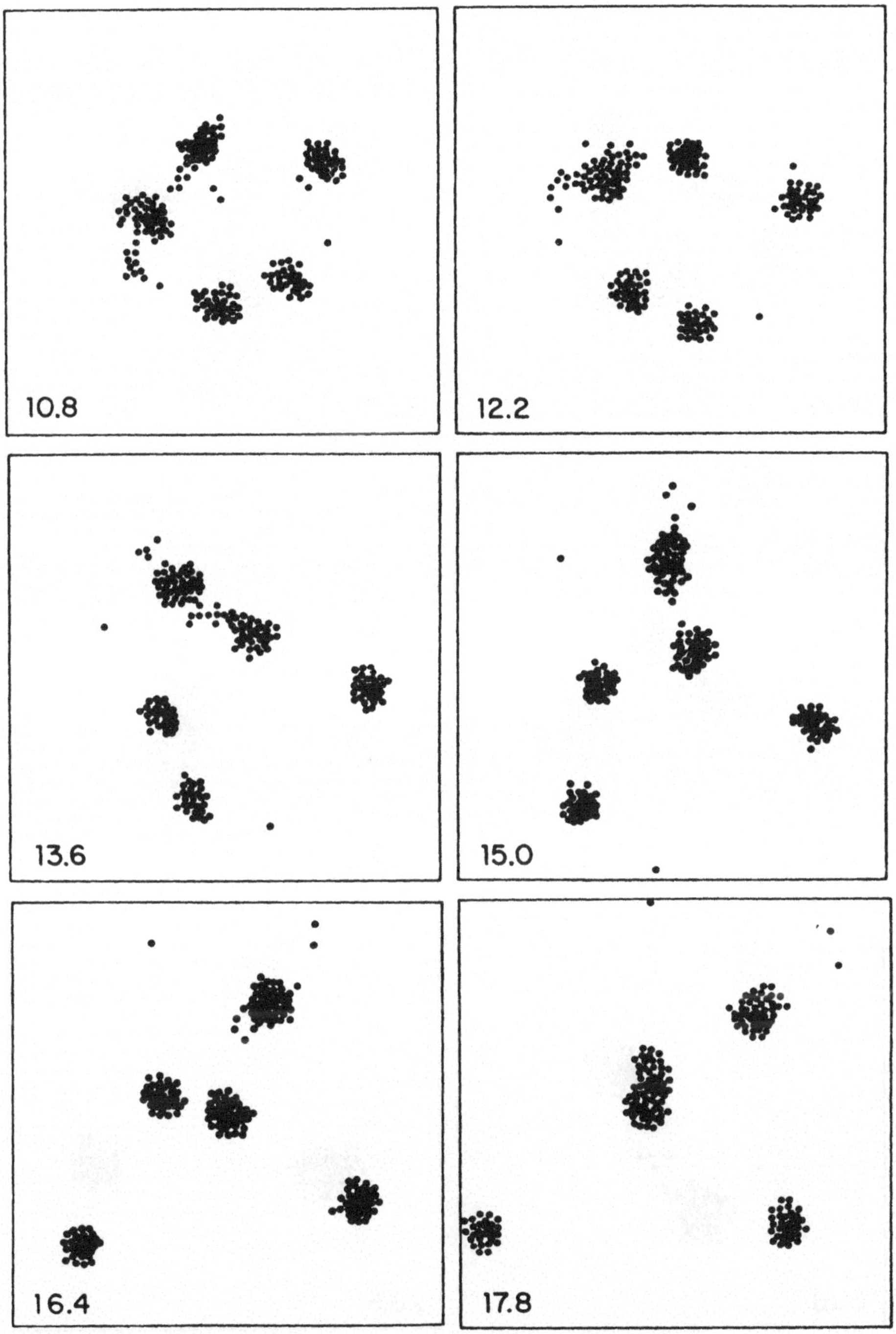

Fig. 2: Evolution of unstable 5-member multiple system formed in Fig.1.

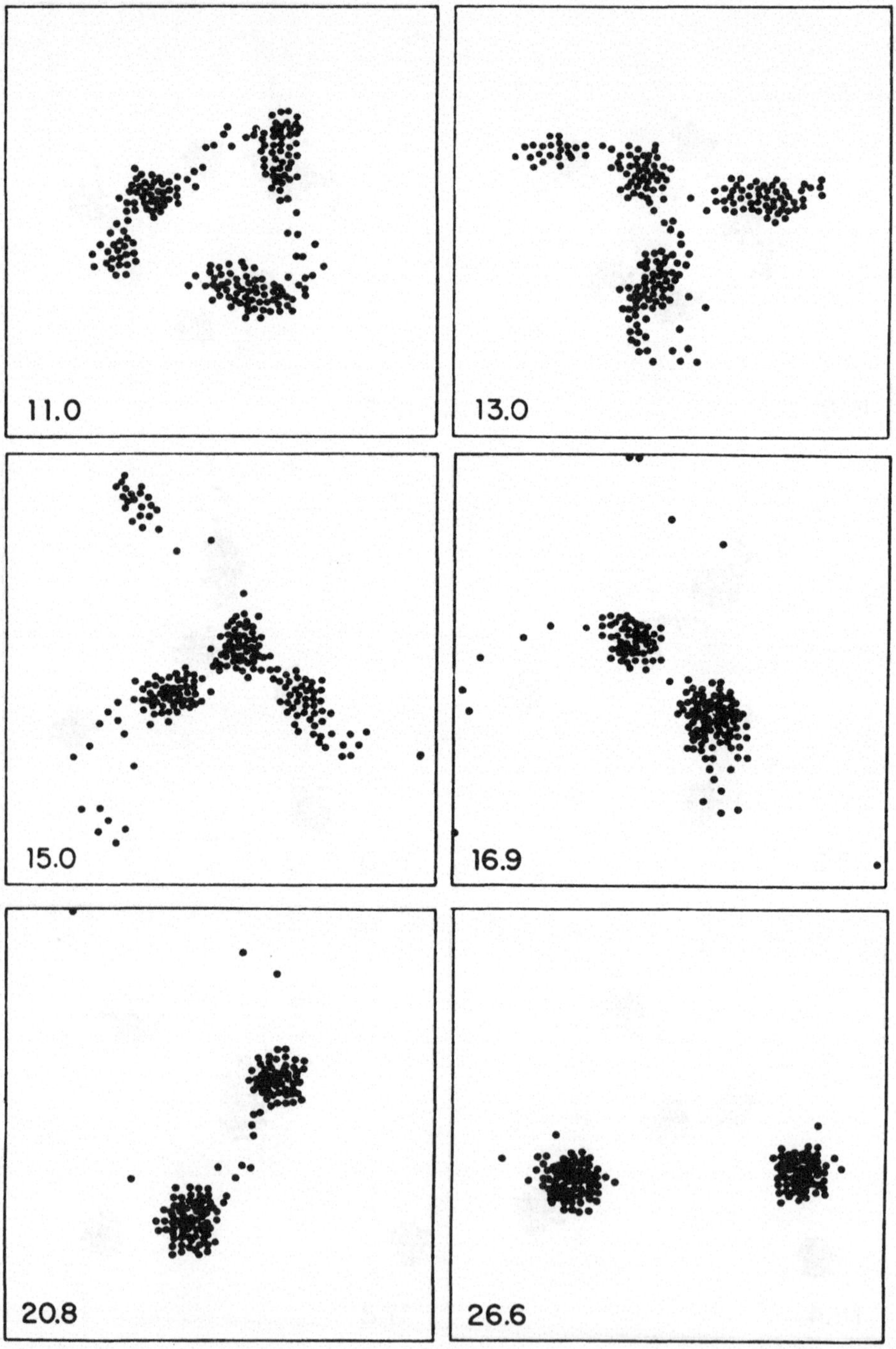

Fig. 3: Evolution of fragments of toroidal protostar having t_T=0.12 initially.

with initial conditions that avoid essentially pre-selecting the fragmentation mode. This is achieved by allowing numerical noise to provide seed amplitudes for unstable modes; a model's fragmentation should then, like that of a protostar, be determined by the non-linear growth of the most unstable mode. Such calculations will now be briefly reported; full details will be published elsewhere.

In these calculations, the initial models are highly flattened, slightly toroidal adiabatic configurations that are symmetric about the invariable plane, uniformly rotating, and in virial equilibrium. Subsequent changes are taken to be isentropic ($\Gamma = 5/3$), with a bulk viscosity term included to simulate the dissipation that would be expected to damp the oscillations of newly formed stars. Because of the simplicity of these assumptions, these models can be scaled to any mass and radius.

Figures 1 and 2 trace the evolution of a model whose initial ratio of thermal to gravitational energy is t_T=0.10; the initial rotation period is 13.5, and the initial time step is 0.35. These diagrams, each of which is an equatorial projection of the coordinates of the 400 particles being followed, show the model evolving into a rather thin torus ($\tau \simeq 5$) and then fragmenting ($\tau \simeq 8$) into an unstable five-component stellar system whose subsequent disruption is complicated by the coalescence of its members (e.g., τ=17.8).

Figure 3 traces the later stages of the evolution of a model with the same initial density structure but now t_T=0.12. After approximately one rotation period (τ=13.0), the model has fragmented into one minor and three major components. Two of the major components coalesce at τ=16.9, so that the final configuration is a binary (q=0.73) with a distant small companion.

Although these calculations have not demonstrated the formation of a q=1 binary, they do suggest the following tentative conjecture: Because of essentially perfect initial conditions, a protostar, following its $\bar{\Gamma}$ < 4/3 collapse, fragments into an unstable multiple system with identical components that then evolves into a binary as a result of the coalescence and occasional ejection of members. A binary with q$\simeq$1 will be a not infrequent outcome of this sequence of events if coalescence occurs with little mass loss.

DISCUSSION

The investigations reviewed and reported here have not yet settled any questions concerning the formation of close binaries. Nevertheless, these early results with 3-D codes are surely encouraging: we can now cleanly follow the growth of the instabilities that presumably convert single protostars into binaries and multiple systems. Provided therefore that our present ideas are basically correct and that effects requiring high resolution are not crucial, numerical simulations should soon decisively advance our understanding of this long-standing problem.

This work has been supported by the National Science Foundation under grant AST 79-13179. A grant of computer time at the Goddard Institute for Space Studies, NASA, is also acknowledged.

REFERENCES

Bodenheimer, P. 1978, Ap.J., 224, 488
Cook, T. L., and Harlow, F. H. 1978, Ap.J., 225, 1005.
Gingold, R. A., and Monaghan, J. J. 1978, M.N.R.A.S., 184, 481.
Gingold, R. A., and Monaghan, J. J. 1979, M.N.R.A.S., 188, 39.
Larson, R. B. 1972, M.N.R.A.S., 156, 437.
Lucy, L. B. 1977, A.J., 82, 1013.
Lucy, L. B., and Ricco, E. 1979, A.J., 84, 401.
Norman, M. L., and Wilson, J. R. 1978, Ap.J., 224, 497.
Tassoul, J-L. 1978, Theory of Rotating Stars (Princeton University Press).

DISCUSSION FOLLOWING LUCY

Nariai: Can you make the mass in your calculation larger and apply your results to the formation of a group of galaxies?

Lucy: Although the mass and radius can be scaled to values appropriate for a group of galaxies or even for a supercluster, the assumption of isotropic changes would not then be so well justified.

Zuiderwijk: What percentage for the thermal energy results in the formation of one single star?

Lucy: The next solution in the sequence has t_T = 14% and ends in a single star surrounded by debris in an equatorial disk.

Morton: I worry about finding such a sharp peak at unity in the mass-ratio distribution. It appears likely that your calculation will give a rather broad distribution.

Lucy: I fear you may be right. My hope, though, is that improved calculations will yield a pure mode, so that the multiple fragments have equal masses; the processes of coalescence and ejection may then end with a q = 1 binary often enough to explain the peak.

Shu: In the spherical calculations, one finds that the first core which forms constitutes only a small fraction of the total mass of the collapsing cloud. Subsequently, there ensues an extended accretion phase. Have you thought about the consequences of such an accretion phase for your models?

Lucy: For rotating protostars, much of the infalling outer envelope will often have too much angular momentum to reach the fragmenting ring.

Rajamohan: Normal single stars on the Main Sequence are essentially fast rotators with a small dispersion in their true rotational velocities, while slow rotators are either peculiar or members of close binary systems. Can this be reproduced by the formation process you have outlined?

Lucy: My calculations indicate that the components of newly-formed binaries are rapidly rotating (t_R = 10 to 15 %). For close systems, tidal dissipation will of course quickly slow down the spin until synchronism is achieved.

STATISTICAL INVESTIGATION OF SPECTROSCOPIC BINARY STARS

A.V. Tutukov and L.R. Yungelson
Astronomical Council,
USSR Academy of Sciences
Mowcos, USSR

ABSTRACT

A catalog of physical parameters of about 1000 spectroscopic binary stars (SB), based on the Batten catalog, its extensions, and newly published data has been compiled. Masses of stars' components (M_1 and M_2), mass ratios of components ($q = M_1/M_2$) and orbital angular momenta are computed, wherever possible. It is probable that the initial mass function of the primaries is non-monotonic and is described only approximately by a power-law. The distribution of q of double-line SB has a sharp maximum near q = 1, that of single-line SB, near q = 0.3. SB with $1.5 \lesssim M_1/M_\odot \lesssim 10$ and semiaxes of orbits $a \lesssim 10\ R_\odot$ probably do not form. A number of assumed "initial" distributions of M_1, q and a were transformed with the aim of obtaining "observed" distributions taking into account the observational selection due to the luminosities of the components, their radial velocities, inclinations of the orbits, and the effects of matter exchange between the components. The best agreement with the observations is obtained for the combination of a power-law initial mass function $\xi(M) \sim M^{-2.35}$, a distribution of q with a maximum near 1 and a distribution of semiaxes with $dN/da \sim 1/a$.

In order to investigate the origin and evolution of close binary stars (CBS) it is necessary to know the distribution of and correlation between the masses of their components M_1 and M_2, semiaxes of their orbits a, mass ratios of the components $q = M_2/M_1$, etc. The only source of this information is the catalogs of spectroscopic binary stars (SB) for which under certain assumptions it is possible to compute M_1, M_2, a and q. We have compiled a catalog of SB, based on the 6th Batten catalog (Batten, 1967), its extensions (Pedoussant and Ginestet, 1971; Pedoussant and Carquillat, 1973), and data published up to the end of 1978. Now it contains data on about 1000 stars. It appeared possible to compute M_1, M_2, a and q for 292 stars with two visible spectra (SB2) and 534 stars with one visible spectrum (SB1). The obtained distributions are by no means the real "innate" distributions of stellar

M. J. Plavec, D. M. Popper and R. K. Ulrich (eds.), Close Binary Stars: Observations and Interpretation, 15–22.

parameters. Because of this we tried to to find the latter by correcting some trial initial distributions for effects of observational selection and stellar evolution.

Referring for computational details and more extensive accounts to our papers (Kraitcheva et al., 1978, 1979; Piskunov et al., 1979) we shall report here some of the most interesting results.

Our results indicate that SB2s are mostly systems prior to mass exchange. Because of this they are well suited for a determination of the initial mass function (IMF) of the primaries of CBS. To get the IMF the observed number of stars with a given M_1 must be divided by the product of the volume of space in which systems with given M_1 are seen and their lifetime. The IMF of 121 SB2s with known inclinations of orbits is shown in Fig. 1. On the average $dN \sim M^{-1.3}\, dM$. The difference from the slope of Salpeter's IMF for single stars is partly due to the selection effects and partly, as it seems to us, to the increased interest of observers in the massive CBS.

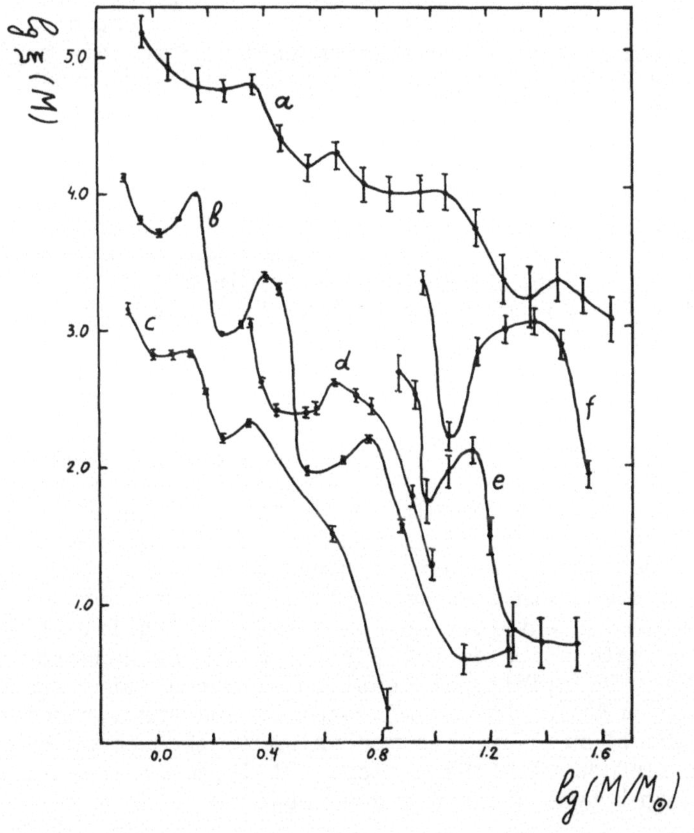

Fig. 1. The initial mass function. a-for double-line SB; b-for stars of the Michigan catalog; c-for stars of b-v catalog; d-for stars of u-b catalog; e-for the brightest stars of LMC clusters; f-for the brightest field supergiants. Dispersion ($\sqrt{n}$) is shown.

Popov (1968) and Svechnikov (1969) on the basis of much more limited data claimed that the IMF may be non-monotonic. With the aim of studying this problem in more detail, we computed the IMF using the spectra of single main-sequence stars (~ 15000 stars from the Michigan catalog by Houk and Cowley, 1975), colors b-v and u-b of main sequence stars (~ 2700 and ~ 1600 stars, respectively, from the catalog of Mermilliod and Nicolet, 1977), luminosities of the brightest stars of the LMC (~500 stars, Hodge et al., 1970), spectra of Ia-Ib field supergiants (~700 stars, Piskunov, 1977). Almost all IMF are non-monotonic with some details recurring in the same mass intervals. Our results are not convincing enough for assertion that the IMF is really non-monotonic. The problem deserves further examination (see also Piskunov et al., 1979).

As SB2s are mostly systems prior to mass exchange, their distribution in the a-M plane (Fig. 2) reflects their distribution at the time of formation. The bottom line in Fig. 2 corresponds to the ZAMS contact systems with components of equal mass. The $a < 10\ R_\odot$, $M_1 > 1.5\ M_\odot$ region is almost empty. There are no selection effects that prevent discovery of stars with such parameters. Three stars that lie in this region, BV 382 Cep, UZ Pup, and V Sge display mass exchange and are probably evolved. To explain the existence of the empty region we may assume that $a \sim 10\ R_\odot$ corresponds to the maximum angular momentum of a protostar in hydrostatic equilibrium on the Hayashi border. If the masses of the components are lower than about $1.5\ M_\odot$, the magnetic braking continuously supported by convection in the envelopes may be effective enough for gradual shrinking of the orbit up to contact. This is similar to the process proposed by Eggleton (1976) as an explanation of the origin of the cataclysmic variables. Detached RS CVn stars, as follows from the time scales (10^7 years) of the decrease of their periods (Hall, 1978), may be in a similar evolutionary stage. Thus the region with $a < 10\ R_\odot$, $M_1 < 1.5\ M_\odot$ is continuously filled by stars, but the region with larger masses remains empty.

A simple estimate shows that the orbital angular momentum of a contact massive ($M_1 > 10\ M_\odot$) binary system with $q \sim 1$ is at least six times higher than the limiting axial angular momentum of a single star with the same mass. We see five possibilities for solving the problem. Angular momentum of a rotating pre-main sequence star may be lost by magnetic braking. A system with $q \ll 1$ or a planetary system or a star with a double core in a common envelope may be formed. Finally, it is possible that massive single stars do not form at all. As concerns double-core stars, they may be stable, at least on a thermal time scale (Tutukov and Yungelson, 1979). Energy of rotation and angular momentum of the cores may be transferred by some mechanism to the common envelope and lost from the surface with the matter. It is possible that some of the Of, Be, or P Cygni type stars are such objects.

We have tried to find "innate" distributions of SBs over their main parameters M_1, a, and q, taking some trivial initial distributions and correcting them for different effects of observational selection and stellar evolution in order to obtain agreement with observed distributions. These effects are: random orientation of orbital planes;

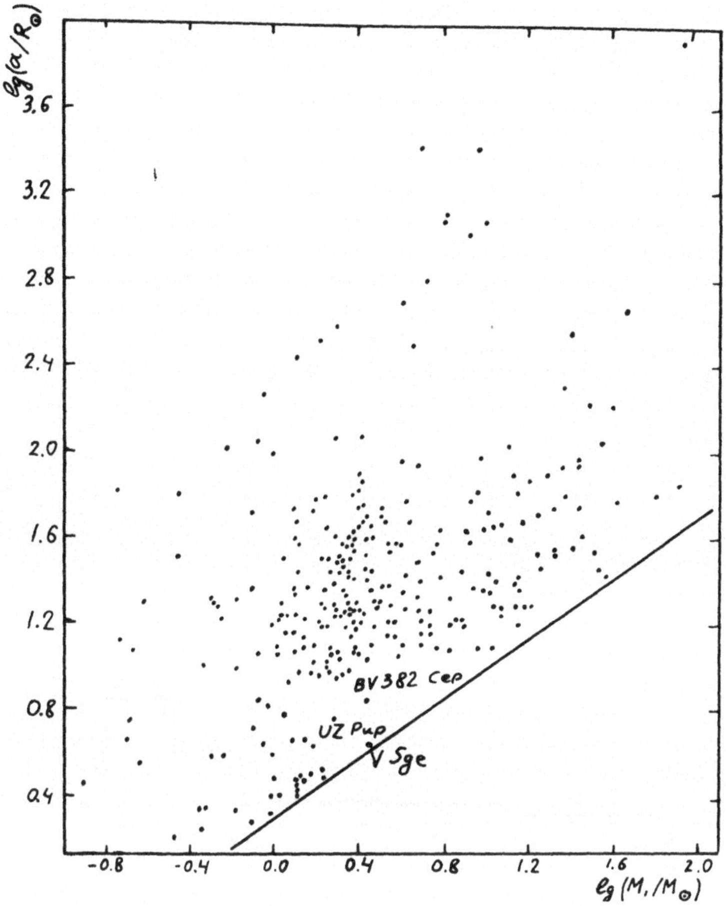

Fig. 2. Double-line SB in lg a-la M_1 plane.

dependence of discovery probability on the semiamplitude of radial velocity K_1 sin i; existence of a lower limit on K_1 sin i which still allows discovery of duplicity by means of spectroscopy and the dependence of this limit on the spectrum of the primary; differences in lifetimes of stars; existence of a limiting visual stellar magnitude for spectroscopic binaries ($\sim 10^m$) and the consequent differences in the limiting distances to observable stars combined with the appropriate spatial distribution (spherical or disc-like); mass exchange between the components leading to changes in their masses, lifetimes, luminosities, semiaxes of orbits; mass loss from the system. In the course of the computation it is possible to separate single- and double-line systems, because for discovery of double spectral lines the semiamplitude of the radial velocity must exceed some limit which depends on the spectrum, and because the difference of visual magnitudes of the components must be less than 1^m.

The theoretical distribution of the mass ratios may be computed as an integral

$$\frac{dN}{dq} = \phi_1(\phi)\int_{M_{1_{min}}}^{M_{1_{max}}} \int_{a_{min}}^{a_{max}} \int_0^{\pi/2} \phi_2(M_1)\ \phi_3(M_1)\ \phi_4(q,M\ ,a,i)\ \phi_5(a)\ dM_1 da\ di$$

where ϕ_1 is the distribution of q, ϕ_2 the IMF, ϕ_3 is the product of the lifetime of the system with given M_1 - T and of volume of space in which is system is seen -V, ϕ_4 describes the dependence of detection probability on the semiamplitude of the radial velocity, ϕ_5 is the distribution of semiaxes ($\int_{a_{min}}^{a_{max}} \phi_5(a)\ da = 1$, and $\int_0^1 \phi_1(q)\ d(q) = 1$). Here functions ϕ_1, ϕ_2, ϕ_5 are our trial functions. Similar expressions could be written for dN/dM_1 and dN/da. Distributions of parameters of systems prior and post mass exchange (we shall name the latter "evolved" systems) can be computed similarly using different integration limits and taking into account changes in the properties of binaries caused by mass exchange/loss from the system. All numerical data necessary for the computations could be found either from the analysis of the catalog of SBs or in the results of evolutionary computations for single and double stars (for more details see the paper by Kraitcheva et al., 1979).

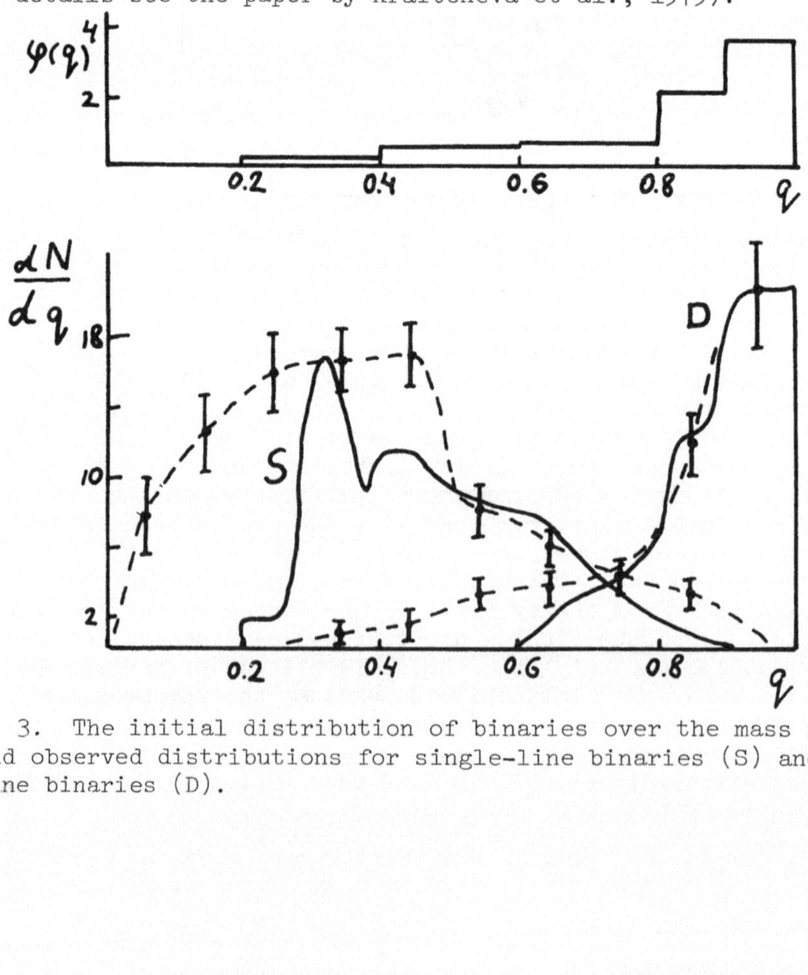

Fig. 3. The initial distribution of binaries over the mass ratio (ϕ) and observed distributions for single-line binaries (S) and double line binaries (D).

We have tried three distributions over q: 1) one with a maximum near q = 1 (Fig. 3a) (Popov, 1970; Lucy and Ricco, 1979); 2) a bimodal one in which the number of systems with $q > 0.4$ is roughly equal to the number of them with $q < 0.4$ (see e.g. Trimble, 1976); 3) one with the number of systems increasing with decreasing q (see e.g. Jashek, 1976). As an initial distribution of masses of unevolved primaries we have used the power law $\phi_2(M_1) = 0.03\,M_2^{-2.35}$ with $M_{1\,min} = 0.8\,M_\odot$, $M_{1\,max} = 30\,M_\odot$; the initial distribution of semiaxes was $\phi_5(a) = 1/a$, with a_{min} corresponding to contact on the ZAMS, $a_{max} = 2000\,R_\odot$. We used a smooth IMF because its nonmonotony is still questionable and because we were interested in the most general properties of distributions.

The best agreement with the observations both in the shape of the curves and in the ratio of numbers of SB1 and SB2 Ns/Nd has been obtained for the case of the initial $\phi_1(q)$ with a maximum near q = 1: see Fig. 3 in which we show both the "theoretical" distribution and the distribution of stars in our catalog smoothed over q; the latter is close to that obtained by Trimble (1976).

Our computations showed that the most severe selection factor is the dependence of the probability of discovery on the semiamplitude of radial velocity. It allows discovery of only about 1/3 of all SBs. This suggests once more that probably almost all stars are members of binaries and that single stars are rather exceptions to the rule. The factor V · T that reaches its maximum value near $\lg M_1 = 0.3 - 0.4$ cuts out low-mass systems because of their weakness and high-mass systems because of their short lifetimes.

It appears that the "theoretical" double-line systems and single-line systems with $q > 0.4$ are mostly unevolved systems; single-line systems with $q < 0.4$ are mainly evolved main-sequence systems. Among the "theoretical" double-line stars about 1/3 are the "A"-systems, 2/3 are "B" and "C"-systems; this proportion is close to that for the observed stars. For unevolved single-line systems this proportion is 1/4 and 3/4. (Here A,B,C are the conventional types of close binaries (see e.g. Paczynski, 1971).) Probably, the initial $\phi_1(q)$ after Popov (1970) with maximum near q = 1 underestimates the number of systems with low q. However, the bimodal initial $\phi_1(q)$ (Trimble, 1976) produces a theoretical value of Ns/Nd twice as large as the observed one. The initial function $\phi_1(q)$ which increases towards small q's produces distributions that do not agree with observations either in their shapes or in Ns/Nd. The "theoretical" number of evolved systems with low q's increases if we take into account possible mass and momentum loss from the system.

The theoretical distributions dN/dM_1 for SB2 and SB1 and dN/da for SB2 agree satisfactorily with the observed ones. The deficit of systems with $\lg a = 0.5 - 0.9$ and $\lg M_1 = 0.0 - 0.3$ may be explained by the existence of an empty "region" in the $a - M_1$ plane (see above). Our trial distribution $\phi_5(a) \sim 1/a$ probably overestimates the number of very close systems. This causes the shift between the observed and computed distributions of single-line systems.

Summarizing, the analysis of the catalog of SB itself and the computation of "theoretical" distributions of CB parameters which take into account the effects of observational selection and stellar evolution, has provided much information on vital properties of CBs. The above described method of computation has shown its great efficiency and will be extended to take into account still unconsidered effects.

REFERENCES

Batten A., 1967. Publ. Dominion Astrophys. Observ. 13, No. 8.

Eggleton P., 1976. In: Structure and Evolution of Close Binary Systems, eds. P. Eggleton et al., Reidel, Dordrecht, p. 209.

Hall D., 1978. Private communication.

Hodge P.W., Welch G.A., Wills R., Wright F.W., 1970. Smithsonian Astrophys. Observ. Rep. 320.

Houk, N., Cowley, A.P., 1979. Michigan Catalog of Two-Dimensional Spectral Types for the HD Stars.

Jaschek, C., 1976. Astron. and Astrophys. 50, 185.

Kraitcheva Z.T., Popova E.I., Tutukov A.V., Yungelson L.R., 1978. Astron. Zh. 55, 1176 (English translation in Soviet Astronomy).

Kraitcheva Z.T., Popova E.I., Tutukov A.V., Yungelson L.R., 1979. Astron. Zh. 56, No. 3 (English translation in Soviet Astronomy).

Lucy L.B., Ricco E., 1979. Astron J. 84, 401.

Mermilliod J.-C., Nicolet B., 1977. Astron. and Astrophys. Suppl. Ser. 29, 259.

Paczynski B., 1971. Ann. Rev. Astron. Astrophys. 9, 183.

Pedoussant A., Ginestet N., 1971. Astron. and Astrophys. Suppl. Ser. 4, 253.

Pedoussant A., Carquillat J.-M., 1973. Astron. and Astrophys. Suppl. Ser. 10, 105.

Piskunov A.E., 1977. Ph.D. Thesis, Tartu Univ.

Piskunov A.E., Tutukov A.V., Yungelson L.R., 1979. Pisma v Astron. Zh. 5, 81 (English translation in Soviet Astron. Letters).

Popov M.V., 1968. Ph.D. Thesis, Moscow University.

Popov M.V., 1970. Peremennye Zvezdy 17, 412.

Svechnikov M.A., 1969. Catalogue of orbital elements, masses and luminosities of close binary stars, Ural Univ. Press, Sverdlovsk.

Trimble V., 1976. Astron J. 79, 967.

Tutukov A.V., Yungelson L.R., 1979. In: Mass Loss and Evolution of O-Type Stars, eds. P.S. Conti and C.W.H. de Loore, D. Reidel, Dordrecht, p. 401.

DISCUSSION FOLLOWING TUTUKOV AND YUNGELSON

Lucy: How do you eliminate evolved--i.e., post-mass-exchange-systems from your sample of SBI's?

Tutukov: We did not make that but our program simulates a simple approximation of evolution of a component. All systems with $\Delta M_b \lesssim 1^m$ are counted by the program as single line binaries if the luminosity and the semi-amplitude of the radial velocity were enough to distinguish the star as a binary. The nature of the hidden component isn't important in this case.

THE PERIOD DISTRIBUTION OF CLOSE BINARY SYSTEMS

L. Mantegazza, P. Paolicchi, P. Farinella and F. Luzny
Osservatorio Astronomico di Brera, Merate, Italy

ABSTRACT. The analysis of the period distribution of eclipsing and spectroscopic binary systems shows the presence of some secundary maxima, which cannot depend on selection effects. These secundary maxima are mainly due to late type stars, as can be seen from the distribution curves for eclipsing binaries of various spectral types. The average separation of the components (in units of the sum of stellar radii) increases with the spectral type from O types to late B types, remaining almost constant for later spectral types.

The period distribution of close binaries (i.e. eclipsing and spectroscopic systems) was first studied by Fracastoro (1954) and then, with more recent catalogue data, by the present authors (Farinella and Paolicchi, 1978, Farinella et al., 1979, hereinafter referred as Papers I and II).

In principle one could represent the distributions either with equal steps in the period P or in the logarithm of the period, resulting in different shapes (the "linear" distribution is monotonically decreasing, while the logarithmic one has a sharp peak for periods of about 3 days and shows also some evidence of other local maxima). In the following, as previously done by Fracastoro and in Papers I and II, we shall employ the logarithmic representation, which seems to be more meaningful when one is concerned with problems of period variations with time, which are often of the form $\Delta P/P = f(t)$.

Another problem refers to the observational selection effect which clearly causes, for geometrical reasons, an overestimate of short-period systems (which more easily produce detectable eclipses). In Paper I the observed distribution for eclipsing binaries (which is similar to the one for spectroscopic systems) has been corrected in various ways, but the most relevant features, i.e. the main peak for periods of about 3 days and the presence of some secundary maxima, were unaffected and must be considered as real features.

M. J. Plavec, D. M. Popper and R. K. Ulrich (eds.), Close Binary Stars: Observations and Interpretation, 23–26.

The third correction method described in Paper I seems to be the most reliable. It employs the known geometrical properties of 140 systems as given by Giannone and Giannuzzi (1974) to evaluate the best-fit probability of observing the eclipse for each period interval. In this way we can estimate that the 3488 observed eclipsing systems could correspond to an available "real" sample of about 12,500 objects, whose distribution is richer of long-period systems than the observed one. This "corrected" distribution is shown in the Figure (case (a)), which allows a comparison with two similar histograms corresponding to systems of a given spectral type (cases (b) and (c); see the following). We remark that the normalization is not the same for the three distributions, which refer to original samples of respectively 3488 (a), 283 (b) and 117 (c) eclipsing systems.

In order to investigate in some more detail how the statistics of period is connected with the formation process and the evolutionary history of close binary systems, various methods are possible. We decided to employ the restricted sample formed by the 1248 systems

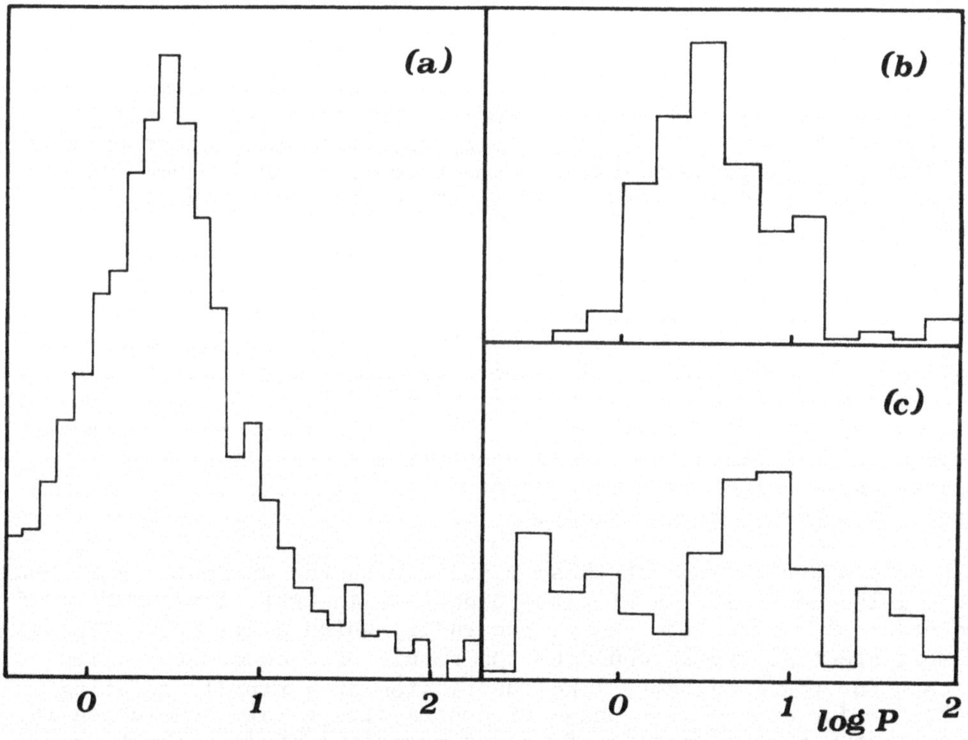

The figure compares the general corrected distribution for eclipsing systems (a) with the corrected distributions for systems containing B (b) and G (c) stars. The step in log P (days) is 0.1 in case (a) and 0.2 in cases (b) and (c).

for which the spectral type of at least one component is known, so that a separate analysis is feasible for different spectral types (see Paper II). Moreover, we tried to isolate the contribution to the distributions both of "evolved" systems (those containing a star of known luminosity class between the I and the IV) and of W UMa systems (whose statistical weight is probably amplified by observational selection). For each spectral class, the "geometrical" selection effect can be taken into account by the same method we used for the general distribution.

An inspection of the results leads to the following considerations. The distributions for early spectral types are rather narrow and show a single marked peak for periods of about 2 - 3 days, while moving towards advanced types the distributions become substantially broader (compare parts (b) and (c) of the Figure, which correspond to B and G stars respectively). Beyond the class F, we get clear evidence of bimodality (or perhaps plurimodality), with the original peak gradually displaced towards shorter periods (less than 1 day) and a new peak for longer periods that increases as the spectral type advances. This latter effect seems directly connected with the increasing contribution of evolved systems, especially for G, K and M classes; on the other hand, the W UMa systems give an important fraction of the short-period peak for F and later spectral classes. It is important to remark that all these conclusions are confirmed if we analyze the data with a smaller subdivision of spectral types (18 classes instead than 6). Moreover, observational selection does not seem to vitiate substantially the results (even if some features may be modified in a relevant way by the correcting method).

An interesting physical interpretation of the statistical results is possible if we study the dependence on spectral type of the average orbital separation of the systems (in units of the sum of stellar radii). If we assume that the systems are formed by two equal main sequence spherical components, we find that this quantity increases from about 1.5 (O types) to about 3 (late B types), then it remains approximately constant for a large range of spectral types (though in this range the absolute quantities - stellar radii and separations - change by a large factor) and finally, beyond early G types, it seems to grow rapidly.

We can try to connect these results with the current ideas about the origin and evolution of close binaries. An orbital separation of the order of 3 times the sum of the radii, which seems to be typical of most spectral types, could be the result of a common formation process for these systems, like the fission of a rapidly rotating protostar during the last phase of contraction : this conclusion is consistent with the results of recent numerical studies about the evolution of rotating collapsing clouds (Bodenheimer 1978). The increase of the average separation for the first spectral types could also be a by-product of the formation process (if the fission of more massive protostars results in smaller average separations), or it may be an evolutionary effect. In this latter case, the effect could be due to the fact that more massive systems evolve more rapidly towards shorter periods, or to different average ages of systems with different spectral type. The timescale of the physical process responsible for the evolution should be of the order of 10^8 yr, i.e. the maximum main sequence lifetime of the systems for which the effect appears.

Finally, the seeming increase of the average separation for systems of advanced spectral types is clearly determined by the high fraction of systems in which at least one of the components is subgiant or giant (so that our assumption about main sequence radii fails). On the other hand, the real increase of the mean period is consistent with the presence of theoretical and observational evidence about drastic changes undergone by "evolved" systems and due to the interaction of components.

REFERENCES

Bodenheimer,P.: 1978, Astrophys.J. 224, pp. 488 - 496.
Farinella,P., and Paolicchi,P.: 1978, Astrophys. Space Sci. 54, pp. 389 - 406.
Farinella,P., Luzný,F., Mantegazza,L., and Paolicchi,P.: 1979, Astrophys. J., in press.
Fracastoro,G.: 1954, Mem. Soc. Astr. Ital. 25, pp. 347 - 382.
Giannone,P., and Giannuzzi, M.A.: 1974, Astrophys. Space Sci. 26, pp. 289 - 304.

DISCUSSION FOLLOWING MANTEGAZZA, PAOLICCHI, FARINELLA AND LUZNY

Tutukov: There are many selection effects more: luminosity, the ratio of luminosities, the lifetime, evolution. What are the influences of those of your results?

Farinella: We agree that there are many selection effects. On the other hand, we think that luminosity and lifetime effects affect mainly the distribution of systems among different spectral types (even if they can lead to some wrong estimate of the percentage of evolved systems). The evolution and ratio-of-luminosities effects are obviously important mainly for the evolved systems, whose properties should be analyzed in much more detail than possible in a statistical survey. These effects probably decrease the statistical weight of late-stage systems, so that the reliability of our conclusions about formation and early evolution could be even strengthened.

ROTATION IN CLOSE BINARIES

R. Rajamohan and P. Venkatakrishnan
Indian Institute of Astrophysics
Bangalore-560034
India

ABSTRACT

It has been demonstrated (Zahn 1966 a, b, c; 1975, 1977) that turbulent viscous dissipation of tides can brake the rotation of stars having convective envelopes, while radiative dissipation can do likewise for those having radiative envelopes. The time scales for the former are very small whereas those for the latter depend sensitively on the ratio r/R where r is the radius of the star and R, the binary separation. For most systems these turn out to be much less than their main sequence lifetimes. This fact affords a natural explanation for the preponderance of synchronous systems.

The parameter which quantifies the tidal braking is the ratio of t_{ms}, the main sequence lifetime (a measure of the star's age) to t_{syn}, the synchronisation time scale (a measure of the rotational deceleration). A plot of the ratio of orbital period P_o to spin period P_s of the star versus such a parameter is shown in figure 1. Rotational velocities of the eclipsing binary systems were obtained from Levato (1974). The fractional radii and inclination of the orbit were taken from the Graded Catalog of Photometric Systems (Koch et al., 1970) while the Seventh Catalog of Spectroscopic Binary Systems (Batten et al., 1978) was our source for the orbital parameters and mass functions. These were utilised to derive the masses, radii and separation of the components of eclipsing binary systems. For Am spectroscopic binaries, the vsini values were obtained from Abt (1975) and Abt and Moyd (1973). The masses of these systems were taken from Allen (1973) which corresponded to their hydrogen spectral types (whenever available) or colours. The synchronisation time scale was calculated according to Zahn's (1977) formula whereas t_{ms} was calculated from Iben (1964).

Figure 1 shows possibly two sequences of stars. The upper sequence behaves "normally" in the sense that it shows asynchronous rotation when the tidal actions are not strong enough to act within its lifetime ($t_{ms}/t_{syn} < 1$). The lower sequence continues to remain near

M. J. Plavec, D. M. Popper and R. K. Ulrich (eds.), Close Binary Stars: Observations and Interpretation, 27–30.

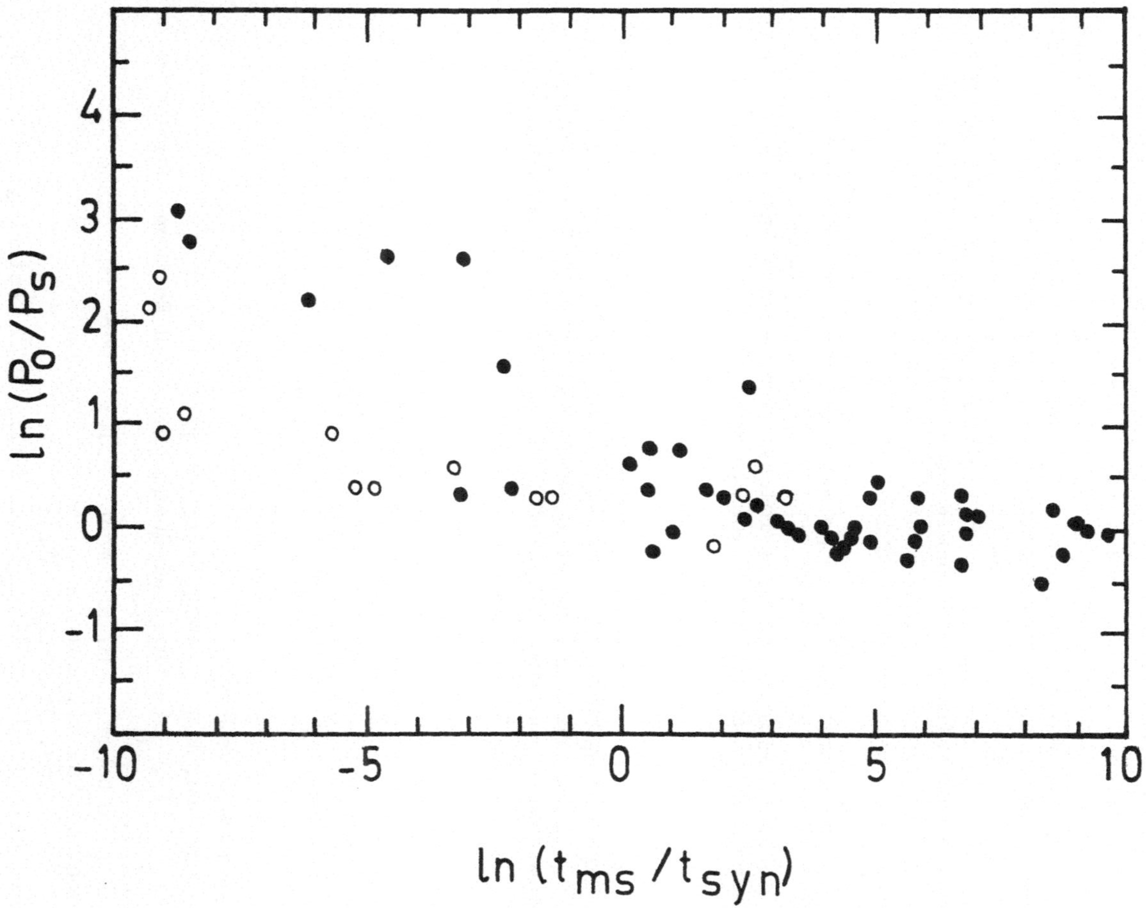

Fig. 1: Plot of the ratio of the orbital to spin period versus the ratio of the main-sequence life-time to the synchronisation time scale. Open circles refer the Am binaries. Based on Iben (1964).

synchronism even when $t_{ms}/t_{syn} < 1$. A majority of these happen to be Am binaries. The possibility of magnetic braking was tested using equation 12 of Joss et al. (1979) but enormous fields ($\approx 10^5$G) seem necessary for this. Such fields could hardly have escaped detection. Angular momentum transport by magneto acoustic waves may not be a good idea purely because of the difficulty of generating these waves in the almost convection-free surfaces of early type stars. A third alternative would be that these had been born with an intrinsically slow rotation. The question which one would now like to ask is whether the Am phenomenon, rather than being isolated among a fraction of the binaries, could be a manifestation of some more general properties of close binaries with the surface peculiarities being wiped out by tidal effects in closer systems. In such a case the equivalent Am stars amongst the close binaries must exist. Then we would be forced to look for two modes of binary star formation.

Finally, when one considers the mass transfer picture for semi-detached systems, one would expect these to be conspicuous in a background of detached systems for the following reasons: Any mass transfer on such large scales as envisaged would, most probably, spin up the recepient. The fact that the donor becomes less massive, coupled with the fact that hardly a fraction of the synchronisation time has elapsed since the mass transfer took place, makes it impossible for the donor to brake the spun up companion. Hence primaries of Algol systems ought to appear with non-synchronous rotation on any plot consisting of a mixture of detached and semi-detached systems. Except for U Cephei, however, the rest of the Algol systems appear quite synchronous in Figure 1.

If one considers the rotational velocities of Algol systems, it is found that, excepting for a few fast rotators, the mean value is about 70 kms^{-1}. Interestingly enough, this value is similar to those of Ap and Am stars.

REFERENCES

Abt, H.A.: 1975, Astrophys. J. 195, 407.

Abt, H.A. and Moyd, K.I.: 1973, Astrophys. J. 182, 809.

Allen, C.W.: 1973, Astrophysical Quantities, 3rd Edition, University of London, The Athlone Press.

Batten, A.H., Fletcher, J.M. and Mann, P.J.: 1978, Publ. Dom. Astrophys. Obs. 15, 121.

Iben, Jr., I.: 1964, Astrophys. J. 140, 1631.

Joss, P.C., Katz, J.I. and Rappaport, S.A.: 1979, Astrophys. J. 230, 176.

Koch, R.H., Plavec, M., and Wood, F.B.: 1970, Publ. Univ. Pennsylvania Vol. XI, 1.

Levato, H.: 1974, Astron. Astrophys. 35, 259.

Zahn, J.P.: 1966a, Ann. Astrophys. 29, 313.

Zahn, J.P.: 1966b, Ann. Astrophys. 29, 489.
Zahn, J.P.: 1966c, Ann. Astrophys. 29, 565.
Zahn, J.P.: 1975, Astron. Astrophys. 41, 329.
Zahn, J.P.: 1977, Astron. Astrophys. 57, 383.

BINARITY AS A FACTOR IN STELLAR RADIO EMISSION

D.M. Gibson
New Mexico Institute of Mining and Technology
Socorro, New Mexico, U.S.A.

1. INTRODUCTION

A simple examination of Table I which lists the 26 positively detected "active" radio stars[1] reveals that 21 are interacting binaries. Thus, on face value, membership in a close binary system would appear to be an important factor in producing conditions which favor stellar radio emission. However, this conclusion ignores the considerable biases which have characterized most of the surveys (c.f., references in Wendker 1978) made to date wherein selection criteria such as binarity itself, evidence for mass exchange, strong magnetic fields, optical activity, and similarity to previously detected systems have been cited. In very few cases (c.f., Gibson 1977) have complete samples (i.e., a uniform survey of all stars in a particular class to a given luminosity limit) been observed. Therefore, it is difficult to examine the role of binarity in stellar radio emission and draw conclusions which have statistical support. However, the body of radio star data does include observations of about 700 objects and thoughtful analyses of a carefully selected subset should reveal trends which, if they do not lead to firm conclusions, should at least point out the direction for future study.

2. SAMPLE SELECTION

The primary reference source for this study was the Catalog of Radio Stars (Wendker 1978) which includes 607 entries from 222 different references published before 1978. Stars classified as normal single stars (class 2), binaries and other multiple systems (class 3), and white dwarfs, magnetic stars, and peculiar stars (class 6) were included in the sample. Cataclysmic variables (a subset of class 4 --

[1]An "active" radio star is herein defined to be one which has two normal stellar components (i.e., components of luminosity classes I-VII) and exhibits microwave emission which is variable on timescales of minutes to a few days.

M. J. Plavec, D. M. Popper and R. K. Ulrich (eds.), Close Binary Stars: Observations and Interpretation, 31–38.

Table I

ACTIVE RADIO STARS

Name	Spectral Types	Period (Days)	Dist. (pc)	S_{max} (mJy)	L_{max} (10^{15} ergs $s^{-1}Hz^{-1}$)
λ And	G8III-IV + ?	20.5	23	20	13
UX Ari	G5V + KOIV	6.4	55	200	720
54 Cam	GOV + G2V	11.1	38	12	21
RS CVn	F4IV + KOIV	4.8	145	7	180
σ CrB	F8V + GOV	1.1	23	57	36
RZ Eri	F5V + G5IV	39.3	105	4	53
σ Gem	K1III + ?	19.6	59	19	79
AR Lac	G2IV-V + KOIV	2.0	47	500	1300
HK Lac	F IV + KOIII	24.4	150	25	670
RT Lac	G9IV + K1IV	5.1	205	35	1800
SZ Psc	G8V + K1IV-V	4.0	100	100	1200
UV Psc	G2IV-V + KeIV-V	0.9	125	14	260
HR 1099	G5V + KOIV	2.8	33	1200	1600
HR 5110	F2IV-V + ?	2.6	52	425	1400
HD 216489	gK1 + ?	24.6	200	33	1600
HD 224085	KOIV-V +?	6.7	29	92	93
CC Cas	O9IV + O9IV	3.4	1000	10	12000
β Lyr	B8IIp + ?	12.9	260	20	1600
β Per	B8V + G5III	2.9	25	1020	760
α Sco AB	M1Iab + B4Ve	?	180	10	390
R Aql	gM5e-8e		150	240	6500
π Aur	M3II		170	31	1100
α Ori	M2Iab		200	110	5300
α Sco A	M1Ib		180	2	89
YZ CMi	dM4.5e		20	120	48
UV Cet	dM5.5e		5	300	8

novae, dwarf novae, and eruptive variables) and those flare stars (class 1) for which cm-wave or interferometric observations had been made were also included. X-ray stars and emission-line stars were excluded.

In addition, recently reported surveys of the RS CVn binaries (Owen and Gibson 1978), related late-type binaries (Gibson and Owen 1979), and single G- and K-subgiants (Gibson 1977) were also included as were observations of individual stars (Davis et al. 1978; Feldman 1978, 1979; Feldman et al. 1979; Gibson 1978; Gibscn and Newell 1979). In total, 469 stars of the aforementioned classes have been observed at least once.

To form a meaningful subset of this sample, two general considerations were applied. Firstly, the sample had to be large enough to assure at least minimal statistical reliability. Since the sample is not complete -- as described above -- "large enough" was a subjective choice which depended on the types of analyses to be done. In our case, we wished to compare the radio emitting properties of the various subclasses of stars, e.g.: are double-lined spectroscopic binaries stronger radio emitters than single-lined systems? or, are giants stronger than dwarfs? Therefore, we adopted the criterion that there be about 10 systems each within luminosity classes III - V for each entry in Tables II and III.

Secondly, the working sample would need to have reasonable source-distance limits so that non-detections would be a true measure of source inactivity and, thus, a meaningul parameter in the analyses. In selecting this limit, we examined both the maximum radio luminosities of the known emitters, L_{max} (see Table 1) and the sensitivity of the radio telescopes used for the surveys. We noted that 14 of the 26 detected systems have maximum radio luminosities $\sim 10^{18}$ ergs/s^{-1}/Hz^{-1}. Assessment of the upper limits obtained in the previous surveys suggests a 20 mJy source could have reasonably been detected in most cases. These two factors were combined to set the distance limit at 200 pc. Distance estimates for the stars were obtained from the original radio surveys (e.g., Spangler et al. 1977) when possible. If not, spectroscopic parallaxes were computed from the spectral type(s), luminosity class(es), and visual magnitude using the absolute magnitudes given by Allen (1973). Applying this distance limit to the original sample resulted in the selection of 201 systems which marginally meet our first criterion.

3. DISCUSSION

The working sample is comprised of 70 double-lined spectroscopic binaries (2SB's), 78 single-lined spectroscopic binaries (1SB's), and 53 single stars. Eighty of the systems have at least one early-type (<F5) component while 131 have at least one late-type (>F5) component. Twenty-three have been detected -- CC Cas, RT Lac, and β Lyr are beyond the distance limit -- including 12 2SB's, 5 1SB's, and 6 singles. We can further break down the sample by luminosity class as shown in Tables II and III below where we tabulate the number of systems detected/number of systems observed.

Table II

SYSTEMS WITH AN EARLY-TYPE COMPONENT

		Luminosity Class				
	Total	I	II	III	IV	V
2SB's	2/26	-	-	0/4	1/3	1/23
1SB's	0/44	-	0/1	0/7	0/10	0/26
Singles	0/10	-	-	0/2	0/4	0/4

Table III

SYSTEMS WITH A LATE-TYPE COMPONENT

	Total	Luminosity Class				
		I	II	III	IV	V
2SB's	15/58	-	0/1	2/8	9/18	6/31
1SB's	5/41	-	-	3/8	2/10	0/23
Singles	6/37	2/2	1/2	1/10	0/15	2/8

Several characteristics of the working sample are immediately obvious. Most outstanding is the fact that all of the detected systems (within the distance limit) have at least one late-type component. There is now strong evidence the radio emission in many of the late-type binaries is one of a number "super solar flare" phenomena which arise near the surface of the active component (c.f., Weiler et al. 1978). The two detected systems which have both an early- and a late type component, Algol and b Per, are characterized by non-thermal radio emission which is similar in its timescales for variability, energies, and spectra to the late-type systems. Thus, there exists the unresolved question whether these systems should be classed with the late-type stars, or perhaps, with the Antares system which is discussed later.

The two other early-type detected systems are CC Cas and β Lyr (both of which are beyond the distance limit). CC Cas may be a thermal emitter. It has a somewhat longer timescale for variability compared to those of the other active radio stars and has never exhibited a non-thermal spectrum. Furtherfore, there is good agreement between its maximum radio luminosity and its excitation parameter (the luminosity of an ionization-bounded HII region). All of this suggests CC Cas may be a stellar-wind type source which undergoes variations in its mass-loss rate. β Lyr, like CC Cas, exhibits some characteristics of a stellar-wind emitter, most notably optical evidence for an expanding envelope (Sahade and Wood 1978) and a radio spectral index which is typical of an optically thick source (at cm-wavelengths) with an inverse square-law density distribution. However, the ionization mechanism for such a wind is probably not radiative since the B8IIp primary is too cool to have the required Ly-α flux to match the source's observed excitation parameter. However, collisional excitation via turbulent mass motions is a possible thermal excitation mechanism. It is also possible for β Lyr to be a non-thermal emitter whose radio spectrum appears opaque due to the large amount of absorbing gas in the vicinity of the binary. Clearly, this puzzling source must be investigated further.

Can the sample give any indication of what are the key characteristics which distinguish the active emitters from the quiscent objects? This is a very difficult question to answer because there is obviously interplay between a number of factors. However, there are some interesting trends in the data. For instance, the only source which clearly shows evidence for the conversion of mass kinetic energy (mass exchange) to radio emission is the wide binary Antares. In this system, particle acceleration as well as magnetic field amplification probably take place at the interface between the winds (or the wind and the magnetosphere, respectively) of the M-supergiant and the B4Ve component (Gibson 1978; Kudritzki and Reimers 1978). The question of how mass exchange plays an active role in the other active binaries is unclear. β Lyr has by far the highest rate of mass transfer of any of the systems investigated and it is a strong radio emitter which may extract its energy from the turbulent motions of mass exchange. However, in a system like Algol, which is known to undergo significant mass transfer, one has the apparent contradiction that the emission is clearly non-thermal and could be attributed to phenomena on the late-type secondary. The picture is further clouded by the fact that many of the binaries surveyed, including other Algol-type systems, early-type systems (e.g., α Vir), and cataclysmic variables show evidence for mass exchange but are not radio emitters, while the RS CVn binaries, which probably have mass exchange rates five orders of magnitude less than β Lyr, have similar radio luminosities to that system. In addition, six of the systems which do exhibit non-thermal emission are single. Thus, at present, mass exchange does not appear to play a primary role in stellar radio emission except in a few special instances. What its secondary roles might be (e.g., as a perturbation on the surface boundary conditions of a member of a close binary) remain to be investigated more fully.

Another factor which does play a role in stellar radio emission is the magnetic field, especially when one considers that 24 of the 26 detected systems are synchrotron or gyrosynchrotron emitters. In this respect, we need to consider the effect of binarity on the magnetic field both in the "passive" (i.e., as a factor in determining the structure of the radio emission region) and the "active" (i.e., as a factor in the possible storage and subsequent release of energy) sense. In examining the passive case, it is interesting to note that there appears to be a tendency for stronger emission from systems which have active components which fill or nearly fill their Roche lobes. Two examples are Algol and HR 1099. Both have been observed by VLBI (Clark et al. 1976; Cohen 1979) and apparently have sizes comparable with the system size. Likewise, observations of AR Lac with the VLA (Owen and Spangler 1977) did not reveal any evidence for radio eclipses which also implies the radio emission region is at least twice as large as the component stars. If we combine these data with the inferred magnetic field strengths determined by observations of circular polarization (c.f., Owen et al. 1976), $B \sim 30$ Gauss, we find that the magnetic field energy density is large and should be a significant factor in the intrabinary gas dynamics. The weakest emission, of the detected cases, comes from dwarf systems (both single and binary) and evolved single-lined systems (e.g., λ And and σ Gem) which are well contained within their Roche lobes. It would seem physically possible for gas to escape from the "active" stars and drag with it magnetic field whose lines are anchored to that star. Thus, in the more active binaries the size of the radio source may correspond to the volume into which "spill-over" from the active star occurs while in the less active binaries or single stars spillover may not occur and the radio emission volume may be characteristic of the volume of the star itself or perhaps something analogous to a magnetic loop anchored to its surface. Clearly there would be great benefit from a concerted effort to observe radio stars with VLBI techniques.

The question of how magnetic fields would play an "active" role in close binaries is confusing. On one hand, there is no evidence for radio emission from early-type stars with large global magnetic fields, e.g., the Ap-stars. On the other hand, relatively strong emission arises from late-type systems which exhibit phenomena usually associated with starspot activity. Thus, at present, it appears the presence of magnetic field is primarily of consequence in those systems which have stars with convective envelopes. The most obvious role it would have in those stars would be in the action of a convective dynamo. Binarity could have several effects on this activity. Firstly, it could significantly affect the rotational velocity of the star. Single late-type stars are magnetically braked as they evolve onto the main sequence and typically have rotation velocities $\sim$ 20 km/s. In close binaries where tidal synchronization takes place, this same late-type star may have a rotation velocity an order of magnitude higher. Perhaps this is one reason why the RS CVn binaries are at least 100 times more luminous than their single counterparts.

A second consequence of binarity -- tidal distortion -- may also have a marked effect on dynamo action. There is a general tendency for enhanced radio activity in systems with distorted components. Systems with subgiant or giant components, especially if the active component is the secondary, tend to be stronger emitters than dwarf systems which have more spherical components. Among the detected single-lined spectroscopic binaries, the weakest are those with primaries which are well contained within their Roche lobes. However, it is confusing at best to distinguish between tidal distortion and its implied consequence, Roche lobe spillover (and its effects) with regard to their relative importantance to the radio emission process.

Based on the available data, two other factors may be important to stellar radio emission but the role of binarity in enhancing these processes is very unclear. There is weak evidence that the maximum radio luminosity depends on the depth of the convective zone. Stars like Betelguese do have very deep convective zones and it is a very strong radio emitter at times. On the other hand, late-F and early-G main sequence stars, which have relatively shallow convective zones, are much weaker. The second factor is that stars which are likely to be in rapid phases of evolution appear to be more active (as measured by flare frequency rather than flare energy) than stars in long-lived states. There is good evidence that the components of the RS CVn systems are rapidly evolving as evidenced by their positions within or astride the Hertzsprung gap (Popper and Ulrich 1977). There is also evidence that the dMe-stars may be in the process of evolving onto the main sequence. Possibly these two factors act to affect the differential rotation, and thus the dynamo action in these systems.

4. SUMMARY

Certainly binarity is important to stellar radio emission. Gibson's (1977) observations that single late-type subgiants are at least a factor of 100 less luminous than binaries with similar components is perhaps the best evidence of this fact. However, our understanding of which specific aspects of binarity affect the radio emission process is still minimal. At present, there appear to be at least three fundamentally different types of radio binaries:

i) CC Cas-type -- thermal emitters possibly having a variable stellar winds,

ii) Antares-type -- non-thermal emitters with energy derived from shock-excitation at the collision of two stellar winds, and

iii) RS CVn-type -- non-thermal emitters with energy derived from dynamo action in the component star's convective zone.

In class i) the effect of tidal interaction on the production of a stellar wind has yet to be addressed. In class ii) the physics of interacting stellar winds including the associated instabilities requires considerable attention because it is important to other branches of astrophysics. In class iii) we have suggested that a number of factors may be important including, rotation, tidal distortion, etc. Whether or not all of these factors can be integrated onto a working model of the source is unclear. However, it is an important, if not an unenviable, task which must be addressed if we are to move to a higher level of understanding of these systems.

REFERENCES

Allen, W.C.: 1973, Astrophysical Quantities, (The Athlone Press: London), 200.

Clark, T.A., Hutton, L.K., Ma, C., Shapiro, I.I., Wittels, J.J., Robertson, D.S., Hinteregger, H.F., Knight, C.A., Rogers, A.E.E., Whitney, A.R., Neill, A.E., Resch, G.M., and Webster, W.J.: 1976, Astrophys. J. Lett. 206, L107.

Cohen, N.L.: 1979, private communication.

Davis, R.J., Lovell, B., Palmer, H.P., and Spencer, R.E.: 1978, Nature, 273, 644.

Feldman, P.A.: 1979, IAU Circ. 3366.

Feldman, P.A.: 1978, Bull. Am. Astron. Soc., 10, 418.

Feldman, P.A., MacLeod, J.M., and Andrew, B.H., 1979, IAU Circ. 3385.

Gibson, D.M.: 1978, Bull. Am. Astron. Soc., 10, 637.

Gibson, D.M.: 1977, Bull. Am. Astron. Soc., 9, 600.

Gibson, D.M., and Newell, R.T.: 1979, IAU Circ. 3337

Gibson, D.M., and Owen, F.N.: 1979, work in progress.

Kudritzki, R.P., and Reimers, D.: 1978, Astron. Astrophys., 70, 227.

Owen, F.N., and Gibson, D.M.: 1978, Astron. J., 83, 1488.

Owen, F.N., Jones, T.W., and Gibson, D.M., 1976, Astrophys. J. Lett., 210, L27.

Owen, F.N., and Spangler, S.R.: 1977, Astrophys. J. Lett., 217, L41.

Popper, D.M., and Ulrich, R.K.: 1977, Astrophys. J. Lett., 212, L131.
Sahade, and Wood, F.B.: 1978, Interacting Binary Stars, (Pergamon Press: New York), 136 ff.
Spangler, S.R., Owen, F.N., Hulse, R.A.: 1977, Astron. J., 82, 989.
Weiler, E.J., Owen, F.N., Bopp, B.W., Schmitz, M., Hall, D.S., Fraquelli, D.A., Piirola, V., Ryle, M., and Gibson, D.M.: 1978, Astrophys. J., 225, 919.
Wendker, H.J.: 1978, Abh. Hamburger Sternw., Band 10, 3.

ULTRAVIOLET SPECTROSCOPY OF BINARY SYSTEMS

A.K. Dupree, L. Hartmann, and J.C. Raymond
Harvard-Smithsonian Center for Astrophysics
Cambridge, MA 02138 U.S.A.

1. INTRODUCTION

The recent launch of the International Ultraviolet Explorer (IUE) satellite and its spectrographic instruments (Boggess et al. 1978a,b) suddenly made accessible large numbers of diverse types of binary systems. These range from the luminous hot supergiants with compact companions that produce copious amounts of X-rays to the dwarf solar like stars of W UMa-type. For systems such as these and many others the IUE is a new and unique tool with which to probe the structure of extended stellar atmospheres, the presence and acceleration of a stellar wind, and the appearance of stellar chromospheres and coronae.

In this review, we shall discuss four typical systems that illustrate some of the major problems in the study of binary stars. They are:

- High Luminosity X-ray sources typified by Cyg X-1 (HDE 226868) and Vela XR-1 (HD 77581).
- Low Luminosity X-ray sources (HZ Her).
- Late-Type Systems of W UMa and RS CVn type.
- Cool Supergiants with a Hot Companion (VV Cephei).

Each of these classes also shows the unique nature of ultraviolet emissions with which to investigate particular aspects of stellar atmospheres.

2. HIGH LUMINOSITY X-RAY SOURCES

These systems typically contain a hot early type supergiant star and a companion compact object. X-rays are produced by accretion of material onto the compact object either from the massive stellar wind or following Roche Lobe overflow. The luminous primary star dominates the ultraviolet spectrum although effects of the X-ray source upon the stellar wind can be found; studies of such effects were not fruitful before the advent of IUE.

M. J. Plavec, D. M. Popper and R. K. Ulrich (eds.), Close Binary Stars: Observations and Interpretation, 39–52.

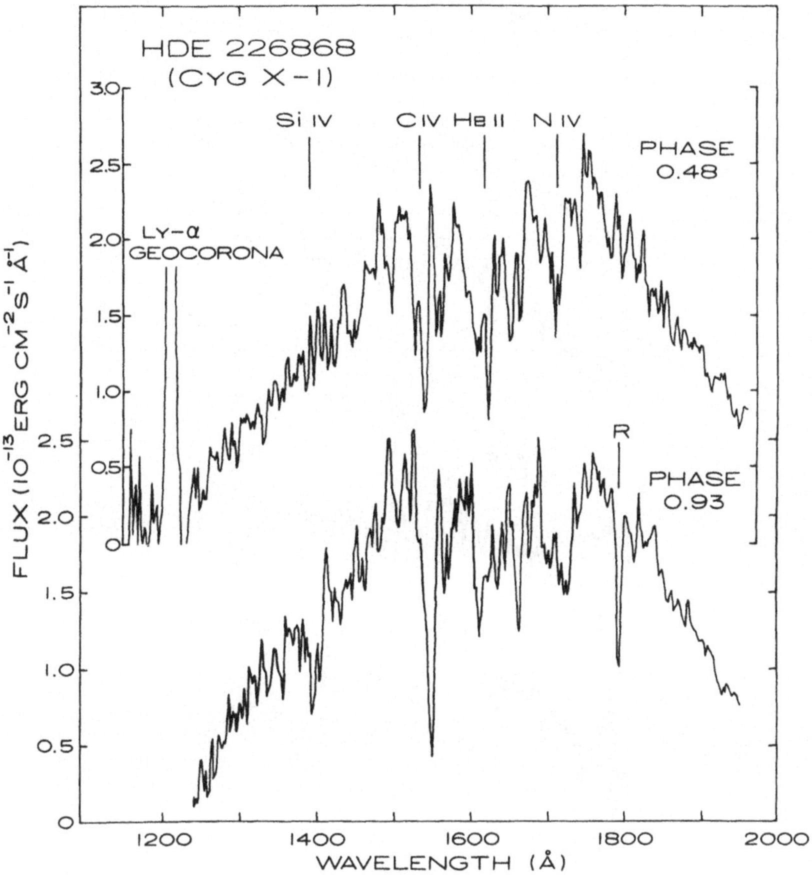

Figure 1. Low dispersion short wavelength exposures from IUE of HDE 226868 (Cyg X-1) at two orbital phases (Dupree et al. 1978). Note the weakening of the Si IV and C IV lines at phase 0.48 (near inferior conjunction of the compact object) as compared to phase 0.93. R denotes a reseau - a fiducial mark.

The first spectroscopic measurements of HDE 226868 (Cyg X-1) (Fig. 1) were taken by the IUE Commissioning Team in the low dispersion mode (Dupree et al. 1978) and appeared to be a typical spectrum of a highly reddened early - B supergiant showing characteristic P Cygni profiles indicating the presence of substantial mass loss. However, variation in the line profiles with orbital phase was apparent. From the low dispersion spectra, one finds the absorption components are strong at superior conjunction of the compact object. They become weaker near phase 0.5 when the compact object lies between us and the supergiant. Although only 2 spectra were available at first, a later more thorough study (Treves et al. 1979) of this source as part of an international collaboration of scientists from Europe, the United Kingdom and the U.S.A. confirmed the earlier result.

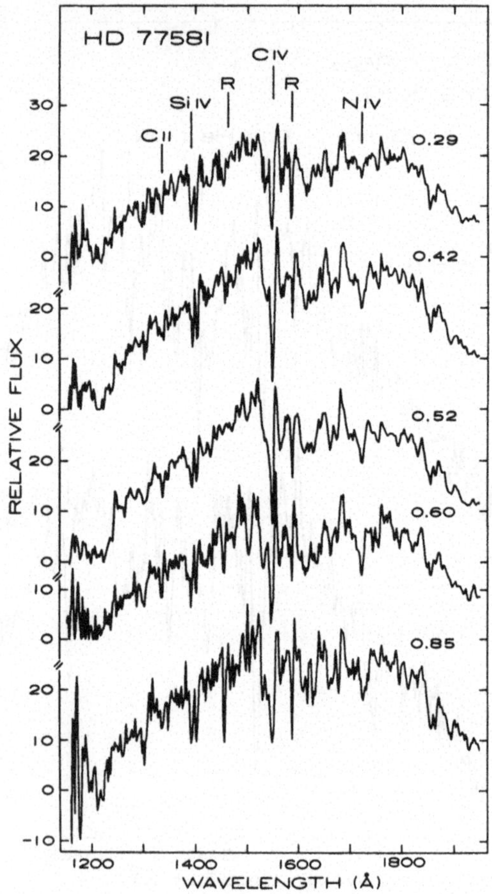

Figure 2. Low dispersion short wavelength exposures from IUE of HD 77581 (Vela XR-1) at many orbital phases (Dupree et al. 1979). Phase 0.0 corresponds to eclipse of the X-ray emitting pulsar.

Another binary system, HD 77581, identified with the X-ray source Vela XR-1 (4U 0900-40) is composed of a hot supergiant (B0.5Ib) and a pulsar with a 283 second period (McClintock et al. 1976) which is orbiting the primary with an 8.96 day period. At low dispersion (Fig. 2) the spectrum appears similar to HDE 226868 and also shows similar phase effects in the P Cygni features (Dupree et al. 1979). Fortunately, this system is relatively bright so that accurate high dispersion measurements are possible with IUE. Fig. 3 shows a portion of the high dispersion spectrum near λ1550. The extended short wavelength absorption at phase $\phi = 0.04$ indicates terminal velocities $V \sim 1700$ km s^{-1}. The most interesting aspect is the dramatic change in appearance of the P Cygni profiles. As the secondary passes in front of the primary star, the edge velocity of the profile decreases and a high velocity absorption feature appears at phase 0.52. A similar phenomenon is found in the Si IV lines.

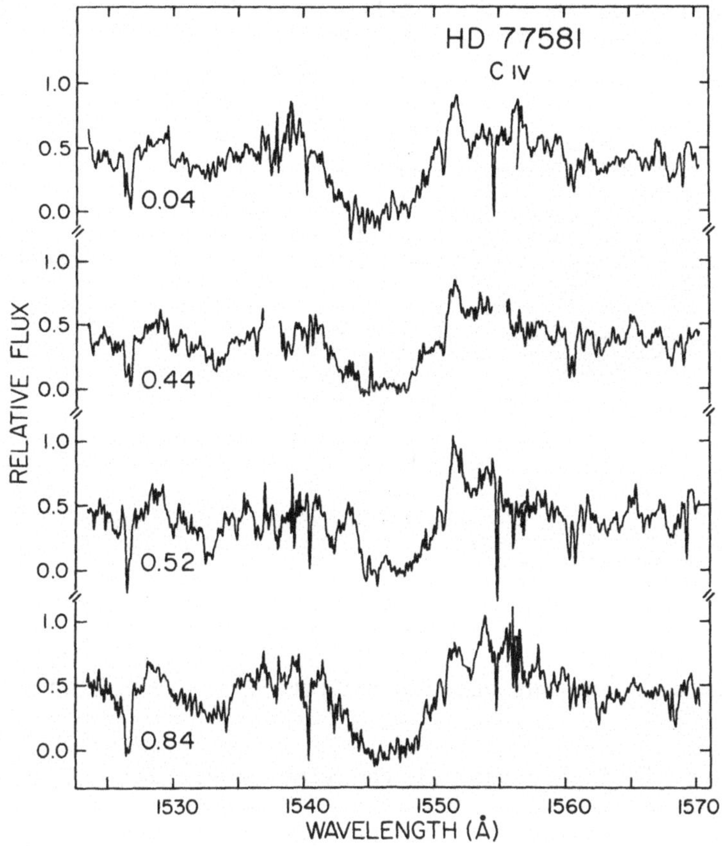

Figure 3. High dispersion IUE observations of the spectrum near λ1550 in HD 77581 showing the C IV P Cygni type profile and its variation with phase. Note the appearance of a high velocity feature at phase 0.52.

Such behavior can be understood in terms of the heating and ionization caused by an X-ray source immersed in a stellar wind as suggested by Hatchett and McCray (1977). Fig. 4 schematically illustrates the volume in which C IV can be ionized by an X-ray source in a system such as HD 77581. Within the broken line surrounding the X-ray source, C IV is assumed to be nonexistent. It is easy to see how the apparent edge velocity of the wind will decrease and an absorption feature can appear at high velocities. Calculations of the expected blue wing of the line profile (Fig. 5) confirms the appearance of an absorption feature as the X-ray source passes in front of the primary star.

It is noteworthy that HD 153919, an O6f star identified with the X-ray source 4U 1700-37 does not show such phase dependent behavior (Dupree et al. 1978) apparently because the large optical depths of the strong resonance lines mask any changing ionization state in the wind.

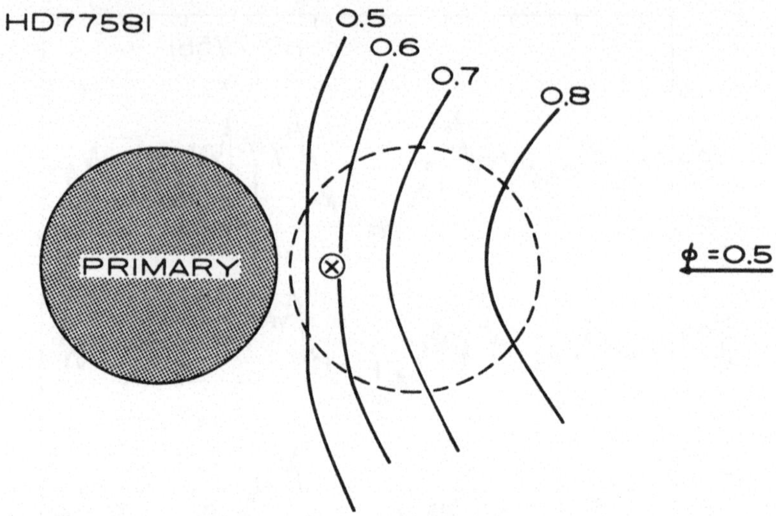

Figure 4. Schematic diagram of the HD 77581 system showing a typical ionization volume (broken line) surrounding the X-ray source marked by X. Within this volume a selected ionic species is absent. The solid lines mark the locus of a constant value of V/V_∞ where V_∞ is the wind terminal velocity

The size of the ionized cavity in the wind sensitively depends upon the X-ray spectrum and the mass loss rate. Thus studies of binaries containing X-ray sources provide criteria to test our understanding of X-ray emission and stellar winds. The detailed structure of the velocity law in the wind can also be evaluated. In HD 77581, the terminal velocity of 1700 km s^{-1} is higher than the 950 km s^{-1} derived on the basis of Abbott's (1978) theory that the terminal velocity should be three times the escape velocity (corrected for continuum radiation pressure). But the velocity agrees with the empirical relation between terminal velocity and escape velocity derived by Lamers, van den Heuvel, and Petterson (1976).

3. LOW LUMINOSITY SOURCES (HZ Her)

The complex and fascinating X-ray binary HZ Her = Her X-1 was also observed with IUE as part of a collaborative effort of American and European astronomers (Gursky et al. 1979). Fig. 6 indicates typical spectra of HZ Her, similar to those reported by Dupree et al. (1978). The very strong Ly α feature is geocoronal in origin. The dominant features are the N V λ1240 and C IV λ1550 emission lines superposed on a relatively flat, featureless continuum. He II λ1640 is also present at a weak level. There is evidence for Si IV λ1394, λ1403 emission as well. O V λ1371, N IV λ1719, and N III λ1750 may be present but are not apparent in all spectra.

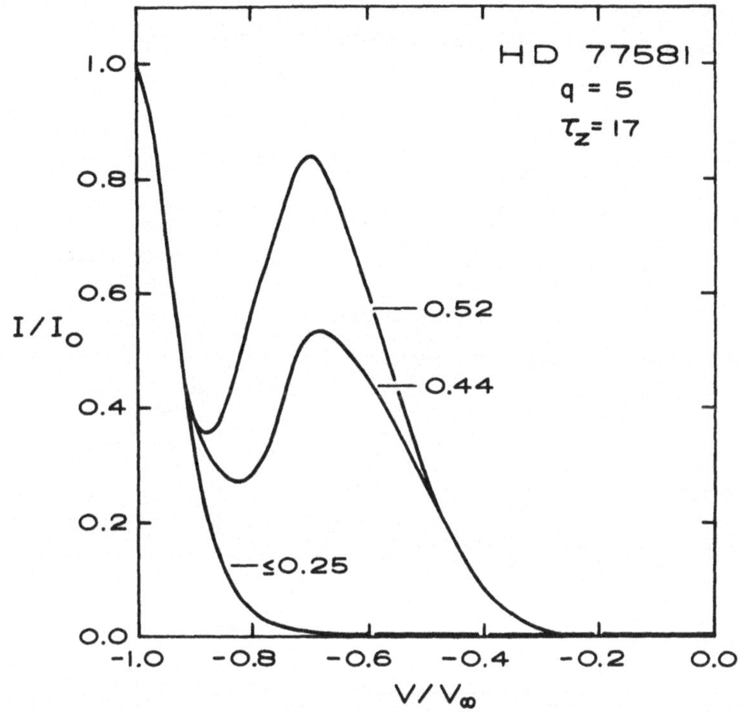

Figure 5. Calculated blue absorption wing of P Cygni profiles at three phases (0.52, 0.44, ≤ 0.25). High velocity absorption appears at phases close to inferior conjunction of the secondary.

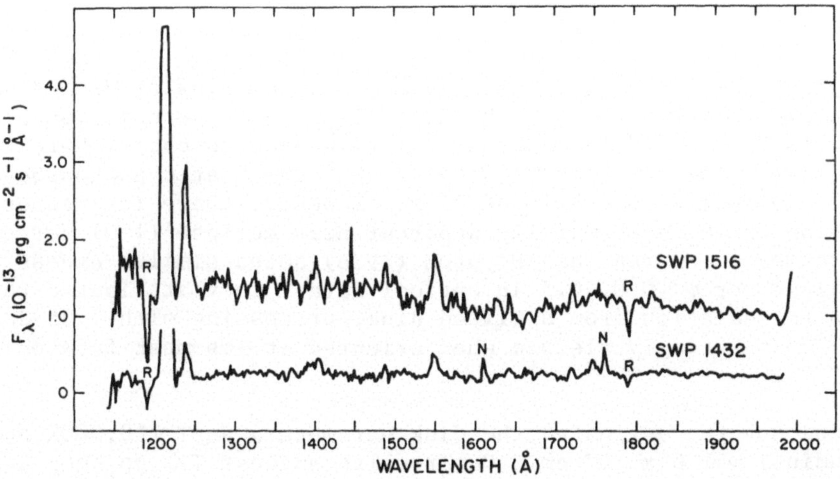

Figure 6. Low dispersion IUE spectra of HZ Her (Gursky et al. 1979) at two orbital phases: $\phi = 0.143$ (SWP 1432) and $\phi = 0.350$ (SWP 1516). Fiducial marks are denoted by R (reseau); noise spikes are indicated by N.

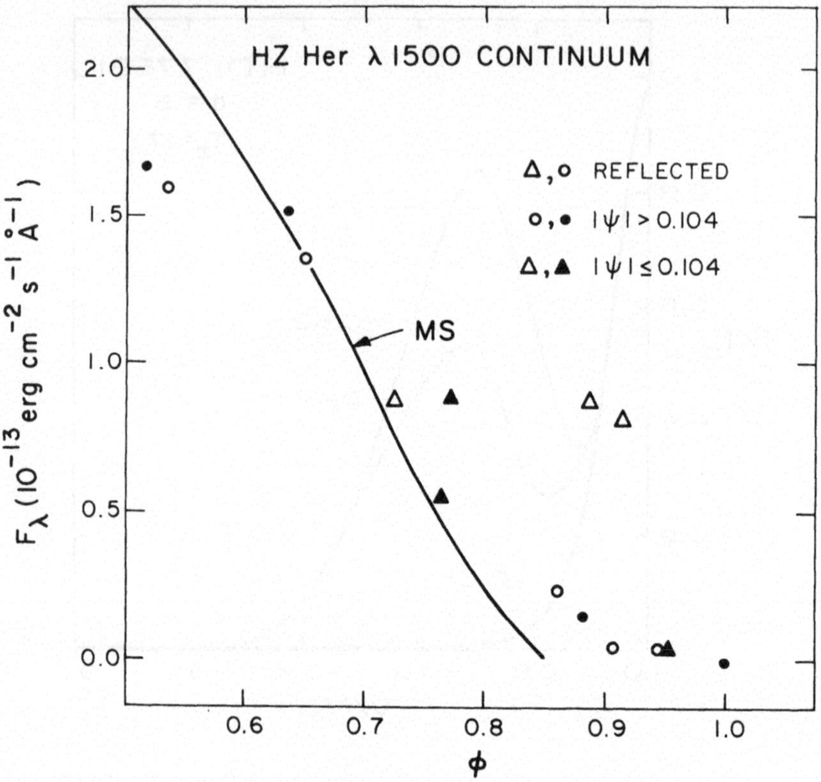

Figure 7. Observed and predicted (denoted MS) flux variation of HZ Her at λ1500 as a function of phase.

Fig. 7 shows the behavior of the 1500Å continuum flux of HZ Her as a function of orbital phase ϕ. The solid line is the prediction of the Milgrom & Salpeter (1975) model of the X-ray heated photosphere of the primary star. The agreement indicates that the heated photosphere is a primary component of ultraviolet flux. However, there is another source of emission which is especially apparent near eclipse ($\phi \sim 0$) for certain 35-day phases ψ. Gerend and Boynton (1976) noted similar excess fluxes near eclipse for $\psi \sim 0.0$, 0.5 in optical data. They attributed this excess emission to radiation from a tilted disk, precessing with a 35-day period, which is most strongly visible when oriented at its most face-on towards Earth.

Gerend and Boynton estimated the disk parameters as $T \sim 10{,}000$K and r (disk radius) $\sim 1.6 \times 10^{11}$ cm. Our results suggest $T \gtrsim 20{,}000$K and r $\sim$ 1×10^{11} cm. The discrepancy in temperature is not surprising, for the ultraviolet spectral region is bound to be more sensitive to high-temperature material. However, it does indicate the importance of acquiring UV data in arriving at a physically meaningful picture of disk properties.

The general behavior of the C IV and N V lines strongly suggests that they are principally formed in or near the heated photosphere as is the continuum, particularly N V, for which the equivalent width is roughly constant as a function of orbital phase. Simple calculations using the results of Hatchett, Buff, & McCray (1976) indicate that the heated photosphere can account for the N V flux; this same model predicts a C IV flux which is of magnitude larger than observed, suggesting that optical depth effects may be important. The C IV flux does not vary as much as N V with its equivalent width increasing as one approaches eclipse. Most of the optical emission lines show no strong correlation with orbital phase (e.g. Crampton and Hutchings 1974; Barbon _et al_. 1974). The N V and C IV line behavior suggests that the phase-dependent modulation increases with increasing temperature of formation, again indicating the importance of UV data in investigating variations in physical conditions.

4. RS CVn AND W UMa BINARIES

Binary systems containing late-type giants or dwarf members are typically strong sources of ultraviolet emission. The RS CVn stars are late-type giant and subgiant binaries whose surface activity has been known from optical studies to be peculiar, in that large spots cover their surfaces (Caston & Hall 1979). Ultraviolet observations with _Copernicus_ and IUE of members of this class such as λ And, HR 1099, and HR 4665, have shown that these stars are powerful emitters in transition-region lines, with much higher surface fluxes than the quiet sun. RS CVn binaries have also been shown to be strong soft X-ray sources (Walter, Charles, and Bowyer 1978). This combination of high chromospheric and coronal activity with spot activity makes it plausible to interpret these phenomena as extreme analogues of solar activity.

The W UMa stars are late-type, dwarf binaries which are orbiting so closely together that they may share a common convective envelope (Lucy 1976; Flannery 1976; Shu, Lubow & Anderson 1976). Two W UMa systems have been observed with IUE: VW Cep (G5) and 44 Boo (G2). The system VW Cep was recently discovered to be an X-ray source by the NRL experiment on HEAO-A (Cruddace 1978). These stars are contact binary systems that eclipse with periods of less than one day. Both stars have ultraviolet spectra that are very similar. The short wavelength spectrum of 44 Boo is shown in Fig. 8. High temperature emission lines, N V, C IV, and Si IV are apparent, as is He II λ1640.

In Fig. 9, the stellar surface fluxes in the binary systems are compared to that of the quiet sun, assuming that the chromospheric activity is uniformly distributed across the stellar surface. The fluxes for λ And and HR 1099 were taken from Linsky _et al_. (1978). The surface fluxes show a general progression of increasing enhancement with increasing temperature much like that found for a solar active region (Dupree _et al_. 1973) and for active dwarfs (Hartmann _et al_. 1979). However, the fluxes of the W UMa stars are an order of magnitude larger than typical solar active regions. The He II λ1640 line is strongly enhanced, which may reflect photoionization by increasing coronal emission measures (Hartmann et al. 1979).

There appears to be a dependence of the surface flux upon orbital period. The W UMa stars 44 Boo and VW Cep have periods $0^{d}.7$; the RS CVn stars have longer periods: HR 1099 ($2^{d}.8$) λ And ($20^{d}.5$); HR 4665 ($64^{d}.4$). The results in Fig. 6 suggest that the flux is enhanced with decreasing orbital period. Because of tidal synchronization, the rotational periods should be close to the orbital periods; this is verified by the spot light curve periods in RS CVn stars. Thus the behavior of emission strength with orbital period may be viewed as an extension of the general increase of chromospheric flux with increasing rotation rate originally found for dwarfs (Kraft 1967; Skumanich 1972). Observations of such a general relationship encompassing stars of very different interior structure provide important constraints on theories of chromospheric and coronal heating.

5. COOL SUPERGIANTS WITH A HOT COMPANION

Binary systems composed of a late-type supergiant with a hot companion offer the unique situation in which to probe the structure of an extended atmosphere with continuous source of radiation. Particularly fortuitous was the egress from eclipse of the secondary of VV Cephei during the first

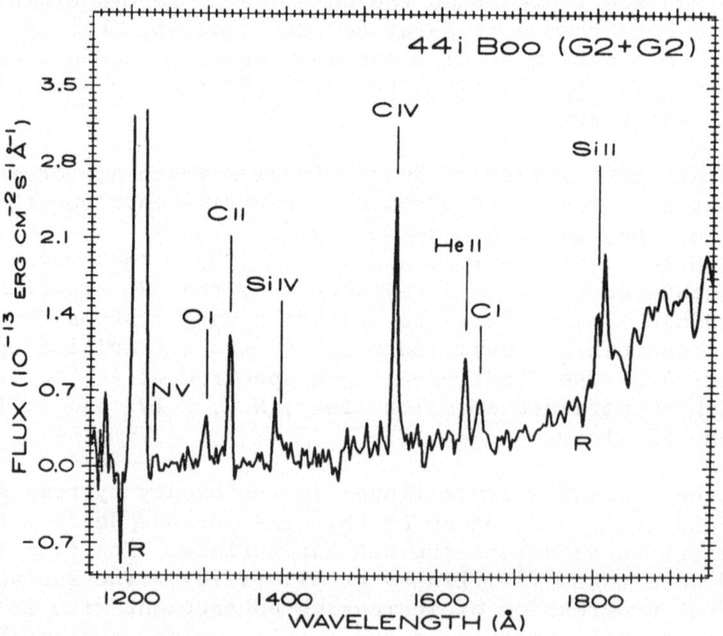

Figure 8. Short wavelength low dispersion IUE spectrum of a W UMa - type star: 44 Boo.

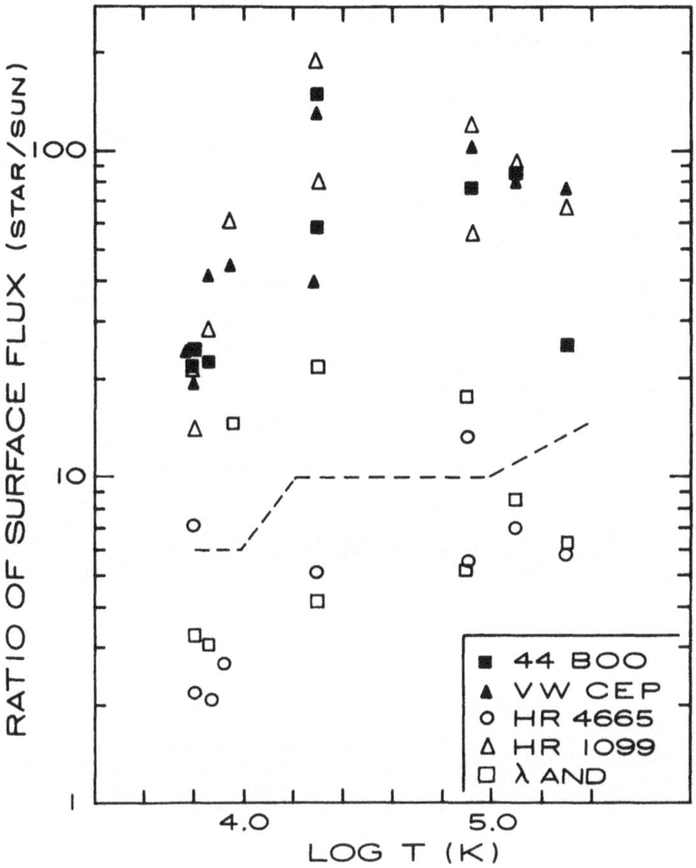

Figure 9. The ratio of stellar surface flux to solar flux in various emission lines as a function of temperature of formation of the ionic species. These stars are all binary systems. The broken line indicates the ratio for a typical solar active region.

months after the launch of IUE. This eclipsing binary system (M2 Iabep + B) was monitored by Faraggiana and Selvelli (1979) and Hagen et al. (1980) during 1978 and 1979. Fig. 10 shows ths appearance of the short wavelength ultraviolet spectrum as the general flux level increases reflecting the changing path length through the atmosphere of the M supergiant. This variation is used by Hagen et al. (1980) to determine the density law in the atmosphere. Egress in the ultraviolet lagged visible egress by several months due to the higher opacity of the M star atmosphere in the ultraviolet. Fluorescent emission lines dientified by Hagen et al. dominated this spectrum. These features result from the excitation of species in the atmosphere of the M supergiant by the ultraviolet continuum of the hot companion. Later, the hot continuum becomes blanketed by absorption lines arising in the extended atmosphere of the supergiant. High dispersion spectra (Fig. 11) show the stratification

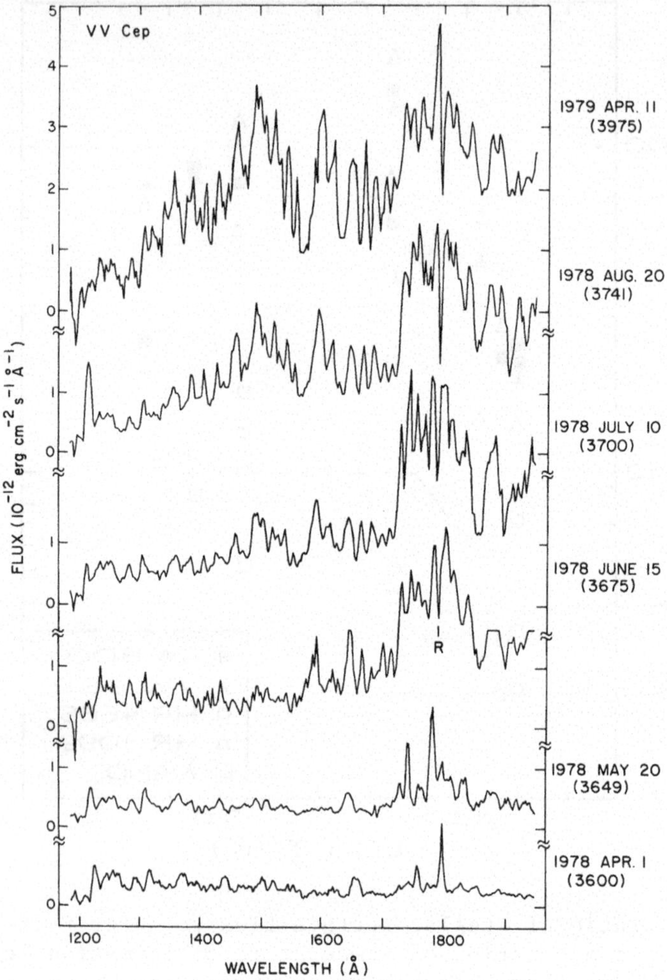

Figure 10. The short wavelength IUE spectra of VV Cephei as the secondary emerged from eclipse (Hagen et al. 1980).

of ionic species as marked by the weakening or disappearance of Fe I absorption features. Apparently, neutral iron does not extend as far out in the envelope as Fe II. The Fe II emission profiles change dramatically with time. For example, between 20 May and 15 July 1978 a sudden enhancement of the blue emission peaks relative to the red peaks of these lines was observed, although the total fluxes of the lines remained roughly constant (Hagen et al. 1980).

The surface fluxes of the chromospheric emission line Mg II and Fe II are substantially enhanced over the values found for single stars of comparable spectral type. Such enhancement may result from the presence of a hot companion. The spectral type of the companion is still indeterminate

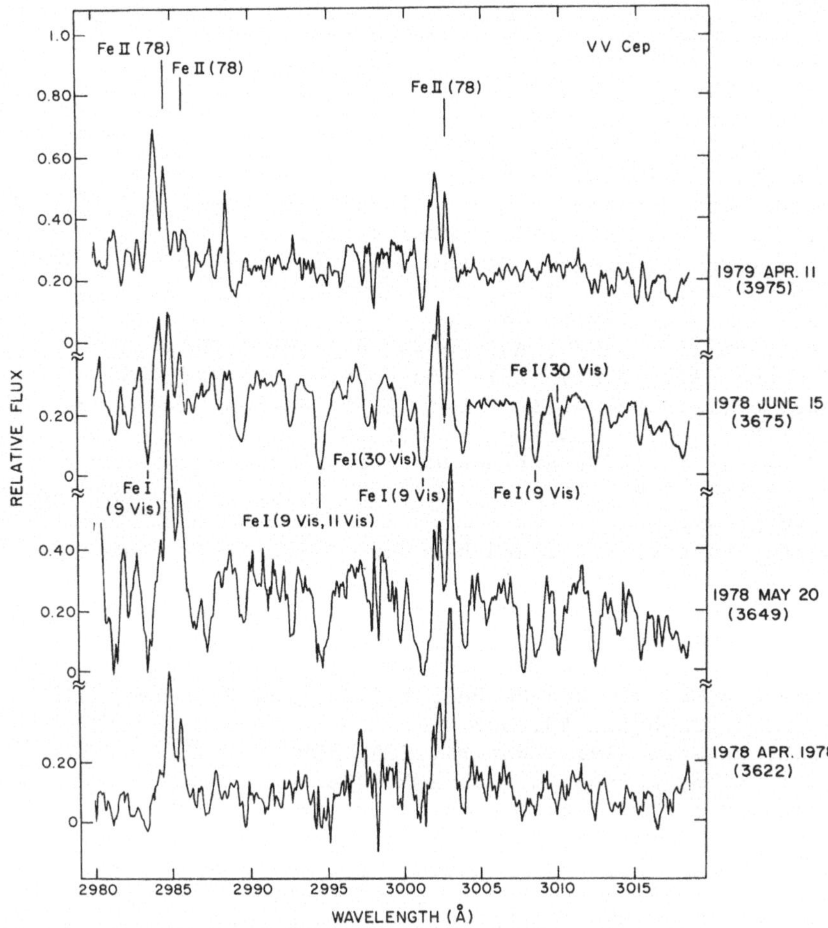

Figure 11. A section of a high dispersion spectrum of VV Cep near λ3000. Note the changing strength of the emission components of the Fe II transitions, and the disappearance of the Fe I multiplets by LL April 1979 (Hagen et al. 1980).

(April 1979) due to the proximity to eclipse and/or the interaction bebween the components of the system. Work is continuing on this complex and important system in order to form a more comprehensive physical picture of the extended atmospheres of late supergiant stars.

ACKNOWLEDGEMENTS

This work was supported in part by NASA Grant NSG 5370 and NSG 7176 to the Harvard College Observatory.

REFERENCES

Abbott, D.C.: 1978, Astrophys. J., 225, 893.
Barbon, R., Benvenuti, P., Bernacca, P.L., and Ciatti, F.: 1974, Astron. and Astrophys., 31, 237.
Boggess, A. et. al.: 1978a, Nature.
Boggess, A. et al. 1978b, Nature.
Crampton, D., and Hutchings, J.B.: 1974, Astrophys. J., 191, 483.
Cruddace, R.: 1978, personal communication.
Dupree, A.K., Huber, M.C.E., Noyes, R.W., Parkinson, W.H., Reeves, E.M., and Withbroe, G.L.: 1973, Astrophys. J., 182, 321.
Dupree, A.K. et al.: 1978, Nature.
Dupree, A.K. et al.: 1979, Astrophys. J., submitted.
Eaton, J., and Hall, D.S.: 1979, Astrophys. J., 227, 907.
Faraggiana, R., and Selvelli, P.L.: 1979, Astron. Astrophys. 76, L18.
Flannery, B.P.: 1976, Astrophys. J., 205, 217.
Gursky, H. et al.: 1979, Astrophys. J., submitted.
Hagen, W., Black, J.H., Dupree, A.K., and Holm, A.V.: 1980, Astrophys. J., in press.
Hartmann, L., Dupree, A.K., and Raymond, J.C.: 1979, submitted to Astrophys. J. (Letters).
Hatchett, S., Buff, J., and McCray, R.: 1976, Astrophys. J., 206, 847.
Hatchett, S., and McCray, R.: 1977, Astrophys. J., 211, 552.
Kraft, R.P.: 1967, Astrophys. J., 150, 551.
Lamers, H.J.G.L.M., van den Heuvel, E.P.J., and Petterson, J.A.: 1976, Astron. Astrophys., 49, 327.
Linsky, J.L. et al.: 1978, Nature, 275, 389.
Lucy, L.B.: 1976, Astrophys. J., 205, 208.
McClintock, J.E.: 1976, Astrophys. J. (Letters), 206, L99.
Milgrom, M., and Salpeter, E.E.: 1975, Astrophys. J., 196, 589.
Shu, F.H., Lubow, S.H., and Anderson, L.: 1976, Astrophys. J., 209, 536.
Skumanich, A.: 1972, Astrophys. J., 171, 565.
Treves, A. et al.: 1979, Astrophys. J., submitted.
Walter, F., Charles, P., and Bowyer, S.: 1978, Astrophys. J. (Letters), 225, L119.

DISCUSSION FOLLOWING DUPREE, HARTMANN AND RAYMOND

Hack: You said that CIV and SiIV in Vela were destroyed by X-ray radiation. What about NV?

Dupree: We expect NV to behave in a similar way as C IV and Si IV However because of its higher ionization potential the volume affected by the X-rays will be smaller than those for C IV and Si IV. The NV transitions at λ1240 are weakly exposed on our IUE spectra and it is not possible to comment on any variation.

Plavec: I noticed in your IUE spectra that the N V line is always much weaker than the C IV line, as usual in stellar transition regions.

I observed similar lines in my long-period interacting binaries, but arising under clearly very different physical conditions. The N V line is actually stronger than the C IV line. Moreover, the power radiated in either line is very much larger than in the systems you have observed.

OBSERVATIONS OF SPECTROSCOPIC BINARIES WITH A SOLID-STATE DETECTOR

Francis Fekel, Jr., Claud H. Lacy, and J. Tomkin
Department of Astronomy and McDonald Observatory
The University of Texas, Austin, Texas 78712

The recent installation of a solid-state 1024-element silicon photodiode array detector (Reticon) at the coude focus of the 2.7 m McDonald Observatory reflector has greatly extended its limits of observation for binary and multiple systems which have weak and/or broad-lined components. This detector can produce extremely high signal-to-noise ratio observations and has high quantum efficiency over the wavelength region 3000-11000Å. The observational programs of three users of this device are described below.

1. RETICON OBSERVATIONS OF ECLIPSING BINARIES BY J. TOMKIN

Secondaries of Algol-type eclipsing binaries are being observed by J. Tomkin. Most of the stars in his program have evolved secondaries although a few main-sequence pairs are also included. The sensitivity of the Reticon in the red and near-infrared--where the secondaries are relatively brighter than in the photographic part of the spectrum and blending problems are less severe--are advantages. The outstanding advantage, however, is the large signal-to-noise ratio that can be obtained. This is particularly true of bright objects. The detection of the NaD lines of the secondary of Algol (Fig.1) is an example.

Studies of δ Lib, Algol and U Sge have been published. A preliminary analysis of U Cep gives a K_2 of 200 km s^{-1}. Batten had suggested a tentative value of 180. The secondaries of RZ Cas, S Cnc, U CrB, Alpha CrB, AI Dra and EK Cep, have also been detected and further observations of these systems are underway.

J. Tomkin, D. Lambert and M. Parthasarathy are studying abundances in eclipsing binaries. An investigation of the metal abundances of the secondary of U Cep shows that they are not significantly different from solar. A preliminary analysis of the secondary of U Sge shows that it also has solar metal abundances. These results do not support the conclusions of earlier investigations by narrow-band photometry which had indicated large metal-deficiencies in evolved Algol-type secondaries. An investigation of carbon, nitrogen and oxygen abundances in primaries and secondaries is planned.

M. J. Plavec, D. M. Popper and R. K. Ulrich (eds.), Close Binary Stars: Observations and Interpretation, 53–57.

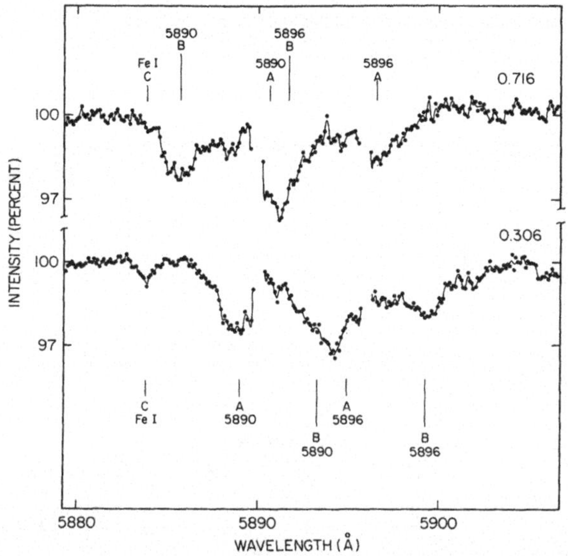

Figure 1. Two spectra of NaD lines in Algol are plotted on an expanded intensity scale. At phase 0.716 the 5890 Å NaD line of Algol B is resolved and the 5896 Å line is blended with the 5890 Å NaD line of Algol A. At phase 0.306 the situation is reversed. Lines from Algol C are also weakly present.

2. ECLIPSING BINARY PROGRAM OF CLAUD LACY AND DAVID EVANS

This program is restricted mainly (but not exclusively) to those systems which already have accurate photometric elements, but lack double-lined RV curves. Prime targets (listed in Table 1) are bright systems with large light-ratios and fainter systems with light-ratios nearer unity. The ultimate goal is to obtain accurate RV curves, and thus masses and absolute dimensions for these systems.

TABLE 1. ECLIPSING SYSTEMS IN WHICH MEASURABLE LINES OF SECONDARY HAVE BEEN OBSERVED.

*	Sp	V	l_1/l_2	Remarks
KP Aql	A2(K)	9.7pg	1	Sp looks like F-stars.
CW CMa	A4	8.58	1	
YZ Cas	A2V + F5V	5.65	9(Red)	Blue light ratio is 15.
V442 Cyg	F4	10.0pg	1	Period is double that of GCVS.
TT Hya	A3e + (G5)	7.5pg	8(Blue)	Si II broad, Fe I sharp.
FL Lyr	G5	9.30	2	
EE Peg	A4mV + FV	6.97	9(Blue)	
IQ Per	B7V + A2V	7.72	10(Blue)	Detection tentative.
V906 Sco	B9V + B9V	5.96	1	Triple-lined system.
TX UMa	B8V + gF2	7.06	4(Yellow)	Broad sec. lines ∿1% deep.

A pair of representative spectra are shown below:

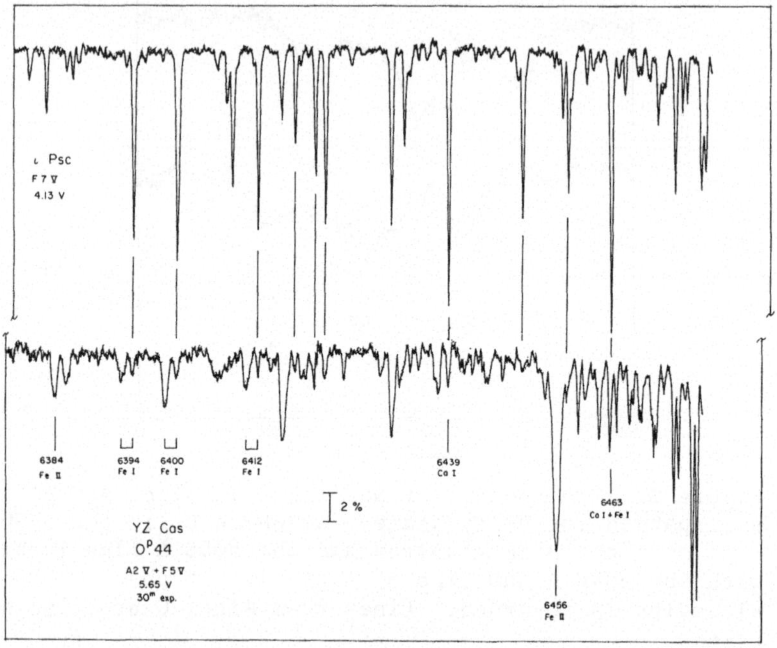

Figure 2--YZ Cas and ι Psc in the red. The upper comparison spectrum has been aligned with features due to the faint secondary in the lower spectrum. The strongest unblended lines have been marked. Some doubled lines are indicated. The strongest features of the secondary are ∿ 2% deep. The signal-to-noise ratio of this spectrum is about 500 to 1. The light-ratio is 9 to 1.

3. MULTIPLE STAR PROGRAM OF F. FEKEL

For the past several years Francis Fekel has obtained Reticon observations of short period (P $\leq$ 20 years) visual binaries, one or both of whose components are short period spectographic binaries, making the systems triple or quadruple. These spectroscopic observations, combined with visual and speckle interferometric data, enable the parallax and individual masses to be determined.

Masses have already been determined for the Ψ Sgr system which contains an early F III and a G 5III component. In collaboration with A. Batten, C. Morley and J. Tomkin, observations of HD 165590 have been completed. Work will shortly be completed on HD 203345 after it passes through nodal passage in its 6 year visual orbit later this year. It had been suggested that this system might contain a neutron star. Reticon observations showed the third component to be an early K dwarf. The visual secondary of μ Ori was thought to violate the mass-luminosity relation. Reticon observations (below) showed that the secondary is not a single star but a double-lined binary, thus solving this problem.

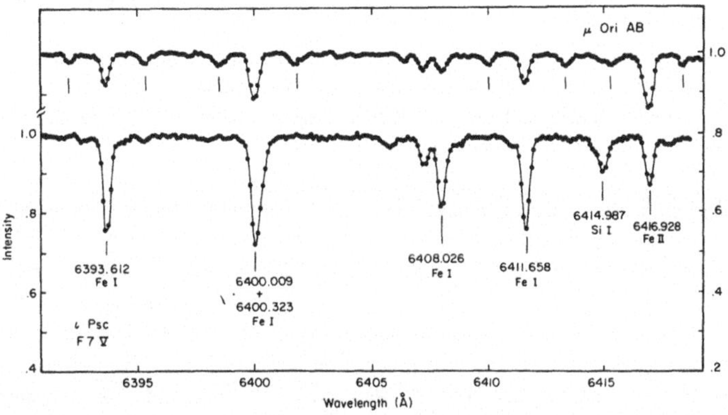

Weak broad-lined features can also be detected in the blue. Examples of this are HR 2130 and HR 266 (below). The broad

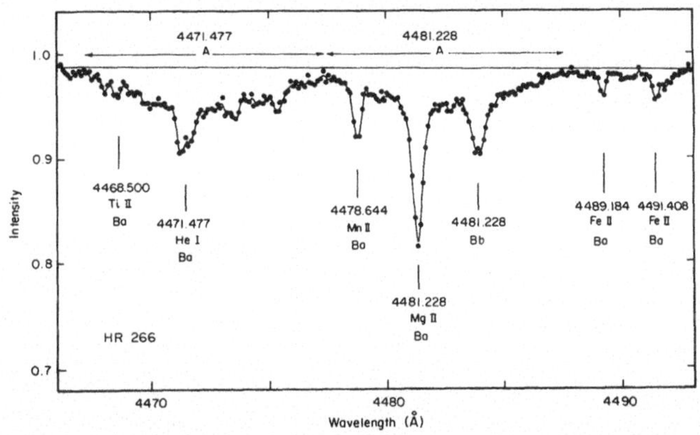

lines of HeI and MgII were not detected on well exposed IIaO plates but are obvious in the Reticon observations. Finally, in collaboration with George O'Brien, the mass ratio and rotational velocity of Capella A, B are being redetermined because of the very broad-lined nature of Capella B.

From the systems presently observed with the Reticon it appears the mid and late type systems can be detected as double-lined binaries if they have mass functions $\gtrsim 0.1\ M_{\odot}$. These systems have mass ratios as small as 0.6.

DISCUSSION FOLLOWING FEKEL, LACY AND TOMKIN

Popper: If I were much younger and was considering going into this field with the facilities now available to me, I would withdraw in favor of the superior capabilities of techniques such as that employed at McDonald. I hope adequate observing time will be made available for this fundamental work.

Budding: Does your list of stars include R CMa? It would be interesting to have independent confirmation of the existence of a definite class of "underluminous subgiants".

Lacy: I believe Joc Tomkin has it on his list.

Mayo: You have shown us some very impressive detections of the lines of faint components in binary and triple systems. Although Cygnus X-1 is usually considered as a binary system containing a black hole, the alternative hypothesis of replacing the black hole by a 'normal' star with a neutron star in orbit around it cannot be unambiguously ruled out. The question is: have you tried looking for lines of the unseen star? If not I strongly recommend doing so.

Fekel: To the best of my knowledge no one at Texas has looked at Cygnus X-1 with the coudé Reticon system. If lines of a possible third component are at least several percent deep they might be detected with our system.

HYDROGEN-LINE PHOTOMETRY OF CLOSE BINARIES

J.D. Dorren and M.J. Siah
Biruni Observatory, Shiraz University, Shiraz, Iran
E.F. Guinan and G.P. McCook
Villanova University, Villanova, PA 19085, USA

A program of Hα and Hβ narrow- and intermediate-band photometry of close binaries has been carried out at Biruni Observatory since 1977 and at Villanova University Observatory since 1971. The advantages of this type of photometry for close binary work are discussed. Stars which have extensive coverage are listed and results for Cyg X-1, WZ Sge, W Ser and HR 1099 are briefly discussed.

1 INTRODUCTION

Hα and Hβ photometry of close binaries is a powerful technique which can be used to extract useful information about stellar temperatures, or Hydrogen-line emission strengths arising from the presence of gas streams, envelopes and disks, or about centers of surface activity in RS CVn-type stars. The method, developed by Strömgren (1958) and Crawford (1958) is to employ pairs of intermediate and narrow band interference filters centered near the rest wavelengths of the Balmer lines Hα (6563A) and Hβ (4861A). The narrow-band filters are sufficiently wide (FWHM ~ 25-35A) to allow for the Doppler shifts due to binary motion, yet narrow enough to be sensitive to the line feature. On the other hand, the intermediate bandpass filters are sufficiently broad (FWHM ~ 150-300A) to remain little affected by the presence of the line. They therefore provide a measure of the continuum. A photoelectric Hα or Hβ index is defined in the following way:

$$\text{H-index} = -2.5 \log (F_N/F_I) + \text{constant}$$

where F_N and F_I are the stellar fluxes measured through the narrow and intermediate bandpass filters respectively. The method of Hβ photometry has been fully described by Crawford and Mander (1960), and used by Crawford and others chiefly in

M. J. Plavec, D. M. Popper and R. K. Ulrich (eds.), Close Binary Stars: Observations and Interpretation, 59–63.

the study of O- to F-type stars which are cluster members. Empirical calibrations of the Hβ system have been carried out (Crawford, 1975 and refs. therein) which relate the β index to such parameters as M_v, T_{eff} and to standard color indices such as c_1, b-y, and m_1. A comprehensive bibliography of Hβ photometry (as well as uvby photometry) is given by Philip and Perry (1979). Similar systems have been developed for Hα (Abt and Golson 1966, and Baliunas et al. 1975). Andrews (1968) has also obtained Hα indices for 976 B- and A-type stars by spectrophotometry. These studies show that the α index is more sensitive to emission than the β index. Additional advantages of Hα and Hβ photometry are the following: speed of measurement of the index and accuracy of line strength determination; the adequacy of a modest aperture telescope; results which are practically independent of atmospheric extinction and are not appreciably affected by interstellar reddening. On the other hand this method, unlike spectroscopy, does not provide line profiles or equivalent widths directly.

2 OBSERVATIONS AND RESULTS

Table I lists the stars for which we have extensive Hα or Hβ photometry. This encompasses stars of a wide range of dimensions, masses and spectral types, and periods from 83 min. to over 20 years. Except for the star R CMa which was observed in New Zealand in 1968, all the data was obtained using the 51-cm reflector at Biruni Observatory or the 38-cm reflector at Villanova Observatory, and in several cases the same stars were observed at both sites. The observations at Biruni were carried out by Dorren, Guinan and Siah, while the data at Villanova were obtained by Guinan and McCook or by students under their direction. The Biruni observations commenced in 1977 and cover a total of about 200 nights while the Villanova Hα /Hβ program began in 1971 and covers about 425 nights. Descriptions of the filters and instrumentation at Biruni and Villanova are to be found in Guinan et al. (1979)

TABLE I STARS WITH EXTENSIVE Hα AND/OR Hβ PHOTOMETRY.

RS CVn-type Binaries and Related Objects	- λ And, ζ And, RS CVn, BY Dra, Z Her, HR 1099, HR 4665, HR 5110, HD 224085
Detached and Semi-Detached Systems	- R CMa, δ Cap, RZ Cas, V380 Cyg, AI Dra, DI Her, TX Leo, β Per, λ Tau, TX Uma, ε UMi
Contact-Over-Contact Binary Systems	- VW Cep, AM Leo, V566 Oph, V1010 Oph, W UMa, AW UMa
Strongly Interacting Systems With Hot Components	- UW CMa, AO Cas, IM Mon, HD 47129
Binaries in the Rapid Phase of Mass Transfer and Systems with Disks	- SX Cas, Cyg X-1, TT Hya, β Lyr, ν Sgr, μ Sgr, W Ser
Systems With Cool Supergiant Components	- ζ Aur, 31 Cyg, 32 Cyg, VV Cep, θ Her, KQ Pup
Cataclysmic and Symbiotic Variables	- Nova Cyg 1975, Nova Cyg 1978, P Cyg, CH Cyg, WZ Sge
Be Stars	- γ Cas, 28 Cyg, φ Per, 48 Per, Pleione
Surveys (Indices only)	- Brighter members of the Pleiades, Bright Be stars, Bright Northern Hemisphere Close Binaries

and Guinan and McCook (1974) respectively. The observing procedure was the usual one of measuring the light of the variable relative to a suitably chosen comparison star. The intermediate bandpass measures yield essentially continuum measures. Light curves and Hα , Hβ and color indices were extracted from the observations, except for the surveys for which only the Hα and Hβ indices were measured. Detailed discussions for the following stars are available: R CMa (Guinan 1976), γCas (Baliunas and Guinan 1976), 32 Cyg (Guinan and McCook 1974), Cyg X-1 (Guinan et al., 1979) β Per (Guinan et al., 1976) and a survey of bright Be stars (Baliunas et al., 1975). Here we select a few stars to illustrate the types of results obtainable.

<u>HR 1099 (V711 Tau)</u>: The Hα index reveals strong, variable Hα emission which does not appear to be phase dependent. Sharp Hα enhancements (flares) of duration a few days to a few weeks have been detected. Smaller variations also appear to be present for the OI λ7774 index. As shown in fig.1, a large increase in Hα emission was observed around 22/23 October, 1977 (JD 2443440); several smaller Hα "flares" are also seen as well as a clear build-up of Hα emission prior to the large radio flare of March 1978.

<u>Cygnus X-1</u>: In 1977 the Hβ index showed a systematic variation with phase, with increased Hβ absorption (or decreased emission) near phase 0.4. A possible mechanism for the variation was suggested (Guinan et al., 1979) in which it was attributed to the projection of a gas stream on the O-star. The data

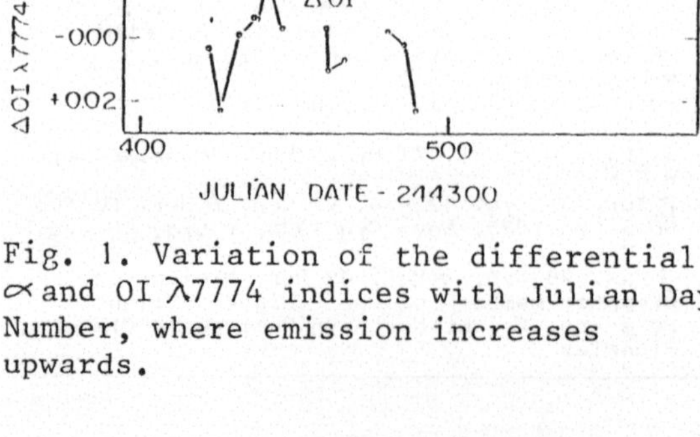

Fig. 1. Variation of the differential α and OI λ7774 indices with Julian Day Number, where emission increases upwards.

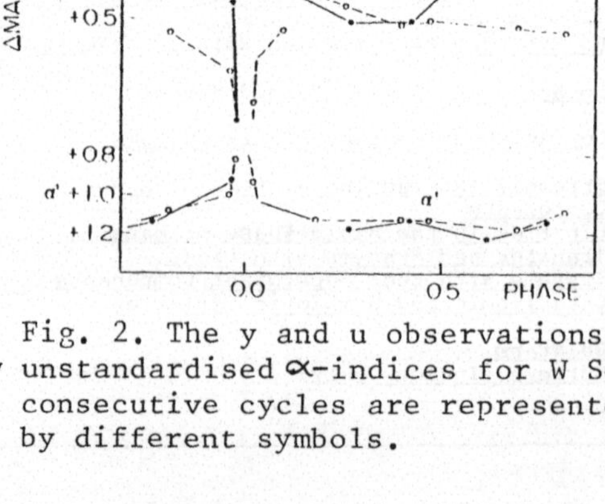

Fig. 2. The y and u observations and unstandardised α-indices for W Ser consecutive cycles are represented by different symbols.

obtained in 1978 indicate that both the λ4870 light curve and the β-variation have changed. The light curve appears more symmetrical with maxima of nearly equal heights, while the β-index shows little marked phase dependence.

W Ser: Of all the eclipsing systems studied so far, including those of the surveys, W Ser shows the strongest net Hα emission. During May and June of 1978 uvby and Hα observations of W Ser were made at Biruni in which two cycles of the orbit were covered. The y and u observations and the α indices are shown in fig.2. The light curves are peculiar, showing one minimum which is deeper in y than u. During the first cycle a large increase in light (particularly in u) is seen before the eclipse. The α index appears to vary smoothly with phase, with greatest emission occurring during the light minimum, and appears to be little affected by the light changes from cycle to cycle.

WZ Sge: This star was observed with Hα narrow- and intermediate band filters during its recent outburst in Dec. 1978. An approximately sinusoidal light variation was seen at λ6585 with a total range of 0.3 mag and a phase-dependent variation in α was found with an increase in net emission during the light minimum.

REFERENCES

Abt, H.A. and Golson, J.C.: 1966, Astrophys. J. 143,pp.306-312.
Andrews, P.J.: 1968, Monthly Notices Roy. Astron. Soc. 72, pp.35-99.
Baliunas, S.L.,and Guinan, E.F.: 1976, Publ. Astron. Soc. Pac. 88, pp.10-12.
Baliunas, S.L., Ciccone, M., and Guinan, E.F.: 1975, Publ. Astron. Soc. Pac. 87, pp.969-974.
Crawford, D.L.: 1958, Astrophys. J. 128, pp.185-206.
Crawford, D.L.: 1975, in B. Hauck and B.E. Westerlund (eds.) Problems of Calibration of Absolute Magnitudes and Temperatures of Stars, pp.93-113.
Crawford, D.L., and Mander, J.: 1966, Astron. J. 71, pp.114 118.
Guinan, E.F., and McCook, G.P.: 1974, Publ. Astron. Soc. Pac. 86, pp.947-951.
Guinan, E.F., McCook, G.P., Bachmann, P.J., and Bistline, W.: 1976, Astron J. 81, pp.57-66.
Guinan, E.F.: 1977, Astron. J. 82, pp.51-66.
Guinan, E.F., Dorren, J.D., Siah, M.J., and Koch, R.H.: 1979, Astrophys. J. 229, pp.296-303.
Philip, A.G.D., and Perry, C.L.: 1979, Publ. of Dudley Obs. and Louisiana State University (preprint).
Strömgren, B.: 1958, in D.J. O'Connell (ed.), Stellar Populations, Interscience Publishers, New York, p.385.

DISCUSSION FOLLOWING DORREN, SIAH, GUINAN AND MC COOK

Oswalt: For the RS CVn and BY Dra systems on your program have you noted any correlations between variations in the Hα index and the distortion wave often present in the light curves? For example, has there been any tendency for strongest emission to occur near times of wave maximum or minimum?

Guinan: Of the stars listed in Table I, RSCVn, Z Her, V711 Tau (HR 1099), BY Dra, and HD 224085 show emission at Hα. Only two--RS CVn and HD 224085 show evidence of an emission variation which is phase-locked. In the case of RS CVn this variation is associated with the eclipses and for HD 224085 the Hα variation is shifted approximately a quarter period from the light variation.

Rucinski: The star HD 224085 has recently been named II Pegasi.

Garmany: Have you looked at HD 108? It is an O supergiant which Hutchings found to have about a 4-day period but Conti and Vreux have since cast doubt on. It has strong Hα emission from which the mass loss is determined, but this loss is not borne out by recent IUE observations. You could enter the controversy and perhaps say something about the nature of the variability.

Guinan: HD 108 has not been observed by us yet. I thank you for your suggestion and I will place the star on our 1979/1980 observing program.

Scarfe: Have you observed any long-term variations in the emission from β Lyrae?

Guinan: Yes. Long term as well as short variations in the net Hα and Hβ line strengths appear to have occurred. Only the 1971 and 1972 data have been reduced and when the 1973 to 1975 observations are processed more can be said.

Rodono: How long have you been observing RS CVn and BY Dra? Since their brightness variations undergo periodic short- and long-term changes it would be interesting to have your observations extended for a long period of time.

Guinan: Hα wide and narrow-band and uvby photometry of RS CVn was obtained on about 25 nights during May and June of 1977. The observations were made at Biruni Obs. by Dorren, Siah and myself. The program on BY Dra was just initiated this year.

POLARIMETRY OF CLOSE BINARY SYSTEMS; OBSERVATIONS AND INTERPRETATIONS

Ian S. McLean
Steward Observatory and Lunar & Planetary Lab.,
University of Arizona

ABSTRACT: Polarization caused by scattering of starlight on gaseous extrastellar material in close binary systems is reviewed. A simple physical derivation is given to illustrate how in principle, variations synchronous with the orbital period of the Stokes parameters of the linear polarization can yield the orbital inclination and other parameters. High resolution multichannel spectropolarimetry across the emission line profiles of binaries is discussed as a new technique in studying the physics and kinematics of gaseous streams and stellar winds. The methods have application to a range of binary objects including systems like Algol and Beta Lyrae, X-ray binaries, Of and Wolf-Rayet binaries, VV Cephei stars and symbiotic stars. Some new observational results are presented.

1. INTRODUCTION

There are many classes of objects now known to exhibit measurable intrinsic polarization at optical wavelengths. Among the most important of these are close binary systems undergoing mass loss and/or mass transfer. Whether we consider semi-detached, detached, contact or X-ray binaries, one property common to many of them is a gaseous extrastellar 'envelope'. Typically, in any such nonspherically-symmetric distribution of material or even in the component stars themselves, processes such as scattering, reflection and light emission in a magnetic field can produce partial linear and/or circular polarization of the emergent radiation. The polarization may be apparent in the continuum flux or in spectral lines or both. This intrinsic polarization will vary with the binary phase and will be dependent on the polarizing mechanism, the distribution and physical state of the material, and geometric factors such as the aspect of the system as seen from Earth. In practice the observed polarizations are small but the rapid development of new types of instruments over the past decade has led to a growing understanding of the role of polarimetry in astronomy. Most polarimeters are now capable of reliably measuring degrees of polarization as small as 10^{-4} (i.e. a polarized flux equal to 0.01% of total flux). Of particular

M. J. Plavec, D. M. Popper and R. K. Ulrich (eds.), Close Binary Stars: Observations and Interpretation, 65–70.

importance is the application of polarimetry to complement or even supplement other means of deriving vital quantities such as the orbital inclination. Below is a brief review in which the aim is to indicate the current status and the changing trends in this field. Our emphasis is on hot (early-type) systems and on two different methods viz., (i) those using periodic variations in the polarization parameters and (ii) those employing polarimetry across certain spectral lines.

2. PERIODIC VARIABILITY OF POLARIZATION IN CBS

Lists of binaries already found to have variable, and hence, intrinsic polarization have been compiled by Kruszewski (1974) and by Pfeiffer and Koch (1977). More recently, it has been demonstrated by systematic observation that in many CBS the variations are indeed repetitive and phase-locked to the orbital period (see, for example, Rudy 1979, Rudy and Kemp 1978). There has also been a concurrent and rapid development in theoretical models, especially concerning analytic treatments of polarization due to single scattering in optically thin (ionized) circumstellar material (cf. Brown and McLean 1977, Brown, McLean and Emslie 1978 and Rudy and Kemp 1978). The essential physics relating polarization and binary parameters can be understood using the simple account given below. Here, the scatterers are free electrons.

Basically, expressions for the linear Stokes parameters ($Q = p \cos 2\theta$ and $U = p \sin 2\theta$) as a function of orbital phase are required. Consider a localized optically thin scattering volume, a 'blob', of N electrons moving in a circular orbit of radius R and inclination i about an isotropic (point) source star (O) as shown in Fig. 1. The stellar intensity is I_o, the scattered intensity is $I_1 = I_o \tau_B (1+\cos^2\chi)$ and its degree of polarization is $P_1 = \sin^2\chi/(1+\cos^2\chi)$. Here χ is the scattering angle and $\tau_B = 3N\sigma_T/32\pi R^2$ is an effective optical depth. For $I_1 \ll I_o$, the net (diluted) polarization is $P_R \simeq I_1 P_1/I_o = \tau_B \sin^2\chi$. When $i=0^o$ (pole-on), χ is always 90°, so $P_R = \tau_B$ independent of phase and the polarization vector (defined by the normal to the scattering plane) rotates uniformly. When i=90° this vector lies constantly along the normal to the orbit while χ varies through 360°. In the Stokes parameter plane, these two cases correspond to motion on a circular locus and to harmonic linear motion during each HALF of the orbital period. For general i the locus is an ellipse. When the blob is nearest (or farthest from) Earth U=0, $Q_1 = -\tau_B \cos^2 i$ while at quadratures U=0, $Q_2 = \tau_B$, hence the (Q,U) ellipse has eccentricity (e) defined by $(1-e)/(1+e) = |Q_1/Q_2| = \cos^2 i$; a result derived independently by J. C. Kemp and the author. In fact, the derivable properties in this case may be summarized as follows; (i) shape of ellipse defines i; (ii) its size defines τ_B; (iii) orientation gives projection of orbit normal on the sky; (iv) phases of extrema in P_R locate longitude of the blob; (v) the sense of circulation defines the sense of orbital revolution. Note that since we deal with the (Q,U) locus, any interstellar polarization is by-passed and merely introduces a displacement of the origin.

In a real binary system the situation is more complex, yet, the above canonical result can be maintained for an envelope of entirely arbitrary shape provided we have corotation, circular orbits and isotropic intensity distributions from both stars. Any absorption effects must not be phase-dependent. Material symmetric about the orbital plane yields SECOND harmonic (half-period) behaviour as described above while material asymmetric about the plane introduces FIRST harmonic terms in Q and U resulting in loci which are general Lissajou-type figures (cf. Fig. 4). Note that the material need only appear to be asymmetric about the plane due to absorption in the envelope. Often, however, such effects lead to a wavelength dependence of polarization which is separately detectable.

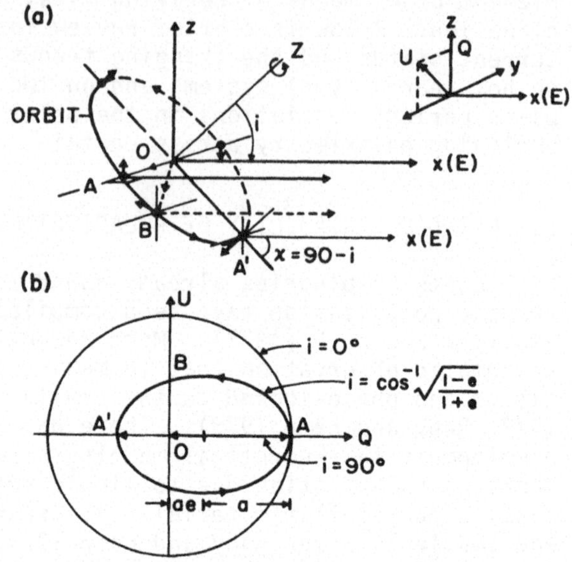

Fig. 1. Scattering geometry and (Q,U) loci for an orbiting blob of electrons or any Rayleigh particles.

Departures from the above assumptions generally destroys the canonical relation (cf. Brown et al. 1978 and Milgrom 1979) by addition of other harmonics and uneven scaling of the 2nd harmonics in Q and U. In mitigation, however, corotation is a good initial assumption and the effects due to eclipses can be recognized and smoothed-out or separately modelled to yield a further independent estimate of i (Piirola 1980). Furthermore, it is for non-eclipsing systems that the polarimetric method is most needed. For an envelope symmetric through the orbital plane, the effect of a mildly eccentric orbit mainly scales the second harmonics in Q and U by the same factor (ignoring terms of order e^4) and introduces 1st and 3rd harmonic terms which can be isolated by Fourier analysis and we again obtain the canonical result. Probably the most serious source of confusion could come from phase dependent absorption effects but in this case spectroscopic data could at least provide a warning. Recently, methods of statistical analysis have been developed (Simmons, Aspin and Brown 1980) which enable one to quantify the significance of the canonical model fit to observational data. Of course, in practice, a good model must also involve external constraints set by other observational methods and each system must be taken on its own merits.

A typical example of observational data is shown in Fig. 2; the variation of Q and U as a function of phase is given for Algol. The 2nd harmonic modulation is clear despite the rather small level of polarization (due mainly to the thinness of the gaseous stream), and the polarimetrically derived inclination of i=81° is in good agreement with the photometric value. The effect of the eclipse is quite marginal.

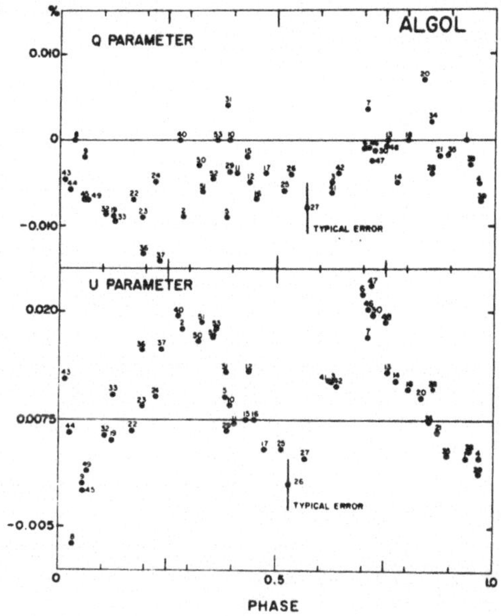

Fig. 2. Polarization (Stokes parameters) versus orbital phase for Algol (β Persei).

Figures 3 and 4 show respectively, (Q,U) versus phase and a (Q,U) locus for the Wolf-Rayet object HD 50896 (WN5) from McLean (1980). Recent observations by Firmani et al. (1979) of this apparently "single" WR star have revealed periodic variations in radial velocity and light on a period of 3.76 days; they interpret these variations in terms of orbital motion. The strong second harmonic modulation in the polarization gives some support to this contention and permits a rough estimate of i=71°. For a $10M_{\odot}$ WR star, this inclination together with the observed mass function implies a companion mass of $1.3M_{\odot}$. HD 50896 may be a member of a new class of WR stars having low mass evolved companions, probably neutron stars, with orbits essentially within the WR star's atmosphere. This is a fine example of the interplay of polarimetry and other methods and is one result in an extensive program being carried out at the University of

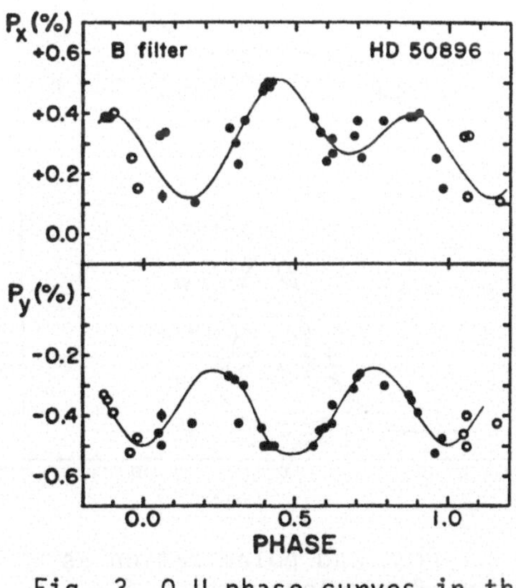

Fig. 3. Q,U phase curves in the B filter for HD 50896 (WN5).

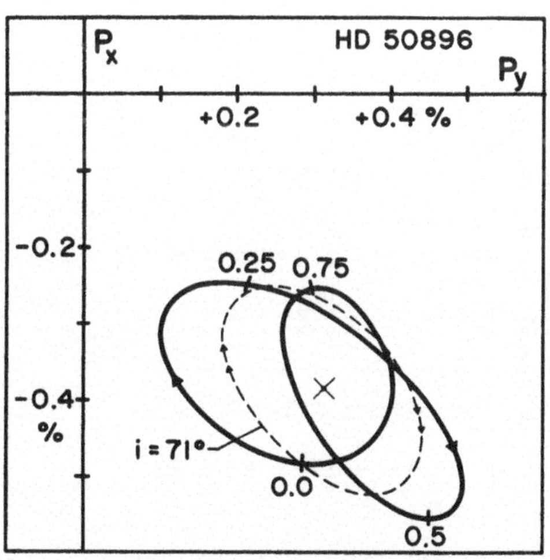

Fig. 4. The smoothed (Q,U) locus derived from the data of Fig. 3.

Arizona. Some of the principles discussed here are applied in the paper on Cygnus X-1 by Dr. Kemp.

3. Polarimetry Across Doppler Broadened Emission Lines

It has recently become possible to measure polarization across wide, Doppler broadened spectral lines with high precision using multichannel echelle spectropolarimetry. The only detailed work concerns the broad Balmer emission lines of Be stars, some of which are certainly binaries, but the same technique can be applied to numerous emission line binaries. Lines Doppler broadened by rotation and/or expansion in gaseous extrastellar material effectively provide us with a means of spatially resolving the envelope. Consequently, the polarization which we see is no longer that of an unresolved, axisymmetric envelope but rather that from non-axisymmetric portions moving toward (or away) from us. Hence, for arbitrary i, the amount and direction of the intrinsic polarization can change rapidly with wavelength across the line (McLean 1977,1979; McLean, Coyne, Frecker and Serkowski 1979; Poeckert and Marlborough 1978). For a simple case, consider an envelope which has no motion except rotation; the scattered flux at wavelengths in the blue wing of the line is more attenuated for the approaching

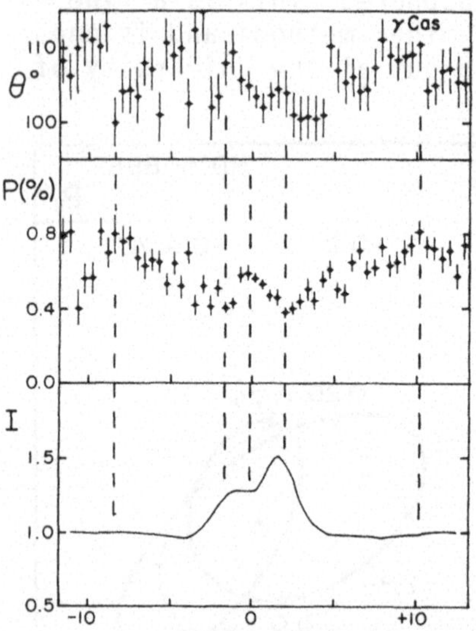

Fig. 5. Hβ flux and polarization with 0.5A resolution for the Be star γ Cas.

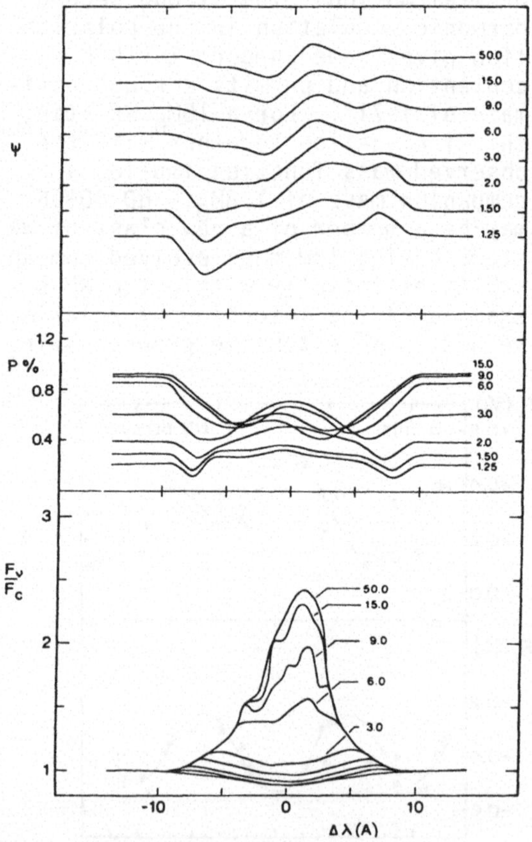

Fig. 6. Hα flux and polarization as a function of envelope extent for a stellar wind model of γ Cas.

part of the envelope and less attenuated for the receding part. The opposite is true for wavelengths in the red wing. Since for each wing the integration of the scattered flux is not symmetrical with respect to the rotation axis, the polarization plane for the polarized flux in each wing will differ from that in the continuum, provided $i \neq 90$. For pure rotation, the variation in θ is of the same amplitude but opposite sense for the two wings. In practice, random motions and expansion dilute this simple effect and cause further variations in θ. Figure 5 shows the remarkably detailed polarization structure across Hβ in the weak X-ray source γ Cas obtained with a Digicon spectropolarimeter (McLean et al. 1979). The oscillations in θ imply that the Be star disk is not viewed edge-on and from the stellar wind model shown in Fig. 6 Poeckert and Marlborough (1978) derive an inclination in the range 45°-50°.

4. Summary

Intrinsic polarization should be relatively common in close binary systems and, at least in principle, polarimetry can provide three independent ways of obtaining the orbital inclination (and other parameters) viz., (i) periodic variations of polarization synchronous with the orbital motion, (ii) polarization changes during eclipses if these occur and (iii) polarization structure across emission lines formed in the envelope and Doppler broadened by mass motions. These new methods carry great potential and over the next several years a rapid development trend is expected.

References

Brown, J. C., McLean, I.S.: 1977, Astron. Astrophys. 57, 141.
Brown, J. C., McLean, I.S., Emslie, A.G.: 1978, Astron. Astrophys. 68, 415.
Firmani, C., Koenigsberger, G., Bisiacchi, G.F., Ruiz, E., Solar, A.: 1979, IAU Symp. No. 83, Mass Loss and Evolution of O-type Stars, ed. P.S. Conti and C. de Loore, Reidel, Dordrecht.
Kruszewski, A.: 1974, IAU Coll. No. 23, Planets, Stars and Nebulae Studied with Photopolarimetry, ed. T. Gehrels, Univ. of Arizona Press.
McLean, I.S.: 1977, Astron. Astrophys. 55, 347.
________: 1979, Monthly Notices Roy. Astron. Soc., 186, 265.
________: 1980, Astrophys. J. submitted.
McLean, I.S., Coyne, G.V., Frecker, J.E., Serkowski, K.: 1979, Astrophys. J. 228, 802.
Milgrom, M.: 1979, Astron. Astrophys. 76, 338.
Pfeiffer, R.J., Koch, R.H .: 1977, Publ. Astron. Soc. Pac. 89, 147.
Piirola, V.: 1980, IAU Symp. No. 88, Reidel. Dordrecht (this volume).
Poeckert, R., Marlborough, J.M.: 1978, Astrophys. J. 220, 940.
Rudy, R.J., Kemp, J.C.: 1978, Astrophys. J. 221, 200.
Rudy, R.J.: 1977, Ph.D. Dissertation, University of Oregon.
Rudy, R.J.: 1979, Monthly Notices Roy. Astron. Soc. 186, 473.
Simmons, J., Aspin, C., Brown, J.C.: 1980, Monthly Notices Roy. Astron. Soc. submitted.

THE EFFECT OF ORBITAL ECCENTRICITY ON POLARIMETRIC BINARY DIAGNOSTICS

C. Aspin, J.C. Brown, J.F.L. Simmons
Dept. of Astronomy, University of Glasgow, U.K.

ABSTRACT. The polarimetric variation from a binary system with an eccentric orbit, thus non-corotating, are calculated and the effect on determining the system parameters is discussed, relative to the circular case.

We relax the assumption of corotation implicit in the canonical model of polarimetric binaries (Brown et al 1978) by considering a localized scattering region in an orbit of eccentricity e about a single light source. This scattering 'blob' has N electrons and can be considered in physical terms as similar to a small accretion disk around a compact star i.e. neutron star or black hole.

The introduction of the eccentricity results in the model predicting variations in the Stokes Parameters Q and U with binary phase of harmonics, in addition to the basic second harmonics which characterise circular orbits when the scatterers are symmetric about the orbital plane. The variation of the source to scatterer distance can be treated purely geometrically in terms of binary phase and the harmonic phase dependence of Q and U can be found exactly so yielding the form of the locus described in the Q,U plane, once per orbital period. Since we take our Q,U measurements in time rather than phase we have re-expressed the eccentric orbit model's predictions in this form by a solution of Kepler's equation to order e which will suffice for all practical cases.

It is found that (Q,U) time harmonics up to and including third are present in this approximation, the first and third harmonic coefficients being respectively of order e and 3e relative to the dominant second harmonics. The resulting harmonic coefficients are:

$$\begin{aligned}
p_1 &= \tau_* e \cos \Lambda p(1+\cos^2 i) & u_1 &= 2\tau_* e \sin \Lambda p \cos i \\
q_1 &= -\tau_* e \sin \Lambda p(1+\cos^2 i) & v_1 &= 2\tau_* e \cos \Lambda p \cos i \\
p_2 &= -\tau_* (1+\cos^2 i) & u_2 &= -8\tau_* e \sin \Lambda p \cos i \\
q_2 &= 4\tau_* e \sin \Lambda p(1+\cos^2 i) & v_2 &= -2\tau_* \cos i \\
p_3 &= -3\tau_* e \cos \Lambda p(1+\cos^2 i) & u_3 &= 6\tau_* e \sin \Lambda p \cos i \\
q_3 &= -3\tau_* e \sin \Lambda p(1+\cos^2 i) & v_3 &= -6\tau_* e \cos \Lambda p \cos i
\end{aligned} \quad (1)$$

M. J. Plavec, D. M. Popper and R. K. Ulrich (eds.), Close Binary Stars: Observations and Interpretation, 71–72.

where e = eccentricity; Λ_p = longitude of periastron and τ_* = weighted optical depth of the scatterer.

These coefficients are therefore functions of e, i, Λ_p, τ_* and Θ the orientation of the binary axis relative to the (Q,U) lab frame. It is shown how, given accurate data, one can solve for these variables in terms of the harmonic coefficients, together with several consistency checks on the results.

The possibility is then discussed that a wrong determination of i (or any other parameter) could be made if a canonical model with circular orbit, but asymmetric envelope geometry, were applied to diagnosing a truly eccentric orbit situation. It is shown that the model predictions are in fact mutually exclusive and so the correct model centre found if the data are accurate enough (eg. the canonical model predicts no third harmonics).

At present we are extending this analytic treatment to the more realistic case of noisy data by employing the optimization technique of Simmons et al (1979). Our aim is to obtain the best fit eccentric orbit parameters and their associated confidence intervals for interesting systems such as Cygnus X-1 using data of (eg. Kemp et al, 1979).

We acknowledge receipt of an S.R.C. travel grant to enable C.A. and J.F.L.S. to attend the Symposium.

References.

Brown, J.C., McLean, I.S. and Emslie, A.G.: 1978, Astron. Astrophys. 68, 415.

Kemp, J.C., Barbour, M.S., Parker, T.E. and Herman, L.C.: 1979, Ap.J. Letters, 228, L23.

DISCUSSION FOLLOWING McLEAN and ASPIN et al.

Massey: Have you combined your ephemeris with Moffat's proposed "velocity curve", or is the period too uncertain? Did you derive the period independently?

McLean: The zero phase shown in the Figure is based on Moffat's ephemeris (WR star in front). Since the polarimetry was obtained a year or so after the spectroscopy, and the period still is a little uncertain, then there is an uncertainty of about 0.05 in the location of zero phase. We do derive the period independently but only to an accuracy of about 0.01 day.

A NEW METHOD OF LIGHT CURVE ANALYSIS: APPLICATION TO V444 CYGNI

S.A.H.Smith,Dept. of Computer Science,
University of York

A.C. Theokas, Dept. of Astronomy,
University of Manchester

ABSTRACT-A new method of light curve analysis is introduced for systems involving an atmospheric eclipse.Occultation or partial eclipses may be treated, with or without the presence of transparency effects. Besides admitting an arbitrary degree of transparency to the eclipsing component, an arbitrary law of limb darkening may also be given the eclipsed star.

The method has successfully been applied to the study of narrow band continuum light curves of V444 Cygni obtained by Cherepashchuk and Khaliullin. Primary and secondary minima are examined separately.

1.INTRODUCTION

A novel approach to the determination of geometrical elements of eclipsing binary systems is here outlined. A more complete presentation will be given elsewhere. The method involves the solution of a system of non-linear, transcendental equations relating the morphology of eclipse minima and the geometrical elements. These include the relative radii, brightnesses, angle of incination, and coefficients of limb darkeming and transparency. The last parameter arisos from the assumption that the eclipsing star can not be considered a wholly opaque disc.

The method is checked by using the solutions to the equations as parameters which generate a synthetic light curve. The model assumes spherical stars, circular orbits, spherically symmetrical, homogeneous extended atmospheres, no phase effects and symmetrical minima.

2. THE BASIC EQUATIONS

Let the total amount of light of the system be

M. J. Plavec, D. M. Popper and R. K. Ulrich (eds.), Close Binary Stars: Observations and Interpretation, 73–78.

$$\ell = L_1 + L_2 - \int_A J(r)\,dS \qquad (1)$$

where J is the brightness per unit area, dS the infinitessimal element of eclipsed area and A is the eclipsed area. If the brightness is uniform then $J(r)=L_1/\pi r_1^2$ where the subscripts refer to the eclipsed star. Following this we can say

$$1-\ell(r_1, r_2, \delta, J) = \alpha L_1$$

where

$$\alpha(r_1, r_2, \theta, i) = 1/\pi r_1^2 \int_A dS$$

and δ is the apparent separation of the two stars, θ the phase in degrees, and i the angle of inclination. Subsequently a plot of ℓ against $\sin^{2m}\theta$ may be represented by

$$A_{2m}(r_1, r_2, i, L_1) = \int_0^{\pi/2} \{1-\ell(r_1, r_2, \delta, J)\} \frac{d(\sin^{2m}\theta)}{d\theta} . \qquad (2)$$

The A2m's on the LHS may be empirically determined from the data and m equations of the form of eq.(2) may be inverted to determine r_1, r_2 and i.

Using a similar approach we develop a system of equations of the form

$$A_{2m} = A_{2m}(r_1, r_2, i, J, F)$$

where F is the transparency function involved by letting the eclipsing star vary in opacity across the disc much the same way the brightness is allowed to vary due to limb darkening. An expression of the form

$$A_{2m}(r_1, r_2, i, F, J) = \int_0^{r_2} F(s) \frac{\partial}{\partial s}(A_{2m}(r_1, s, i, J))\,ds \qquad (3)$$

is derived and solved numerically to obtain the elements. Note that s in F(s) refers to the radius of the eclipsing star while dS represents an element of eclipsed area.

To solve equations of the form of eq.(3) the transparency function F(s) is adopted while further expressions for $\partial/\partial s\, A_{2m}(r_1, s, i, J)$ are derived. In the course of these derivations successive terms arise defined as

$$I_m = I_m(r_1, J) = \int_0^{r_1} \frac{J(r)}{r_1^{2m-2}} \frac{\partial}{\partial r}(\pi r^{2m})\,dr$$ where $I_m = L_1$ for a non-limb darkened star

$$R_m = R_m(r_2, F) = \int_0^{r_2} \frac{F(s)}{r_2^{2m}} \frac{\partial}{\partial s}(s^{2m})\,ds$$

where $R_m = 1$ for an opaque star

$$Q = Q(r_1, r_2, J, F) = -\int_0^{\min(r_1, r_2)} F(s)J(r)\, 2\pi s\,ds\ .$$

Ultimately, expressions of the following form are determined.

$$A_2(r_1, r_2, i, J, F) = I_1 R_1 r_2^2 \csc^2 i + \cot^2 i\, Q - \psi \qquad (4)$$

for m=1

where the ψ term is treated as an error term. A similar system of equations may be derived using the Kernel $\cos^{2m}\theta$ which weights the data near mid-minima more heavily.

APPLICATION TO V444 CYGNI

Solutions have been obtained for narrow band continuum light curves of the Wolf-Rayet eclipsing binary V444 Cygni. The observations were obtained by Cherepashchuk and Khaliullin(1972). Primary and secondary eclipse were examined separately. For the primary eclipse (that of the O9 by the WN6) the adopted transparency function was

$$F(s) = (1 - v(s/r)^2) \qquad (5)$$

where v is the coefficient of transparency and $0 < v < 1$. The behavior of this function is shown in fig.1 below. The adopted law of transparency determines the coefficients R_m and here $R_m = 1 - v/(m+1)$, e.g. $R_1 = 1 - 0.5v$ for m=1. The loss

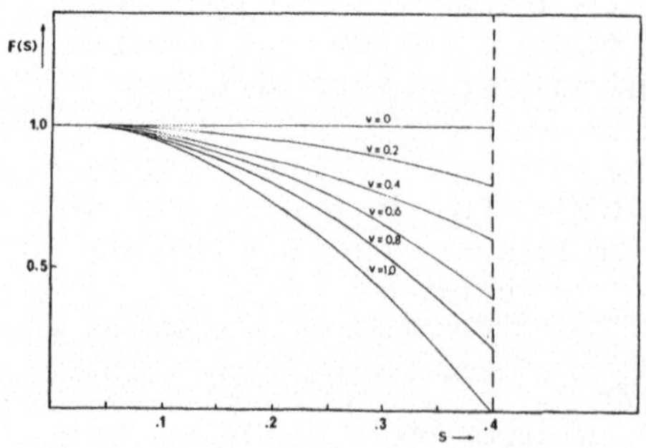

Fig.1 Law of transparency

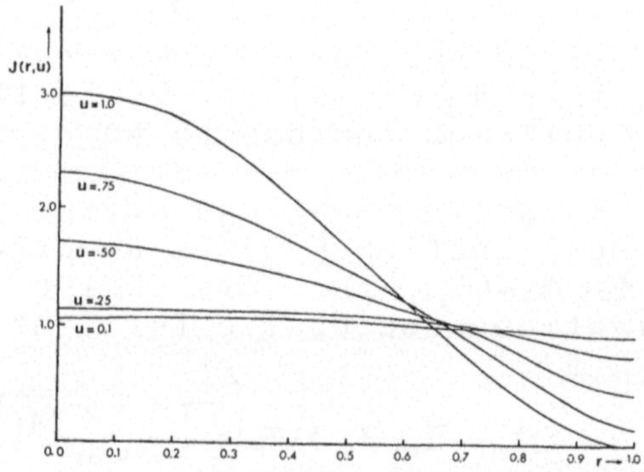

Fig.2 Law of limb-darkening

of light for a concentric eclipse, Q, is similarly dependent on F(s) and is derived as

$$Q = L_1 (1 - 0.5v (r_1/r_2)^2) \qquad (6)$$

while all I_m are equal to unity, the O star treated as non limb darkened.

For the secondary eclipse the O star is treated as opaque and the WR component is assigned a non-linear law of limb darkening(see Fig.2). This will determine the values of the coeffients I_m while all R_m=1. Due to lack of space solutions for the light curve taken at λ4789 are given below. In Fig. 3 the solid line represents the synthetic curve fit to the primary eclipse while the dotted lines represents the region in which all reasonable solution sets may lie. The parameters which generated this curve are the solutions to the primary eclipse listed in Table 1, along with solutions to the secondary eclipse. Although the solution for

	primary	secondary
WR	0.430±.045	0.134±.016
09	0.260±.026	0.268±.005
u	-	0.69±.04
v	0.39±.37	-
i	79.7° ± 1	80° adopted
L_1	0.252± .026	0.154±.098

Table 1.

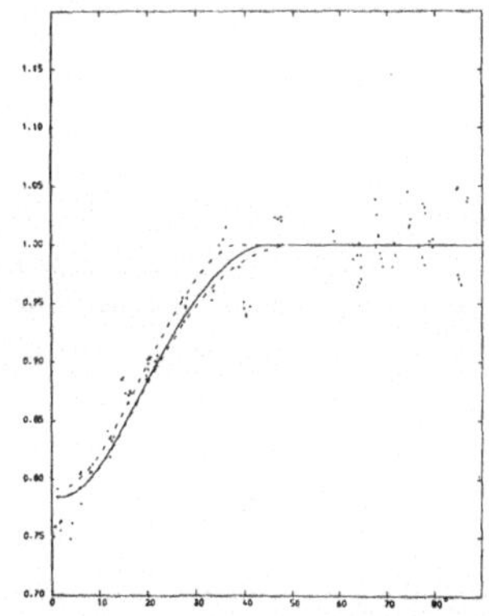

Fig.3

coefficients of transparency has an unacceptably large error it does indicate, as do the solutions for the other light curves, the presence of transparency effects. It is felt that, although mathematically sound, the method requires the use of highly accurate data before the application can be reliable to the degree required.

REFERENCE

Cherepashchuk, A.M. and Khaliullin, Kh.F.: 1972, Peremmenye Zvezdy, 18, 321.

DISCUSSION FOLLOWING SMITH AND THEOKAS

Rucinski: The method by Cherepaschuk (applied by him most extensively to V444 Cyg) does not require any knowledge of shapes of functions describing the transparency or surface brightness versus radius; it is assumed that the functions are monotoneous, however. Could you comment how your method is related to the approach by Cherepaschuk?

Theokas: First the model does not require rectified data, simultaneous solution of the light curve minima or any precondition such as $r_1 > \cos\theta$ as does Cherepaschuk's method. Secondly the adoption of the law of transparency need not be completely arbitrary but other considerations may be taken into account such as those of Cherepaschuk you mention. This method in a sense thus may compliment his.

Mook: V444 Cyg has a phase-dependent linear polarization. Can your model predict this? (Your initial response that your model does not speak to conditions outside eclipse is not responsive to my question. You could write your transfer equations for polarized components and solve for the variation of polarization during the eclipse. That is, just as you solve for overall flux as a function of time--your synthesized light curve--you could also predict % polarization as a function of time and use that as an independent check.)

Theokas: Yes, in theory it would be possible to expand the model to examine polarization rather than flux as a function of time. Within the context of this model however, the mathematics would become considerably more complex diminishing its practicality.

Smak: What is the maximum photometric phase of eclipse, or, in other words, how close is it to totality?

Theokas: These are very nearly complete eclipses 70 - 80% or more. If they were total we would have the unusual occurrence of an occultation eclipse at primary and secondary.

van Paradijs: I hope that your model will allow one to determine the radial density distribution of stellar winds.

Theokas: Not directly. Some distributions may be ruled out by considering what laws of transparency will fit the data and which give poor approximations. The model is probably best applied to those systems where stellar winds are low as in ζ Aur.

Linnell: The semi-transparent layers are subject to substantial distortion by the companion, while your idealized model assumes spherical symmetry. How wide latitude of assumed model characteristics do you feel is possible which will produce a light curve in good agreement with the observations?

Theokas: V444 Cygni perhaps represents the limit to that latitude as represented by the scatter in the light curve shown. This is due to variations in both the rate and amount of mass outflow so the assumption you mention is not well founded here and we should expect better results with less scatter. In other words an extended but quiescent atmosphere.

Al-Naimy: Why is there a great difference in the radius of the WR star?

Theokas: The fractional radius of the WR component is not the same for secondary and primary eclipse due to the presence of the highly extended atmosphere of the WR component. It behaves differently as a luminous object and does not take part in the secondary eclipse. This is just a confirmation of an old result.

Al-Naimy: How do you trust your results according to your assumption?

Theokas: I'm satisfied with the results for the fractional radii and brightnesses but do not trust the coefficients of transparency or limb darkening to tell us a great deal physically about the stars. They were very sensitive to changes in initial conditions. The accuracy of data is a problem here and perhaps satellite-based observations would overcome this.

II. DETACHED SYSTEMS

DETERMINATION OF ABSOLUTE DIMENSIONS OF MAIN SEQUENCE BINARIES

J. Andersen, J.V. Clausen, and B. Nordström
Copenhagen University Observatory, Denmark

ABSTRACT

Determinations of absolute dimensions of main sequence binaries are no longer useful unless random and systematic errors are kept appreciably below 10%. Some essential conditions for achieving this are reviewed.

1. INTRODUCTION

Determining absolute dimensions of main sequence stars from spectroscopic and photometric observations of double-lined eclipsing binaries has been a standard technique for several decades. Hundreds of determinations have been published and are summarized in standard compilations of, e.g. mass and radius vs. spectral type.

It is therefore necessary to ask ourselves whether it is still worthwhile, or even reasonable, to continue such "routine" work. Are we just beating a dead horse and piling up determinations of something that has already long been known to everybody's satisfaction? In other words, aren't we wasting valuable telescope time and journal space?

A devastating answer to this question is closer to the truth than generally recognized. Further work can in fact only be justified on the following two conditions:

1 - More accurate data really are needed, and
2 - The new data really have the required precision.

As just one example of the need for better data, one of the popular uses of such empirical determinations in recent years has been to compare with theoretical computations of various degrees of refinement of the evolution of single or double stars. Our own experience in determinations of the highest accuracy indicates that these "empirical data",

M. J. Plavec, D. M. Popper and R. K. Ulrich (eds.), Close Binary Stars: Observations and Interpretation, 81–88.

whether taken from simple literature surveys or from standard compilations of M and R vs. spectral type, may be in error by as much as 10-40% in M and a factor 2 in R. This is quite inadequate for any critical test of the theoretical models: Little meaningful information can be extracted on chemical composition, age, or the importance of any processes of mass loss. What is worse, previous mass determinations for early-type stars are found to have serious systematic errors (see 2.3). In fact, use of the older data would lead one to derive a helium abundance Y = 0.33-0.40 for the (young) B stars as compared to Y = 0.20 for the (older) F and G stars (Popper et al., 1970), a degree of helium enrichment difficult to explain theoretically--and entirely spurious.

We conclude that a real accuracy in M and R of 1-2% or at least appreciably less than 10% is required. Only very few systems have yet been analyzed to this precision, but it is essential for future work to result in new information. New determinations should also be accompanied by photometric indices on a system providing quantitative and sensitive reddening-free parameters related to effective temperature and gravity. The parameters supplied by even a good MK class are simply too coarse to resolve moderate evolutionary effects easily discerned in the binaries. Stars of a given MK class may differ by as much as 30% in mass and a factor 2 in radius.

Experience from more than a dozen high-precision determinations has taught us that the precision limit set above will not be reached in practice without careful attention to a number of questions of method. These are reviewed below, hopefully for the benefit of colleagues interested in using and/or producing data of this kind.

2. NECESSARY CONDITIONS FOR ACHIEVING HIGH ACCURACY

2.1 Choice of System

Not all double-lined eclipsing systems are suitable for precision determination of stellar dimensions. They must have well-defined eclipses of appreciable and comparable depth and well-resolved spectral lines of both components. In too strongly interacting systems, the dimensions cannot be very accurately determined, and in addition their validity for single stars becomes doubtful.

Obviously, the system must be bright enough to yield observations of good quality with the available instrumentation (see 2.2). If a third component contributes appreciably to the total light, it must be sufficiently well separated from the main pair to allow an accurate determination of its contribution (area scanner, electrography).

2.2 Choice of Observational Material

The spectra used for radial velocity determination must be of sufficiently high resolution to produce no appreciable degradation of the line profiles, so that remaining line blending problems are inherent in the stellar spectrum and not caused by the instrument. Particularly for early-type stars, the signal-to-noise ratio of the spectra must be high enough to allow measurement of those, mostly relatively weak lines that yield velocities free of systematic error (see 2.3). The usual precautions in radial velocity work must be taken, including observations of standard stars.

The orbit should of course be well covered, but avoid phases when the lines of the two components are not clearly separated. In our view, judgment of such line blending can only be done reliably when the actual line profiles can be inspected during measurement, e.g. on an oscilloscope screen.

The light curves should be obtained in two or more colours in a homogeneous well-defined and well-calibrated system (i.e. intermediate or narrow-band). Avoiding wide-band systems eliminates troublesome bandwidth effects in extinction corrections and transformations, apart from the difficulty of interpreting wide-band observations with the modern, essentially monochromatic light curve simulation models. The photometric indices needed for both components were specified above (see 1).

The light curves should be well covered with observations of sufficiently long integration time (on both the variable and the two carefully selected comparison stars) to ensure good accuracy of each magnitude difference--about $0\overset{m}{.}005$ or less. In order to detect any non-orbital light variations, all phase intervals should be observed on at least two different nights.

2.3 Choice of Spectral Lines

Methods for selecting spectral lines for the determination of spectroscopic orbits have been described by Popper (1967, 1970) and Andersen (1975a). They will not be repeated here, but we stress the importance of an objective method to test the consistency of the velocity measures, both for each line by itself and from line to line (Andersen 1975a). Without a description of such tests--or even a list of the lines used--one cannot judge the quality of a published analysis.

A particularly common--and disastrous--mistake is to ignore the effects of blending of the Stark broadened wings of the double Balmer and diffuse He I lines in early-type spectra, unfortunately the strongest lines in these spectra. Figure 1 (based in part on unpublished data) illustrates the systematic underestimate of the masses resulting from this, found for luminosity ratios ranging from one (Andersen 1975a, b) to about 0.05 (Hill et al. 1976). The empirical mass scale for B stars requires significant revision.

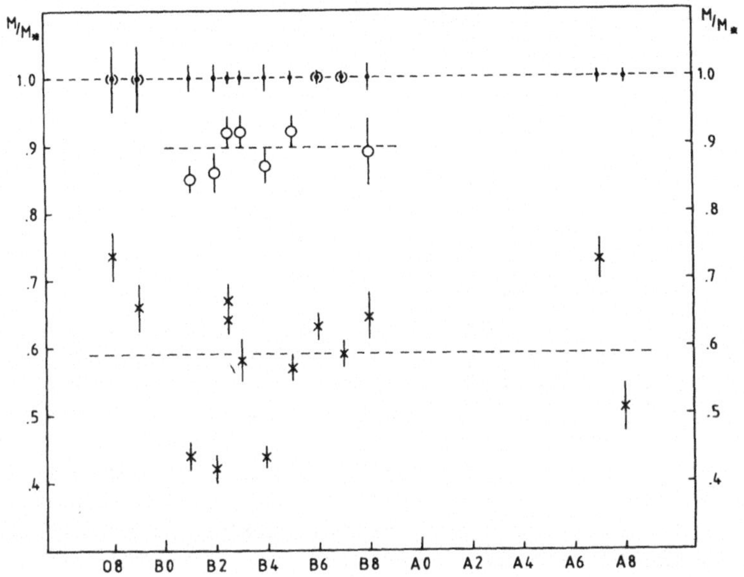

Fig. 1. The ratio between the masses derived from Balmer lines (crosses) or diffuse He I lines (circles), and the true masses derived from sharp lines (dots), plotted against spectral type. Mean errors are shown for each point. The average systematic error is 10% for He I and 40% for Balmer lines.

2.4 Choice of Photometric Model

The analysis of the light curves must be done with a physically realistic model. A few systems with circular orbits and very small fractional radii yield consistent results with almost all methods, including those involving rectification, etc. For most carefully observed systems, however, a physically realistic model is the way to accurate results - as distinct from small formal errors.

The computation of tidal and rotational deformation should take into account both circular and eccentric orbits, and non-synchronous as well as synchronous rotation. Reflection, gravity brightening, and limb darkening should be included in a realistic manner. Further, it must be possible to fix certain parameters like mass ratio or axial rotation, if they are better known from the spectroscopic analysis.

It is not advisable to attempt to solve simultaneously for all parameters defining the light curve. Some fundamental parameters (e.g. limb darkening or gravity brightening coefficients) should be fixed at values known to be applicable for the stars under study. One should then solve for the parameters peculiar to the system (i, k, r_A, T_B/T_A, e, ω). Subsequently, the effect upon the derived elements of likely errors in the fixed parameters should be investigated and taken into account in the discussion of the probable overall errors of the results.

2.5 Consistency Between Photometric and Spectroscopic Data

Traditionally the determination, if not the discussion, of photometric elements has proceeded without consideration of the spectroscopic data, or vice versa, usually because they were simply not available. In our program, photometric and spectroscopic observation and analysis normally proceed hand in hand, and we have become convinced that this is indeed essential to satisfy ourselves that our results are really correct.

First of all, the mass ratio cannot be used as a free parameter in fitting the computed light curve to the observations, when an accurate spectroscopic value is known. Next, not only should the photometric elements derived from light curves in different colours agree among themselves, but the computed luminosity ratio must be consistent with the observed line ratios. This condition sometimes offers the only possibility of fixing k in partially eclipsing systems of closely similar stars (see Fig. 2). A particular combination of assumed initial parameters and distribution of the observations may therefore lead even a good program completely astray on even the best observations, if no independent check is possible.

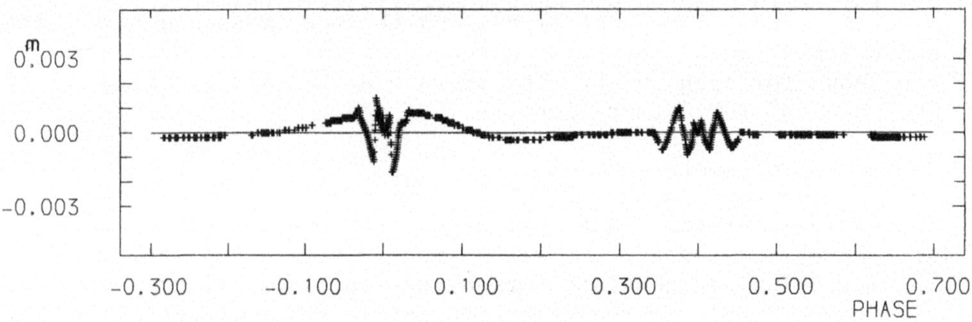

Fig. 2. The difference between two simulated light curves for QX Car, the first with k as a free parameter, with second with fixed k = 0.94, reproducing the spectroscopic line ratios. The curves are photometrically indistinguishable, yet the luminosity ratios differ by 40%.

Next, the computed rotations should reproduce the observed line widths when compared with spectra of rotation standards (Clausen et al. 1977). Finally, the observed colour changes in eclipse should be consistent with the temperature difference found in the analysis, and the colour indices themselves should be consistent with the spectral types. The latter constraint is a rather weak one, since spectral types are difficult to determine accurately in double-lined spectra, particularly from the high-dispersion spectrograms needed to determine accurate radial velocities.

3. CONCLUSIONS

The basic result of our work is a set of accurate values of mass, radius, and colour indices for the components of the binary systems studied. There are good reasons to believe that they are valid also for single stars. Interesting comparisons with stellar structure models have already been possible, leading to determinations of helium abundance also in later-type stars without observable spectral lines of helium. Some improvement in the determination of the basic parameters may still result from better treatment of, e.g. limb darkening and reflection. However, although eventually a purely empirical calibration of colour indices in terms of mass and radius is possible, the greatest uncertainty in the comparison with theory, and thus in the derivation of age and chemical abundance, stems from the calibration of colour indices into effective temperature and other intrinsic parameters.

We conclude that it is still useful to continue our work, provided we take it quite seriously and produce results of the quality needed: "Routine" work all too easily results in useless numbers.

ACKNOWLEDGEMENTS

This program was inspired by the work of D.M. Popper, and we thank him for numerous invaluable discussions. The work was done in a group including also H.E. Jørgensen, K. Gyldenkerne, B. Reipurth, and the late B. Grønbech. It has been generously supported by ESO observing time, by several grants from the Danish Natural Science Research Council and by Julie M. Vinter Hansen travel grants.

REFERENCES

Andersen, J.: 1975a, Astron. Astrophys. 44, 445.
Andersen, J.: 1975b, Astron. Astrophys. 44, 355.
Clausen, J.V., and Grønbech, B.: 1977, Astron. Astrophys. 58, 131.
Clausen, J.V., Gyldenkerne, K., and Grønbech, B.: 1977, Astron. Astrophys. 58, 121.
Hill, G., Aikman, G.C.L., Cowley, A.P., Bolton, C.T., and Thomas, J.C.: 1976, Astrophys. J. 208, 152.
Popper, D.M.: 1967, Ann. Rev. Astron. Astrophys. 5, 85.
Popper, D.M.: 1970, IAU Colloq. No. 6, Eds. K. Gyldenkerne and R.M. West, University of Copenhagen, p. 6.
Popper, D.M., Jørgensen, H.E., Morton, D.C., and Leckrone, D.S.: 1970, Astrophys. J. 161, L57.

DISCUSSION FOLLOWING ANDERSEN, CLAUSEN AND NORDSTRÖM

Popper: I endorse the comments of Dr. Andersen. There are five additional points I would like to make briefly about the treatment of photometric data of close binaries. Three of them are pleas to those discussing light curves of relatively simple systems with one or the other of the available computer programs. The output of these programs tends not to give some of the important quantities derivable most directly from the photometric observations. First is the ratio of the surface fluxes of the two stars. This basic result is obtainable directly from the ratio of the depths of the minima, without reference to the ratio of the radii or to the ratio of light, which may be poorly determined in systems with partial eclipses that are not very deep. Second, the color index of a component undergoing eclipse is obtainable directly from the color index of the light lost, again without reference to the photometric solution. Third: important information about the system is contained in the light variation between eclipses. In the bad old days one used this variation to "rectify" the light curve within minima. While rectification is rarely carried out now, the coefficients of the variation between minima, a_0, a_1, and a_2, are very useful in understanding properties of the system. My fourth point concerns use of the spectrographic luminosity ratio, as discussed by Dr. Andersen, to rescue an otherwise indeterminate photometric solution, or at least to improve the solution. It is not safe to assume that the luminosity ratio is the same as the ratio of like intensities if the two components differ in temperature or gravity. Lastly, it is not valid, in my opinion, to derive "mean errors" of photometric elements by the usual least squares treatment of the observations. The conditions required for this treatment to be valid are not generally satisfied.

Andersen: It is very much our own philosophy to give the fundamental data obtained directly from the observations, so that anyone may rediscuss the interpretation with improved temperature scales, etc., but without having to reanalyze the observations themselves. We will be glad to discuss directly with you how to best do this in practice. With regard to your fourth point, the line ratio is always determined from several unblended lines varying differently with spectral type (the gravity differences are always quite small in the cases when the photometric indeterminancy requires additional spectroscopic data). We quite agree with your final point and base our own estimates more on interagreement between solutions in different colours, with different assumptions on limb darkening, gravity brightening, reflection, agreement with spectroscopic data, etc.

Smak: Did you try to use a profile-synthesis technique to extract information from blended hydrogen lines?

Andersen: Not yet. We are planning to do it, but at the moment we concentrate on getting a number of reliable orbits from the sharp lines, so we can check the results of such a correction procedure.

Wilson: You imply that parameters are more easily determined for well-detached binaries than for those with larger components. However, I would say that the reverse is often the case, especially in regard to the mass ratio, because light curves of very close binaries provide information from the proximity effects as well as from the eclipses.

Andersen: That's right, of course, but it puts you into the hands of your model, and we have seen "determinations" of mass ratio with less refined models than yours, which we found to be completely up in the air when we got the radial velocities. So I must confess a preference for getting the mass ratio from spectra, and to the precision we are aiming at, that is best done in detached systems.

Van Paradijs: It would be interesting to compare your accurate gravity values with spectroscopic gravities as obtained from the ionization equilibrium or the profiles of spectral lines. Have such comparisons been made?

Andersen: No. This paper is on methods, not on results, but it is certainly one of our hopes in doing this work that it would interest some of the spectrum-synthesis groups to have a number of stars for which log g is known to 0.01 or even better.

OBSERVED PERIOD CHANGES IN DETACHED BINARY SYSTEMS

Tibor J. Herczeg
Department of Physics and Astronomy, University of Oklahoma

A comprehensive study of period changes in eclipsing binaries is being carried out at the University of Oklahoma. The survey of available information is supplemented by observations obtained at the University Observatory, at Kitt Peak and Cerro Tololo observatories, and, as part of a cooperative program, at Ankara University. The present review reports the behaviour of 20 systems classified as detached, some of which do show noticeable variations of the period.

We would not expect detached systems of binary stars to show appreciable variations of the period, apart from geometric effects such as apsidal motion or light-time shifts. But are their periods really constant?

This question was first systematically investigated by Plavec, Pěkný and Smetanová [1960], and the present survey is a direct-line continuation and extension of that study. It became obvious very soon, however disappointing, that the main obstacle in 1960, the lack of suitable observations (regular timings of minima) remains the same today. Observers of detached systems are more interested in orbital parameters; time and again, complete and accurate light curves are observed and used for many years of orbit calculation. In the meantime, the O-C curve of the minimum epochs is reduced to hardly more than a few isolated points. Further, since the minima are usually not as deep as primary minima of semi-detached systems, much of the earlier visual or photographic observations is of limited usefulness.

Nonetheless, a new comprehensive study of the literature, with some additional observations, is needed if we want eventually to reach a clearer understanding of how detached pairs behave. Twenty systems have been selected with which to begin this study:

RT And, V805 Agl, RX Ari, WW Aur, AR Aur, β Aur, SS Cam, AW Cam, RS Cha, RX Her, TX Her, VZ Hya, YY Gem, CM Lac, UV Leo, U Oph, WZ Oph, V451 Oph, EE Peg, and RS Sgr.

M. J. Plavec, D. M. Popper and R. K. Ulrich (eds.), Close Binary Stars: Observations and Interpretation, 89–93.

The following discussion, hardly more than an extended abstract, offers a progress report of this investigation.

Two problems present somewhat unexpected difficulties:

1. In quite a number of cases the nature of the binary system (detached or semi-detached) is not easy to decide upon; also, several systems considered detached are now reclassified as semi-detached. For these reasons ZZ Cyg, MR Cyg, DI Peg and a few other binaries have not been included; moreover, the classification of one object or other on the above list remains in contention.

2. In some cases the O-C time residuals may be rendered spurious or, at least, markedly influenced by distortions in the light curve, notably by asymmetries of the minima. If the eclipsed component has, for instance, a large asymmetrically placed starspot on its surface turned toward the earth, the time of minimum light will not necessarily coincide with the moment of inferior conjunction. A discussion of the system XY UMa by Geyer [1977] is very instructive; clearly, this problem deserves much careful attention. I am convinced that we can decide upon the reality of the O-C values and hence the period changes, provided the photometric observations are accurate enough and the individual data points are made accessible. Available evidence seems to indicate that the majority of the observed time residuals are not spurious.

A good fraction, about one-third, of the stars in our list show no sign of period variations. This finding, however, is based on observations so few in number or encompassing such a short time span that the constancy of the period cannot be definitely established as yet. In three further cases, constant periods appear to be far better documented, for time intervals of 30-60 years; these systems are: β Aur, VZ Hya and CM Lac.

Ten systems out of the twenty show period variations or, at least, are suspected of period changes. In this latter group of suspected "period variables" we find WW Aur, AR Aur, UV Leo, WZ Oph and EE Peg; some of these belong to the well-behaved systems which supply us with the best data on masses and radii. The evidence is in the form of a few, more or less isolated, minimum epochs which show deviations (usually of the order of 0.01-0.02 days) from linear ephemeris formulae. When backed by photoelectric observations, these O-C values are certainly significant. On the other hand, the linear ephemeris formulae themselves are sometimes unreliable; at any rate, the fragmentary O-C diagrams do not even indicate the mode of the period variations.

Two of the systems, YY Gem and V451 Oph, had earlier been expected to show apsidal rotation; recent data seem to, if not prove, at least support this hypothesis. The case of the late dwarf binary YY Gem (Castor C) may prove particularly interesting. The light curve exhibits, however, quite conspicuous distortions, thus, if anywhere,

here is a cautious treatment of possibly spurious residuals advisable. The apparently opposite shift of primary and secondary minimum rests entirely with recent observations made by Leung and Schroeder [1978] and by Budding [1975]. If real, this shift would indicate an apsidal period of some 200 years or possibly much longer. The corresponding apsidal motion coefficient would be close to the lowest values found: $\log k \lesssim -2.6$, as crude estimates indicate.

Finally, three systems show marked changes of the period: RT And, TX Her and U Oph. In the case of RT And the time residuals grow to rather substantial values (well over 1 hour) and, quite significantly, the period variations seem to be discontinuous, similar to those shown by many Algol-type semi-detached systems [Williamon, 1974]. Here the fact of a variable period is demonstrated beyond doubt, yet the detached nature of the system is sometimes questioned. It appears, however, that a recent photometry of RT And by Dean [1974] demonstrated that the primary eclipse is a transit; also, "both stars are well within their contact surfaces."

TX Her is the only system among those considered where the time residuals may indicate a light-time effect in a triple system [Vetešnik and Papoušek, 1973]. Between 1972 and 1976, eleven additional photoelectric minima have been observed and later published; they do not contradict the periodic representation of the paper quoted, although the residuals turn out predominantly negative.

U Oph is another very interesting case. The system consists of two early-type stars and has the relatively short period of 1.677 days. Yet U Oph is certainly a detached system, although a very close one with fractional radii $r_1 + r_2 \approx 0.52$. The period is known to vary and attempts have been made to represent the observed O-C values by a periodic curve, see recent discussions of the period by Frieboes-Conde and Herczeg [1973] and by Koch and Koegler [1977]. A periodic representation of the residuals, however, remains somewhat unconvincing, apparently due to the inacceptably low accuracy of the early measurements but also to conflicting recent data [Batten and Scarfe, 1977].

This star is not neglected by photoelectric observers and a glance at the diagram of the photoelectrically determined O-C values for primary minima reveals immediately that the system very probably exhibits irregular changes of the period (Fig. 1). Secondary minima are in good agreement with this O-C diagram. The period increased steadily since about 1963 and its present value is 1.677355 days. Just before 1963 it underwent a sharp decrease, by about 1.0 sec; prior to this decrease, the period was for years apparently constant, at 1.677344 days.

One has to point out the possibility of an alternate interpretation: A short period sine curve (approx. 28 years) would fit most of these data equally well; the decision depends entirely on which of two conflicting early photoelectric determinations (near 2434500) is more accurate.

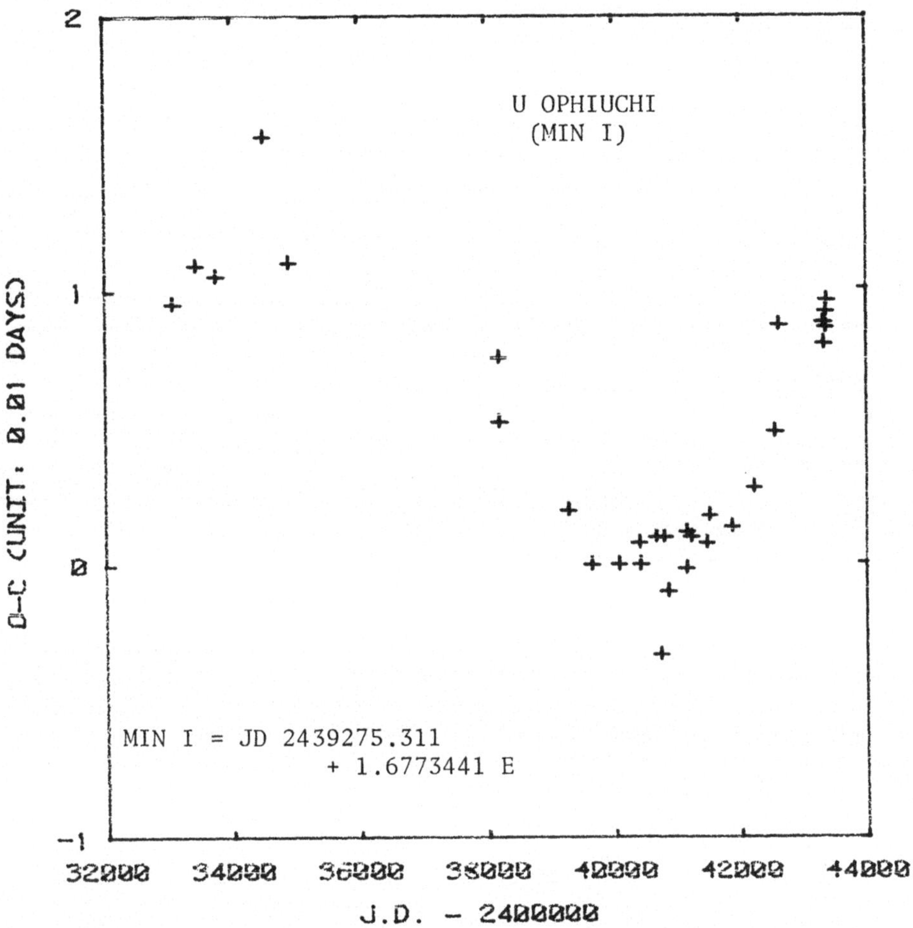

This investigation is supported in part by a grant from the National Science Foundation.

References

Batten, A.H., Scarfe, C.D. 1977: Rev. Mex. Astron. Astrofis. 3, 21
Budding, E. 1975: Tokyo Obs. Bull. 2nd Ser. No. 240
Dean, C.A. 1974: PASP 86, 912
Frieboes-Conde, H., Herczeg, T. 1973: Ast. Aph. Suppl. 12, 1
Geyer, E.H. 1977: Ap.Sp.Sc. 48, 137
Koch, R.H., Koegler, C.A. 1977: Ap J 214, 423
Leung, K.C., Schroeder, D.P. 1978: AJ 83, 618
Plavec, M., Pěkný, Z., Smetanová, M. 1960: BAC 11, 180
Vetešnik, M., Papoušek, J. 1973: BAC 24, 57
Williamon, R.M. 1974: PASP 86, 924

DISCUSSION FOLLOWING HERCZEG

Smak: What is the percentage of detached binaries with a third body?

Herczeg: Surprisingly low, if we consider only the information from period changes, the light-time effect. I do not know of a single case where the existence of a third body in the system was discovered in this way, although it was occasionally demonstrated later, for instance in the case of Algol, after the spectroscopic or astronomic discovery, by observing the time resuduals. Counting the few uncertain cases, I would estimate the percentage values at 1 or perhaps 2 but this certainly reflects mainly an unfavorable observational selection.

Budding: I have a couple of remarks to make about RT And. Daniel Jassur and I have been looking carefully at the system recently. In some work which we hope to publish soon we should show good reasons why the system can be taken to be detached. But the other point I would like to make is that the system does exhibit fairly substantial irregular photometric fluctuations. In view of what you said already about the effects of this on apparent period variation, how reliably would you say a genuine period variation has been shown for this system?

Herczeg: Obviously RT And is still comewhat controversial; I have based in classifying it a detached system on the papers of Gordon and Dean. As to the O-C curve of RT And in particular, its type suggests that the period changes are not spurious: the period is constant for years then suddenly changes, to a new value; this is typical also for Algol-type stars or W UMa systems. It is difficult to imagine that a certain deformation stays unchanged for such a long time and goes unnoticed. Available evidence, for instance with YY Gem, suggests that the effect of light curve variations is rather an increased scatter of data in the O-C diagram.

Andersen: RS Cha figures in your list. In a paper in press by J.V. Clausen and B. Nordström discussing new light curves of it, there is no indication of period change. But it contains at least one δ Scu star, so here is another source of scatter in the times of minimum (though hardly any systematic variation over longer periods).

Herczeg: Yes, RS Cha is one of those cases in my list where presently we cannot say more than this period seems to be constant, at least for the relatively short time interval of reliable observation.

THE SURFACE DISTRIBUTION OF CHEMICAL ANOMALIES OF Ap COMPONENTS IN DETACHED CLOSE BINARIES

M. Kitamura
Tokyo Astronomical Observatory
University of Tokyo, Mitaka
Tokyo 181, Japan

ABSTRACT

By estimating the orbital inclinations of non-eclipsing detached close binaries with Ap spectra, a marked statistical preference is obtained on the frequency distribution of the inclination which suggests that the abundance anomalies of Ap components tend to concentrate towards the stellar polar region.

1. INTRODUCTION

It is well known that no eclipsing system has been discovered in spectroscopic binaries with Ap spectra, except for AR Aur, for which Dworetsky (1975) reported that it is an Ap star of the Hg-Mn type.

Suppose that the orientations of the orbital planes of the binary systems are at random in space; then the relative frequency ratio of observable eclipsing and spectroscopic binaries may be written as

$$R = f_{ecl}/f_{sp} \tag{1}$$

where f_{ecl} is the probability of discovery of eclipsing binaries and f_{sp} the probability of discovery of spectroscopic binaries. The Seventh Catalog of spectroscopic binaries (Batten et al. 1978) lists 978 systems, of which 378 are eclipsing binaries. Kitamura and Kondo (1978) made a statistical study of spectroscopic binaries of Am spectral features with periods less than 12 days and showed that 11 of the 45 systems are eclipsing. Thus, the probability of discovery of eclipsing binaries in spectroscopic ones should about to 25 $\sim$ 40% generally.

Therefore, the actual lack of eclipsing systems in spectroscopic binaries with Ap spectra might suggest any particular reason.

M. J. Plavec, D. M. Popper and R. K. Ulrich (eds.), Close Binary Stars: Observations and Interpretation, 95–99.

2. Ap CLOSE BINARIES WITH PERIODS LESS THAN 13 DAYS AND THEIR ORBITAL INCLINATIONS

It is customary to divide the Ap stars into two groups, the Hg-Mn type and the magnetic type. So far, 14 close binaries of the Hg-Mn type and 5 of the magnetic type are known. Of these, 9 are double-lined and the remaining are single-lined (Kitamura, 1980).

For a non-eclipsing double-lined binary, the orbital inclination can be estimated from the $m_1 \sin^3 i$ value with the assumed m_1 value corresponding to its spectral type as the main sequence. For a non-eclipsing single-lined binary, the orbital inclination can be given from the mass function as

$$\sin^3 i = (m_1/m_2 + 1)^2 \cdot f/m_2 \qquad (2)$$

In view of the invisibility of the fainter component on spectrograms, we may put a condition that

$$m_1/m_2 > 1.5 \qquad (3)$$

Also, the lowest mass for the secondary component may be put as

$$m_2 > 0.3 \qquad (4)$$

Substituting conditions (3) and (4) into (2), it follows that the orbital inclination of a single-lined binary should fall within the two limiting values as

$$(10\, m_1 + 3.3)f > \sin^3 i > 9.4\, f/m_1 \qquad (5)$$

From expression (5) we can find the value of orbital inclination of a non-eclipsing single-lined binary with a certain limit, which enables us to make a statistical study of the frequency distribution of the inclinations.

3. STATISTICS

In order to find the frequency distribution of the orbital inclinations of 19 spectroscopic binaries with Ap features, numbers of binaries within intervals of 0°-15°, 15°-30°, 30°-45°, 45°-60°, 60°-75°, and 75°-90° in the orbital inclination were counted. If the range of the inclination deduced for a single-lined binary was spread over more than one interval, a weight reduced reciprocally to the number of the relevant intervals was applied. For example, the orbital inclination of 21 Her is given to be 10°-21° spreading over two intervals of 0°-15° and 15°-30°. In such a case a reduced weight of 0.5 was applied to each of the relevant intervals. The results of the counts are given in the second line of Table 1.

Table 1. The Frequency Distribution of the Orbital Inclinations

i	90°-75°	75°-60°	60°-45°	45°-30°	30°-15°	15°-0°
No.	2.28	3.28	2.78	3.95	3.20	3.50
f_{Ap}	0.120	0.173	0.146	0.208	0.168	0.184
f_{Am}	0.312	0.196	0.188	0.133	0.107	0.064
f_{Ap}/f_{Am}	0.38	0.88	0.78	1.56	0.57	2.88

In the third line of Table 1, the corresponding fraction is given. F_{Am} in the fourth line is taken from Kitamura and Kondo's statistical study (1978) on Am spectroscopic binaries. According to Abt and Bidelman (1969), the A2-F0 type stars as the components of close binaries are all Am stars and therefore the frequency distribution of orbital inclinations of Am binaries may be considered to represent the ordinary A-type spectroscopic binaries. The values of f_{Am} and f_{Ap} are shown in Figure 1 as well.

3. DISCUSSION

In general the discovery of spectroscopic binaries is easier the larger their velocity variations are. As the orbital inclination is larger, the velocity variations should be larger. For the random orientations of the orbital planes of binary systems in space, the frequency distribution of the orbital inclination for observable spectroscopic binaries should increase with increasing orbital inclinations, as in the case of Am spectroscopic binaries. However, as is evident in Figure 1, the distribution of f_{Ap} and f_{Am} are quite different.

Next we shall correct for the discovery probability in order to find out the real relative frequency distribution of the orbital inclinations of Ap binaries in space. This can be done by dividing the observed frequency of Ap binaries with the relative frequency for Am binaries, because f_{Am} may be regarded as representing the relative discovery rate of the general spectroscopic binaries. The ratio f_{Ap}/f_{Am} should represent the relative frequency distribution of the orbital inclinations of the 19 Ap binaries in space, corrected with the discovery probability.

Figure 2 shows the distribution of f_{Ap}/f_{Am} with the inclination. As is evident from this figure, the frequency distribution in space of orbital inclinations of Ap spectroscopic binaries has a considerable preference towards the smaller orbital inclinations. What does this mean?

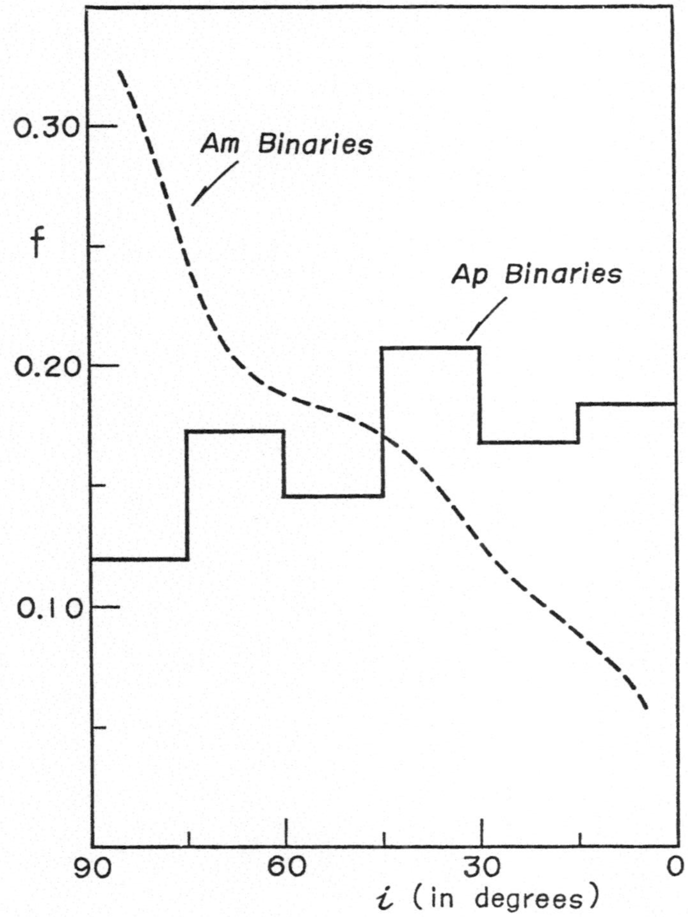

Figure 1. Histogram for the frequency distribution of the orbital inclination for the 19 Ap binaries of the Hg-Mn type and the magnetic type. The dotted curve represents the corresponding smoothed frequency distribution for the Am spectroscopic binaries with periods less than 12 days.

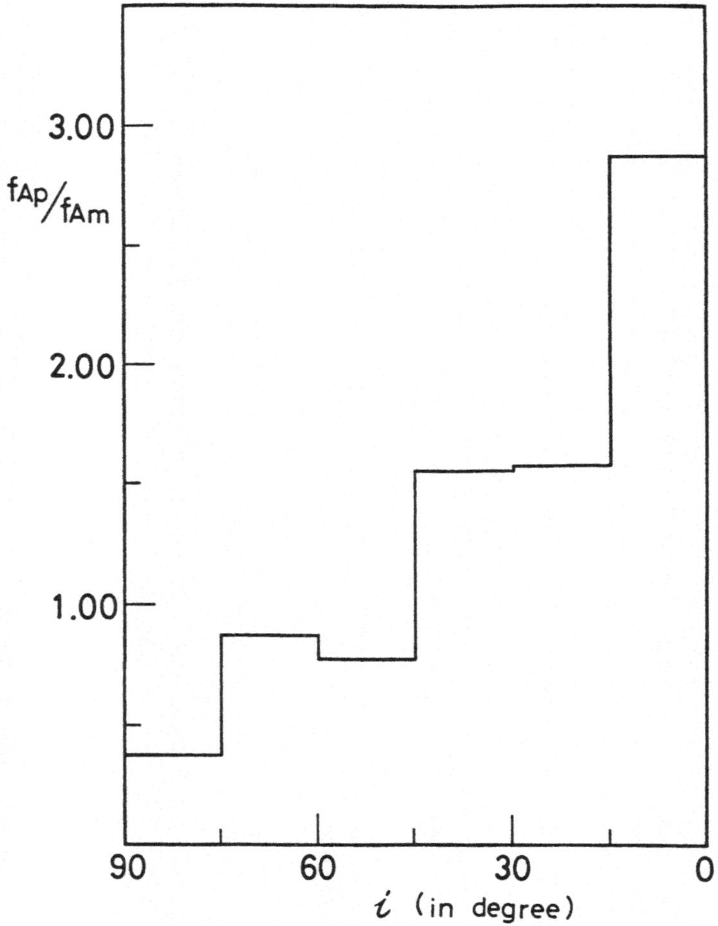

Figure 2. Relative frequency distribution of the orbital inclinations of the 19 Ap binaries, corrected for the discovery probability.

In close binaries, the components may be considered to rotate around the axes perpendicular to their orbital planes, and therefore the inclinations should correspond to the angle between the line of sight and the rotation axis of either component. Therefore, the present result indicates that the surface distribution of Ap chemical anomalies should have a marked preference towards the pole of the Ap star.

It has been conventional to divide the Ap stars into two distinct groups, the Hg-Mn type and the magnetic type. However, the present statistics reveal that both types have a similar nature. The full discussions will be published in Ap. Sp. Sci. soon.

REFERENCES

Abt, H.A. and Bidelman, W.P.: 1969, Astrophys. J., 158, 1091.
Batten, A.H., Fletcher, J.M. and Mann, P.J.: 1978, Publ. Dominion Astrophys. Obs., 13, No.8, 119.
Dworetsky, M.: 1975, In W.W. Weiss, H. Jenkner and H.J. Wood (ed.), Physics of Ap Stars (IAU Coll. No. 32), Wien Univ. Obs., p. 228.
Kitamura, M. and Kondo, M.: 1978, Astrophys. Sp. Sci., 56, 341.
Kitamura, M.: 1980, Astrophys. Space Sci., 68, 283.

COMMENTS ON KITAMURA

Nariai

The paper puts the Hg-Mn stars into the group of Ap-type stars. It is probably better to separate the Hg-Mn stars from the Ap stars, because the Hg-Mn stars seem to form the hot extension of the Am-type stars (cf. S. Wolff, IAU Coll. No. 32, Vienna, 1975, p. ...).

III. MASS TRANSFER AND MASS LOSS - THEORY

CLOSE BINARIES BEFORE AND AFTER MASS EXCHANGE: A COMPARISON OF OBSERVATIONS THROUGH THEORETICAL COMPUTATIONS(1)

J.P. De Grève(2) and D. Vanbeveren
Astrophysical Institute, Vrije Universiteit Brussel, Brussels, Belgium.

From a search through the literature 174 close binaries with completely known absolute dimensions are sampled. Distinction is made between systems before and after mass exchange, giving resp. 100 and 40 systems (a third group contains the systems not definitely belonging to these two). Mass, period and mass ratio distributions and relations of the group of "unevolved" binaries (i.e. prior to mass exchange) are transformed into corresponding distributions and relations of evolved binaries. The transformations are based upon the $M_{1f}=g(M_{1i})$ relation derived from an extended set of published theoretical computations on the evolution of close binaries. Final masses resulting from the same initial mass are averaged. Equations are derived for the cases A (for all masses), B1 ($M_{1i}/M_{\odot}\leq 2.8$), B2 ($2.8<M_{1i}/M_{\odot}\leq 9$) and B3 ($M_{1i}/M_{\odot}>9$). For the changes of the period due to angular momentum loss the formalism of Vanbeveren et al. (1979) was adopted. The following characteristics of the system after mass exchange are computed: M_{1f}, M_{2f} (and q_f), P_f. Three different modes were applied for the mass loss from the system:

a) conservative case (mass and angular momentum of the system remain constant), called C.

b) non conservative case with 50% of the transferred mass leaving the system with a small or a large angular momentum loss (resp. called NC51 and NC53).

c) non conservative case with 100% of the transferred mass leaving the system with a small or a large angular momentum loss (resp. called NC101 and NC103).

The distributions of the parameters M_1, q^{-1} and P of the transformed unevolved systems are compared to the corresponding distributions of observed evolved systems. The most important features of the comparison are given in tables 1, 2 and 3. From this comparison the following conclusions are derived:

1. The mass distribution of the unevolved systems gives a fair representation of reality in the range $1.6M_{\odot}-15M_{\odot}$,

M. J. Plavec, D. M. Popper and R. K. Ulrich (eds.), Close Binary Stars: Observations and Interpretation, 103–107.

Table 1. Comparison of the mass distribution of evolved primaries and transformed ones in the range $-0.4<\log M_1\leq 0.6$. The transformed systems have been normalized to the total number of evolved systems in that mass range.

$\log M_1$-bin	-0.4 -0.2	-0.2 0.0	0.0 0.2	0.2 0.4	0.4 0.6
nr of evolved systems	12	10	5	3	3
nr of transformed systems	13	11	5	2	2

Table 2. Comparison of characteristics of the q^{-1} distributions of set II', with the distribution of observed evolved systems (set II).

Group	Maximum $q_1<q^{-1}\leq q_2$	% of max.	% in interval $0.1<q^{-1}\leq 0.3$	% systems with $q^{-1}>0.6$
Observations	0.2 - 0.3	30	57	0
Conservative	0.1 - 0.2	43	50	3
NC50	0.1 - 0.2	55	74	3
NC100	0.2 - 0.3	29	57	11

Table 3. The same as Table 2, but for the period distribution.

Group	Maximum $\log P_1<\log P\leq \log P_2$	% of max.	% in interval $-0.2<\log P\leq 1.0$	% systems with $\log P>1.4$
Observations	0.2 - 0.6	37	84	3
Conservative	1.4 - 1.8	26	18	62
NC51	1.4 - 1.8	32	22	59
NC53	1.0 - 1.4	28	45	17
NC101	1.4 - 1.8	32	20	54
NC103	-0.2 - 0.2	27	54	2

the low mass stars being slightly underabundant; for the evolved system good correspondence is found in the range $0.4M_{\odot}-2.5M_{\odot}$.

2. The theoretical formalism of M_{1f} depending on M_{1i} results in a striking correspondence between set II and set II' (= the transformed set I).

3. The best quantitative correspondence for the q^{-1}

distribution is found with mode NC100; the appearance of systems with mass ratios $q^{-1}>0.6$ for that mode may indicate that the external mass loss is influenced by the parameters M_1, q, P.

4. The best correspondence in the period distributions is obtained when a large amount of angular momentum is removed from the system; a still better agreement results if one assumes that not all, byt say 80% of the transferred mass is leaving the system. This also causes a strong decrease of the excess of transformed short period systems, appearing with mode NC103.

5. Only a very small fraction of the evolved systems can be transformed into unevolved by using the conservative mode. The average value of the maximum fraction of mass transferred to the secondary (in order to obtain a non-negative initial mass of the secondary) is ~0.6.

The different distributions were analysed against selection effects. It was found that the selection effects in the distributions of M_1, q and P do not alter the conclusions at all.

REFERENCE

Vanbeveren, D., De Grève, J.P., van Dessel, E.L., de Loore, C.: 1979, Astron.Astrophys. 73, 19.

Notes

(1) For the details of this investigation we refer to De Grève, J.P. and Vanbeveren, D., Astrophys.Space Sci. (submitted).

(2) Research Fellow, F.K.F.O., Belgium.

COMMENTS FOLLOWING DE GREVE AND VANBEVEREN

Meyer-Hofmeister: For a comparison we need complete samples of observed and computed evolved binaries. Initial distributions of binary masses and separations are needed to determine the distribution of evolved binaries. How did you determine the sample of evolved (computed) binaries?

De Grève: Computed evolved binaries (called "transformed") are constructed by transforming each observed unevolved system in the following way. The mass of the loser is transformed into a remnant mass $q(M_i)$. The mass of the gainer is determined by the assumption on the mass leaving the system (0, 50 or 100%), resulting in a value q_f^{-1}. The final period is computed with the formalism quoted, adopting assumptions on the angular momentum loss. Thus several sets of transformed systems are constructed. These sets are compared to the set of observed evolved binaries. The influence of selection effects is discussed in the comparison, but herefore I refer to the extended mass paper. For example for the mass distributions the ranges where the distributions are close to reality are determined as well as the proper fractions of systems, taking into account evolutionary aspects, such as different lifetimes for different masses.

Smak: I am still under an impression that, because of the severe selection effects affecting your main sequence input sample and the semi-detached sample, no meaningful comparison can be made.

De Grève: It is so that for the mass distributions of unevolved and evolved systems selection effects were examined by constructing different mass distributions within well defined space volumes from the catalogue of Batten and supplements to it. The comparison shown in table 1 refers only to the mass range where the mass distributions of both set I and II are close to the volume distributions. For the period and mass ratio distribution also arguments are given (in the extended paper) to show that the selection effects in the distributions do not alter the comparison made in tables 2 and 3.

Sugimoto: Since the physical situations are very complicated and many parameters are involved, it is better to ask the following questions. How much uncertainty, do you think, is involved in your conclusion of 80 percent mass loss? Do you think it possible to obtain any quantitative result concerning the amount of the angular momentum loss from such considerations?

De Grève: I do agree that many parameters are involved and that precise quantitative results on mass and angular momentum loss vary from system to system. The 80 percent quoted in my talk is to be interpreted as an average result. It is difficult to say something about the uncertainty but I could mention two aspects that give at least an idea:

1 - If one assumes that less than 50 percent of the transferred mass leaves the system, the resulting q^{-1} and P-distributions differ rather strongly from the observed.

2 - The fact that the average maximum fraction of the mass accreted by the secondary, as derived from the determination of initial parameters of evolved systems, is 0.6. I think that a detailed analysis of these minima for an extensive set of evolved systems can give more information.

Plavec: Typical detached systems have periods of a few days, and the relative dimensions of the components are so large that when the more massive one overflows, the stream hits the other star. It is therefore rather surprising to hear that almost all the material should leave the system. This would imply that these stars probably are not the progenitors of Algols, and that the real progenitors are wider systems that we probably do not observe.

De Grève: The fact that most of the material should leave the system is not so surprising. Theoretical computations have shown that either the stream hits the gainer or falls on it through a ring structure; the gainer swells up rapidly and fills its own critical lobe, forcing the material to leave through L_1. Only in the last stage of mass transfer when the loser has already lost the largest fraction of the material, the gainer reacts less violently to accretion. Whether or not the progenitors of Algols should be searched among more wide systems is not so easy to determine. From the comparison of the distribution values for the parameters determining the mass and angular momentum loss result in a ratio of initial to final period near or larger than unity, indicating that the Algol-progenitors may still result from systems with periods of a few days.

de Loore: I want to make a comment concerning the value of the mass and angular momentum losses. It should be kept in mind that these results are global results, and the conclusion is that mass and angular momentum losses are very large. The values of $\sim$100% quoted within this paper should be interpreted in this sense.

Most probably these losses are different from system to system; it should not be concluded that none of the secondaries can accrete a part of the expelled matter.

EVOLUTION OF INTERMEDIATE MASS CLOSE BINARIES

Th.J. van der Linden
Astronomical Institute, University of Amsterdam

Abstract

Numerical simulations of close binary evolution were performed for five binary systems, using a newly developed evolutionary program. The systems have masses 3+2, 4+3.2, 6+4, 9+6, 12+8 $M_{\odot}$ and periods 2^d, 1^d78, 3^d, 4^d, 5^d respectively. The primary component was followed from the zero-age main sequence through the mass transfer phase to core-helium burning. Special care was given to the self-consistent determination of the mass transfer rate and the detailed treatment of composition changes. After the mass transfer phase the resulting systems consist of a main sequence star with a helium star companion of mass 0.36, 0.46, 0.82, 1.48, 2.30 $M_{\odot}$ for the five systems respectively. Interesting "thermal pulses" were found in the 3+2 $M_{\odot}$ system at the onset of helium burning.

Introduction

Most theoretical studies of the evolution of close binaries have so far been restricted to systems with high ($M > 15$ $M_{\odot}$) and low-mass ($M < 3$ $M_{\odot}$) primaries. The high-mass systems evolve into Wolf-Rayet binaries, while the low-mass systems lead to Algol-type systems and further to systems containing a white dwarf (cf. Paczynski 1971; Thomas 1978). The intermediate mass (3-15 $M_{\odot}$ primaries) close binaries are of particular importance, as they cover the entire B spectral range, and so far were studied in only a very few cases (Kippenhahn and Weigert 1967; Harmanec 1970; Plavec, Ulrich and Polidan 1973). As still little is known about the final evolutionary stages of these systems, a new evolutionary program was constructed which enables us to use as much (nuclear) input physics as is necessary to carry out reliable computations for the advanced evolutionary stages. Our calculation include all the effects of mass transfer, nuclear burning and composition changes in a completely self-consistent way. Evolutionary sequences were calculated for five systems, which cover the mass range of 3-12 $M_{\odot}$ for the primary, assuming conservative mass transfer and chosing periods such that mass transfer starts as Case B and at a time when the outer layers are still in radiative

M. J. Plavec, D. M. Popper and R. K. Ulrich (eds.), Close Binary Stars: Observations and Interpretation, 109–114.

equilibrium. Mass ratios were chosen such that no problems were expected with the accreting companion star (cf. Kippenhahn & Meyer-Hoffmeister 1977). The initial composition used was X = 0.70 and Z = 0.03.

Input physics and numerical techniques

For the equation of state the contributions due to electrons ions and photons are taken into account. The thermodynamic properties of the electrons gas are used in tabular form, allowing for partial as well as relativistic degeneracy. The contributions from the ions and photons are calculated analytically. The cross-sections used for the nuclear reactions were taken from the expressions given by Fowler et al. (1975). Absorption coefficients were interpolated from the tables of Cox and Stewart (1969), using 2 point interpolation. The program used obtains a fully implicit solution of the structure equations. Composition changes are calculated algebraically by solving a diffusion equation (which includes a complete thermonuclear reaction network) simultaneously with all other structure equations. Implicit boundary points are used for all convective zones (central as well as surface and intermediate convection zones). An abbreviated nuclear network was used for PP and CNO-cycles, except for the 12+8 $M_\odot$ system where all of the 14 reactions in the CNO-bicycle were treated seperately. All equations were solved algebraically with a Newton-Raphson scheme using a tridiagonal (Jacobi) matrix system devised by R.J. Takens. The central boundary conditions used are described by Savonije and Takens (1976). Surface boundary conditions used ar similar to those described by Savonije (1978). A more detailed description of input physics and numerical tehcniques will be given elsewhere.

Discussion

The mass loss rates as determined by using implicit boundary condition are shown in figure 2. All curves show a double peaked structure. The second peak in these curves is caused by the onset of convection in the outer layers. In the 3 $M_\odot$ primary and the 4 $M_\odot$ primary deep surface convection occurs when the stars have almost reached their minimum luminosity in the HR-diagram, giving rise to an increased mass transfer rate during the second part of the mass transfer. In the 6,9,12 $M_\odot$ system no deep surface convection occurs during mass transfer. The increase of the mass transfer rate here is due to the increase in size of the small convection zone present in the atmosphere, the extend of which is controlled by the luminosity of the star. In the time following minimum luminosity the extending atmospheric convection gives rise to an extension of the rapid mass transfer phase.

The tracks of the mass losing star in the HR-diagram show some systematic trends. They all show the familiar drastic decrease in luminosity during the rapid mass transfer phase due to the absorption of radiation energy in the outflowing atmosphere. Mass transfer terminates at the onset of core helium burning after which the star relaxes to its thermal equilibrium position near the helium main sequence. The low mass remnants are very faint in comparison to their newly formed companion. This is clearly demonstrated in Table 1, which shows that for the lower-mass systems the resulting helium star companion will be practically unobservable. A particularly interesting phenomenon occurs at the onset of helium burning in the (original) 3+2 $M_\odot$ system. As the mass-losing star in this system becomes degenerate before helium burning, this remnant reacts differently to the onset of helium burning than the other four systems. The mass fraction of the star which becomes convective is shown as a function of time in figure 4. The convective core shows three thermal oscillations before it reaches its final thermal equilibrium state, each oscillation removes a part of the degeneracy. During the oscillations over 50% of the star becomes convective which is about 3 times the extent of the core in thermal equilibrium. The remnant has a hydrogen content $X \approx 0.50$ in its outer layers after mass transfer, in contrast to the other systems which have $X \approx 0.20$. This difference in hydrogen content is responsible for the increase in luminosity by more than one magnitude before settling on the helium main sequence.

Examination of Table 1 shows that in each case the resulting system consists of a newly formed bright main sequence B-star accompanied by a faint helium star in a rather wide ($P > 60^d$) orbit. The difference in luminosity between the two stars is about 5 magnitudes for the low mass-systems and 2 magnitudes for the heavier ones. Due to the faintness of the companion the systems are expected to be unrecognized as a binary in the visible part of the spectrum. Most of the flux of the helium star is emitted in the UV part of the spectrum due to its high surface temperature ($\log T_{eff} \sim 4.7$). Because of this high UV flux it has been suggested that these systems might appear as Be stars (see Križ and Harmanec 1975 and references therein). It should be pointed out that this kind of evolution has significant influence on the statistics of B-stars, as every B-star in a binary system produces another B-star of somewhat earlier spectral type, thereby increasing the lifetime of the system as a binary with the primary on the main sequence by about a factor of two. We are presently continuing the evolution of the helium stars in these five systems in order to study the type and composition of their final remnants.

TABLE 1

M_1, M_2	P_i	P_f	A_f	M_{1f}	M_2	Δm
3+2	2^d	93	147	.36	4.64	5.0
4+3.2	1.78	11.8	195	.46	6.74	4.0
6+4	3	96	191	.82	9.18	3.5
9+6	4	78	190	1.48	13.52	2.7
12+8	5	65	184	2.30	17.70	-

Acknowledgements

It is a pleasure to acknowledge the inspiration and guidance of E.P.J. van den Heuvel throughout this work. I am indebted to R.J. Takens for the introduction to and the use of his algebraic scheme and numerical routines. This work was supported by the Netherlands Organization of Pure Research (Z.W.O.) under projectnumber 78-110.

References

Cox, A.N. and Stewart, J.N. 1969, Sci. Infor. Astr. Council USSR Acad. Sci. Vol. 15.
Fowler, W.A., Caughlan, G.R. and Zimmerman, B.A. 1975, Ann. Rev. Astr. Astrophys. *13*, 69.
Harmanec, P. 1970, Bull. Astron. Inst. Czech. *21*, 113.
Kippenhahn, R. and Weigert, A. 1967, Z. Astrophys. *65*, 251.
Kippenhahn, R. and Meyer-Hoffmeister, E. 1977, Astron. Astrophys. *54*, 539.
Kriz, S. and Harmanec, P. 1975, Bull. Astron. Inst. Czech. *26*, 65.
Paczynski, B.E. 1971, Ann. Rev. Astron. Astrophys. *9*, 183.
Plavec, M., Ulrich, R.K. and Polidan, R.S. 1973, Publ. Astron. Soc. Pac. *85*, 769.
Savonije, G.J. 1978, Astron. Astrophys. *62*, 317.
Savonije, G.J. and Takens, R.J. 1976, Astron. Astrophys. *47*, 231.
Thomas, H. 1977, Ann. Rev. Astron. Astrophys. *15*, 127.

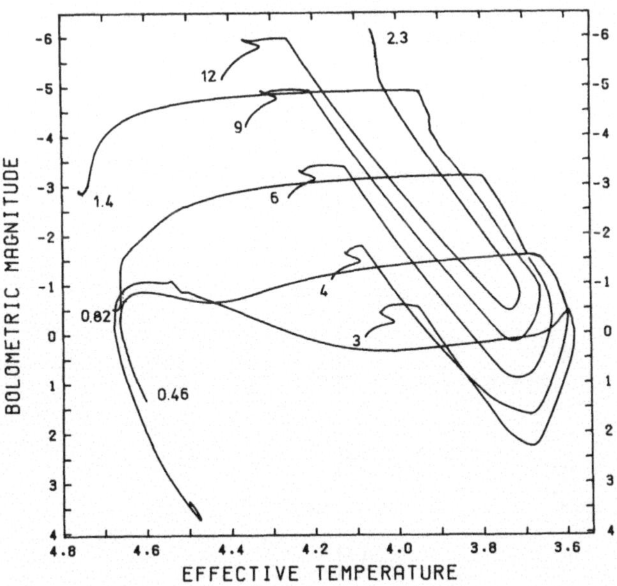

Fig. 1. Evolutionary tracks of the primary component, from the ZAMS through mass exchange phase to He-main sequence, except for the 12 $M_\odot$ track which is terminated at He-ignition.

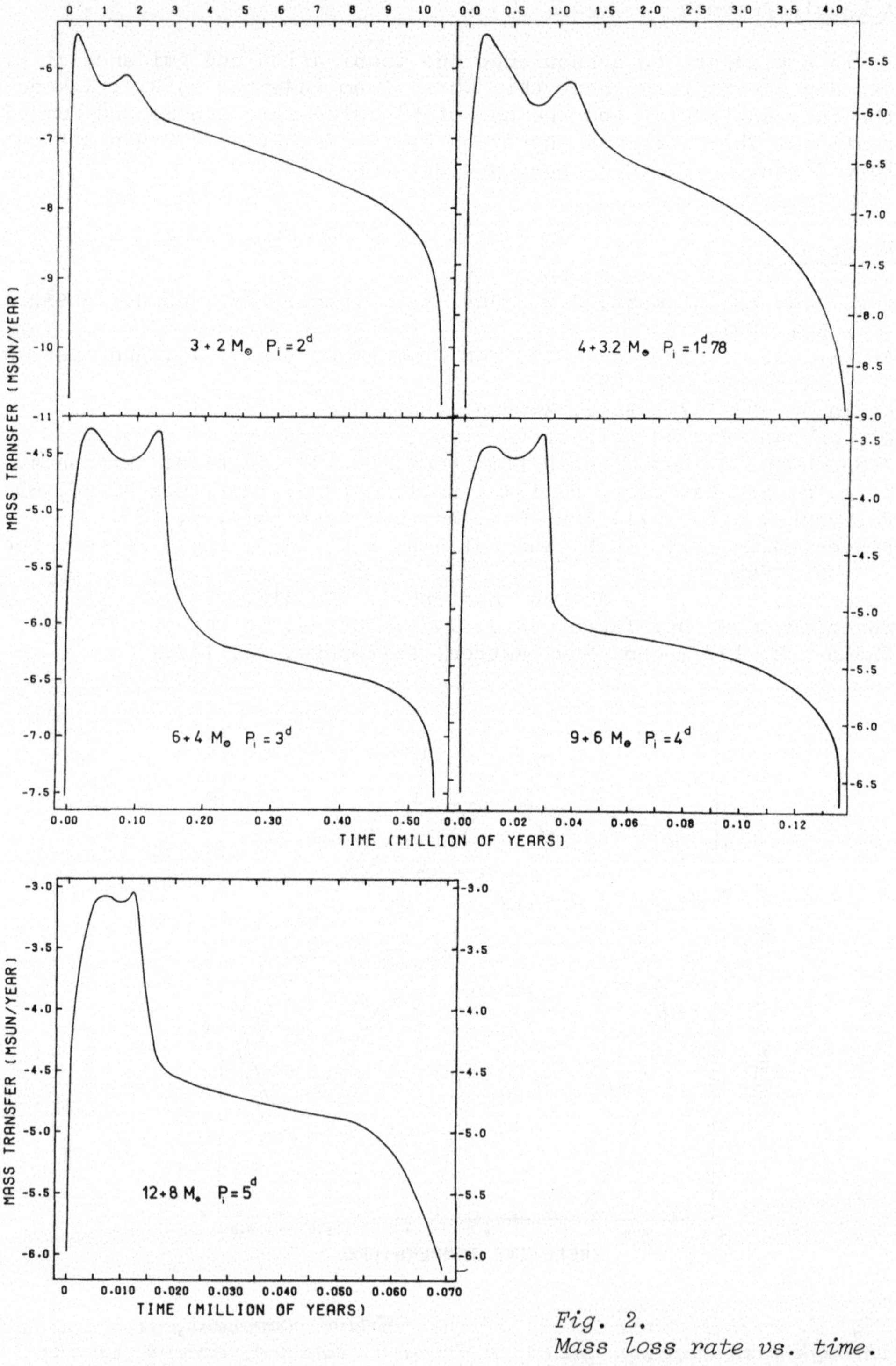

Fig. 2.
Mass loss rate vs. time.

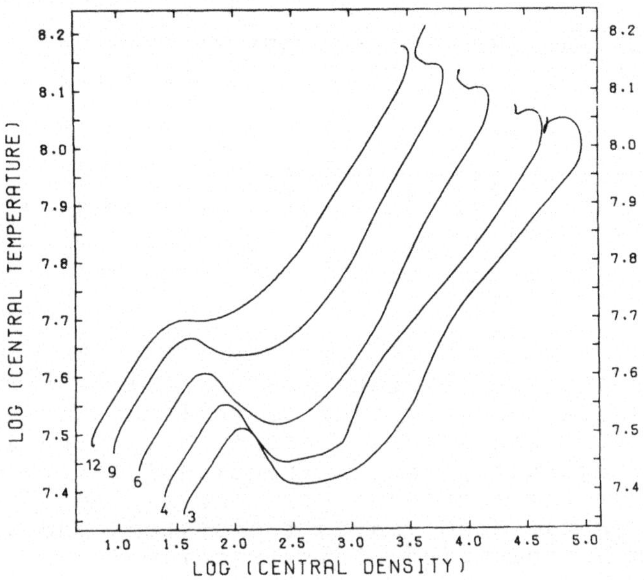

Fig. 3. Central values of density and temperature.

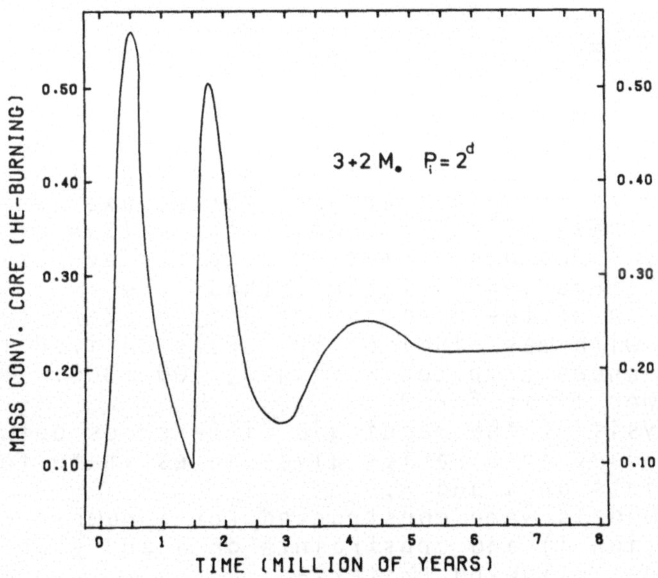

Fig. 4. Relative mass fraction of convective core vs. time, in the remnant of original 3 $M_\odot$ primary, after the onset of He-burning.

THE EVOLUTION OF MASSIVE CLOSE BINARY SYSTEMS : ESTIMATES OF THE PARAMETERS GOVERNING MASS AND ANGULAR MOMENTUM LOSS

D. Vanbeveren and C. de Loore
Astrophysical Institute, Vrije Universiteit Brussel, Brussels, Belgium

It becomes more and more evident that for close binary evolution during Roche lobe overflow as well mass transfer as mass loss occurs. When a mass element ΔM is expelled from the primary during this phase, a fraction β is transferred to the secondary; the remaining part leaves the system. Moreover, angular momentum leaves the system, and also this fraction has to be specified; this fraction is related to a parameter α (Vanbeveren et al., 1979). For the computation of the evolution of massive close binaries also mass loss due to stellar wind of both components, prior to the Roche lobe overflow has to be taken into account. The mass loss rate $\dot{M}$ due to radiation driven stellar winds can be expressed as

$$\dot{M} = -N \frac{L}{c^2}$$

An estimate of the N-value for single O-stars has been made by Lamers, Paerels and de Loore (1979); they found that N is of the order 100 to 200. The N-value for binaries could be higher, due to the gravitational forces exerted by the secondary. We have tried to specify the values of N, β and α, using various methods, relating computed evolutionary models with observations. These methods are :

1. comparison of the position of O-type stars sufficiently well known with mass-luminosity curves derived from evolutionary sequences computed with N=0, 100, 300. This leads to a lower limit for N.
2. An analysis of the magnitude differences between primary and secondary in binaries gives an estimate for N.
3. Constraints on α and β.

Evolutionary models were constructed for a number of systems (see section 1) and constraints on α and β were determined using the following criteria :

a) Vanbeveren and Packet (1979) determined upper and lower limits for the amount of mass that still has to be removed after RLOF in order to obtain an early WN star or a WC star.

M. J. Plavec, D. M. Popper and R. K. Ulrich (eds.), Close Binary Stars: Observations and Interpretation, 115–121.

Using the observed masses of WR systems we can use these results to point out that in massive systems the mass ratio $(\frac{\text{secondary}}{\text{primary}})$ just after RLOF is less than 2.

b) The mass ratio of the system (secondary to primary) will increase during core hydrogen burning due to stellar wind mass loss, and come nearer to 1. The results of Garmany (1979) and the computations of Vanbeveren et al. (1979) reveal that we can expect an overabundance of systems just before Roche lobe overflow with a mass ratio larger than 0.7.

c) A large majority of massive close binaries evolves through a case B of mass exchange. An analysis of the observed periods for systems after the mass exchange and the minimum periods before RLOF in order to have a case B system reveals that the period ratio (after to before) should be of the order 1 or smaller.

1. SUFFICIENTLY WELL DEFINED SYSTEMS EVOLVED OR NOT

We considered a system as evolved, if the mass of the secondary exceeds that of the primary, hence the system is in its post mass exchange stage (PRLOF). The parameters of these systems are shown in Table 1 (evolved systems) and in Table 2 (non evolved systems). Table 1 contains the OB components of 6 X-ray binaries, with well defined parameters (mass, luminosity).

Table 1. Observed parameters for evolved systems.

System	q	$M_1\sin^3 i$	$M_2\sin^3 i$	i	P(d)	$-M_{bol_1}$	$-M_{bol_2}$	Spectral types 1	2	References
HD90657	2.01	6.8	13.6		6.5			WN5	O6	(3)
HD152270	2.68	1.8	4.9		8.9			WC7	O5	(5)
HD168206	4.35	8.1	35.2	≥60°	29.7			WC8	O	(6)
HD186943	2.38	3.36	7.94		9.6			WN4	B	(7)
HD190918	3.70	0.21	0.78		85		9.8 (CA) 10.1 (W)	WN4	O9I	(7)
HD193576	2.32	8.4	19.5		4.2			WN5	O6	(2)
HD211853	2.87	11.5	33	≥60°	6.7		10 (CA) 10.7 (W)	WN6	O6I	(1)
γ Velorum	1.88	17	32		78.5		9.8 (CA) 10.1 (W)	WC8	O9I	(4)
HD1337	1.37	19	26	~90°	3.5	8.6	8.6	O9III	O9III	(11)
HD57060	1.21	24	29	~90°	4.4	9.6 (CA) 10.3 (W)		O7f	O7	(11)
		23	30			9.7 (CA) 10.15 (W)		O8Ia	O7	(12)
HD149404	1.68	1.6	2.7			9.65	8.77	O8.5I	O7III(f)	(13)
HD163181	1.74	12.5	21.8	~73°1	12.01	8.8	6.5	BO.5Ia	B	(10)
HD166734	1.07	29	31		34.54	9.6 (CA) 10.7	9.6 (CA) 10.3	O7f	O7f	(16) (W+16)

Table 1 (continued)

System	q	$M_1\sin^3 i$	$M_2\sin^3 i$	i	P(d)	$-M_{bol_1}$	$-M_{bol_2}$	Spectral types 1	Spectral types 2	References
HD190967	1.28	17.5	22.4	~90°	6.5		7.2(CA 7.0(W)	B1Ib	O9.5V	(5)
HDE228766	1.48	23	34		10.74	10.3		O5.5f	O7.5	(9)
BD+40°4220	3.44	9	31	≥60°	6.6	10.8	10.5	O6f	O7f	(8)(9)
HD153919	20.69	1.31	27.1	~90°	3.41		10.2		O6f	(14)
Cen X-3	12.11	1.42	17.2	~80°	2.09		9.0		O6.5III	(14)
HDE226868	1.61	1.48	2.39	~29°	5.60		8.7		O9.7Iab	(14)
SMC X-1	15.86	0.79	12.53	~70°	3.89		8.1		BOI	(14)
HD77581	15.01	1.42	21.32	~74°	8.97		9.6		BO.5Iab	(14)
LMC X-4	9.00	2.5	22.53	~75°	1.41		8.4		O7	(15)

(1) Stempien (1970)
(2) Ganesh et al. (1967)
(3) Niemela (1976)
(4) Niemela and Sahade (1979)
(5) Seggewiss (1974)
(6) Cowley et al. (1971)
(7) Bracher (1967)
(8) Bohannan and Conti (1976)
(9) Massey and Conti (1977)
(10) Woodward and Koch (1975)
(11) Stothers (1972)
(12) Hutchings (1977)
(13) Massey and Conti (1979)
(14) Conti (1978)
(15) Hutchings et al. (1978)
(16) Conti,P.S.,Ebbets,D.,Massey,P.,Niemela,V.,1979,private communic.

Table 2. Observed parameters for non evolved systems

System	q	$M_1\sin^3 i$	$M_2\sin^3 i$	i	P(d)	$-M_{bol_1}$	$-M_{bol_2}$	Spectral types 1	Spectral types 2	References
HD19820	0.48	18.9	9.2		3.4			O8	O8	(3)
HD34333	1	19.7	19.7	~64°	4.1	5.8	5.3	B3III	B3III	(3),(4),(5)
HD36486	0.36	28	10	~67°	5.7	9.1	5.5	O9.5	B	(3),(4),(5)
HD57060	0.83	23	19	~90°	4.4	9.57(CA) 10.27(W)		O7f	O7	(1)
HD93205	0.38	39	15	<50°	6.08	10.0 10.5	7.4	O3	O8	(11)
HD190967	1.00	23	22.4	~90°	6.5		7.2(CA)	B1Ib	O9.5V	(5)
HD191201	1	13.9	12.9		8.3	8.5(CA)	8.5(CA)	O9.5III	O9.5III	(3)
HD152218	0.80	13.4	10.7		5.4	8.5(CA) 7.9(W)		O9.5III		(9)
HD152248	0.92	24.4	22.5		5.97	9.6(CA) 10.3(W)		O7f		(9)
HD159176	0.96	11.4 14.7	10.8 14.1	<50°	3.4	8.5-9.1	8.5-9.1	O7	O7	(6) (7)
HD165052	0.87	2.5	2.2	<25°	6.1	8.6(CA)	8.6(CA)	O6.5	O6.5	(10)
HD167771	0.87	2.7	2.3	<28	3.97	9.4(CA)	8.4(CA)	O7 ((f))	O9	(10)
HD206267	0.33	18.7	6.5	<55°	3.7			O6	O9	(8)
HD228854	0.89	37	33	~90°	1.9			O7	O8	(2)

(1) McCluskey and Kondo (1972
(2) Batten (1968)
(3) Batten (1967)
(4) Koch et al. (1970)
(5) Stothers (1972)
(6) Conti et al. (1975)
(7) Evans (1979)
(8) Crampton & Redman (1975)
(9) Hille et al. (1974)

2. ESTIMATES FOR N DURING CORE HYDROGEN BURNING DERIVED FROM THE OVERLUMINOSITY OF O TYPE COMPONENTS IN BINARIES

a) Comparison of the O star in Figure 1 with the mass-luminosity curves computed for N=0,100,300 gives a lower limit for N. Indeed, the considered systems have not necessarily finished core hydrogen burning. In order to explain the observed overluminosities for O components in binaries, large N-values have to be involved. For the indicated supergiant components in X-ray binaries, which are most probably at the end of the core H-burning, a comparison with the theoretical curves allows a real estimate for N. From the figure may be derived that N~400.

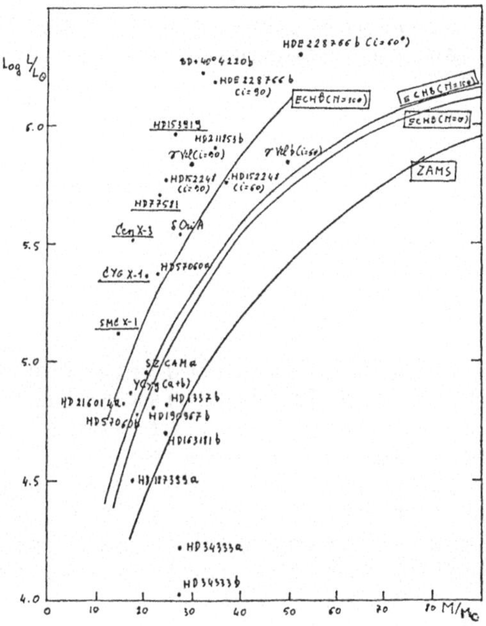

Figure 1. Comparison between the theoretical M-L relations at the end of core hydrogen burning for different N-values with observed luminosities of binary components.

b) We consider now the systems with sufficiently accurate determined spectral type. Using the evolutionary sequences of de Loore et al. (1977,1978) with the inclusion of mass losses and selecting these models with effective temperature corresponding with these spectral types, we can plot the luminosity of these O components as a function of their mass

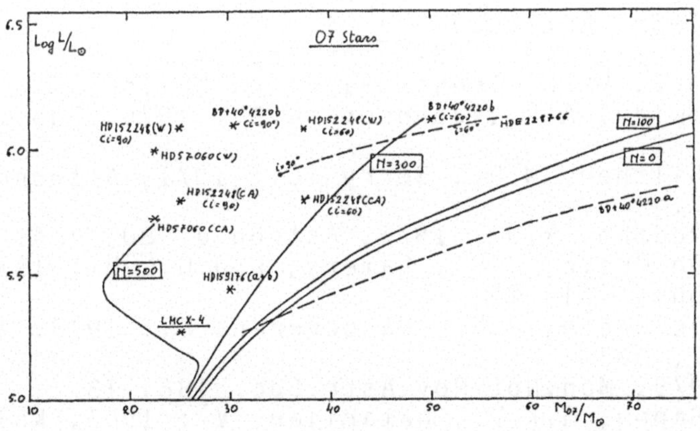

Figure 2. The M-L relations for O7 stars for different N-values: the positions of observed O7 stars, members of close binaries are indicated.

for N=0,100,300,500. The observed position of the O6, O7 and O9 stars in this diagram corresponds with large stellar wind mass loss rates ($N \geq 300$)(as an example we have chosen the O7 spectral type in Figure 2; the conclusion for N holds for O6 and O9 as well).
Conclusion :

$$300 \leq N < 500$$

3. CONSTRAINTS ON α AND β

For evolved systems it is possible to calculate their original parameters at the ZAMS, for different assumptions on α, β and N. This was done for evolved O and Of stars by using interpollation formulas derived by Vanbeveren and De Grève (1979) for N=0 and 300, α=1,2 and 3 and β=0, 0.5, 0.8 and 1. For unevolved binaries the future of these systems can be evaluated by calculating evolutionary sequences, including stellar wind mass loss (N=300) for the same values of α and β as before. Inspection of the mass ratios and the periods (or period ratios, final to initial) and comparison with the constraints,concerning N and the mass ratio, determined in the previous sections lead to a large mass and angular momentum loss from the system during RLOF, i.e. at least 50% of the mass lost by the primary should leave the system taking with typically some 50% of the orbital angular momentum.

REFERENCES

Batten, A.H.: 1967, Sixth Catalogue of the Orbital Elements of Spectroscopic Binary Systems, Publ.Dom.Astrophys.Obs., Vol.XIII, No.8.

Batten, A.H.: 1968, Astron.J. 73, 1.

Bohannan, B., Conti, P.S.: 1976, Astrophys.J. 204, 797.

Bracher, K., 1967, Ph.D. Thesis, Indiana University.

Conti, P.S., Cowley, A.P., Johnson, G.B.: 1975, Publ.Astron. Soc.Pac. 87, 327.

Cowley, A.P., Hiltner, W.H., Berry, C.: 1971, Astron.Astrophys. 11, 407.

Crampton, D., Redman, R.O.: 1975, Astron.J. 80, 454.

de Loore, C., De Grève, J.P., Lamers, H.J.G.L.M.: 1977, Astron.Astrophys. 61, 251.

de Loore, C., De Grève, J.P., Vanbeveren, D.: 1978, Astron. Astrophys. 67, 373.

Evans, T.L.: 1979, Mon.Not.Roy.Astr.Soc. 186, 13.

Ganesh, K.S., Bappu, M.K.V., Natarijan, V.: 1967, Kodaikanal Obs.Bull. No. 184.

Garmany, K.: 1979, this symposium.

Hill, G., Crawford, D.L., Barnes, J.V.: 1974, Astron.J. 79, 1271.

Hutchings, J.B.: 1977, Publ.Astron.Soc.Pac. 89, 668.
Hutchings, J.B., Crampton, D., Cowley, A.P.: 1978, Astrophys. J. 225, 548.
Lamers, H.J.G.L.M., Paerels, F., de Loore, C.: 1979, preprint.
Koch, R.H., Plavec, M., Wood, F.B.: 1970, Univ. of Pennsylvania, Pub. 10.
Massey, P., Conti, P.S.: 1977, Astrophys.J. 218, 431.
Massey, P., Conti, P.S.: 1979, in "Mass Loss and Evolution of O-Type Stars", eds. P.S. Conti and C. de Loore, Reidel, Dordrecht.
McCluskey, G.E., Kondo, Y.: 1972, Astrophys.Space Sci.17,134.
Niemela, V.S.: 1976, Astrophys.Space Sci. 45, 191.
Niemela, V.S., Sahade, J.: 1979, in "Mass Loss and Evolution of O-Type Stars", eds. P.S. Conti and C. de Loore, Reidel, Dordrecht.
Seggewiss, W.: 1974, Astron.Astrophys. 31, 211.
Stempien, K.: 1970, Acta Astron. 20, 117.
Stothers, R.: 1972, Astrophys.J. 175, 431.
Vanbeveren, D., De Grève, J.P., van Dessel, E.L., de Loore,C. 1979, Astron.Astrophys. 73, 19.
Vanbeveren, D., De Grève, J.P.: 1979, Astron.Astrophys (in press).
Vanbeveren, D., Packet, W.: 1979, Astron.Astrophys.(in press)
Woodward, E.J., Koch, R.H.: 1975, Astron.J. 87, 901.

DISCUSSION FOLLOWING VANBEVEREN AND DE LOORE

Sugimoto: Is the separation between the components decreasing or increasing during the mass loss before filling their Roche lobes?

de Loore: For the orbital period variation, the equation for isotropic mass loss of Hadjidemetriou (Advan. Astron. Astroph. 5, 131 1967), was used

$$P/P_o = \left(\frac{M_1^o + M_2^o}{M_1 + M_2}\right)^2 \quad \text{(subscript o refers to initial situation)}$$

Hence the period, and also the separation between the components during the phase before RLOF is increasing.

Massey: How are the luminosities plotted in your slides derived? Several of these stars are not known to be members of clusters (HDE 228766, HD 211853 for example). If the luminosities come from the spectral types and not from distance moduli, aren't they highly uncertain for the evolved stars? How did you correct for the continuum contribution of the WR star in the WR + OB systems?

de Loore: The luminosities are partly determined from the distance of the systems. If spectral type and luminosity class are sufficiently well known the bolometric magnitude is found from the studies of absolute magnitude of Conti and Altschuler (Ap. J., 170, 325, 1971) and Walborn

(Astron. J., 77, 392, 1972). Bolometric corrections are from Morton (Ap. J., 158, 629, 1968).

I agree that uncertainties remain due to the continuum contribution of the WR star in WR systems.

DISCUSSION FOLLOWING VAN DER LINDEN

Sugimoto: Are you planning to compute the evolution of the mass-accreting companion?

van der Linden: Until now we limited ourselves to evolution of the loser. But we are certainly planning to calculate also the mass-gaining star. Work on this is in progress.

MASS TRANSFER BETWEEN BINARY STARS

Jerry L. Modisette
Houston Baptist University

Youi Kondo
NASA-Goddard Spaceflight Center

The transfer of mass from one component of a binary system to its companion can involve ejection of the mass by radiation pressure, by Roche lobe overflow, by centrifugal action through the coupling of the upper atmosphere to the stellar rotation through magnetic fields, and by the dynamics of the atmospheric fluid under the action of an energy source at the photosphere or lower chromosphere. We will primarily address the latter mechanism, which commonly goes under the title of stellar winds. The major determinant of a stellar wind is the energy source. In the case of a binary system, the dynamic effects of the system on the flow pattern, particularly for close binaries, is a second major determinant, once an energy source is available to drive the wind. Roche lobe overflow might be thought of as a limiting case in which the system dynamics predominate.

In addition to ejection, the transfer of mass requires capture of the ejected matter by the companion. The first issue is free particle vs. continuum flow. We will only consider continuum flow, as the applicability of the free particle concept to any binary system with significant mass transfer is doubtful, because of the scale height and because of the likely presence of magnetic fields in the interstellar plasma. The need to consider continuum flow in the capture process complicates the analysis considerably, particularly for close binaries where the coriolis force enteracts strongly with the collective motion of the plasma. We have not solved this problem, which requires multi-dimensional flow analysis with only one simplifying symmetry. However, for more distant binaries, rather simple considerations show that the capture of matter for continuum flow can be enhanced over that for free particle flow.

Now for the principal topic, mass ejection through fluid dynamics. As shown by Parker years ago, a static stellar atmosphere is unstable, provided the temperature is high enough. The corona of the sun, $>1,000,000^{\circ}$K, flows outward in a continuous, though highly variable wind as predicted by Parker and discovered by the Mariner II Spacecraft. It seems clear from the work of Parker and others that any star with a hot corona will have a stellar wind.

M. J. Plavec, D. M. Popper and R. K. Ulrich (eds.), Close Binary Stars: Observations and Interpretation, 123–126.

The existence of the hot corona itself once appeared paradoxical, since its energy presumably comes from the solar interior through the relatively cold photosphere and chromosphere to the corona. The spontaneous flow of heat from the cold photosphere to the hot corona directly violates the Second Law of Thermodynamics. The resolution of the paradox, of course, is that the energy does not flow as heat, but rather as mechanical energy that is dissipated into heat in the chromosphere and corona. In the solar corona only a fraction of the mechanical energy appears as heat--most shows up in the kinetic energy of bulk motion of the solar wind. The stagnation temperature of the relatively cold solar wind is $\approx$30,000,000^{o}K.

For the sun, the mechanical energy flow is in the form of acoustic waves generated by the deep convection zone. As the waves move outward their amplitude increases until dissipation occurs. This phenomenon of the increase in amplitude of waves traveling from heavier to lighter media is widespread. It accounts for the breaking of waves on a beach, the heating of the ionosphere by thunder, and the cracking of a whip. In each case the wave, traveling with constant energy, is propagated by decreasing mass, which can only contain the energy by increasing the amplitude of its motion. A sufficient amplitude leads to dissipation so that small disturbances in a dense medium can produce great heating in a contiguous medium of lesser density.

For single main sequence stars, acoustic waves from deep convection zones have had great success in explaining heating of the upper atmosphere. However, stellar winds can only be observed for the sun, so that the accepted model, based on the systematic decrease in chromospheric emissions and increase in stellar rotation from late to early type stars is inferential.

In studying binaries we have a richer field. It is possible to observe the effects of winds, if they exist, as they impinge on companions. There are sources of disturbance in the lower atmosphere other than acoustic noise associated with deep convection zones that can drive stellar winds, for example, g-mode oscillations and x-ray heating. We will conclude by describing some results of model calculations for binary systems showing how relatively small disturbances in the lower chromosphere or photosphere can cause substantial mass ejection in binary systems.

The model is a straightforward numerical integration of the equations of motion, including the energy equation, with an initial static atmosphere and with various temperature fluctuations imposed at the base of the corona. The initial static atmosphere is isothermal in the corona, with a linear temperature profile in the chromosphere. The flow is calculated along the line of centers between the two binary components, in the rotating frame of the system, so that the centrifugal force is included. The coriolis force is not included, so that the present model is not valid for close binaries with large orbital angular velocities. Energy flows by convection, by thermal conduction, and by propagation of longitudinal pressure waves.

There are two ad hoc assumptions: The flow is one dimensional in a diverging/converging flux tube configured to produce radial flow at each star, and dissipation of waves is included as a coefficient multiplied by the velocity gradient. The flux tube parameters can be adjusted to correspond to the capture cross-section, once that problem is solved. Meanwhile, it permits the examination of the effect of constricting the flow on the ejection of mass. The dissipation coefficient is set so that waves from the base of the atmosphere are dissipated totally in the lower corona, thereby providing the energy to heat the corona and drive the stellar wind.

The model has been applied to several situations: using a linear temperature pulse it has been shown that a substantial rise in the temperature at the base of the corona can account for the mass ejections needed to cause the x-ray turn-ons observed or HZ-Herculis. It is hypothesized that G-mode oscillations may account for the temperature fluctuations through the amplitude amplification mechanism described above. By starting the disturbance further down in the atmosphere the model shows that rather small temperature fluctuations lead to large mass ejections.

An application that tends to validate the model itself is the quiet sun. Beginning with a static chromosphere and corona, a periodic oscillation in the lower corona of about 10% of the static value initiates mass ejection that grows into a typical quiet-time solar wind in a few hours.

If the oscillations were initiated as temperature fluctuations, the fluctuation amplitude in the photosphere would be $\pm$ 500°K. The corresponding velocity field is $\pm$ 4 km/sec., about what is observed in the photosphere. The fluctuating quantity used to initiate the oscillations is not very important, only introducing a phase shift in the overall motion.

For sources of disturbance other than convection zones (say, x-ray heating), the altitude of the initiating disturbance may vary. This altitude is very important in determining the disturbance amplitude required for a given velocity of ejection, since for disturbances higher in the chromosphere, the amplification with outward propagation is reduced.

There is also a considerable effect of altitude on mass loss, both because of the amplification and because there is less mass in the atmosphere at high altitudes. For the HZ-Herculis x-ray turn-ons, a disturbance at the base of the corona must be of the order of the coronal temperature in amplitude, whereas at lower altitude 10% disturbances will suffice.

The dissipation also has a significant effect on mass loss. For disturbances originating at low altitudes, a small dissipation coefficient permits waves to travel upwards until conduction in the corona provides damping of the waves. The result will be a thin chromosphere and low mass loss. Dissipation in the chromosphere heats and expands it, thereby increasing the mass loss.

The principal result of the modeling study is that any modest disturbance at low altitudes can cause mass loss through a stellar wind mechanism, due to the amplification of the disturbance as it propagates into the thinner atmosphere. The possible sources of the disturbance are many, so that this approach is capable of explaining many mass ejection phenomena. Binary stars offer a good application to such an approach, since it is possible to observe the effects of mass ejection. For a better model, the ad hoc assumptions of dissipation coefficients and flux tubes need replacing by careful analysis of decay mechanisms and the fluid dynamics capture problem.

ON THE ONSET OF CONTACT IN EARLY-TYPE BINARIES

R.F. Webbink
University of Illinois

ABSTRACT

A previously published model of a detached binary evolving into deep contact, under the assumption that both components fill the same equipotential surface, is tested for self-consistency. Hydrodynamic equations describing the flow between components at their interface indicate an enormous discrepancy in the magnitude and direction of the flow in the evolutionary calculation. This behavior strongly supports the view that binaries reaching contact in this way undergo a cyclic thermal instability as in many models of W Ursae Majoris systems, rather than overflowing an outer Lagrangian point.

1. INTRODUCTION

It was first shown by Benson (1970) and Yungel'son (1973) that an accreting secondary with a radiative envelope grows rapidly out of thermal equilibrium during the initial, rapid (thermal) time scale mass transfer in intermediate- and high-mass binaries. This rapid expansion typically brings the binary into contact when the secondary mass has increased only a few percent. This phenomenon has since been confirmed by many investigators (Mullen 1974; Ulrich and Burger 1976; Webbink 1976b; Kippenhahn and Meyer-Hofmeister 1977; Flannery and Ulrich 1977; Neo, et al. 1977; Packet and De Greve 1979), and is easily understood in terms of the marked increase with radius of the specific entropy within the envelope of a radiative star. When such a star accretes matter, high entropy material at the surface is buried, and must rid itself of its energy excess for the star to reach thermal equilibrium; if accretion is rapid compared with thermal relaxation in the envelope, the star cannot radiate this energy excess away, and becomes bloated and superluminous.

Several of these studies of accreting stars have attempted to follow evolution through the contact phase, imposing only the condition that the binary now fill a common equipotential surface, but neglecting

M. J. Plavec, D. M. Popper and R. K. Ulrich (eds.), Close Binary Stars: Observations and Interpretation, 127–132.

the possibility of energy exchange independent of net mass transfer, although such energy transfer is clearly of profound importance to the structure of zero-age contact binaries, the W Ursae Majoris systems. In this contribution, one such calculation of accretion by a star with a predominantly radiative envelope is examined in detail, and the need for a non-equilibrium model for the structure of common envelopes exposed.

2. THE MODEL

A feature common to models of W Ursae Majoris systems which appeal to a large-scale circulation for energy transfer (Hazelhurst and Meyer-Hofmeister 1973; Nariai 1976; Webbink 1977b) is the existence of a potential surface within the common envelope on which pressure forces between components are balanced. A slight difference in density (or entropy) between components then leads to a slight pressure imbalance of opposite signs above and below this surface, creating conditions for a closed circulation. Whether or not this sort of circulation accounts for energy exchange in W Ursae Majoris systems, it is certainly true that the absence of pressure balance on equipotential surfaces will lead to a hydrodynamic mass flow between components.

As a test of whether the failure to insist on at least approximate local pressure balance is a serious oversight, an evolutionary sequence by the author (Webbink 1976b, first model with stream effects included) was adopted in which deep contact phases had been computed imposing only equality of surface potentials. Insistence on that condition is sufficient to determine the mass flow rate as an eigenvalue of the evolutionary problem, and it is this net mass transfer rate which we wish to compare to that obtained by an approximate solution to the hydrodynamic flow, given the common envelope structure at each state as determined by the evolutionary calculations. The model of the flow adopted is a modification of one previously employed by Webbink (1977a; Appendix), itself a variant of one by Jedrzejec (Paczynski and Sienkiewicz 1972). In the present modification, the mass flux on any equipotential surface was computed by identifying the component whose envelope pressure was higher on that surface, and allowing that material to expand freely and adiabatically to the sonic pressure, or to the pressure at the same equipotential in the companion's envelope, whichever was greater. The net mass flow is then obtained by integrating this flux, with the appropriate sign, over a surface normal to a line connecting the component centers.

The results of the hydrodynamic approximations to the mass transfer rate are compared in Figure 1 with those obtained in the simplified evolutionary treatment, Note the difference in scales. If the evolutionary calculations were self-consistent, the two estimates should agree tolerably well; in fact, there is an enormous discrepancy not only in the <u>magnitude</u> of the flow, but in its <u>direction</u> as well. Apparently,

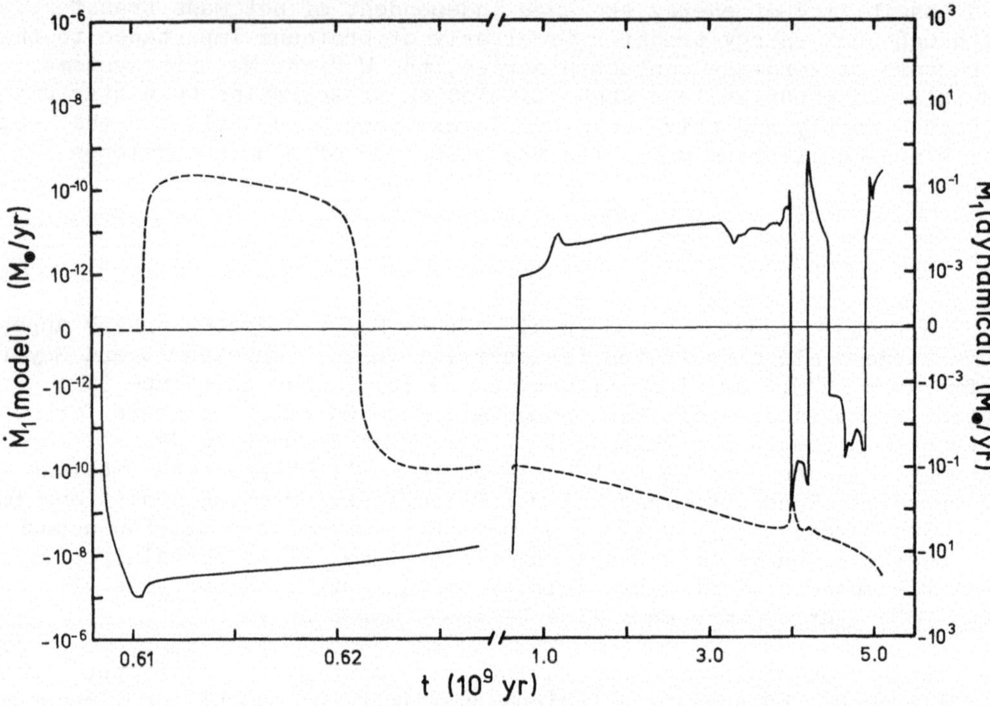

Figure 1. Comparison between mass transfer rates deduced for a binary (Webbink 1976b) evolving under the condition that both components fill the same equipotential surface (Solid line, scale at left) with rates computed for these models with a hydrodynamic model of the flow between components (broken line, scale at right).

the cooler, denser envelope of the secondary almost immediately reverses the mass flow once it expands beyond the inner critical potential. Even in the absence of any closed circulation or energy exchange, it is obviously unphysical to insist that the two components in these non-equilibrium contact systems have equal surface potentials. Even a slight difference in envelope structure between components is then adequate to produce an enormous mass flux in deep contact: in the evolutionary sequence used here, for instance, the two components never differed by more than 15 percent in effective temperature. Strong support is thus given to the stability arguments of Hazlehurst (1974). Under these circumstances, the entire subject of contact evolution in massive systems bears reconsideration.

3. CONCLUSIONS

The calculations presented here were a prime motivation for the original caution (Webbink 1976b) that the published models beyond establishment of contact were unphysical. As noted then, the structures of the two components as contact is attained bear a substantial resemblance to the corresponding phase in cyclically unstable contact binaries (Lucy 1976; Flannery 1976), and we should therefore anticipate a similar cyclic instability in these systems. With the possible exception of very massive contact binaries (Webbink 1979), their ultimate fate is probably the same (Webbink 1976a). Certainly, we should expect contact to be prevalent among evolved Case A systems, but remnants of these binaries do not appear to be present among Algol-type systems in the numbers to be expected if they did not coalesce (Plavec 1973; Ziolkowski 1976). On the other hand, if binaries evolving in Case B were initially near enough unit mass ratio that contact ensured after mass reversal, there is some possibility that these systems could yet emerge as Algol-type binaries, even if they were forced to evolve in marginal contact. This could result if the primary were actually the cooler component when contact was established, in which case energy exchange would proceed from secondary to primary, suppressing the expansion of the former while stimulating mass loss from the latter.

This research was supported by National Science Foundation Grant AST 78-12309.

REFERENCES

Benson, R.S. 1970, Ph.D. Thesis, University of California, Berkeley.
Flannery, B.P. 1976, Astrophys. J., 205, 217.
Flannery, B.P., and Ulrich, R.K. 1977, Astrophys. J., 212, 533.
Hazlehurst, J. 1974, Astr. Astrophys., 36, 49.
Hazlehurst, J., and Meyer-Hofmeister, E. 1973, Astr. Astrophys., 24, 379.
Kippenhahn, R., and Meyer-Hofmeister, E. 1977, Astr. Astrophys., 54, 539.
Lucy, L.B. 1976, Astrophys. J., 205, 208.
Mullen, E.F.F. 1974, Ph.D. Thesis, University of Florida.
Naraia, K. 1976, Publ. Astr. Soc. Japan, 28, 587.
Neo, S., Miyaji, S., Nomoto, K., and Sugimoto, D. 1977, Publ. Astr. Soc. Japan, 29, 249.
Packet, W., and De Greve, J.P. 1979, Astr. Astrophys., 75, 255.
Paczynski, B., and Sienkiewicz, R. 1972, Acta Astr., 22, 73.
Plavec, M. 1973, in "Extended Atmospheres and Circumstellar Matter in Spectroscopic Binary Systems", IAU Symposium No. 51, ed. A.H. Batten (Dordrecht: D. Reidel), p. 216.
Ulrich, R.K., and Burger, H.L. 1976, Astrophys. J., 206, 590.
Webbink, R.F. 1976a, Astrophys. J., 209, 829.
Webbink, R.F. 1976b, Astrophys. J. Suppl., 32, 583.
Webbink, R.F. 1977a, Astrophys. J., 211, 486.
Webbink, R.F. 1977b, Astrophys. J., 215, 851.
Webbink, R.F. 1979, review presented at IAU Colloquium No. 53.

Yungel'son, L.R. 1973, Nauchn. Inf. Akad. Nauk S.S.S.R., 27, 93.
Ziolkowski, J. 1976, in "Structure and Evolution of Close Binary Systems", IAU Symposium No. 73, ed. P. Eggleton, S. Mitton, and J. Whelan (Dordrecht: D. Reidel), 321.

DISCUSSION FOLLOWING WEBBINK

Budding: Can W UMa systems live as long as 10^{10} years?

Webbink: This is a question I attempted to address in a paper this January in the Astrophysical Journal. Certainly we do see W UMa systems in the very old open clusters, for example, AH Cnc in M 67, which is roughly half the age of the globular clusters. One cannot exclude that some contact binaries may arise out of the evolution of initially detached systems, but apart from that possibility, my inclination is that, though primordial contact binaries may survive 10^{10} years, the mass ratio will necessarily have become so extreme in that time (to accommodate growth of the primary), that their low amplitudes and short periods will make them very difficult to detect.

Sugimoto: Consider the case when gas is being transferred almost dynamically from the secondary star to the primary. When the gas contained in the common envelope has been transferred, new gas has to be pumped up from the secondary in order to continue the mass transfer. It requires much energy and makes the radius of the star smaller. Therefore, the mass transfer in this direction seems to stop. How far did you follow this mass transfer in your numerical computations, or in other words, how long the mass transfer continued at the rate as high as $10^{-3} \sim 10^{-1}\ M_{\odot}\ yr^{-1}$ and how much mass was transferred in your computations? Did not you solve the initial-boundary value problem, or did you solve only the boundary value problem?

Webbink: Your points are well-taken. I should make perfectly clear that the hydrodynamic estimates of mass flow rates are indeed made by treating the so-called evolutionary sequence of models as a sequence of initial-value problems; these estimates do not themselves form a self-consistent time sequence. Clearly, it is absurd to suppose that mass transfer rates as large as the estimates you cite can be sustained for 10^7 years, as portrayed in Figure 1; personally, I find it difficult to envision net mass transfer rates in either direction much exceeding the thermal rate, $\sim 10^{-7}\ M_{\odot}/yr$. Rather, when contact is first reached, the primary is very undersized and underluminous for its mass, while the secondary is oversized and overluminous for its mass; this is precisely the same condition of thermal disequilibrium as in the TRO models of W UMa systems at the beginning of their good thermal contact phase. In those models, the energy supply needed to keep the secondary in contact, to which you refer, is provided by energy transfer from the primary, rather than from the core of the secondary. This dams up the energy outflow from its interior, permitting it to remain oversized and lobe-filling, even though it is losing mass. My feeling is that the same circumstances are appropriate here, but as you imply, this should be demonstrated by calculating a self-consistent evolutionary sequence using a detailed model of mass flow such as the modified Jedrzejec model I have used here. This remains to be done.

Linnell: Perhaps I can ask Ron whether there is a problem concerning the total mass flux rate through L_1. In his paper on mass circulation, the total mass flow is 2 or 3 orders of magnitude larger than the net mass transfer between components. If the solutions indicating extremely marginal contact are correct, is there any problem with the rate of mass flow near L_1 which would be required on the mass circulation model?

Webbink: In that paper, I tried to estimate the minimum depth of contact at which mass flow at sonic velocities could carry the required net energy flux. It is roughly this state about which systems should oscillate in the TRO models, and the data I had at hand at that time seemed to indicate the depths of contact are indeed more or less evenly apportioned to either side of this marginal state. Systems in extremely marginal contact, i.e., having smaller depths of minima than this threshold, cannot then be presently in good thermal contact: although possibly still in physical contact, they correspond more nearly to the semi-detached phase of the TRO model.

Shu: If this scenario is right, what are Algols?

Webbink: That is precisely the reason why this problem needs so desperately to be resolved. I think the tendency of theoretical studies to show formation of contact systems is to a large extent an artifact of choosing initial systems with rather extreme mass ratios, which tends to aggravate the problem. It is embarrassing that none of the theoretical calculations including evolution of the secondary has succeeded in producing an Algol system. I have offered some speculations on how this dilemma might be resolved, but though I would like to think these are educated guesses, they should be recognized as nothing more.

EVOLUTION OF EARLY-TYPE CONTACT BINARIES

Daiichiro Sugimoto and Shigeki Miyaji
Department of Earth Science and Astronomy, College of General Education, University of Tokyo, Tokyo 153, Japan

1. SV CENTAURI AS AN EVIDENCE OF MASS EXCHANGE

SV Centauri is an early-type contact binary for which detailed physical data have been obtained from observations (Wilson and Starr 1976). The orbital period is $p = 1.659$ days and the separation is $a = 16.1\ R_\odot$. The radii of component Stars A and B are, respectively, $R = 7.2$ and $6.9\ R_\odot$ so that they are in overcontact. The masses and the luminosities are $M = 11.1$ and $9.3\ M_\odot$, and $L = 2830$ and $10900\ L_\odot$ for Stars A and B, respectively. The more massive Star A is called the primary star but its luminosity is fainter than Star B. This implies that it is in the phase of rapid mass transfer and mass is outflowing from Star A. Indeed one of the most interesting and important points

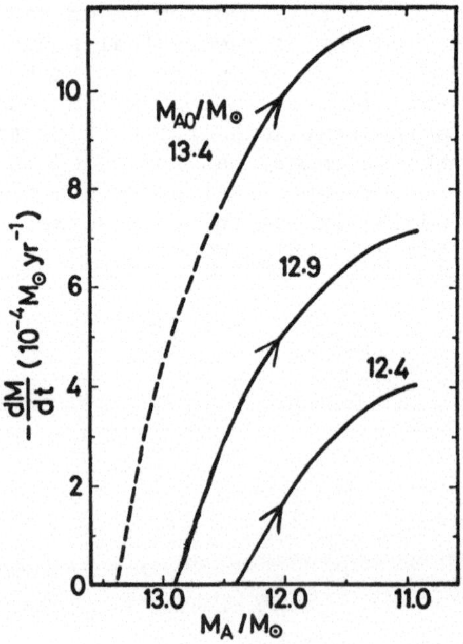

Fig. 1. —— Changes in the rate of mass transfer are shown against the mass of Star A. Three different cases for its initial mass. For the dashed part of the curve, the computed results are not smooth enough, because advance guesses for the rate were not sufficiently accurate. However, it seems to affect but little the behavior in later stages.

M. J. Plavec, D. M. Popper and R. K. Ulrich (eds.), Close Binary Stars: Observations and Interpretation, 133–137.

of this system is that the rate of period change is measured to be dP/dt = -9.4 x 10-8 (Irwin and Landolt 1972) which corresponds to the mass transfer rate of $dM_A/dt = -4 \times 10^{-4}\ M_\odot\ yr^{-1}$.

Recently, Nakamura, Saio, and Sugimoto (1978) computed a model for this binary system by using a stellar evolution code in which included are the effects of mass loss as well as mass acceptance. They assumed the conservations of mass and angular momentum during the mass exchange. Only one parameter of the model is at our disposal, which they chose to be the initial mass of Star A, i.e., M_{A0}. Assuming its value they followed evolution of both stars until the mass of Star A was reduced down to its present value, i.e., to $M_A = 11.1\ M_\odot$, as a result of overflow from its critical Roche lobe.

In constructing such models Nakamura et al. (1978) computed three cases for the initial mass of Star A, i.e., for $M_{A0} = 12.4$, 12.9, and 13.4 $M_\odot$. The last one was not published in their paper, because it resulted in too high rate of mass transfer. In Figure 1 we reproduce the time changes in their mass transfer rates for all of the three cases. The case of $M_{A0} = 12.4\ M_\odot$ fits well with observations both in the HR diagram and in the rate of mass transfer, as was discussed in detail by Nakamura et al. (1978). This implies that SV Cen can be a compelling evidence supporting general idea of mass transfer during the evolution of binary stars.

2. INCREASE IN RADIUS OF MASS ACCEPTING STAR

Of particular importance and interest is the evolution of the mass accepting Star B. As a result of rapid mass acceptance in large amount, the radius and the luminosity of Star B increase. Nakamura et al. (1978) computed evolution of the mass accepting Star B only for the cases of $M_{A0} = 12.4$ and $12.9\ M_\odot$. At the stage of $M_B = 9.3\ M_\odot$, its radius is 6.0 and 8.1 $R_\odot$, respectively. For the case of $M_{A0} = 13.4\ M_\odot$, we estimate its radius to be 10 $R_\odot$ by extraporating in log R - log (dM/dt) plane. Such large difference in the radii of Star B is ascribed mainly to the difference in the rates of mass transfer as discussed by Kippenhahn and Meyer-Hofmeister (1977) and by Neo, Miyaji, Nomoto, and Sugimoto (1977). The difference in the rates of mass transfer is ascribed to the difference in the sizes of the initial critical Roche lobe, which are $R_{cr,A0} = 7.3$, 7.8, and 8.5 $R_\odot$, for $M_{A0} = 12.4$, 12.9, and 13.4 $M_\odot$, respectively. Here we see how large difference in the mass transfer rates results from a relatively small difference in the initial Roch lobes.

The size of the critical Roche lobe for Star B is $R_{cr,B} = 5.9\ R_\odot$ at the present stage of $M_B = 9.3\ M_\odot$. In the case of $M_{A0} = 12.4\ M_\odot$ Star B will be marginally accommodated within its critical Roche lobe, and the mass transfer will continue conservatively. In the case of $M_{A0} = 13.4\ M_\odot$, on the other hand, Star B will not accept matter any more after its radius exceeds $R_{cr,B}$ appreciably. Then the mass overflowing from Star A must escape from the binary system and the mass transfer will become non-conservative. Here we notice that only a small difference in the initial conditions leads to large difference in the mode of mass transfer, conservative or non-conservative.

3. FATE OF CONSERVATIVE MASS EXCHANGE

Conservative mass exchange in a binary system has been studied extensively in these ten years. In our case of $M_{A0} = 12.4\ M_\odot$ (or less), the mass transfer commences when hydrogen is still left in the convective core. It continues until a stage when the mass ratio is almost reversed. After that the mass transfer continues with the nuclear time-scale as Star A evolves and expands. Finally almost all of the hydrogen-rich envelope is transferred from Star A to Star B. Now Star A is almost a helium star of mass about $M_A \simeq M_{A0}/4$. Then it evolves, makes a supernova explosion, and leaves a neutron star.

The separation between the components increases after the reversal of the mass ratio. If the initial mass ratio was unity, the separation and the period before the supernova explosion is $a \simeq 5.2a_0$ and $P \simeq 12P_0$, where a_0 and P_0 are their initial values. For our model of $M_{A0} = 12.4\ M_\odot$ we have $a \simeq 60\ R_\odot$ and $P \simeq 12$ days, when $M_A = 3.1\ M_\odot$. When a supernova explosion takes place its separation becomes somewhat wider. The system would become an X-ray binary. Thus X-ray binary of relatively long orbital period seems to have been born as a result of the conservative mass exchange.

4. NON-CONSERVATIVE MASS EXCHANGE AND SHRINKAGE OF SEPARATION

When the initial separation is somewhat large as in the case of our model of $M_{A0} = 13.4\ M_\odot$, the mass transfer will become non-conservative. Such situation has already been considered by Flannery and Ulrich (1977). However, their discussions were limited only to early stages of non-conservative mass exchange during which the mass of the primary star is reduced from 14.4 to 13.5 $M_\odot$. Therefore, it is well worth extending them into the later stages.

For such discussion, we have to know how much angular momentum is carried away with a unit mass of escaping matter. We have such information only for particles which are ejected from the second Lagrangian point. Though, the escaping matter is not the particle but the gas, its flow will soon become supersonic, which could be well assimilated by motion of a particle. The angular momentum carried away with a unit mass of matter is denoted by $\ell\omega a^2$, where ω is the orbital angular velocity. According to computations of particle trajectories (Nariai 1975; Flannery and Ulrich 1977), the value of ℓ is about 1.7 irrespective of mass ratio of the binary system. The orbits of the binary stars are assumed to be circular, because the circular orbits will be recovered by tidal interaction.

The total orbital angular momentum is ωa^2 times the reduced mass of the system, and the loss of the angular momentum is described by

$$\frac{d}{dM_A}\left(\frac{M_A M_B}{M_A+M_B}\,\omega a^2\right) = \ell\omega a^2. \qquad (1)$$

Here M_A is decreasing by mass loss while M_B stays constant. With the help of the Kepler's third law, equation (1) is integrated to yield (Nariai and Sugimoto 1976)

$$\frac{a}{a_0} = \frac{M_A+M_B}{M_{A0}+M_B}\left(\frac{M_A}{M_{A0}}\right)^{2\ell-2} \exp\left(-2\ell\,\frac{M_{A0}-M_A}{M_B}\right). \qquad (2)$$

Here we notice that the effect of angular momentum loss brings the component stars very close each other when (M_{A0} $-M_A$) is comparable to or larger than M_B. Let us consider the case of $M_{A0} \simeq M_B$. Because the mass loss from Star A is unstable, it will continue until the hydrogen-rich envelope is almost stripped off. If we use the approximate relation of $M_A \simeq M_{A0}/4$, equation (2) yields the value as small as $a/a_0 = 7.0 \times 10^{-3}$ for $\ell = 1.7$. A system similar to SV Cen but with a little longer orbital period, i.e., the case of $M_{A0} = 13.4\ M_\odot$ corresponds to such case. For a system considered by Flannery and Ulrich (1977), i.e., for $M_{A0} = 14.366\ M_\odot$ and $M_B = 5.634\ M_\odot$, we obtain $a/a_0 = 1.54 \times 10^{-4}$ for $M_A = 4\ M_\odot$. If these are the case, Star B can not be accommodated within its critical Roche lobe and will be dissipated by or will coalesce with a compact star.

Thus our conclusion is somewhat paradoxical. When the initial separation between the component stars is relatively narrow, they will be well separated even after the mass transfer. On the contrary, when it is relatively wide, they will coalesce each other.

This research is supported in part by the Scientific Research Fund of the Ministry of Education, Science and Culture (464080).

REFERENCES

Flannery, B.P., and Ulrich, P.K.: 1977, Astrophys. J., 212, pp. 533-540.
Irwin, J.B., and Landolt, A.U.: 1972, Publ. Astron. Soc. Pacific, 84, pp. 686-696.
Kippenhahn, R., and Meyer-Hofmeister, E.: 1977, Astron. Astrophys., 54, pp. 539-542.
Nakamura, Y., Saio, H., and Sugimoto, D.: 1978, Publ. Astron. Soc. Japan, 30, pp. 649-655.
Nariai, K.: 1975, Astron. Astrophys., 43, pp. 309-311
Nariai, K., and Sugimoto, D.: 1976, Publ. Astron. Soc. Japan, 28, pp. 593-598.
Neo, S., Miyaji, S., Nomoto, K., and Sugimoto, D.: 1977, Publ. Astron. Soc. Japan, 29, pp. 249-262.
Wilson, R.E., and Starr, T.C.: 1976, Monthly Notices Roy. Astron. Soc., 176, pp. 625-632.

DISCUSSION FOLLOWING SUGIMOTO AND MIYAJI

Shu: An important effect left out in your calculations may be the back pressure of the common envelope. Have you considered the modulation on the mass transfer/loss rate due to this back pressure? In particular, have you considered the possibility that it might stop the mass transfer/loss altogether?

Sugimoto: I have not computed the effect of the back pressure, but I think it is negligible because of the following reasons. Almost all of the nuclear energy released in the mass losing star is absorbed in the deep interior where the temperature is about 1 x 10^7K. It is consumed in pushing out the outflowing matter against the entropy gradient, and it is necessary to keep the stellar radius as small as its Roche lobe. This implies that the interior of the star is greatly out of thermal equilibrium. If the mass outflow should stop, the star would expand greatly and it would overcome the back pressure. The energy involved in the mass outflow of the present model is of the order of the thermal energy of the whole star, which is much greater than the energy contained in the common envelope. Therefore, it will not be the common envelope but the thermal disequilibrium of the interior that determines the mass outflow.

Wilson: I suggest RT Sculptoris and V 701 Scorpii as other systems whose evolution would be interesting to model. RT Scl is in the rapid phase before mass ratio reversal. V 701 Sco is also in the rapid phase, only slightly after mass ratio reversal.

Sugimoto: Thank you for this information. It will be interesting to try to construct their models. From the standpoint of computational techniques, it is now easy to do so.

Webbink: I would like to endorse the remark by Frank Shu regarding the importance of back pressure once the secondary reaches contact. The presentation I have scheduled for Friday morning illustrates just this point.

Sugimoto: The effect of the back pressure may be important when the degree of thermal disequilibrium is very small in the deep interior of the star as in the case of W UMa stars. If you insist it to be important even in our case of a large thermal disequilibrium of the interior, please make clear the physics involved, when you read your paper.

FORMATION OF LUMINOUS CONTACT BINARIES BY RAPID ACCRETION ONTO WHITE DWARFS

Ken'ichi Nomoto, Kyoji Nariai*, and Daiichiro Sugimoto**
Department of Physics, Ibaraki University
*Tokyo Astronomical Observatory, University of Tokyo
**College of General Education, University of Tokyo

1. INTRODUCTION

During the evolution of a close binary system, there is a phase of mass exchange between its component stars. When the mass-losing star is an evolved star off the main sequence, the mass exchange is very rapid. Its rate may well exceed the critical accretion rate corresponding to the Eddington limit of the mass accreting star. Effects of such a rapid accretion were studied for the mass-accreting main-sequence stars (Flannery and Ulrich 1977; Kippenhahn and Mayer-Hofmeister 1977; Neo et al. 1977). The mass-accreting stars become quickly overluminous and their radii increase even to a red-giant size as mass is accumulated. Then the both component stars will fill their Roche lobes to form a contact binary system.

In some cases, the mass-accreting star is a white dwarf. The accreting white dwarf will be rekindled. Though such a rekindling was studied by several authors, they considered only the cases of slow accretion (Giannone and Weigert 1967; Taam and Faulkner 1975). The rekindled hydrogen shell-burning is unstable to develop a shell flash or even to grow into a nova-like explosion.

It will be interesting to investigate what happens in the case of extremely rapid accretion onto a white dwarf. Which is a more important effect, the rapid mass accumulation or the shell flash? Will it become a red giant star? We have computed the whole processes of mass accretion starting from its onset through the shell flash and further mass accumulation. Throughout our computation the effect of gravitational energy release has been correctly taken into account.

The initial model was chosen to be a helium white dwarf, whose mass, luminosity and radius are 0.4 $M_\odot$, $1.7 \times 10^{-4} L_\odot$, and $1.6 \times 10^{-2} R_\odot$, respectively. The mass accretion rate dM/dt was assumed to be 2×10^{-5} $M_\odot yr^{-1}$, which is the critical accretion rate for the initial radius of the model. Assumptions and more details of the results are given by Nomoto et al. (1979).

M. J. Plavec, D. M. Popper and R. K. Ulrich (eds.), Close Binary Stars: Observations and Interpretation, 139–143.

2. NUMERICAL RESULTS

The evolutionary stages will be characterized by the mass ΔM_H contained in the hydrogen-rich envelope. In our case of rapid accretion, ΔM_H is practically equal to the amount of mass that has been accreted since the onset of the accretion. It is related with the evolutionary time by $t = 5 \times (\Delta M_H/10^{-4} M_{\odot})$ yr. We have found the following sequence of events takes place.

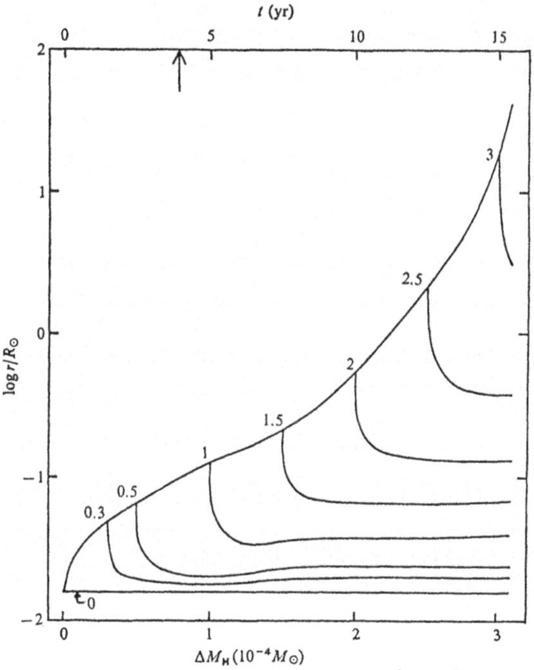

Fig. 1

a) Increase in Luminosity

Figure 1 shows evolutionary changes in the radial distance $\underline{r}$ of Lagrangian shells against ΔM_H or t. (For each shell attached is the value of $M_r - 0.4 M_{\odot}$ in units of $10^{-4} M_{\odot}$.) As the weight of the overlying layers increases, the accreted matter contracts and releases the gravitational energy. In the deep layers of the envelope, the contraction is almost adiabatic because the time scale of contraction is shorter than that of the heat transport In the relatively outer layers, on the contrary, an appreciable fraction of the gravitational energy release is radiated away. Figure 2 shows evolutionary path in the HR diagram. (The values of ΔM_H in units of

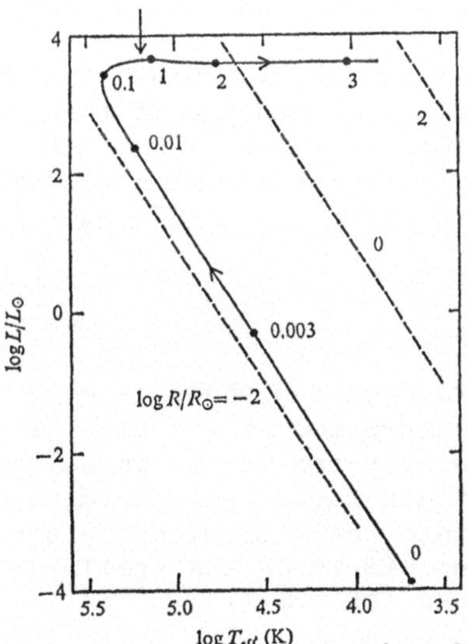

Fig. 2

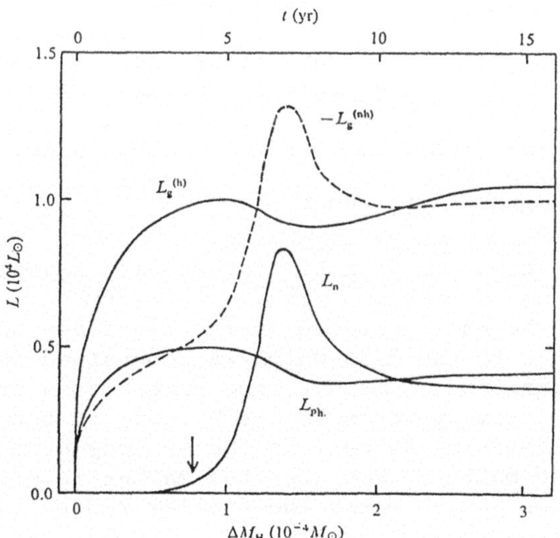

Fig. 3

$10^{-4}M_\odot$ are attached.) The luminosity of the white dwarf increases quickly in 0.2yr, though the stellar radius increases only slightly during these initial stages. Afterwards, the radius increases as matter is accumulated.

Changes in the luminosity L_{ph}, the rate of the so-called gravitational energy release L_g, and the nuclear energy generation rate L_n are shown in Figure 3. Here, L_g is divided into two terms, one ascribed to a homologous contraction $L_g^{(h)}$ and the other ascribed to non-homologous contraction $L_g^{(nh)}$ (see Nomoto et al. 1979). We find that $L_g^{(h)} \simeq 2L_{ph}$ and $-L_g^{(nh)} \simeq L_{ph}$ before the ignition of the flash. This implies that much entropy is brought into the star. The resultant non-homologous change in the stellar structure is important to increase the stellar radius as will be described below.

Though dM/dt is just equal to the critical rate at the initial stage of accretion, L_{ph} reaches only $5 \times 10^3 L_\odot$, i.e., 0.3 times the Eddington limit. This is explained by the fact that the contraction of the lower layers is suppressed as seen in Figure 1 because of the trapped entropy within the star.

b) Hydrogen Shell-Flash

As ΔM_H increases, the bottom of the hydrogen-rich envelope is compressed. When ΔM_H becomes $7.8 \times 10^{-5}M_\odot$, the hydrogen shell-burning is ignited. In Figures 1-3, the ignition time is indicated by the arrow. The shell burning is thermally unstable and makes a relatively weak shell flash as seen in Figure 3. The peak energy generation rate amounts only to $L_n = 8.3 \times 10^3 L_\odot$.

The reason why the shell flash is so weak can be understood as follows. The shell flash is stronger when the pressure of the shell is higher at the stage of the ignition, i.e., the value of ΔM_H and/or the mass of the white dwarf are larger (Sugimoto et al. 1979). Since the accretion is very rapid in our case, much entropy is retained in the accreted mass element. Consequently, the hydrogen burning is ignited at a relatively smaller ΔM_H, which is only 3.8% of that for much slower accretion onto the same 0.4 $M_\odot$ white dwarf with dM/dt = $1 \times 10^{-13}M_\odot yr^{-1}$ (Sugimoto et al. 1979).

c) Formation of a Supergiant-like Envelope

The mass accreted during the flash and its decaying phase is comparable to the mass which was contained in the envelope at the time of the ignition. However, the contraction of the accreted matter is halted in relatively inner layers because of the heat generated by the hydrogen shell-burning as well as the entropy brought into the star with the accreted matter. The accreted matter is piled up simply on the stellar surface, which makes the stellar radius increase (Figure 1).

Thus the gravitational contraction takes place only in the relatively outer layers of large radial distance r so that the gravitational energy release $L_g = L_g^{(h)} + L_g^{(nh)}$ is small as compared with L_{ph}. This fact implies also that the matter is accumulated almost adiabatically in the relatively inner layers. In other words, the small L_g implies that $L_g^{(h)} \simeq - L_g^{(nh)}$. Such non-homologous changes of the outer layers are the main cause of the formation of a supergiant-like envelope (Neo et al. 1977).

3. DISCUSSION

As a result of further mass accumulation, the envelope of our accreting white dwarf will become of a red giant size. Such formation of a red giant-like envelope takes place when the accretion rate is higher than the critical one $(dM/dt)_N$. This is the rate of core growth if the white dwarf were immersed deep in the hydrogen-rich envelope of a red giant star, and is given by

$$(dM/dt)_N = L_H / X_e \cdot E_H \simeq (2 \times 10^{-8} - 6 \times 10^{-7}) M_\odot yr^{-1}, \quad (1)$$

for $M = (0.4 - 1.4) M_\odot$. Here X_e and E_H are the concentration of hydrogen and the nuclear energy release from unit mass of hydrogen, respectively, and the value of the hydrogen burning rate L_H is taken to be that of the corresponding red giant star which has the same core mass as the mass of the white dwarf. When $dM/dt > (dM/dt)_N$, all of the accreted matter can not be processed into helium. Then the accreting gas is piled up to form an extended envelope of a red giant size.

In a close binary system, the radius of such an accreting envelope exceeds its Roche lobe to form a contact system. After that the mass overflowing from the companion star will not be swallowed any more but eventually escape from the binary system. The following sequence of events is very likely. The escaping mass carries away the angular momenta, the separation between the stars becomes smaller and smaller, and finally the white dwarf falls into the companion star.

REFERENCES

Flannery, B.P., and Ulrich, R.K.: 1977, Astrophys. J., 212, pp. 533-540.

Giannone, P., and Weigert, A.: 1967, Zs. für Astrophysik, 67, p. 41.

Kippenhahn, R., and Meyer-Hofmeister, E.: 1977, Astron. Astrophys., 54, pp. 539-542.

Neo, S., Miyaji, S., Nomoto, K., and Sugimoto, D.: 1977, Publ. Astron. Soc. Japan, 29, pp. 249-262.

Nomoto, K., Nariai, K., and Sugimoto, D.: 1979, Publ. Astron. Soc. Japan, 31, pp. 287-298.

Sugimoto, D., Fujimoto, M.Y., Nariai, K., and Nomoto, K.: 1979, IAU Colloquium No.53, <u>White Dwarfs and Variable Degenerate Stars</u>.

Taam, R.E., and Faulkner, J.: 1975, Astrophys. J., 198, pp. 435-438.

DISCUSSION FOLLOWING NOMOTO, NARIAI AND SUGIMOTO

Friedjung: Can you relate your theoretical predictions with a class of observed objects?

Nomoto: I would like to know the candidates.

Smith: How would your results be affected by including the angular momentum of the accreted material?

Nomoto: Since the stellar radius is very large, effects of angular momentum are small in this case.

DO BLUE STRAGGLERS FORM BY COMMON ENVELOPE EVOLUTION?

F. Meyer and E. Meyer-Hofmeister
Max-Planck-Institut für Physik und Astrophysik,
München, FRG.

COALESCENCE OF CLOSE BINARIES

All proposals to explain the formation of blue stragglers advanced up to now seem to have inherent difficulties. One appealing possibility, the formation by coalescence of close binary stars, appears to be ruled out by the existence of blue stragglers with more than twice the turn-off mass of their cluster. We suggest that this contradiction does not exist, since the binary system on its way to coalescence passes through a phase in which a very extended common envelope with a size of several hundred solar radii forms. In this stage a third companion in the range of this envelope will be captured and finally merge to form a blue straggler that can have more than twice the turn-off mass.

COMMON ENVELOPE EVOLUTION

The processes leading to coalescence of two stars are complicated. Rapid mass transfer in a close binary system where the primary is still in the stage of core hydrogen burning ("case A") induces the secondary to expand until a common envelope around the two stars is formed. It seems plausible that corotation cannot be maintained in the envelope. The tidal interaction between common envelope and interior binary core produces so much frictional heat that the common envelope expands until it reaches an equilibrium "red-giant stage". In this equilibrium the density in the envelope has decreased to such a low value that the generation of frictional heat is just limited to the amount needed to support the envelope. Due to the low pressure in the envelope it is possible that all unprocessed material of the primary will flow into this common envelope and only the He-enriched core remains together with the secondary as the inner binary system. This inner binary shrinks by frictional transfer of angular momentum to the extended envelope. A model for common envelope evolution is developed in Meyer and Meyer-Hofmeister (1979).

M. J. Plavec, D. M. Popper and R. K. Ulrich (eds.), Close Binary Stars: Observations and Interpretation, 145–148.

As an example we have chosen a system with $M_1 = 1.6\ M_\odot$ and $M_2 = 1\ M_\odot$, an initial distance $a_o = 5\ R_\odot$ and a period of $P_o = 0.8^d$. After about $1.3\ 10^9$ years the primary fills its critical Roche lobe and initiates rapid mass transfer. Assuming that synchronism is lost the common envelope formed will expand due to the frictional heat generated. Our computations (chemical composition (X = 0.7, Z = 0.0004) for the common envelope stage show that during this evolution luminosity and radius stay at almost constant high value, $\log L/L_\odot = 4.8$, $R/R_\odot = 570$. Comparison with computations for a population I mixture show that the chemical composition does not essentially influence this result. After a time of about 100 yrs the distance of the inner binary has shrunk that much that the secondary fills its critical Roche lobe initiating the final merging of the stars, possibly via a second mass transfer process. In any case only a finite amount of potential gravitational energy is available due to the finite extent of the internal cores, and this will be exhausted by the high frictional luminosity of the common envelope on a timescale comparable to the shrinkage time of a hundred years. As the interior mass distribution approaches a rotational symmetric configuration the frictional heat generation switches off and the matter in the extended envelope settles on the core. For some time this may keep up the high luminosity. Finally it will relax to a configuration where the He-enriched core of the former primary is surrounded by the rest of the mass.

ANGULAR MOMENTUM

During the common envelope evolution most of the angular momentum was in the outer extended region. This will remain so as the mass in the envelope settles gradually down towards the core while the envelope itself stays fairly extended, since convective "viscous" friction will tend to achieve corotation. This means that most of the mass in the envelope settles down at very low rotational speed. The small remaining fraction containing most of the angular momentum cannot settle in this way and will form a disk as the luminosity decreases. Disk evolution brings down most of this fraction with Keplerian velocity while the angular momentum is removed together with less and less mass towards infinity. We therefore expect a relatively slowly rotating blue straggler that might contain a rapidly rotating core depending on whether or not magnetic or other forms of coupling play a role.

CAPTURE OF A THIRD STAR

During the stage of the extended common envelope any companion directly caught will spiral down in the way discussed in Meyer and Meyer-Hofmeister (1979). This process occurs on a timescale which is at first short compared with the shrinking time of the interior binary. The additional frictional luminosity helps to keep up the extended envelope and will lead to a slight reduction of the interior luminosity and to a reduced shrinking rate in the core. Finally the mass of the three originally separated stars will form one blue straggler. Masses above twice the turn-off mass are indicated for a few observed blue stragglers. In Wheeler's (1979) compilation there are 4 cases out of about 100 blue stragglers.

Besides direct capture the mechanism of tidal interaction between a more distinct companion and the slowly rotating extended envelope can degrade the orbit of a companion so much that it enters the envelope. In this way all those companions are captured for which the tidal interaction time is shorter than the lifetime of a few hundred years for the common envelope stage. Applying a rough formula for the tidal interaction one finds that stars within about twice the radius of the envelope should be captured.

The observations of progenitor systems with a third companion require high resolution spectroscopy. Recently Fekel (1979) employing a solid state spectroscopy technique found a system consisting of GO dwarf + MO dwarf binary with a period of $0\overset{d}{.}88$, orbited by a G5 dwarf in highly eliptic orbit with a period of 20 years.

EXOTICA

(a) Wheeler has pointed out that one blue straggler in Hagen's (1970) catalogue in NGC 6940 appears to have a mass about 3.9 times the turn-off mass of the cluster. Masses up to 4 times the turn-off mass could come about if the companion captured is itself a close binary system. Though the incidence of such quadrupole systems is quite high with stellar systems of larger extent (Abt and Levy, 1976) little is known about the occurrence of such systems within separations of a few astronomical units.

(b) Could the occurrence of a blue straggler in a close binary system be understood in a common envelope evolution model? Such a result must occasionally occur when the timescale for capture of a third companion becomes sufficiently large. Though the companion might still enter the common envelope and at first rapidly reduce its separation from the binary core its shrinking time increases about inversely proportional to the separation. If during this time the interior system already coalesces and the interior frictional luminosity is switched off the mass of the envelope might settle while the companion star is still orbiting at some distance and leave a system in which a blue straggler is orbited by the original companion, but at a much decreased separation.

CONCLUSION

The model of an extended common envelope evolution can resolve apparent contradictions in the suggestion that blue stragglers are the result of stellar coalescence. The structure of such common envelopes is relatively well understood, whereas a better quantitative understanding of the processes occurring during the first phase of contact and rapid mass transfer are needed to decide under what circumstances synchronism of a not yet extended common envelope is maintained or lost and the phase of extended envelope supported by frictional heat is initiated.

REFERENCES

Abt H.A., and Levy S.G., 1976, Ap. J. Suppl. 30, 273
Fekel F., 1980, IAU Symp. No. 88, 53.
Hagen G.L., 1970, Publications of the David Dunlop Observatory, Vol. 4
Meyer F. and Meyer-Hofmeister E., 1979, Astron. Astrophys., in press
Wheeler J.C., 1979, Ap. J., in press

COMMENTS FOLLOWING MEYER AND MEYER-HOFMEISTER

Shaviv: How were the masses of the blue stragglers obtained? Did you assume the blue stragglers to be normal MS stars with normal evolution?

Meyer-Hofmeister: I took the distribution of blue straggler masses from a paper of Craig Wheeler (to appear in Ap. J.)

Shu: I would like to emphasize the distinction between your kind of common envelope binaries and ours. We assume synchronism whereas nonsynchronous motion of the cores through the common envelope is crucial to your model. I have no doubt that there is a regime of parameter space where synchronism cannot be maintained, and your type of model is more appropriate. However, I would hate to think that all close binaries evolve to coalesce into single stars shortly after leaving the main sequence since then we would be in danger of having no interesting close binaries to study like Algols, cataclysmic variables, etc.

Meyer-Hofmeister: I think it is important to note that rapid case A mass transfers as is needed for this kind of common envelope evolution only happens for very close systems. I quite agree that there is a wide range of parameters (for example case B mass transfer) where synchronous evolution could occur.

THE EVOLUTION OF PRE-MAIN SEQUENCE CLOSE BINARIES

Jon K. West and Kwan-Yu Chen
Department of Astronomy, University of Florida

1. INTRODUCTION

Conventional computer models of close binary star systems usually start with at least one component on the main sequence. Models of pre-main sequence binaries have been computed to study mass transfer (Yamasaki 1971). However no pre-main sequence computations have been published that follow the evolutionary tracks of a binary system onto the main sequence, even though some observed systems appear to be pre-main sequence (Field 1969). The main purpose of this investigation is the evaluation of individual close binaries with a pre-main sequence model. The evaluation will be accomplished by comparing the positions of the observed binary on the Hertzsprung-Russell diagram with the evolutionary tracks generated by the pre-main sequence model. If both components appear to have the same age and fall near the tracks of the model, then the system is possibly pre-main sequence. Eleven semi-detached binaries were considered, each with a total mass between 2.5 and 6 solar masses and with a period between 0.9 and 3.4 days.

2. MODEL AND RESULTS

The stellar structure is represented by a polytrope of index $n = 3$ for the radiative solution. A polytrope of index $n = 1.5$ is used for the Hayashi solution (Hayashi 1961). The initial configuration introduced is constrained by the Roche lobe. This limits the radius of each component of the binary system. The basic assumptions are as follows: 1) Mass and energy are conserved both for the individual components and the binary system as a whole; 2) the binary system obeys conservation of angular momentum, and circular orbits are assumed; 3) the stars are spherically symmetric; 4) the stars are in quasi-hydrostatic equilibrium; and 5) the chemical composition of Population I is chosen to be the same for each star (X:Y:Z: = 0.750:0.224:0.026).

With the masses of each component and their periods being free parameters, a computer program calculated the sequence of contraction

M. J. Plavec, D. M. Popper and R. K. Ulrich (eds.), Close Binary Stars: Observations and Interpretation, 149–153.

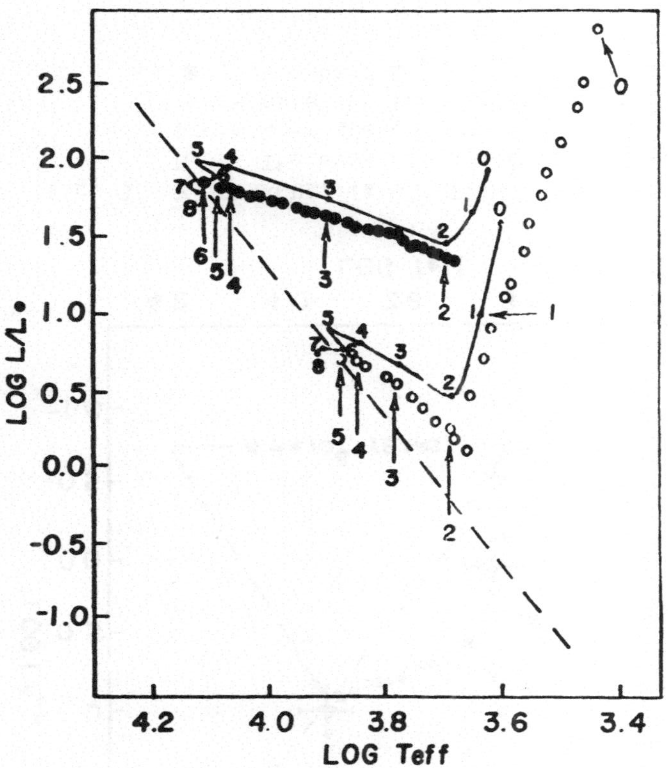

Fig. 1. Evolutionary tracks of a pre-main-sequence model for a binary (● = 3 $M_\odot$, ○ = 1.5 $M_\odot$) and the evolutionary tracks for single stars published by Iben (1965).

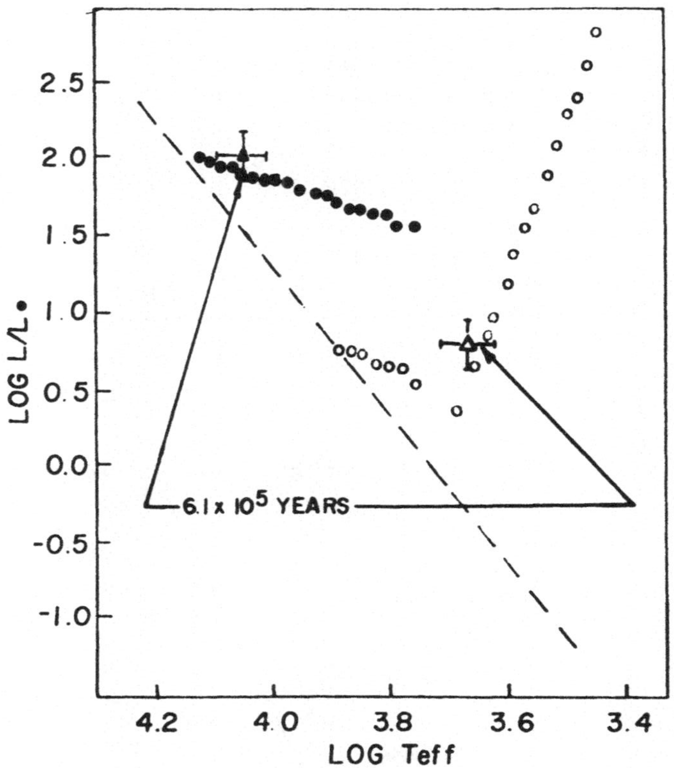

Fig. 2. U CEPHEI (▲ = 3.19 $M_\odot$, Δ = 1.53 $M_\odot$). The evolutionary tracks of the pre-main-sequence binary model is represented by the circles (primary ●, secondary ○).

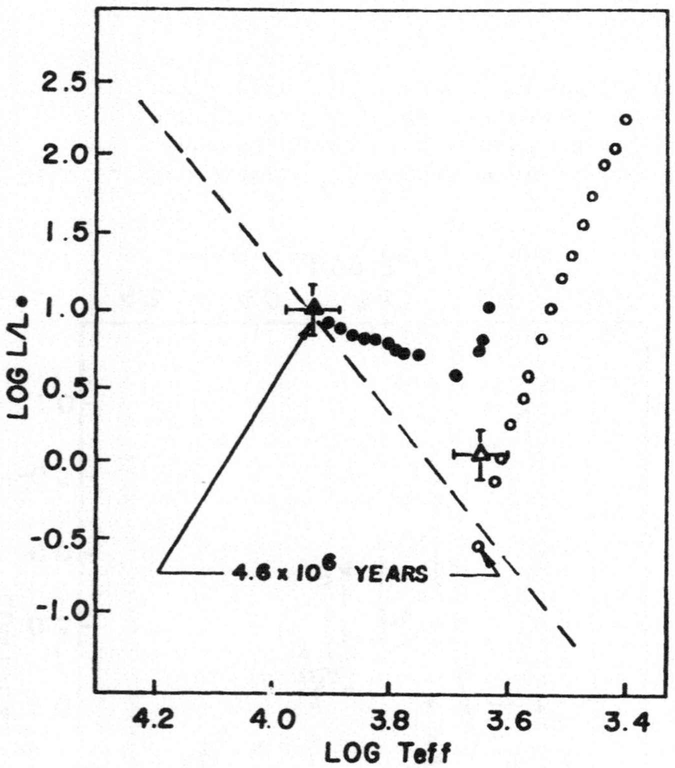

Fig. 3. X TRIANGULI (▲ = 1.72 $M_{\odot}$, Δ = 1.00 $M_{\odot}$). The evolutionary tracks of the pre-main-sequence binary model is represented by the circles (primary ●, secondary ○).

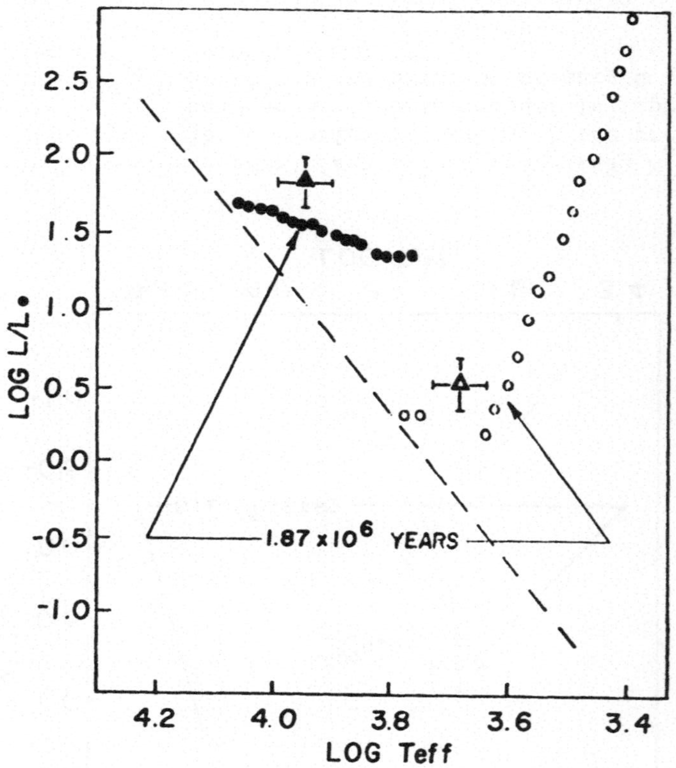

Fig. 4. W URSAE MINORIS (▲ = 2.68 $M_{\odot}$, Δ = 1.19 $M_{\odot}$). The evolutionary tracks of the pre-main-sequence binary model is represented by the circles (primary ●, secondary ○).

models. It calculated both sequences simultaneously which allowed comparison of the age and structure of each component. A test model is compared with the evolutionary tracks published by Iben (1965). The primary for the test model is 3 solar masses, the secondary is 1.5 solar masses, and the period is 3 days. Initially both test stars are filling their Roche lobes and are on the Hayashi Track. Figure 1 shows the plots of the test model and Iben's pre-main sequence tracks for single stars on the Hertzsprung-Russell diagram. The initial point for Iben's models is labeled "0". The points 1 through 5 can be used for age comparison with the test model. The same numbering is used for the test model. The following table shows the ages (10^6 years) at these points for each star.

	Primary = 3 $M_\odot$		Secondary = 1.5 $M_\odot$	
Age Point	Iben	Test Model	Iben	Test Model
1	0.034	---	0.23	0.205
2	0.208	0.06	2.36	1.391
3	0.763	0.747	5.80	3.483
4	1.135	1.202	7.58	4.962
5	1.250	1.351	8.62	5.949
6	1.465	1.432	10.43	---
7	1.741	---	13.39	---
8	2.514	---	18.21	---

The absolute dimensions of the eleven systems are published by Giannone and Giannuzzi (1974). Figures 2, 3, and 4 show the three systems found to fit the pre-main sequence model best: U Cephei (6.1 x 10^5 years), X Trianguli (4.6 x 10^6 years), and W Ursae Minoris (1.87 x 10^6 years). Four systems are found not to fit the pre-main sequence model: TV Cassiopeia, AI Draconis, β Persei, and TX Ursae Majoris. Finally, four systems could fit the pre-main sequence model with modifications: IM Aurigae (3.5 x 10^5 years), δ Librae (8.2 x 10^5 years), U Sagittae (3.78 x 10^5 years), and V505 Sagittarii (1.89 x 10^6 years). These modifications would require differing chemical compositions for each component of a binary. This investigation shows that it is possible to represent some semi-detached close binaries by a pre-main sequence model. More detailed study of individual systems is needed in order to adapt any model to the observed data and inferences therefrom.

The funds for computing time were made available through the Battery Business Department of General Electric Company.

REFERENCES

Field, J.V.: 1969, Monthly Notices Roy. Astron. Soc. 144, 419.
Giannone, P., and Giannuzzi, M.A.: 1974, Astrophys. Space Sci. 26, 289.
Hayashi, C.: 1961, Publ. Astron. Soc. Japan 13, 450.
Iben, I. Jr.: 1965, Astrophys. J. 141, 933.
Yamasaki, A.: 1971, Publ. Astron. Soc. Japan 23, 33.

DISCUSSION FOLLOWING WEST AND CHEN

Anderson: How accurately determined are the observed masses, which you assume for your theoretical models? In many of these systems, they must be exceedingly uncertain.

West: These data were taken from Giannone and Giannuzzi (1974). The error bars shown were calculated from their paper. They estimated the maximum errors caused by inaccuracy in the mass determination.

Zuiderwijk: Do you have a physical reason for starting your computations with two Roche lobe filling stellar models?

West: Typically, pre-main sequence models published by Yamasaki (1971) terminated at the end of the mass transfer phase in a few hundred years. Therefore, this assumption is good. Also, starting at this point fills a theoretical gap between conventional post-main sequence binary evolution (which typically starts with a binary on the main sequence), and the work of Yamasaki.

Shu: Do any of the systems you discuss exhibit the characteristics of T Tauri stars which might indicate that they are young?

West: For the cases which fit the pre-main sequence model, we would expect T Tauri type spectra. However, the luminosity of the primary would maks these characteristics.

Whelan: Are you really saying that U Cep is a pre-main sequence object? It is known as a classical post-main sequence one.

West: It appears to fit this model well.

Wilson: Are there not too many of these binaries for them to be pre-main sequence objects? Why should the low mass components accurately fill the Roche lobes so often, and where are the intermediaries between lobe-filling cases and fully contracted cases?

West: The model fits three of these eleven close binary systems. It is not a thorough survey.

Smak: I am old enough to make two historical remarks. First, it was in the 30s that Kurt Walter originally suggested that the secondaries of Algols would be in the contraction phase. And secondly, it was in the 50s, when--mostly due to the work of Otto Struve and Su-Shu Huang--this hypothesis was found incompatible with the observational data. Namely, it was found that the secondaries are greatly overluminous for their masses, etc., etc.

West: This is what the pre-main sequence model shows. Only three systems, selected for their mass range, appear to fit this pre-main sequence contraction model. All other Algol-type systems evaluated show secondaries too over-luminous to be pre-main sequence.

CATACLYSMIC VARIABLES, HUBBLE-SANDAGE VARIABLES AND η CARINAE

G.T. Bath
Department of Astrophysics, Oxford

The Hubble-Sandage variables are the most luminous stars in external galaxies. They were first investigated by Hubble and Sandage (1953) for use as distance indicators. Their main characteristics are high luminosity, blue colour indices, and irregular variability. Spectroscopically they show hydrogen and helium in emission with occasionally weaker FeII and [FeII], and no Balmer jump (Humphreys 1975, 1978). In this respect they closely resemble cataclysmic variables, particularly dwarf novae. In the quiescent state dwarf novae show broad H and HeI, together with a strong UV continuum. Weak FeII in emission has been observed in U Gem and SS Aur (Warner 1976). The Balmer jump is either not present, or weakly in emission. The principal spectroscopic difference is the increased breadth of the occasionally doubled emission lines.

In contrast to the spectroscopic similarities, the luminosities could hardly differ more. Rather than being the brightest stars known, quiescent dwarf novae are as faint or fainter than the sun.

I suggest that the close correspondence between the spectral appearance of the two classes combined with the difference in luminosity is well accounted for by a model of Hubble-Sandage variables in which the same physical processes are occurring, but on a larger scale. The spectral characteristics of cataclysmic variables are generated by infall of gas through an accretion disk onto a white dwarf (Bath et al. 1974). The outer disk radii are $\sim 5 \times 10^{10}$ cm. The Hubble-Sandage variables would result from similar physical conditions (temperature, density etc.) being produced in an accretion disk which was geometrically larger, and therefore radiating a larger total flux. The occurrence of occasional nova-like behaviour in Hubble-Sandage variables is some support for this identification (see below).

I assume the Hubble-Sandage variables contain a main sequence star in a mass transferring binary with a separation $> 10^{12}$ cm. This separation allows an accretion disk to be formed around the main sequence star, and allows a more massive, evolving primary to be accommodated

M. J. Plavec, D. M. Popper and R. K. Ulrich (eds.), Close Binary Stars: Observations and Interpretation, 155–160.

within the companion's Roche lobe. The major differences affecting accretion result from the much-reduced gravitational potential at the stellar surface, and the possibility of central masses greater than $\sim 2M_{\odot}$.

At the accretion rate $\dot{m}$ gs^{-1}, the luminosity due to accretion is $L = GM\,\dot{m}/R$ and the characteristic source temperature, $T = (L/4\pi\sigma\, R^2)^{1/4}$. To have the same spectral appearance as that produced by an accretion disk around a white dwarf, the accretion rate must be increased by a factor $\dot{m}_{MS}/\dot{m}_{WD} = (R_{MS}/R_{WD})^3\,(M_{WD}/M_{MS}) \sim 10^5$ to 10^7 (for masses $1 < M_{MS}/M_{\odot} < 100$). The ratio of luminosities is then $L_{MS}/L_{WD} = ([\dot{m}_{MS}\,R_{WD}\,M_{MS}]/[\dot{m}_{WD}\,R_{MS}\,M_{WD}]) = (R_{MS}/R_{WD})^2 \sim 10^4$ to 10^6. The obvious conclusion is demonstrated that the accretion luminosity of a main sequence star with a disk structure similar to that of a quiescent dwarf nova is very much brighter. It will have a luminosity $\sim 10^{36}$ to 10^{39} erg s^{-1}.

This model may be examined in greater detail using steady disk models (Shakura and Sunyaev 1973). The consequences of time-dependence in the disk may be shown to be unimportant. Assuming the disk is optically thick, which is valid so long as $T > 10^4$, then the temperature distribution as a function of radius and accretion rate is given by:

$$T(R) = (3GM\dot{m}/8\pi\sigma\, R_{MS}^3)^{1/4}\,(R/R_{MS})^{-3/4}\,(1 - [R/R_{MS}]^{-1/2})^{1/4} \qquad (1)$$

The temperature has a maximum at a radius $\sim 1.4\ R_{MS}$, and then decreases outward. The maximum disk temperature is given by:

$$T_{MAX} = 0.488\ (3GM\dot{m}/8\pi\sigma\, R_{MS}^3)^{1/4} = 8.3 \times 10^3\ \dot{m}_{20}^{1/4}\,(M/M_{\odot})^{1/4}\,(R/R_{\odot})^{-3/4} \qquad (2)$$

where $\dot{m}_{20} = \dot{m}/(10^{20}\ gs^{-1})$.

In Figure 1a the variation of T_{MAX} with $\dot{m}$ is shown for main sequence stars in the range 0.5 to 121 $M_{\odot}$. Radii are from Schwarzschild and Härm (1958), Iben (1965), and Stothers (1963). The associated luminosity changes are shown in Figure 1b. The variation of T_{MAX} and L is also shown for white dwarfs (cataclysmic variables) and a 1 $M_{\odot}$ neutron star (binary X-ray sources) for comparison.

Cut-offs exist at both high and low temperatures. Disk temperatures fall below 10^4 K with low accretion rates and the rapid fall in opacity leads to the disk becoming optically thin. In these conditions present models are not applicable. Radiation will be emitted in the lines with no strong continuum. Stars in this condition are likely to be found amongst the Stephenson-Sanduleak emission line objects, and symbiotic stars. For massive main sequence accretors a second cut-off exists, at low temperatures. For $M > 9\ M_\odot$ the nuclear luminosity is greater than the accretion luminosity at low accretion rates. This region is shown by heavy shading.

At high temperatures disk solutions break down when the luminosity becomes supercritical and radiation pressure disrupts the disk (Shakura and Sunyaev 1974). This bounday occurs at $L = L_{ed} = 4\pi\ GMc/\kappa$ and is shown in Figure 1 for a scattering opacity with $X = 0.7$. The behaviour of sources in the supercritical regime is discussed below.

It is evident from Figure 1 that accreting main sequence stars give a satisfactory model of Hubble-Sandage variables, simply through scaling accretion disk structure from the white dwarf to the main sequence case. With accretion rates 10^{20} to 10^{24} gs^{-1}, maximum disk temperatures are the same (10^4 to 4×10^4 K) as quiescent dwarf novae with accretion rates 10^{14} to 10^{17} gs^{-1}. The disk structure will be similar (Equation 1), simply scaled to larger radii. This larger disk has a luminosity $\sim 10^{38}$ erg s^{-1}. To generate the highest luminosities ($\sim 10^{39}$ erg s^{-1}) massive main sequence stars ($> 9\ M_\odot$) are required, with accretion rates $\sim 10^{23}$ g s^{-1}. Accretion disks around massive main sequence stars will form the brightest Hubble-Sandage variables.

It is easy to show (Bath, in press) that the other physical properties apart from the temperature will be similar in the two cases at these rates. The disk optical thickness is $\tau_{MS} \sim 10^3$ and $\tau_{WD} \sim 10^2$. Both are well into the optically thick regime. The surface density differs somewhat, with $\Sigma_{MS} \sim 10^{3-4}$ g cm^2 and $\Sigma_{WD} \sim 10^2$ g cm^2. Density differences may affect the spectrum and could be responsible for the strong FeII lines observed in some Hubble-Sandage variables.

The differences in line broadening are easily accounted for. The broad, doubled lines in cataclysmic variables are caused by emission from material in the disk moving with Keplerian velocities. The maximum disk velocity is $V = (GM/R)^{1/2}$, with an orbital period $P = 2\pi\ (R^3/GM)^{1/2}$. In the mass range $0.5 < M/M_\odot < 121$ the periods range from $1 < P < 14$ hours and velocities $400 < V < 1200$ km s^{-1}. In the white dwarf case $10 < P < 60$ sec and $2{,}000 < V < 5{,}000$ km s^{-1}. Thus, the line broadening in main sequence accretion disks will be 2 to 10 times less than that observed in cataclysmic variables, in agreement with that observed in Hubble Sandage variables.

Both classes lie in roughly the same region in the two-colour diagram, U-B ~ -1.0 and B-V ~ 0.0. The absence of the Balmer jump in both cases is due to the spectrum being composed of contributions to the integrated light from radii at a range of temperatures and luminosities in the disk.

The required mass transfer rates are possible in principle, particularly if the companions are giants. Observed variability (Rosino and Bianchini 1973) is due to fluctuations in the mass transfer rate, possibly caused by instabilities associated with destabilization in the ionization zones of the companion's envelope, as in dwarf novae.

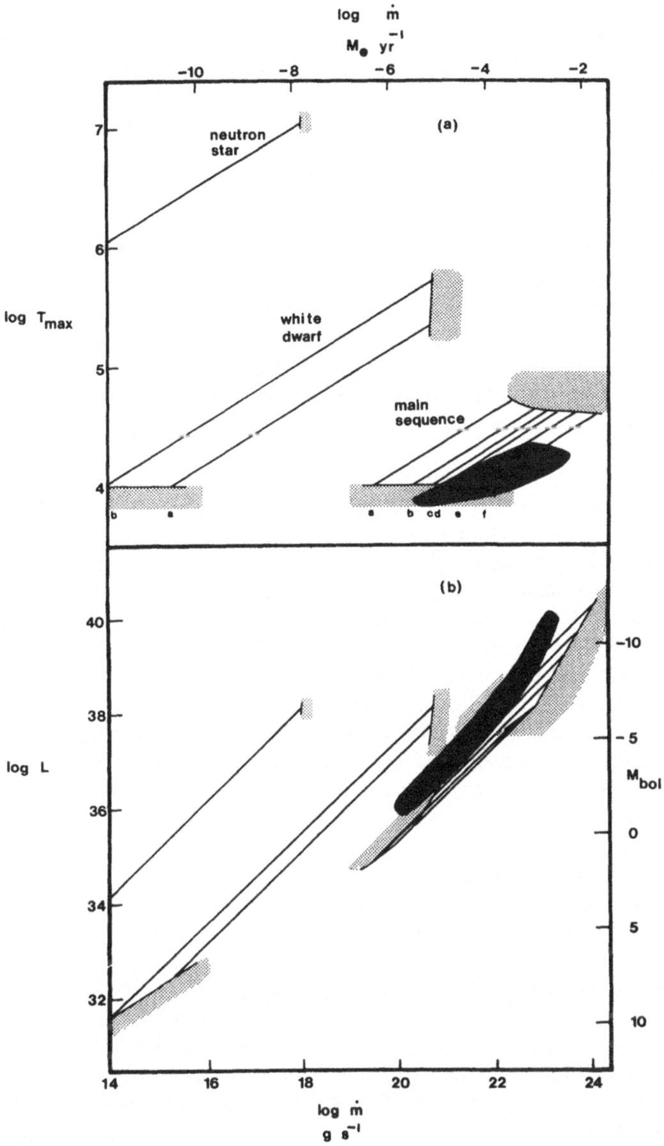

Figure 1. Variation of maximum disc temperature (Figure 2a) and disk luminosity (Figure 2b) with accretion rate. Accretion tracks are shown for 0.5(a), 1(b), 3(c), 9(d), 30(e) and 121(f) solar mass main sequence accretors. The region in which the accretion luminosity is less than the nuclear luminosity is indicated by heavy shading. At the upper boundary the disk luminosity is equal to the Eddington limit and the disk disrupts. At the lower boundary disk temperatures are less than ~ 10^4 °K and the disk becomes optically thin. Tracks for white dwarfs 0.4(a) and 1(b) $M_\odot$ and a neutron star of 1 $M_\odot$ are also shown.

In the supercritical regime, at accretion rates ~ 10^{22} to 10^{24} g s^{-1} (depending on M_{MS}), the inner region of the disk is disrupted by radiation pressure. This leads to a fraction of the accreting gas being ejected in the form of an outflowing wind, which because of the low gravitational potential is likely to be optically thick. It will have the same general properties as the optically thick winds at Eddington luminosities that are generated in classical nova explosions. Increases in the mass ejection rate will produce expansion of the photospheric radius, a decrease of photospheric temperature and of the bolometric correction, and an increase in visual luminosity at approximately constant bolometric luminosity. The emission line spectrum will be replaced by an F or A absorption line spectrum similar to that observed in classical novae at maximum light. The rapid optical brightness changes of M33 Var A (Hubble and Sandage 1953) seem to agree with this interpretation.

One of the closest related stars in our own galaxy to the Hubble-Sandage variables is η Car. η Car exploded in the optical between 1836 and 1856, followed by a subsidiary outburst in 1889. In 1893 the spectrum was in absorption and early F. This was slowly replaced by the characteristic emission line spectrum with strong H, HeI, FeII and [FeII] (Humphreys 1975). The absolute magnitude at outburst was brighter than -11, and may have been as bright as -14 (Warren-Smith et al. 1979). At present the bulk of the emission is radiated in the infrared by a surrounding dust cloud, but the bolometric luminosity is still high, possibly still -14. All the characteristics are in agreement with that expected for a massive interacting binary which underwent a burst of accretion in the 19th Century. If correct the outburst of η Car was an accretion-driven nova explosion.

η Car is the only member of this class for which evidence of binary structure exists. Warren-Smith et al. (1979) find that the dust is concentrated in an annulus of size ~ 10^4 AU, and interpret this as evidence for binary structure. This is the first suggestion of binary structure in these variables. Binary periods must be greater than several days, most probably weeks to months. Doppler shifts will be small (orbital velocities ~ 10 km s^{-1}). They would not be detected through radial velocity changes but a search for eclipses with periods between several days and several months is an outstanding task with exciting prospects.

REFERENCES

Bath, G.T., Evans, W.D., Papaloizou, J. & Pringle, J.E. 1974, Mon. Not. R. astr. Soc., 169, 447.
Bath, G.T. Nature, submitted for publication.
Hubble, E. & Sandage, A. 1953, Astrophys. J., 118, 353.
Humphreys, R.M. 1975, Astrophys. J., 200, 426.
Humphreys, R.M. 1978, Astrophys. J., 219, 445.
Iben, I. 1965, Astrophys. J., 142, 993.
Rosino, L. & Bianchini, A. 1973, Astr. Astrophys., 22, 453.

Schwarzschild, M. & Härm, R. 1958, Astrophys. J., 128, 348.
Shakura, N.J. & Sunyaev, R.A. 1973, Astr. Astrophys., 24, 337.
Stothers, R. 1963, Astrophys. J., 138, 1074.
Warner, B. 1976, I.A.U. Symp. 73, Dordrecht: Reidel (eds. P. Eggleton, S. Mitton & J.A.J. Whelan), 85.
Warren-Smith, R.F., Scarrott, S.M., Murdin, P., & Bingham, R.G. 1979, Mon. Not. R. astr. Soc., 187, 761.

IV. MASSIVE BINARY SYSTEMS

ANOMALOUS MASS RATIOS IN O-STAR BINARIES

Catharine D. Garmany and Peter S. Conti
Joint Institute for Laboratory Astrophysics, University of Colorado and National Bureau of Standards, Boulder, Colorado 80309

ABSTRACT

A study of the binary frequency among the O-stars finds that only 40% are binaries. They show a strong tendency to have mass ratios near unity and none with values less than 0.3. We discuss the evidence that mass loss is responsible for the observed mass ratios.

I. INTRODUCTION

Among the O-type stars stellar winds are widely-observed phenomena which are responsible for continuous mass loss at a rate that is significant on the star's evolutionary time scale (deLoore, DeGreve and Vanbeveren, 1978). In the case of close binaries, Hutchings (1976) has shown that the mass loss rate prior to Roche lobe overflow seems to be even higher than that found for similar single stars, but the binary nature of many O stars is not clear. In one of a series of papers on O stars, Conti has discussed the binary frequency of O stars by comparing new coude spectra radial velocities with previously published values (Conti, Leep and Lorre, 1977). Atmospheric motions in O stars can masquerade as orbital motion, so we have extended this survey by obtaining additional spectra of all O stars brighter than $m_V = 7$ and north of -50° (Garmany, Conti and Massey, 1980). There are 67 stars in this sample, and 40% of them are binaries with either orbits already computed or clear evidence for orbital motion. This figure of 40% is complete for any binary with a period less than about 10 days and a semi-amplitude greater than 15 km s^{-1}, so our results represent a complete sample of massive close binaries and can be compared with the predictions of Vanbeveren and deLoore (1979).

M. J. Plavec, D. M. Popper and R. K. Ulrich (eds.), Close Binary Stars: Observations and Interpretation, 163–167.

II. MASS RATIOS OF THE O-TYPE BINARIES

Compared to later type stars, 40% is a rather low binary frequency, but it is not surprising when one considers the complete lack of binaries with mass ratios, Q, less than 0.3 in our sample. Previous evidence that low amplitude systems with $Q < 0.3$ may not exist has been given by Bohannan and Garmany (1978). The overabundance of double line systems (Garmany 1979) also suggests this, although it could be partly due to a flattening of the mass-luminosity relation for O stars. Figure 1 shows Q vs. binary period for all known O-type binaries. The binaries in our survey are underlined, showing that there is no major change in the distribution of Q with the addition of fainter systems. Obviously Q can only be directly determined for double line binaries, but we have estimated Q for single line systems from the spectroscopic mass of the primary and from an assumed $\sin^3 i$ = 0.67. There is a cutoff at Q = 0.3 and a strong peak around Q = 1.

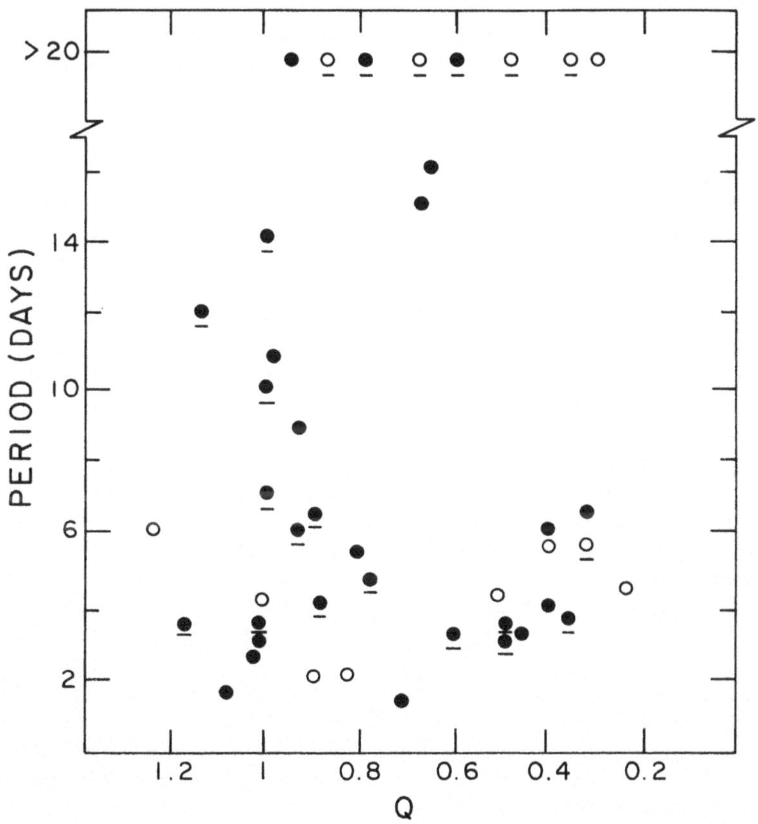

Fig. 1. The mass ratio, Q, vs. period for all presently known O-type binary systems. Underlined systems are included in our survey (• = double line, o = single line binaries).

Let us first consider the reality of this cutoff: could it be a selection effect caused by our inability to detect low amplitude binaries? The answer is probably not, because unless the inclination of the system is very small, mass ratios between 0.1 and 0.3 are not too difficult to detect for short period binaries. For example, O-type binaries with periods less than 6 days and mass ratios of 0.15 will have velocity semiamplitudes of 50 km s^{-1}. At coude dispersions, typical internal mean errors are 3-5 km s^{-1} and plate-to-plate errors are 2-3 times the internal errors, so any real variation greater than about 20 km s^{-1} has been detected. This includes all systems with inclination greater than 30°. Furthermore, the cutoff in Q is not caused by the different evolutionary time scales of the stars in question. From the evolutionary models of deLoore, DeGreve and Lamers (1977) and Iben (1965) an O-type star will still be in core hydrogen burning when a companion as small as 4 $M_{\odot}$ arrives on the ZAMS. Therefore systems with mass ratios as small as 0.1 to 0.2 are to be expected unless they are somehow prevented from forming. Lucy and Ricco (1979) have discussed formation by fragmentation, which produces binaries with Q = 1, and by fission for which Q is typically 0.2. Perhaps fission is somehow prevented in massive stars, or perhaps after reaching the main sequence the primary inhibits further formation of the secondary.

In their evolutionary computations for massive close binaries including mass loss by a stellar wind, Vanbeveren and deLoore (1979) predict that the mass ratio of pre-Roche lobe overflow binaries will tend towards unity as the primary star loses mass faster than the secondary. The number of systems in Figure 1 with Q close to one supports Vanbeveren and deLoore's calculations, although mass loss is not the only mechanism which would produce this distribution. It should be noted that their calculations are based on the assumption that mass loss is proportional to luminosity but most of the available data concern only supergiant O stars.

Recently, Conti and Garmany (1979) have computed mass loss rates from observations taken with the IUE satellite of the ultraviolet P Cygni lines in six main sequence stars. We have fitted the observed profiles with a grid of theoretical profiles computed by Castor and Lamers (1979) and have derived mass loss rates which are much lower than predicted. Figure 2 shows $\dot{m}$ vs. M_{BOL} for all O stars whose status as either single or binary is known. Two things are apparent from this figure: First, the addition of main sequence stars with log $\dot{m} < -6$ destroys any linear relation between $\dot{m}$ and M_{BOL}, and second, the binaries do not show any clear pattern of greater mass loss. The missing factor in Figure 2 may be the age of the star, such that the star's mass loss rate increases as it evolves away from the main sequence.

This work was supported in part by the National Science Foundation, most recently under grant AST78-20135 to the University of Colorado.

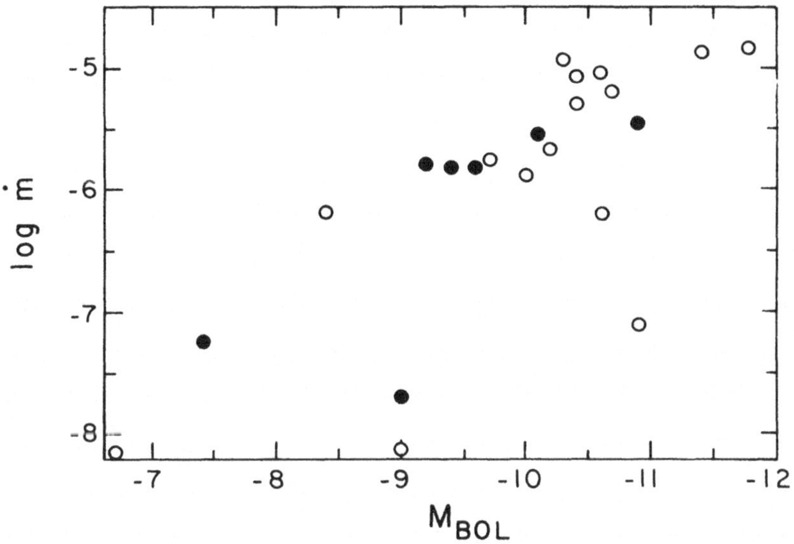

Fig. 2. Mass loss vs. M_{BOL}. Filled circles are binaries, open circles are single stars.

REFERENCES

Bohannan, B. and Garmany, C.D.: 1978, Astrophys. J. 223, p. 908.
Castor, J.I. and Lamers, H.J.G.L.M.: 1979, Astrophys. J. Suppl. 39, p. 481.
Conti, P.S. and Garmany, C.D.: 1979, submitted to Astrophys. J.
Conti, P.S., Leep, E.M. and Lorre, J.J.: 1977, Astrophys. J. 214, p. 759.
deLoore, C.W., DeGreve, J.P. and Lamers, H.J.G.L.M.: 1977, Astron. and Astrophys. 61, p. 251.
deLoore, C.W., DeGreve, J.P. and Vanbeveren, D.: 1978, Astron. and Astrophys. 67, p. 373.
Garmany, C.D.: 1979, in Mass Loss and Evolution of O-Type Stars, IAU #83, eds. P.S. Conti and C. DeLoore (Dordrecht, D. Reidel).
Garmany, C.D., Conti, P.S. and Massey, P.M.: 1980, in preparation.
Hutchings, J.B.: 1976, Astrophys. J. 203, p. 438.
Iben, I.: 1965, Astrophys. J. 141, p. 993.
Lucy, L.B. and Ricco, E.: 1979, Astron. J. 84, p. 401.
Vanbeveren, D. and deLoore, C.W.: 1979, preprint.

DISCUSSION FOLLOWING GARMANY AND CONTI

Nariai: What is the time scale of your possible evolution on the Q-P diagram?

Garmany: It is comparable to the time the stars spend in core hydrogen burning, about 5 x 10^6 years. The rate of change of the period and Q are given, for example, by Vanbeveren (1979).

Vanbeveren: Stellar wind values are determined by using spherical symmetric models for the outflowing atmospheres. These models may not be applicable in binaries where deformations become important. On the other hand, it can be that the underlying hot coronal layers emit soft X-rays changing the ionization equilibrium in the stellar wind and thus changing the mass loss rate determinations.

Garmany: We don't yet know mass losses for enough binary systems to say if the mass loss rate is dependent on the inclination, but as an example HD 48099 probably has an i of about 30° and its rate is low. The evidence so far is more in favor of the mass loss rate being a function of evolutionary age. And yes, any change in our ionization model is directly related to a corresponding change in the derived mass loss rate.

Lucy: Since the Q = 1 peak also exists at later spectral types where it cannot be attributed to mass loss, it seems somewhat unreasonable to invoke mass loss to explain the peak for the O-type binaries.

Garmany: Yes, we could certainly explain the shape of the Q-period diagram by initial formation, although some binaries certainly have a high mass loss rate which would alter their Q.

OBSERVATIONAL AND EVOLUTIONARY ASPECTS OF WOLF-RAYET STARS

Dany Vanbeveren
Astrophysical Institute, Vrije Universiteit
Brussel, Brussels, Belgium

Some of these results are part of a more extended investigation on WR stars that will appear in four papers, i.e. Vanbeveren and Packet (1979), Vanbeveren and Conti (1979), Vanbeveren and Doom (1979), Vanbeveren and de Loore (1979).

1. THE BINARY STATUS OF WR STARS

Many authorities proclaimed that all WR stars are members of close binary systems with OB type companions (see e.g. I.A.U. Symp.n°43). Arguing that fainter stars were less well studied Kuhi (1973) only considered northern systems brighter than m_V=10m. In this sample 73% of the WR stars are binary components. Due to the very low disruption probability during the supernova explosion of the low mass component in a massive binary (De Cuyper et al., 1976) classical evolutionary theory predicts that there should be as many WR+OB type systems as there are WR+compact star systems (de Loore et al., 1975). If 73% WR+OB systems would be real, the evolutionary predictions would give serious problems. Fortunately, the proposed 73% is probably largely overestimated. P.S. Conti made spectra with the 4m telescopes at the observatories of Kitt Peak and Cerro Tololo of nearly all the galactic WR stars and those in the LMC studied by Smith (1968). One of the important results to remember here is that very few additional stars were found to have absorption lines that had not been listed as such by Smith already. On the other hand, the m_V of many of the brightest WR stars is dominated by a companion. In the fainter WR stars the continuum is dominated by the WR star itself. Therefore if one decides only to consider systems brighter than m_V=10m one preferentially includes the binaries and this biases the statistics. The new measurements show a binary frequency ~40(±10)% for the Galaxy and for the LMC considering the whole sample now and using the criterion : binary means absorption lines present.

M. J. Plavec, D. M. Popper and R. K. Ulrich (eds.), Close Binary Stars: Observations and Interpretation, 169–175.

It can be remarked that these numbers do not exclude the possibility of the existence os a low percentage of single WR stars.

2. THE EVOLUTIONARY STATUS OF WR STARS

According to Willis and Wilson (1978) early WN and WC stars show pure (He,Z) layers. This indicates that these stars should be in the He burning stage. Stellar wind mass loss during H burning is not enough to bring these pure (He,Z) layers to the surface. In WR binary systems the remaining hydrogen will be removed during the Roche lobe overflow process (RLOF). In Figure 1 the position in the HR diagram of early WN and WC stars is shown according to Willis and Wilson (1978 and private communication). The position of the late type WN stars is taken according to P.S. Conti (I.A.U. Symp.n°83). On the other hand also evolutionary tracks are shown for a $40M_0$, $60M_0$ and $100M_0$ star. During core hydrogen burning mass loss by stellar wind is taken into account (N=300, cf. Vanbeveren and de Loore, this symposium). As most of the massive binary systems will evolve through a case B of mass transfer (Vanbeveren et al., 1979), the primary reaches its Roche lobe after core hydrogen burning. We remark that the remnant mass after RLOF is largely independent from the period and the mass ratio of the system

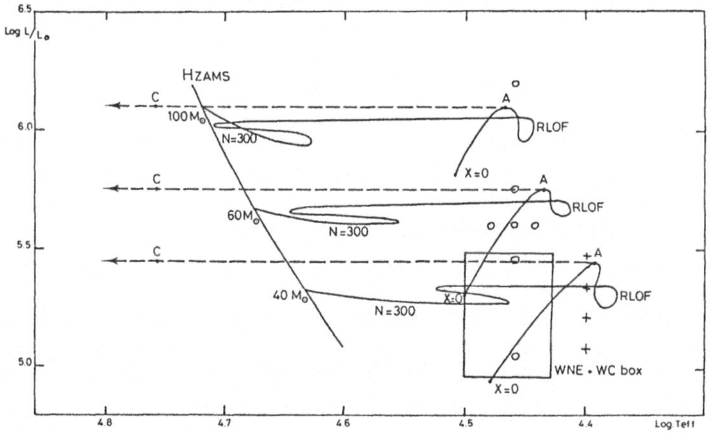

Figure 1. Evolutionary tracks for primaries of binary systems with initial ZAMS masses of $40M_0$, $60M_0$ and $100M_0$. During the hydrogen core burning, mass loss by stellar wind is taken into account (N=300). Point A denotes the end of the Roche lobe overflow. The dashed lines are the tracks with a non-extended, hydrostatic radius determination. The observed WN7 stars (o), the WN8/9 stars (+) and the WNE+WC box are indicated.

(De Grève et al., 1978) and of the mass and angular momentum loss from the system during RLOF (Vanbeveren et al., 1979). After RLOF the surface H abundance ≃0.2 almost independent from the stellar mass (Vanbeveren and De Grève, 1979) so that this stage may account for the late WN stage. The observations reveal that WR stars lose mass by stellar wind at very high rates (typically a factor 10 higher than in O and Of stars). After RLOF the evolution was further carried on including mass loss rates of this order. A mass loss rate formula proportional with luminosity was used. The remaining hydrogen is removed and pure He layers appear at the surface: an early WN star is formed. The hydrogen burning shell dies out very fast and the luminosity decreases, explaining the observed luminosity difference between late type and early type WN stars.
There remains however a large discrepancy: after RLOF the evolutionary tracks cross the HR diagram towards the very high temperature range of the He main sequence in a time scale of the order of 10^4 years. The number of WR stars relative to the number of OB type stars indicates that the WR stage should last for the whole He burning phase. So, according to evolution we would expect to see most of the early WN stars and WC stars in a temperature range around ~90 000K and this is much higher than observed (T≃30 000K). A possible reason for this discrepancy may be the treatment of the atmosphere in evolutionary programs. The outer layers are treated as a planparallel, non extended, gray and hydrostatic atmosphere. In WR stars none of these assumptions are correct. Although the energy production in the core (and thus the luminosity) is nof affected, a correct treatment would result in larger radii and thus lower T_{eff} values. Considering the position of the early WN stars and WC stars in the HR diagram, it cannot be excluded that the real He main sequence for massive stars is situated at lower T_{eff} values than the hydrogen ZAMS.
The evolution after core hydrogen burning was performed including different stellar wind mass loss rates. It is possible to put some limits on these rates. First the mass loss rate may not be too high because in this case we would see layers that were in the He burning core in a previous

$M(M_\odot)$	$t_{He}^{min}(10^5 yrs)$	$t_{He}^{max}(10^5 yrs)$	$\Delta M_{min}(M_\odot)$	$\Delta M_{max}(M_\odot)$
10	3.4	5.3	4	6
17	2.2	3.1	5	7
32	1.0	2	6	10

Table 1. Minimum and maximum values for the total mass removed from the star during core He burning. The corresponding He burning lifetimes are indicated.

stage containing too much carbon as observed in WC stars. On the other hand, in order to see carbon enhanced layers at least once during He burning and taking into account with the WR distribution, i.e. n(late WN)/n(WR)~0.1 (Moffat and Seggewiss, 1979) and n(WN early)≃n(WC) we can determine a lower limit. It is encouraging that both numbers generally do not differ by more than a factor 2. In Table 1 some of these mass loss rates ar shown.

3. THE CHEMICAL ABUNDANCES

In this section we try to determine theoretically the CNO abundances in the surface layers of the remnant stars after RLOF (corresponding with late type WN stars) and in the surface layers of the early WN stars. Therefore the whole CNO cycle was followed in detail in the layers of a 40, 60 and 100 $M_\odot$ star. This mass range covers the WR mass range $8M_\odot \leq M_{WR} \leq 35M_\odot$. The results are shown in Figure 2. As can be noticed the results for N/He, C/He and O/He are almost independent from the stellar mass, the difference being

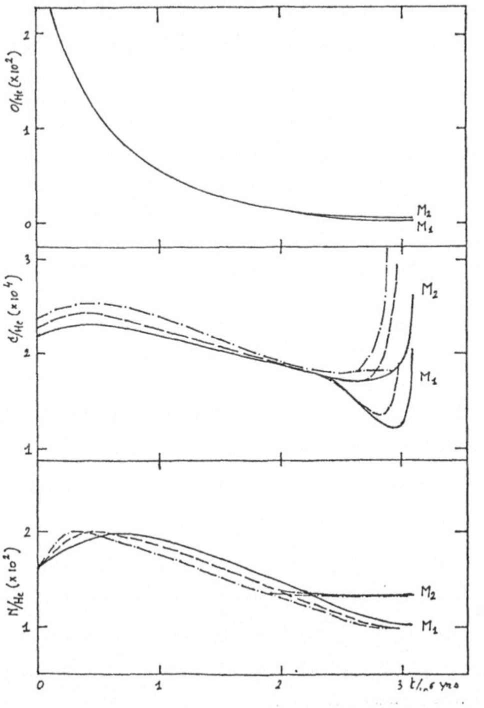

Figure 2. The variation of the N/He, C/He and O/He ratios in the mass layers M_1 and M_2 corresponding resp. with the layer X=0.2 (just after RLOF) and with the layer X=0 (early WN). The end of each curve indicates the N/He ratio at the moment the layer appears at the surface. Different stellar masses are considered, i.e. $40M_\odot$ (full line), $60M_\odot$ (dashed line) and $100M_\odot$ (dashed-dotted line).

largest for C/He because C is most sensitive to temperature variations in the layers considered. The CNO abundances in WR stars should therefore be largely independent from their mass. The correspondence of the theoretical calculations with the abundances in 3 early WN stars and 1 WC star determined by Willis and Wilson is amazingly good if their model with $T_{eff} \simeq 30\,000K$ is considered (which I recall is the most probable one).

4. EVOLUTIONARY MODELS FOR SOME KNOWN WR SYSTEMS

For a number of WR systems where estimates of the masses are available evolutionary models were constructed using the stellar wind mass loss rates obtained in section 3. Using interpollation formulas given by Vanbeveren and De Grève (1979) one can calculate the mass a star had just before RLOF. More details are given in a paper by Vanbeveren and de Loore (1979). In order to be consistent with the overabundance of unevolved systems with large mass ratios, a large mass loss from the system during RLOF is necessary. On the other hand, as it may be expected that more massive systems will evolve through a case B system (Vanbeveren et al., 1979) also a large angular momentum loss is needed. Striking also is the fact that all the WR stars considered here result from initial ZAMS masses larger than $\sim 50M_{\odot}$. This is quite different as what has been found by Smith (1973). From a statistical investigation she found that all OB type stars with masses larger than $25M_{\odot}$ should evolve into a WR star. However this conclusion was based on the use of the initial mass function of Sandage (1957) and very crude estimates of hydrogen and helium burning lifetimes. Since that time things have been improved. Using the initial mass function of Lequeux (1979) one finds that only 1/5 of the stars with masses larger than $25M_{\odot}$ should evolve into WR systems. Only by using $50M_{\odot}$ as a minimum mass one obtains that all OB stars with masses larger than $50M_{\odot}$ should evolve into WR stars.

REFERENCES

De Cuyper, J.P., De Grève, J.P., de Loore, C., van den Heuvel, E.P.J.: 1976, Astron.Astrophys.

De Grève, J.P., de Loore, C., van Dessel, E.L.: 1978, Astrophys.Space Sci. 53, 105.

de Loore, C., De Grēve, J.P., De Cuyper, J.P.: 1975, Astrophys.Space Sci. 36, 219.

Kuhi, L.V.: 1973, in "Wolf-Rayet and High Temperature Stars", eds. M.K.V. Bappu and J. Sahade, Reidel, Dordrecht, p.205.

Moffat, A.F.J., Seggewiss, W.: 1979, Astron.Astrophys. (in press).

Smith, L.F.: 1973 in "Wolf-Rayet and High Temperature Stars", eds. M.K.V. Bappu and J. Sahade, Reidel, Dordrecht, p.15.
Vanbeveren, D., De Grève, J.P., van Dessel, E.L., de Loore, C.: 1978, Astron.Astrophys. 73, 19.
Vanbeveren, D., De Grève, J.P.: 1979, Astron.Astrophys. (in press).
Vanbeveren, D., Packet, W.: 1979, Astron.Astrophys. (in press).
Vanbeveren, D., Conti, P.S.: 1979, preprint.
Vanbeveren, D., de Loore, C.: 1979, Astron.Astrophys. (in press).
Vanbeveren, D., Doom, C.:1979, preprint.
Willis, A.J., Wilson, R.: 1978, Mon.Not.R.Astr.Soc. 82, 55.

DISCUSSION FOLLOWING VANBEVEREN

Friedjung: I wish to make a somewhat provocative remark. Wolf-Rayet stars are very badly understood--we do not understand the nature of the wind. The significance of T_e is not simple if one has in many layers optically thick winds, rather than a hydrostatic photosphere. How can one then make evolutionary calculations for objects so badly understood? Are you sure your calculations refer to Wolf-Rayets? It should also be noted that the University College London group obtained a high Wolf-Rayet mass loss rate from IUE observations.

Vanbeveren: The origin of the high stellar wind is indeed badly understood whereas the determination of an effective temperature is extremely difficult as well. However these problems are not so important for evolutionary calculations of WR stars. As has been done for the evolution of O stars including stellar wind, we have used a parameter equation for the stellar wind in WR stars in order to determine general evolutionary characteristics during the burning. On the other hand, a different treatment of the outer layers in evolutionary computations will not alter the behavior of the core of the star determining the whole evolutionary pattern. Thus although the effective temperature is badly known, this does not affect general evolutionary characteristics as there are the surface CNO abundances, the behavior of the luminosity, the timescales, the calculations of the masses before RLOF, the mass of the star itself so that a comparison with observations is very reliable. The determination of a maximum mass loss rate after RLOF is independent of the atmospheric treatment as well. The reason for the difference between these maximum values and the values obtained by the London group is unclear to me for the moment. If the values of the London group are correct then there should exist a large number of stars having carbon abundances in their outer layers of the same order as the He abundance. This is not observed.

Morton: It is strange that the Willis-Wilson effective temperatures are essentially the same for all WN types. The Zanstra calculations I made show that there are some WN5 stars with surrounding nebulae where $T_{eff} \approx 50000K$ consistent with what we have just heard about the central star of NGC 6888.

Vanbeveren: I would be very happy if the observed T_{eff} values for WR stars were higher than the values obtained by Willis and Wilson. There is something in the determination of T_{eff} for WR stars that we do not yet understand. Even the 50000K for some WN5 stars would not resolve the problem I mentioned i.e. from evolutionary calculations we would expect to see most of the WR stars at $T_{eff} \approx 90000K$. Therefore I considered different parameters to make a comparison, parameters that are only marginally affected by the treatment of the atmosphere.

Holm: A couple of years ago Joe Cassinelli and I compared UV and optical spectrophotometry of HD 50896 with a model extended Wolf-Rayet atmosphere having a temperature of $50,000^{o}K$. The agreement was quite good. The temperature of this atmosphere is about $20,000^{o}K$ hotter than a plane-parallel atmosphere which also fits the observed energy distribution.

OBSERVATIONS OF WOLF-RAYET BINARIES

Virpi S.Niemela[1]
51 esq.11,Villa Elisa,Buenos Aires,Argentina

In many recent papers dealing with computations of evolutionary sequences for massive stars with mass loss,the evolutionary status of Wolf- Rayet stars has been discussed. When one whishes to compare these theoretical computations with observations,it is of great importance to know the masses of Wolf-Rayet stars,which unfortunately are well determined only in a few cases.Estimates of the masses are provided by binaries with WR type components.I will report here preliminary results of spectrographic observations for 4 WN+OB systems,with new information about the masses of their components.

The observations of the southern binaries,namely HD 90657,HD 94546 and HDE 311884 (=MR 42),were obtained during 12 consecutive nights in February 1979 with the Cassegrain spectrograph with IT attached to the 1.0m reflector of the Cerro Tololo Interamerican Observatory.These spectra have dispersion 42 Å/mm and are on Kodak IIIaJ plates. The spectrograms (15 and 30 Å/mm on IIaO plates) of the northern WN+OB binary HD 214419 (=CQ Cep) were obtained at the Dominion Astrophysical Observatory by Dr.A.Batten and myself.

Radial velocity curves have been derived for all 4 binary systems from the NIV λ4058 emission line,assumed to represent the motion of the WN star;and from the Balmer series, HeII and HeI absorption lines of the OB companion.Preliminary orbital elements,listed in Table 1,were obtained by graphical solution from these velocity curves.I have chosen the NIV λ4058 emission for the WN components,because this line was shown to represent closely the stellar radial velocity in WN stars of subclasses WN7/WN8 (Moffat and Seggewiss 1979). However,for the present WN stars of earlier types this may

[1]Visiting astronomer,CTIO,which is supported by NSF under Contract No.AST74-04128.

M. J. Plavec, D. M. Popper and R. K. Ulrich (eds.), Close Binary Stars: Observations and Interpretation, 177–180.

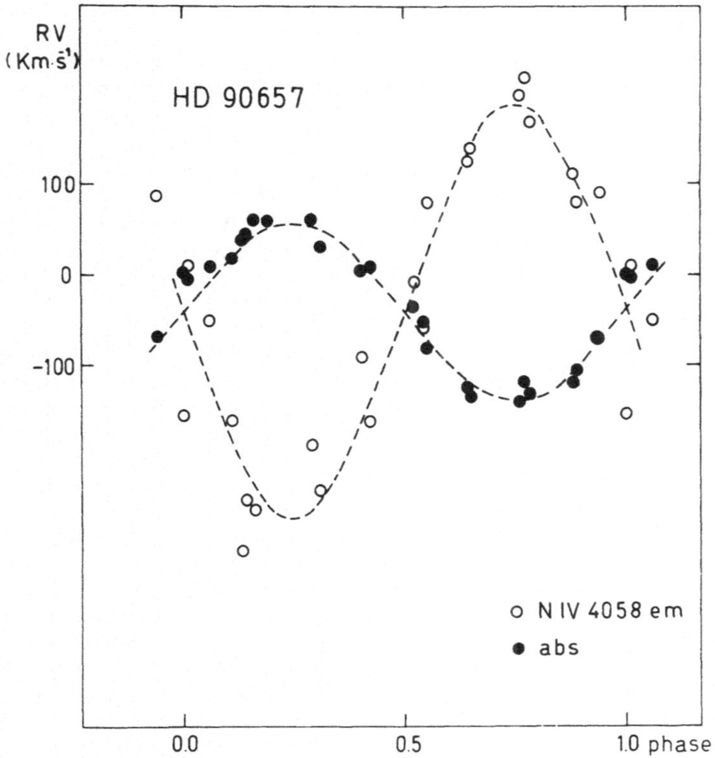

Fig.1. Radial velocities for HD 90657.
Filled circles:mean absorption.
Open circles:NIV λ4058 emission.
Dashed curves:theoretical velocities defined by orbital elements in Table 1.

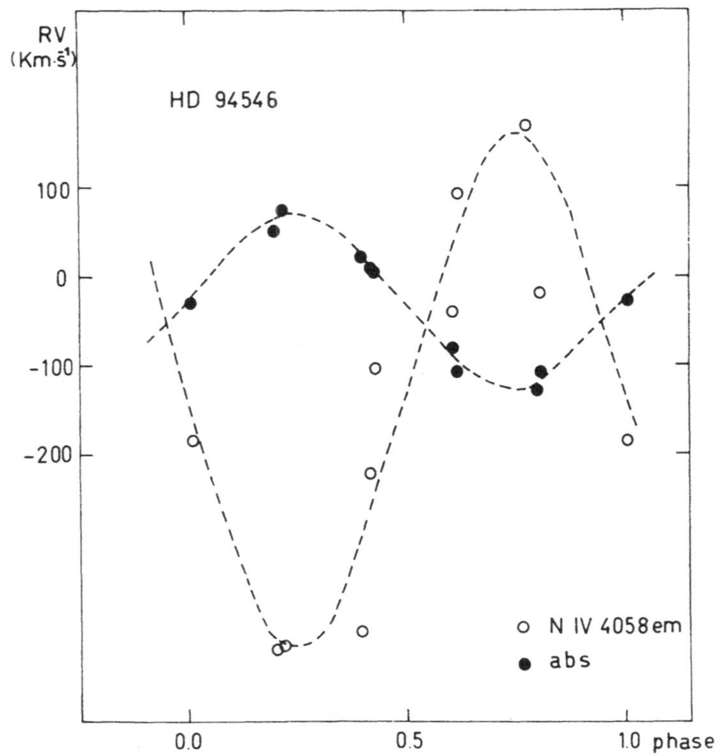

Fig.2. Radial velocities for HD 94546.
The symbols as in Fig.1.

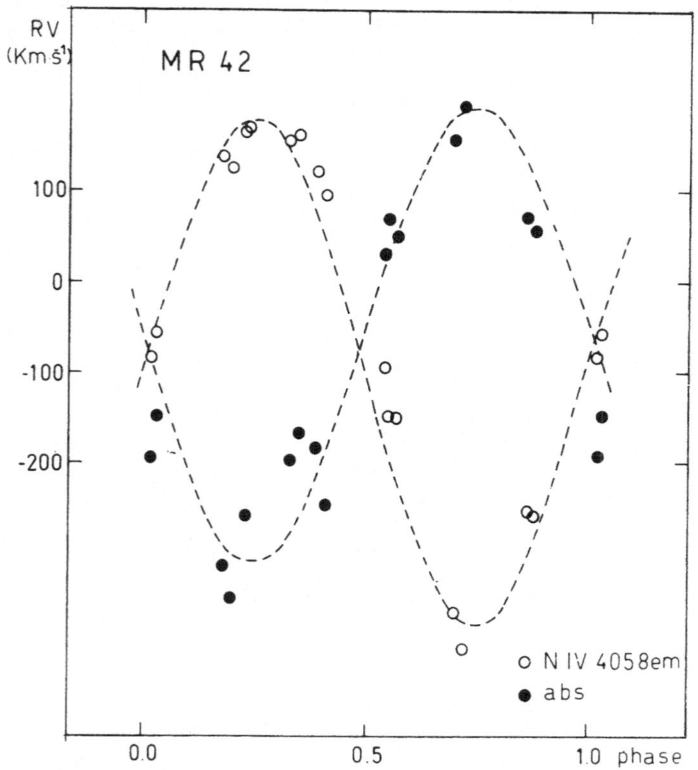

Fig.3. Radial velocities for MR 42.
The symbols are as in Fig.1.

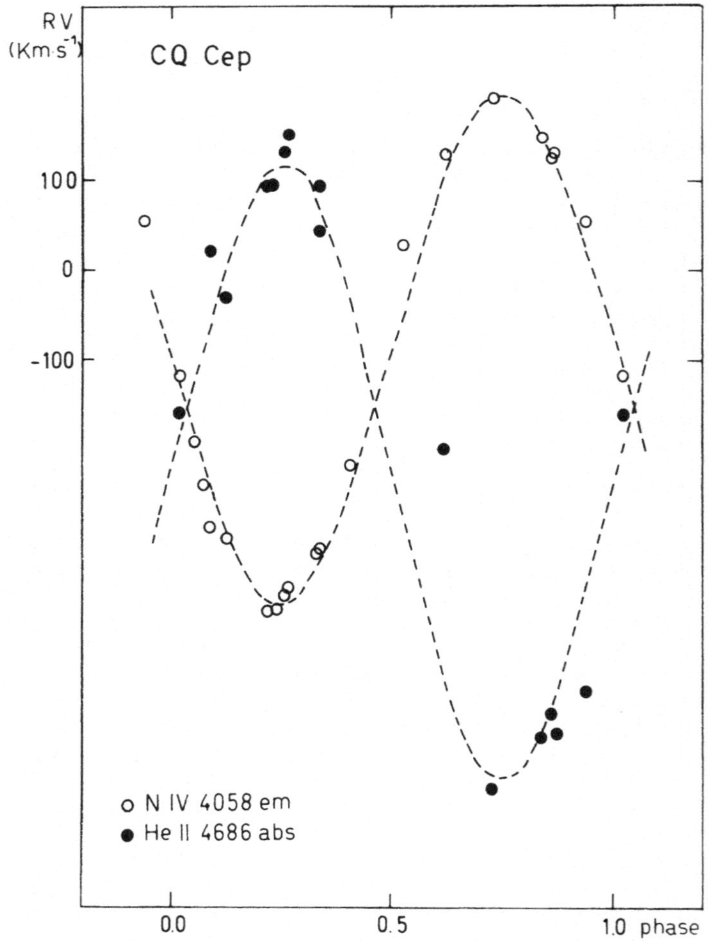

Fig.4. Radial velocities for CQ Cep.
Filled circles:HeII 4686 abs.
Other symbols are as in Fig.1.

Table 1. Preliminary Orbital Elements for 4 WN+OB Systems

	HD 90657	HD 94546	MR 42	CQ Cep
Period (d)	8.2	4.9	6.3	1.6
e=0 (assumed)				
$K_{NIV\ em.}$ (km/s)	230	290	280	285
$K_{abs.}$ (km/s)	100	100	240	340
$\gamma_{NIV\ em.}$ (km/s)	-40	-130	-100	-85
$\gamma_{abs.}$ (km/s)	-40	-30	-50	-220
$M_{OB}\sin^3 i$ ($M_\odot$)	22	23	50	19
$M_{WN}\sin^3 i$ ($M_\odot$)	9.5	8	43	23
Spectral type	WN4+OB	WN4+OB	WN6+OB	WN7+O?

not be true,as the systemic velocities from the NIV emission do not always agree with that of the OB companion.

HD 90657:The velocity curves for this system from 25 spectra obtained in February 1979 are shown in Fig.1.The ellipticity of a previous orbit (Niemela 1976) is found to be spurious by the present observations covering one and a half cycles of the binary.The values of the minimum masses for this system,with the WN star being the less massive,are very similar to those of the "classical" WN+O system V 444 Cygni (Munch 1950).

HD 94546:No previous radial velocity curves for this system are known.A radial velocity variation with a period of 4.9 days is shown from 10 spectra obtained in February 1979 (Fig.2).Spectroscopically this binary resembles HD90657 and also seems to have similar minimum masses.

MR 42:This star has not previously been known as double-lined spectroscopic binary.17 spectra obtained in February 1979 show faint absorption lines from an OB companion to be present.Both components of this system seem to be quite massive (see Fig.3 and Table 1),unlike most other WR binaries.

CQ Cep:The absorption lines previously attributed to the OB companion belong to the WN7 star.Only two absorption lines,the HeII λ4686 superimposed to the WN emission,and one component of the violet shifted HeI λ3888,seem to arise from the companion.

REFERENCES

Moffat,A.F.J. and Seggewiss,W.:1979,in IAU Symp.No.83,eds. P.S.Conti and C.W.H.de Loore,p.447.

Munch,G.:1950,Astrophys.J. 112,p.226.

Niemela,V.S.:1976,Astrophys.and Space Sci. 45,p.191.

WOLF-RAYET BINARIES

A.F.J. Moffat[1*] and W. Seggewiss[2]
1 Département de Physique, Université de Montréal
2 Obs. Hoher List, Universitäts-Sternwarte Bonn

ABSTRACT. An updated list of known spectroscopic double-line and single-line Galactic Wolf-Rayet binaries is presented. From this we discuss e.g. mass ratios, the binary frequency and the evidence for low-mass unseen (compact?) companions. Spectroscopic binary orbits of several WR stars in the Magellanic Clouds are noted for the first time.

I. Galactic WR stars with orbits

Our present knowledge of WR-star masses is based exclusively on an incomplete sample of twelve double-line systems brighter than v = 10.5. Another eight single-line binaries and an effectively single-line star in a triple system complete the list for the Galactic WR stars. All 21 are listed in Table 1. It should be noted that the only really conclusive proof of duplicity comes from the observation of periodic orbital phenomena such as radial velocity (RV) variations. The presence of both emission lines and O-star like absorption lines in the same spectrum does not necessarily guarantee a binary nature for those two components. This is manifested by at least two of the stars in Table 1 (B), MR 25 and CQ Cep, whose respective emission and absorption lines mutually follow the same orbit. For this reason, they are assigned single-line spectral types.

From Table 1 which contains the best values form the literature we note the following:

- No orbits (SB1 or SB2) are available for the hot or cool extremes of the nitrogen or carbon sequences (WN3, WN8; WC5, WC9). This is probably due to selection.
- The eccentricity e is significantly different from zero for binaries with long periods ($P > 50^d$) and binaries with suspected compact companions.
- M_{WR}/M_{OB} for the double-line stars appears to be relatively constant, independent of spectral subclass. Mean values are:

0.44 ± 0.08 m.e. for WN,
0.42 ± 0.07 m.e. for WC (excluding θ Mus).

* Visiting astronomer Cerro Tololo Interamerican Observatory, Chile

M. J. Plavec, D. M. Popper and R. K. Ulrich (eds.), Close Binary Stars: Observations and Interpretation, 181–185.

These imply that the mean masses of WN stars (at least those of lower luminosities: WN3-6) differ insignificantly from those of WC stars, assuming the OB companions to have similar mass in the mean. This puts constraints on the hypothesis that WN stars evolve into WC stars through additional mass loss and exposure of processed core material.

- M_{WR}/M_{OB} is independent of the orbital period, in conflict with the idea that mass tranfer should be most efficient for the closest systems (cf. Vanbeveren and Conti, 1979).
- Among the eight single-line binaries, none is WC, whereas all are WN5-7 with WN7 dominating. In some cases this is probably the result of the drowning out of the OB companion by the luminous WN7 star. In the less luminous WN5,6 stars MR 114 and MR 108 it might be particularly worthwhile looking more carefully with higher signal-to-noise ratio for the presence of an OB companion.
- The deduced mass ratio M_{WR}/M_2 for the five single-line stars MR 114, 108, 25, 111, 118 is slightly larger (0.60 ± 0.10 m.e.; 0.51 ± 0.03 m.e. without MR 25) than for the double-line binaries, subject to assumptions concerning the orbital inclination and WR mass.

Table 1. Galactic Wolf-Rayet binaries with orbits, sorted according to spectral duplicity and WR subclass

MR	HD	v	P	e	Sp	f(m)	m_{WR}/m_2
A *Double-line binaries*						$M_\odot$	
31	94546	$10^m.69$	$4^d.9$	0.0	WN4 + O	12.4	0.34
94	186943	10.36	9.55	0.0	WN4 + B	3.9	0.42
99	190918	7.48	112.8	0.43	WN4.5 + O9.5 Ia	1.24	0.26
106*	193576 V444 Cyg	8.27	4.21	0.0	WN5 + O6	12.4	0.39
23	90657	9.80	8.2	0.0	WN5 + O6	10.4	0.23
116*	211853 GP Cep	9.20	6.69	0.0	WN6 + O6 I	16.1	0.35
42	311884	11.09	6.3	0.0	WN6 + O	14.4	0.86
105	228766	9.33	10.74	0.0	WN7 + O7.5 I	12.7	0.67
43	113904 θ Mus	5.69	18.34	0.0	WC6 + O9.7 Iab	9.9	(<0.03)
65	152270	6.95	8.89	0.0	WC6-7 + O5	2.7	0.36
36	97152	8.25	7.84	0.0	WC7 + O6	3.7	0.55
12	68273 γ^2 Vel	1.74	78.50	0.40	WC8 + O9 I	13.7	0.53
85*	168206 CV Ser	9.43	29.71	0.0	WC8 + O8-9	15.0	0.24
B *Single-line binaries (orbit of WR component only)*							($i=60^\circ$ $m_{WR}=10$)
6†	50896 EZ CMa	6.94	3.76	0.34	WN5	0.015	6.9
114*	- CX Cep	12.40	2.13	0.0	WN5	5.4	0.52
108	193928	10.15	21.64	0.0	WN6	4.9	0.55
102†	192163	7.73	4.5	?	WN6	0.00024	29
25	92740	6.44	80.35	0.55	WN7	1.67	0.98
111	-	14.0	22:	?	WN7:	7.7:	0.42:
113†	197406	10.50	4.32	0.11	WN7	0.32	2.1
118*	214419 CQ Cep	8.94	1.64	0.0	WN7	5.1	0.54

Notes: * eclipsing binary

† compact companion?

- The mass ratio M_{WR}/M_2 for the stars MR 6, 102 and 113 is significantly larger than 1.0. Combined with other evidence below, these stars are prime candidates for possessing relatively low-mass, compact companions.
- The ratio of stars with known orbits (SB2 and SB1) to the total number of WR stars decreases steadily from 71% (5/7) for $v \leq 7.0$ to 43% (10/23) for $v \leq 9.0$, where detailed spectroscopic searches for orbits are less complete. From a study of the statistics of double-line WR stars and deduced WR+compact binaries, Vanbeveren and Conti (1979) estimate a frequency of 40 + 40 = 80%, in fair agreement with the value obtained here for brighter stars with known orbits. It may be inferred that the binary nature appears fundamental to the formation of WR stars, except possibly for the more luminous WN7/8 subgroup.

II. Runaway WR stars

From a theoretical point of view, it appears quite plausible that a massive binary can pass through two WR phases of essentially equal duration (Tutukov and Yungelson, 1973; van den Heuvel, 1976). The first corresponds to the WR + OB (SB2) phase observed unequivocally in ~40 - 50% of the cases. This is followed by a supernova explosion of the WR star, leaving in the majority of cases, a compact + OB recoil system which eventually evolves into a runaway single-line WR (SB1) binary corresponding to the second WR phase. Several properties, summarized below can be expected to characterize such WR stars with collapsed companions:
- high peculiar (radial) velocity and/or large separation $|z|$ from the Galactic plane, depending on the direction of the runaway space velocity vector.
- single-line, short period spectroscopic binary with low mass function; such systems will be more difficult to detect than WR + OB binaries. The orbit may be eccentric.
- expanding gas disc ejected during the phase of rapid, non-conservative mass transfer from the evolving OB star. If this occurred not too long ago, one would expect to see it in form of an expanding H II region, otherwise it may be too dispersed to be seen.

Table 2. Runaway WR candidates

MR	type	MR	type	MR	type
3	4	51	4	100	2
6	1,2	53	4	102	1,2,3
7	2	58	4	113	1,3,4
21	2	60	4	M1-67	2,3
34	2	67	4	LS 8	4
49	2	91	4	LS 11	4
50	4	97	2		

Type of peculiarity: (1) low mass function, (2) H II ring, (3) peculiar RV, (4) $|z| > 400$ pc.

From Table 1 (B) we see three stars which satisfy at least two of the above criteria: MR 6, 102 and 113. In Table 2 we present a summarizing list of probable runaway WR candidates based on these criteria. The fact that there may be a significant fraction of runaway stars (up to ~50%, comprising most of the single-line WR stars) is strongly supported by the statistical analysis of the z-distribution of Galactic WR stars by Moffat and Isserstedt (1979). The mean absolute separation $|z|$ for WR + OB stars is ~80 pc, like the normal population I objects, while for single-line stars it is ~130 pc, much like the well-known runaway OB stars. Some of these stars may turn out to be X-ray sources but of considerably reduced X-ray flux due to the high efficiency of photoelectric absorption in the dense WR envelopes (cf. Moffat and Seggewiss, 1979).

III. Search for binary orbits among extragalactic WR stars

Beyond the Galaxy, WR stars are known to abound in the Magellanic Clouds: there are now 101 known in the LMC (Azzopardi and Breysacher, 1979b) and 8 in the SMC (Azzopardi and Breysacher, 1979a). Many of these stars are bright enough to make a velocity orbit search feasible. During 10 nights of Nov/Dec 1978 one of us (AFJM) obtained repeated image tube spectra at CTIO, Chile, of 17 stars of magnitude V ≤ 12.5 at 46 Å/mm.

Whereas about half of the luminous WN7/8 stars appear to be truly single stars (constant RV) as is the case in our Galaxy (Moffat and Seggewiss, 1979) all the rest except FD 68 (WN6) are WR + OB binaries similar to their Galactic counterparts in Table 1 (A). As in the Galaxy, many of these OB companions are supergiants.

From preliminary analysis of the radial velocities, six (35%) double-line spectroscopic binaries have been found (AB 5,6; FD 21,28,29,67). Except for FD 67, which contains the more luminous WN8 subclass, the mass ratio M_{WR}/M_{OB} of the other five stars appears to be significantly lower on the average than for their Galactic equivalents. Although this may be partly due to spurious effects the low mass ratio appears to be intrinsic, possibly the result of differences in chemical composition between the Magellanic Clouds and the Galaxy such that the presently observed OB component in the Clouds has lost less mass on the average.

Furthermore, seven (41%) stars are definite or probable SB1 or possible SB2 (FD 6,20,31,55,62,66,68) while four (24%) are probably single WR stars (all WN7,8) from our sample of 17 stars in the Clouds. Although these fractions compare favourable with those of WR stars in our Galaxy, it is still too early to make confirm conclusions in view of the magnitude selection effect in favour of binary systems with bright OB companions.

REFERENCES

Azzopardi, M., Breysacher, J.: 1979a, Astron. Astrophys. 75, 120

Azzopardi, M., Breysacher, J.: 1979b, preprint

Moffat, A.F.J., Isserstedt, J.: 1980, Astron. Astrophys., in press

Moffat, A.F.J., Seggewiss, W.: 1979, Astron. Astrophys. 77, 128

Tutukov, A.V., Yungelson, L.R.: 1973, Nauchn. Inform. 27, 3, 58, 70, 86

Vanbeveren, D., Conti, P.S.: 1979, preprint

van den Heuvel, E.P.J.: 1976, IAU Symp. No. 73, 35

DISCUSSION FOLLOWING NIEMELA AND MOFFAT-SEGGEWISS

Herczeg: Is MR42 perhaps eclipsing? The high M $\sin^3 i$ values in your table may suggest this.

Moffat: I obtained photometry of MR42 on 2 nights in 1973 (3 hour runs each) with no evidence for eclipses (see Moffatt and Haupt, Astron. Astrophys., 1974).

Sugimoto: For the former two systems the mass of the Wolf-Rayet star is smaller than that of its companion. For the latter two systems both of these masses are comparable to each other. They seem to be in different evolutionary states. Is there any indication reflecting such a difference, for example, in the ratio of element abundances such as He/H?

Niemela: Yes, in fact the envelopes of the late type WN stars show H lines in their spectra, but the early type WN envelopes do not.

Popper: Is it your perception and also that of the previous speaker that the Wolf-Rayet-O star binaries are mostly semi-detached rather than detached?

Moffat: I would say they are quasi-detached. The OB star does not generally fill its Roche lobe, nor does the WR core. However, the WR envelopes often extend beyond the classical Roche lobes especially for short-period systems. In such cases, one must also consider the strong, fast stellar wind for which the companion star only represents a perturbation of the wind particle trajectories amd the concept of a Roche lobe must be modified.

Van Paradijs: Given the large mass-loss rates of WR stars, wouldn't your model for the runaway WR stars predict that these stars are strong X-ray emitters?

Moffat: Yes, except that virtually all X-rays up to energies of $\gtrsim$ 10 KeV will be absorbed in the dense WR envelope (cf calculation for WR runaway candidate HD 197406 by Moffat and Seggewiss, A&A, 1979, in press.)

HD 192641 AND HD 193077: TWO SINGLE, HYDROGEN-RICH WOLF-RAYETS?

Philip Massey*
Department of Astro-Geophysics and Joint Institute for Laboratory Astrophysics, University of Colorado and National Bureau of Standards, Boulder, Colorado 80309

As recently as six years ago it was generally agreed that all Wolf-Rayet (WR) stars were the product of mass loss induced by Roche lobe over-flow (RLOF), and that thus they were all members of binary systems. In particular, the presence of absorption lines in the spectrum of a WR star was taken as a definitive indicator that the star was a binary, as it was well known that emission alone (other than the absorption present in P Cygni profiles) was a WR characteristic. However, Niemela (1973) showed that in the WN 7 binary HD 92740 that the absorption and emission lines move in phase, proving that in at least one case the absorption spectrum originates in the WR star itself. Conti (1976) has meanwhile suggested that WR stars could form by mass loss due to a stellar wind rather than RLOF. Obviously it is well worth examining the belief that all (or even most) WR stars are binaries. The currently popular feeling is that the late WN stars (WN 7, 8, and 9) are the only possible exceptions to the traditional picture, while the members of the other subclasses are all hydrogen-poor and probably members of binaries. I hope to convince you today that this is not the complete story.

For the past two observing seasons I have been obtaining coude spectrograms (10 Å/mm) at KPNO of the northern WR stars with absorption lines but no orbit solutions. The naive motivation for this project was to determine more orbits and hence masses, since the masses of WR stars are poorly determined. The first two stars studied have been HD 192641 (WC 7) and HD 193077 (WN 5). Both show the upper Balmer lines in absorption plus the emission lines appropriate to their subclasses. I obtained ~1-2 plates each night for a week once a month for six months, and have additional material. The plates were all traced on a PDS microphotometer and radial velocities were

*Visiting student, Kitt Peak National Observatory (KPNO), which is operated by AURA, Inc., under contract with the National Science Foundation.

M. J. Plavec, D. M. Popper and R. K. Ulrich (eds.), Close Binary Stars: Observations and Interpretation, 187–191.

produced using an objective criteria to define line-center. The high dispersion material combined with this reduction procedure resulted in very good velocity measurements for even the broad emission lines.

Much to my surprise neither of these stars showed any trace of radial velocity variations. The scatter in the velocities are 20 km s^{-1} for the "best" emission lines, and so any orbital motion must be such that the semi-amplitude, K_{EM}, of the WR star is less than 30 km s^{-1} for periods less than a few years. Bracher (1966 and private communication) reports measurements of plates that she obtained in 1964 and 1966 of HD 193077; her velocities agree with mine.

Low orbital inclinations or extremely long periods cannot be ruled out for HD 192641, but for HD 193077 neither is likely. The absorption lines in HD 193077 are incredibly broad, with v sin *i* = 500 km s^{-1}, greater than that known for any other early-type star. A Fourier analysis of the absorption lines shows that they are broadened by rotation, and not by turbulence or electron scattering. (The strengths of the absorption lines are normal for an O star, however.) Therefore, if these lines originate in an OB companion its rotation axis must be inclined less than 45° to our line of sight or the equatorial rotation velocity would tear the star apart. If the rotation axis of this alleged companion is aligned with the orbital axis, then this constraint on the orbital inclination, combined with the limit on K_{EM}, demands that the period must be greater than ~1000 days if the system consists of a 10 $M_{\odot}$ WR and a 30 $M_{\odot}$ OB star. This is 10 times longer than the longest known WR binary period.

If HD 193077 is single, then the absorption lines must be formed in the WR star. This interpretation is supported by two indirect pieces of evidence. There is hydrogen present in the WR envelope, as inferred by the odd-n *vs.* even-n Pickering emission line fluxes [implying N(H)/N(He) ~ 0.8]. Furthermore, the emission line strengths in HD 193077 are weak compared to some other WN 5 stars. Although small emission line strengths were mentioned by Kuhi (1973) as possible evidence of a companion (which would contribute to the continuum flux), the argument can be made that weak emission lines indicate that the envelope is not very extensive and hence one may actually see down to the absorption-line forming region rather than only part way into the wind.

Vanbeveren and Conti (1979) have argued that *at most* 40% of the galactic WR stars have massive (uncollapsed) companions, where they have accepted the presence of absorption lines as sufficient evidence of duplicity. They suggest that this number possibly should be doubled to account for the WR plus collapsed companion systems, but that 20% must then be single. If more systems like HD 192641 and HD 193077 are found not to have velocity variations then the percentage of actual single WR stars could be substantially higher.

The details on HD 193077 may be found in Ap. J. (in press). The work on HD 192641 is still in progress.

This research has been supported by National Science Foundation grant No. AST78-20135 through the University of Colorado.

REFERENCES

Bracher, K.: 1966, Ph.D. thesis, Indiana University.
Conti, P.S.: 1976, Mem. Soc. Roy. Sci. Liege 9, p. 193.
Kuhi, L.V.: 1968, in IAU Symposium No. 49, Wolf-Rayet and High Temperature Stars, ed. M. K. Bappu and J. Sahade (Boston: Reidel), p. 205.
Niemela, V.S.: 1973, Pub. A.S.P. 85, p. 220.
Vanbeveren, D. and Conti, P.S.: 1979, Astr. Astrophys. (in press).

COMMENTS FOLLOWING MASSEY

Webbink: Geoff Bath isn't here, but I feel confident that if he were, he would express grave doubts that one can attribute the observed spectral features to a rotating hydrostatic stellar surface. Could not the breadth and Fourier structure of the absorption lines be produced in an optically thick outflowing wind?

Massey: Hardly. I would be pretty amazed if you could produce these higher-order Balmer lines (H8-H13) and He I λ3820 in absorption in a stellar wind.

Zuiderwijk: From the upper limit of 30 km/s you can hardly conclude that the star is not a binary; there is still room for a lower mass companion like a neutron star. If this should be the case one could expect some type of periodic variations in several spectral lines. Did you look in your spectra for such variations?

Massey: Yes, there are no profile changes observed in any of the emission lines I studied. The limit that the emission-line K must be less than 30 km s^{-1} places very strong constraints on sin *i* and P for a canonical WR binary ($30M_\odot + 10M_\odot$). While it is true that a low mass companion (or a hundred low mass companions) cannot be excluded, the only reason for believing the star was a binary was the presence of absorption, which cannot be attributed to a low mass companion.

Mayo: *Re* Webbink's comment that these broad (500 km/sec) higher Balmer absorption lines may arise not in stellar photosphere but in optically thick stellar wind. Another non-stellar interpretation may be possible--namely that these lines arise in an optically thick accretion disc. Such a disc can easily give rise to lines rotationally broadened by 500 km/sec, (whereas you say 500 km/sec rotational velocities are not observed in "normal" O-stars). Also in the discs in cataclysmic variables we see higher order Balmer and He I lines in absorption. However the disc would have to be pretty luminous to show up against such a luminous star as the W-R.

Massey: The line strengths of H9, H10, H11 and He I λ3820 are normal for an O star. I am extremely dubious you could get these lines that strong in anything other than a stellar photosphere. That would have to be some disk!

Seggewiss: I like to add to the paper of Massey that there are at least two Wolf-Rayet binaries in which the central absorption lines originate in the WR atmosphere, because the radial velocities of the absorptions vary in phase with the WR emission lines: HD 92740 and HD 197406, both of spectral type WN 7 (Moffat and Seggewiss, Astron. & Astrophys. 70, 69, 1978 and in press).

Massey: Yes, as I mentioned in my talk, Virpi Niemela (PASP, 85, 220, 1973) showed that in HD 92740 the emission and absorption move in phase, and this has now been confirmed by Conti, Niemela, and Walborn (Ap. J., 228, 206, 1979) and your own work with Moffat. I was unaware that HD 197406 has absorption lines: Bracher did not find any in her comprehensive study (Univ. of Indiana thesis, 1966; PASP, 1979, in press), so I look forward to your paper with considerable interest.

We know of several WN 7-8 systems which have absorption and appear to be single--but I hope I've convinced you that this is not the entire story. There are members of the other subclasses which have absorption lines present but appear to be single.

Collins: I would like to second Ron Webbink's concern about interpreting the hydrogen line profiles as arising from uniformly rotating hydrostatic atmospheres since those used to generate the Fourier transform you used for reference are totally inappropriate for such large velocities.

Massey: Since there are no hydrostatic plane parallel non-LTE calculations available (rotating or non-rotating) for the upper Balmer lines on HE I λ3820, I used the observed lines of the slowly rotating O-star 10 Lac convolved with 500 km s^{-1} rotation and 500 km s^{-1} turbulence for comparison. The value 500 km s^{-1} for v sin i is derived from the locations of the cusps in the transform of the observed profiles, and are not "generated" from atmosphere calculations. While the Fourier technique is not strictly applicable since there is no intrinsic profile when there is an expanding, rotating atmosphere (see Mihalas, Ap. J., 1979, submitted) the method certainly is sufficient to show that the dominant broadening mechanism is rotation. There for large values of v sin i, what to use for the "intrinsic profile" matters very little, since broadening swamps the "intrinsic" shape. There is a 5-10% uncertainty in the derived value of v sin i due to the question of how to treat limb darkening (I have used $\varepsilon = 0.6$ for the continuum), and I have ignored any gravity darkening effects. I expect that the value of v sin *i* = 500 km/s is in error by 10-20 %, but the point is that the absorption broadening is due to rotation, not electron scattering or turbulence.

Kuzma: Was the v sin *i* $\approx$ 500 km sec^{-1} obtained from only the hydrogen lines, or were similar values obtained from the other absorption lines?

Massey: The value was derived from H11 and He I λ3820.

THE INTERPRETATION OF LIGHT CURVES λλ2460 : 35000Å OF THE ECLIPSING WOLF-RAYET BINARY V444 CYG

A.M. Cherepashchuck, Y.A. Eaton and Kh.F. Khaliullin
Sternberg Astronomical Institute
Moscow University
Moscow, U.S.S.R.

ABSTRACT

The structure of the extended atmosphere and the physical characteristics of the WN5 star V444 Cyg have been determined from the analysis of light curves of this system in the range λλ2460 - 35000. The radius of the core of the WN5 star at $\tau = 1$ is $r_o = 2.9R_\odot$; the temperature at the surface of the core is equal to 90000°K, the electron density at its surface is $0.9 \cdot 10^{13}$ cm^{-3}. The bolometric luminosity of the WN5 star $L_b = 2.10^{39}$ erg/s is close to the critical Eddington luminosity $2.6 \cdot 10^{39}$ erg/s. The electron temperature decreases monotonically with height through the extended atmosphere from the value T_e = 80000° - 90000°K at $r = 2.9R_\odot$ to the value T_e = 20000° - 30000°K at $r = 8R_\odot$. The outflow velocity at the surface of the core ($r_o = 2.9R_\odot, \tau = 1$) is v_o = 400 km/s; the outflow in the region $r \geq 12R_\odot$ becomes stationary with the constant velocity v = 1000 - 1200 km/s. The total interstellar absorption for the system V444 Cyg is $A_V = 2^m.66$. The results obtained confirm strictly the model of Wolf-Rayet phenomenon proposed by Beals (1944). Chromospheric-coronal effects in the extended atmosphere are absent until $r = 20R_\odot$. Characteristics of the WR star are consistent with the predictions of the evolution theory for massive close binary systems with mass exchange. The WR star is in a late evolutionary stage after the main sequence stage, being a helium remnant that has been formed as a result of the mass exchange.

M. J. Plavec, D. M. Popper and R. K. Ulrich (eds.), Close Binary Stars: Observations and Interpretation, 193.

MASS FLOW ANALYSIS OF THE ULTRAVIOLET SPECTRUM OF UW CANIS MAJORIS

H. Drechsel
Remeis-Sternwarte Bamberg
Astronomical Institute
University Erlangen-Nuremberg, F.R.G.

Y. Kondo*
Department of Astronomy
University of Pennsylvania
Philadelphia, PA, U.S.A.

G. E. McCluskey, Jr.
Department of Mathematics
Lehigh University
Bethlehem, PA, U.S.A.

J. Rahe**
Laboratory for Astronomy and Solar Physics
NASA-Goddard Space Flight Center
Greenbelt, MD, U.S.A.

ABSTRACT

The ultraviolet spectrum of the close binary UW CMa (= HD 57060; O7f Ia + O-B), obtained with the COPERNICUS (OAO-3) and IUE satellites, is analyzed in terms of a comparison between observed P Cygni resonance line profiles and theoretical calculations in an expanding model atmosphere.

The line fit suggests an electron temperature of the circum-stellar gas of $T_e = 2.2 \times 10^5$ K, and a mass loss rate from the O7f component of $dM/dt = 2.3 \times 10^{-6} M_\odot \cdot yr^{-1}$.

* Permanent address: Goddard Space Flight Center, Greenbelt, MD, U. S. A.

** NAS-NRC Resident Research Associate, on leave from University Erlangen-Nuremberg.

M. J. Plavec, D. M. Popper and R. K. Ulrich (eds.), Close Binary Stars: Observations and Interpretation, 195–198.

The far ultraviolet spectrum of UW CMa was scanned with the U2 spectrometer of the Princeton University Telescope aboard OAO-3 in the wavelength region from 1010 to 1510 Å on 1975 January 29 through 31, with a resolution corresponding to 0.2 Å (McCluskey and Kondo, 1976). The IUE observations were obtained on 1978 June 4 between 1200 and 3000 Å with the short and long wavelength cameras, at resolutions of 0.1 and 0.2 Å, respectively.

The dominant features are P Cygni lines of S IV, P V, Fe III, C III, C IV, and N V. A wealth of interstellar lines, and a few photospheric absorption lines of N III and He II originating from the more luminous O7f binary component are also present. Tables 1 and 2 give the identified stellar lines, laboratory wavelengths, and line strengths in terms of peak height for emission (E) or central depth for absorption (A) components, normalized to the local stellar continuum.

Compared with the large terminal velocities of about -1000 km s^{-1}, which by far exceeds the escape velocity of the system, the observed variations of the P Cygni line center wavelengths with time are an order of magnitude lower, with a theoretical maximum value of 220 km s^{-1}, defined by the projected orbital velocity component of the primary. No definite phase correlation of the line shifts with the 4.4 day binary period of UW CMa was found. Application of accretion theory suggests that only a negligible amount of mass lost by the O7f supergiant (~2%) is transferred to the secondary, and thus we are dealing with a continuous stellar wind from UW CMa which is nearly completely lost by the binary system.

The resonance doublets of S IV (1063, 1073 Å) and of Si IV (1394, 1403 Å), the doublet component of P V (1128 Å), and the unresolved multiplet of C III (1176 Å) are used for an investigation of the mass loss process. The observed P Cygni line profiles are compared with theoretically predicted profiles formed by isotropic and coherent scattering in a spherically symmetric expanding circumstellar envelope applying Lucy's (1971) solution of the equation of transfer in a frame comoving with the gas. For the calculation of grids of P Cygni profiles, analytical parametrized expressions for the velocity law and the variation of the ionization fractions with distance from the stellar surface are used for the computation of the optical depth of the envelope. The mass flow is treated in a fully trans-sonic way, i. e. the thermal random motions of the scattering ions are neglected compared to the mass flow velocity (Sobolev approximation). Grids of computed line profiles and description of the free parameters can be found elsewhere (Drechsel et al., 1979).

Regrettably, there is no space for an additional illustration. For comparison of the theoretically computed profile and the actual stellar spectrum observed with Copernicus, the reader is referred to the article by Drechsel et al. (1979).

Table 1: COPERNICUS (OAO-3) UV SPECTRUM OF UW CMa

Ion	λ_{lab} (Å)	I_o/I_{con} *)	Remarks
N III	1006.015	~0.5 A	Photospheric line
S IV	1062.672	0.50 A 1.15 E	P Cygni profile
S IV	1073.045	0.58 A 1.42 E	P Cygni profile
He II	1084.942	0.66 A	Photopsheric line
P V	1117.979	0.45 A 1.24 E	P Cygni profile
Fe III	1122.5	0.80 A 1.22 E	P Cygni profile
P V	1128.006	0.78 A 1.21 E	P Cygni profile
C III	1175.66+)	0.52 A 1.23 E	P Cygni profile
N III	1183.030	0.68 A	Photospheric line
N III	1184.544	0.68 A	Photospheric line
N V	1238.821	~0.0 A	P Cygni profile
N V	1242.804	~0.0 A 1.80 E	P Cygni profile
Si IV	1393.755	~0.2 A ~1.7 E	P Cygni profile with uncertain background
Si IV	1402.770	~0.1 A ~1.8 E	P Cygni profile with uncertain background

Employing electron temperature dependent theoretical ionization fractions, the line fit procedure for the C III, S IV, Si IV, and P V features yields the following values for the mass loss rate dM/dt of UW CMa, and for the electron temperature T_e of the circumstellar gas:

$$dM/dt = 2.3\ (\pm 0.8) \times 10^{-6}\ M_\odot \cdot yr^{-1} \text{ and}$$

$$T_e = 2.2\ (\pm 0.4) \times 10^{5}\ K.$$

The electron temperature $T_e = 2.2 \times 10^5$ K of the envelope is an order of magnitude higher than the stellar effective temperature, and consistent with the observation of nearly saturated P Cygni lines of N V (1238, 1242 Å) which are produced by collisional ionization. This result is in accordance with the warm coronal model first postulated by Lucy and Solomon (1970) who also suggested a non-radiative heating mechanism driving the expanding atmospheres of hot luminous stars.

The deduced mass loss rate $dM/dt = 2.3 \times 10^{-6}\ M_\odot \cdot yr^{-1}$ can be compared with a rough estimate of McCluskey, Kondo, and Morton (1975), who found a value of about $3 \times 10^{-6}\ M_\odot \cdot yr^{-1}$ from an investigation

Table 2: IUE UV SPECTRUM OF UW CMa

Ion	λ_{lab} (Å)	I_o/I_{con} *)	Remarks
C III	1175.66+)	0.17 A	P Cygni profile
		1.16 E	
N III	1183.030	0.50 A	Photospheric line
N III	1184.544	0.69 A	Photospheric line
Si III	1206.533	0.43 A	Uncertain feature
N V	1238.821	~0.0 A	P Cygni profile, blended with emission feature
N V	1242.804	2.20 E	P Cygni profile, blended with absorption feature
Si IV	1393.755	~0.0 A	P Cygni profile
		1.33 E	
Si IV	1402.770	0.06 A	P Cygni profile
		1.69 E	
C IV	1548.188	~0.0 A	P Cygni profile, weak emission
C IV	1550.762	1.25 E	Uncertain feature
N IV	1718.551	0.48 A	Probably no emission component
N III	1747.848	0.80 A	Photospheric line
N III	1751.437	0.80 A	Photospheric line
N III	1885.136	0.80 A	Photospheric line
Mg II	2795.528	A	Mainly interstellar absorption; slightly asymmetric line profile
Mg II	2802.693	A	

*) A = absorption; E = emission;

+) mean wavelength of unresolved multiplet.

of UV absorption line strengths. Clearly the evolution of Of-type stars will be influenced by mass loss rates of the order of $5 \times 10^{-6}\ M_\odot \cdot yr^{-1}$; about 1/4 up to 1/3 of the initial ZAMS mass is lost during the main sequence hydrogen burning phase.

References

Drechsel, H., Kondo, Y., McCluskey, G.E., Jr., and Rahe, J.: 1979, Astron. Astrophys., in press.

Lucy, L.B.: 1979, Ap. J. 163, 95.

Lucy, L. B., and Solomon, P.M.: 1970, Ap. J. 159, 879.

McCluskey, G.E., Jr., and Kondo, Y.: 1976, Ap. J. 208, 760.

McCluskey, G.E., Jr., Kondo, Y., and Morton, D.C.: 1975, Ap. J. 201, 607.

THE Be STAR ϕ PERSEI

R. Poeckert,
Dominion Astrophysical Observatory,
Herzberg Institute of Astrophysics, Victoria, B.C.

ABSTRACT

He II (λ 4686) emission has been detected in the spectrum of the binary Be star ϕ Persei. The He II emission has velocity variations which are 180° out of phase with the velocity variations of the primary. The primary velocity curve is determined from the broad He I and H lines, and the Fe II emission lines. The amplitude of the primary velocity curve (K_1) is 16 ± 2 km s^{-1}. The amplitude of the He II velocity curve is 121 ± 9 km s^{-1}.

If one assumes that the He II emission arises from the secondary, minimum masses for the primary and secondary are 29 ± 14 $M_\odot$ and 3.8 ± 1.7 $M_\odot$, respectively. We suggest that the secondary is a hot compact star, either the remaining helium core of a once more massive star, or a neutron star. The primary is too cool, B1 IIIe, to have any appreciable He II emission.

DISCUSSION FOLLOWING POECKERT

Guinan: Have you found any evidence of short term or long term variations in the H emission line strengths for ϕ Persei?

Shore: There is another possibility for forming He II 4686 emission in close binaries besides the ones you've listed--collision of stellar winds. The region between the components is heated in a lens-shaped region to temperatures which are deduced from thermalizing the shock $T > 10^6$K and He II emission could occur from recombination. The problem, of course, is that you find, as you've pointed out, a double peak and for this system the only way of doing this is with self absorption which should be unlikely--this lozenge is thin. It would be interesting though to look at other systems for He II λ4686.

M. J. Plavec, D. M. Popper and R. K. Ulrich (eds.), Close Binary Stars: Observations and Interpretation, 199.

V. ALGOL AND ALGOLS

Spectroscopic Orbits of Algols--a Progress Report

Daniel M. Popper
University of California, Los Angeles

Preliminary values of masses and radii are presented for twelve Algol systems with cooler components of class F or later. The results are new for nine of the systems and are based on measures of the D lines for the faint cooler components. For most of the systems with periods longer than six days the orbits of the hotter components are uncertain because of absorption by non-photospheric matter in the systems. The masses of the hotter components are approximately normal for their spectral types, while the masses of the Roche-lobe filling components range from 0.2 to 1.4 solar masses.

This discussion is restricted to semi-detached binaries of the "typical" Algol class in which the secondaries are of type F or later and for which radial velocities of the secondaries have been obtained outside eclipses. Velocities have been published for both components of three of these systems: AW Peg (Hilton and McNamara 1961), AS Eri (Popper 1973), and Algol itself (Tomkin and Lambert 1978). Observations at the Lick Observatory making use of an image tube giving a reciprocal dispersion at the D lines of 11 Å mm^{-1} are in various stages of completion for a number of additional systems in which the D lines and, in a few cases, some additional lines of the secondaries can be measured outside eclipses on spectrograms of high contrast. The Texas reticon work on Algol illustrates the potential of modern detectors, at least on bright stars.

To obtain masses and radii, the orbits of the primaries must be obtained as well. For a majority of the relatively bright systems under discussion velocities of the parimaries have been published by others. For the systems with periods less than about 6 days (Algol, TW And, RY Aqr, TW Dra, AS Eri, RS Vul) the orbits appear to be without complications. On the other hand, for most of the systems with longer periods (RY Gem, TT Hya, AW Peg, RY Per, XY Pup), the velocities appear, to a greater or lesser degree, to depart from the sinusoidal variations shown by the secondaries. S Cnc, P=$9^{d}.5$, may be an exception. As in the well-known cases of U Cep and U Sge, the distortions of the velocity

M. J. Plavec, D. M. Popper and R. K. Ulrich (eds.), Close Binary Stars: Observations and Interpretation, 203–205.

curves may be a result of absorption in gas streams in the systems. The distortions are not, however, necessarily to be attributed to alterations of the line profiles. The velocities of TT Hya, for example, are based on internally consistent measures of a considerable number of symmetrical lines with not particularly broad profiles. These lines vary in intensity around the orbit. Additional evidence of extra-photospheric matter is to be found in the strong double Hα emission in the spectra of all the longer period systems except S Cnc and, in some cases, the weakness of Mg II 4481. A study of all these effects may provide information about the physical properties of the extra-photospheric matter as well as of its distribution and motion.

The table gives provisional values of the masses and radii of the components of Algol systems. The uncertainties in the values are consequences of inadequate observational material obtained to date and, in most of the longer period systems, of distortion of the velocity curves of the primaries by non-stellar matter.

			Hotter		Cooler	
	P	Sp	M($H_\odot$)	R($R_\odot$)	M($M_\odot$)	R($R_\odot$)
RY Aqr	$2\overset{d}{.}0$	A5	1.3	1.5:	0.3:	1.7:
AS Eri	2.7	A3	1.9	1.8	0.2	2.2
TW Dra	2.8	A3	1.7	2.4	0.8	3.4
Algol	2.9	B8	3.7	3.1	0.8	3.2
TW And	4.1	A8	1.8	2.2	0.4	3.4
RS Vul	4.5	B5	4.5	4.1	1.4	5.6
RY Per	6.9	B4	5:	3.4	0.8::	7
TT Hya	7.0	A2	2.6:	2:	0.7::	5:
RY Gem	9.3	A2	2.6	3.3	0.6:	6
S Cnc	9.5	A0	2.4	2.1	0.2::	5
AW Peg	10.6	A5	2.0:	---	0.3::	---
XY Pup	13.8	A2	2.3:	---	0.3::	---

The masses and radii of the primary components lie within the range for main sequence stars of the same spectral types with the exception of TW And, which appears cooler than its mass and radius would predict. The secondary masses show no clear correlation with other properties. Masses as low as 0.2 or 0.3 do not appear to be uncommon in Algol systems. These results must play a role in improving our understanding of the evolutionary histories of Algol systems. We heard at considerable length earlier in the symposium of difficulties in understanding their histories.

REFERENCES

Hilton, W.B. and McNamara, D.H.: 1961, Astrophys. J. 134, pp. 839-849.

Popper, D.M.: 1973, Astrophys. J. 185, pp. 265-275.

Tomkin, J. and Lambert, D.L.: 1978, Astrophys. J. Lett. 222, pp. L119-L122.

DISCUSSION FOLLOWING POPPER

Whelan: Have you found any problems with image tube systems when measuring radial velocities?

Popper: No matter what spectrographic configuration is used, standard-velocity stars should be observed frequently; with the Varo image tube at the Lick 120-inch coude', I find no greater systematic or accidental effects in the velocities than when using direct photographic recording.

Wilson: Lately the situation with regard to determing mass ratios from light curves has improved considerably for semi-detached systems, where the size of the secondary tells you the size of the Roche lobe. Although this method does not always give extremely accurate mass ratios, there are also pitfalls in the double-line velocity approach (as you just pointed out) so it should probably be used more than in the past.

Popper: Reliable masses from any source should not be overlooked. As long as one is clearly aware of the potential pitfalls and does not underestimate the uncertainty of the results, a mass ratio obtained through light-curve analysis could be useful in favorable circumstances.

MASS LOSS AND MASS TRANSFER IN ALGOLS

G. Giuricin, F. Mardirossian, and M. Mezzetti
Astronomical Observatory of Trieste, Trieste, Italy

ABSTRACT

We have examined the observational data of 60 Algols in order to check some theoretical views on mass loss and mass transfer in Algols. Not negligible contradictions between some theoretical non-conservative attempts and the observational scenario have been emphasized.

INTRODUCTION

The evolutionary status of Algols has been qualitatively explained in terms of mass exchange, but the agreement between theory and observations is still poor, since the scenario is complicated by possible loss of mass and angular momentum from the binary systems. In this study we have touched on some general aspects of mass loss and mass transfer in Algols by examining the observational data of 60 Algol-type binaries (known as sd or sd-d systems), whose lightcurves have been analyzed by means of modern techniques of lightcurve synthesis. More specifically, from the total mass, the orbital period, and the mass ratio of these Algols, we tried to estimate the original period and total mass of each binary, focusing our attention on the conservative case (i.e., total mass M and total angular momentum J constant) and on the two very different hypotheses of mass and angular momentum loss proposed by Plavec et al. (1973) (PUP) and by Drobyshevski and Reznikov (1974) (DR). The PUP approach takes f (=fraction of the mass exchanged, which is lost from the system) as a free parameter and is characterized by a relatively small ratio (g) between the specific angular momentum of the lost matter and that of the original system; g is defined as follows:

M. J. Plavec, D. M. Popper and R. K. Ulrich (eds.), Close Binary Stars: Observations and Interpretation, 207–210.

$$g=(1-J/J_o)(1-M/M_o) \tag{1}$$

where J_o and M_o are the initial values of J and M. The DR approach is characterized by intermediate values of f ($0.4 \leq f \leq 0.6$) and relatively high values of g ($g \cong 3$).

ORIGINAL PERIODS AND MASSES

Table 1 lists the parameters used in our computations for each binary, i.e., the period P (in days), the mass ratio $q=M_c/M_h$ (M_h=mass of the hotter component, M_c=mass of the cooler component) and the masses M_h and M_c (in solar units) of each component, as found in the literature. The data of the double-lined spectrum binaries (starred in Table 1) have been taken from Batten et al. (1978), except for δ Lib (see Tomkin 1978) and β Per (see Tomkin and Lambert 1978). For the remaining stars, photometrically determined mass ratios along with estimates of the masses have been taken from Cester et al. (1979), Chambliss (1976), Mancuso et al. (1977), Mardirossian et al. (1979a), Mardirossian et al. (1979b), Mezzetti et al. (1979a), and Mezzetti et al. (1979b). In some cases we have evaluated the errors on M_h, M_c, and q, by inspecting the original published data.

In the conservative case the original period P_o and total mass M_o follow from the formulae:

$$M_o=M \tag{2}$$

$$P_o=P(q/q_o)^3(1+q_o)^6/(1+q)^6 \tag{3}$$

where $q_o \geq 1$ is the original mass ratio.

Within the framework of the PUP approach it is easy to work out the initial values for total mass and period:

$$M_o=M(1+q_o)(1+q(1-f))/((1+q)(1+q_o(1-f))) \tag{4}$$

$$P_o=P(q/q_o)^{3-3f}((1+q)/(1+q_o))^2(1+q_o(1-f))^{8-3f}(1+q(1-f))^{3f-8} \tag{5}$$

where f is a free parameter.

For estimating P_o and M_o, DR give the following analytical formulae:

$$M_o=M((1+1/q_o)(1+1/q))((1+1.59/q)(1+1.59/q_o))^{1.02} \tag{6}$$

TABLE 1

N	Star		P	q	M_h	M_c
1	TW	And	4.1228	0.19±0.01	2.4±0.9	0.5±0.2
2	XZ	And	1.3573	0.47±0.03	3.2	1.5
3	RW	Ara	4.3674	0.21±0.01	3.2	0.7
4	IM	Aur	1.2473	0.26±0.01	4.4±0.7	1.1±0.2
5	SU	Boo	1.5612	0.11±0.03	2.7	0.3
6	Y	Cam	3.3055	0.24±0.02	1.7±1.0	0.4±0.2
7	RZ	Cnc *	2.1643	0.18±0.03	3.1±0.2	0.6±0.1
8	R	CMa	1.1359	0.17±0.01	1.1±0.8	0.2±0.1
9	RZ	Cas	1.1953	0.35	1.75	0.61
10	TV	Cas	1.8126	0.41±0.07	3.8±1.5	1.6±0.6
11	TW	Cas	1.4283	0.41±0.02	2.9±0.3	1.2±0.1
12	U	Cep *	2.4930	0.67±0.07	4.2±0.6	2.8±0.5
13	XX	Cep	2.3373	0.17±0.07	1.4±1.4	0.2±0.2
14	U	CrB *	3.4522	0.38	6.7	2.6
15	RW	CrB	0.7264	0.22±0.01	2.0±0.6	0.4±0.1
16	SW	Cyg	4.5730	0.21±0.03	2.5±1.6	0.5±0.3
17	VW	Cyg	8.4303	0.28±0.04	2.1±1.5	0.6±0.5
18	WW	Cyg	3.3178	0.31±0.05	6.2±2.6	1.9±0.8
19	ZZ	Cyg	0.6286	0.63±0.03	1.8	1.1
20	MR	Cyg *	1.6770	0.56±0.06	4.5±1.1	2.5±0.4
21	V548	Cyg	1.8053	0.29±0.01	3.9±0.5	1.1±0.1
22	W	Del	4.8060	0.18±0.02	3.2±1.3	0.6±0.2
23	TW	Dra	2.8069	0.39±0.01	2.2±0.2	0.9±0.1
24	AI	Dra	1.1988	0.43±0.02	2.7±1.1	1.1±0.5
25	S	Equ	3.4361	0.12±0.01	3.1±1.2	0.4±0.1
26	AS	Eri *	2.6642	0.11±0.01	1.92±0.04	0.208±0.009
27	RW	Gem	2.8655	0.29±0.01	5.6±1.1	1.6±0.3
28	RX	Gem	12.2085	0.20±0.05	3.4±2.0	0.7±0.4
29	AL	Gem	1.3913	0.10±0.01	1.3	0.1
30	UX	Her	1.5489	0.21±0.01	2.7±0.6	0.6±0.1
31	AD	Her	9.7666	0.33±0.01	2.9±0.3	0.9±0.1
32	V338	Her	1.3057	0.16±0.01	1.8	0.3
33	u	Her *	2.0510	0.36±0.01	7.7±0.2	2.8±0.1
34	T	LMi	3.0199	0.10±0.03	8.1±4.8	0.8±0.4
35	δ	Lib *	2.3274	0.35±0.01	4.7±0.2	1.7±0.2
36	RW	Mon	1.9061	0.33±0.10	4.0±3.0	1.3±0.9
37	TU	Mon *	5.0490	0.21±0.06	12.7±2.3	2.7±0.6
38	AR	Mon *	21.208	0.30±0.01	2.69±0.10	0.80±0.05
39	RV	Oph	3.6871	0.10±0.01	3.2	0.3
40	AW	Peg *	10.6225	0.16±0.03	2.0±0.1	0.32±0.05
41	β	Per *	2.8673	0.22±0.01	3.7±0.3	0.81±0.05
42	RT	Per	0.8494	0.24	1.71	0.42
43	ST	Per	2.6484	0.16±0.03	3.3±1.9	0.5±0.3
44	IZ	Per	3.6877	0.31±0.10	3.2±2.5	1.0±0.7
45	Y	Psc	3.7658	0.25±0.01	2.8±0.6	0.7±0.2
46	XZ	Pup	2.1924	0.41±0.01	3.0±0.3	1.2±0.1
47	U	Sge *	3.3806	0.33±0.02	5.8±1.0	1.9±0.3
48	XZ	Sgr *	3.2755	0.14±0.02	1.9±0.8	0.3±0.1
49	V356	Sgr *	8.8961	0.38±0.05	12.2±1.0	4.7±0.4
50	v505	Sgr	1.1829	0.49±0.01	2.7±0.3	1.3±0.2
51	μ^1	Sco *	1.4463	0.66±0.01	13.6±0.4	9.0±0.2
52	λ	Tau *	3.9530	0.27±0.02	6.8±0.3	1.8±0.1
53	RW	Tau	2.7688	0.19±0.01	5.5±0.9	1.0±0.2
54	X	Tri	0.9715	0.51±0.06	2.3±0.7	1.2±0.3
55	TX	UMa *	3.0632	0.30±0.03	3.6±1.1	1.1±0.3
56	VV	UMa	0.6874	0.21±0.01	2.4±0.5	0.5±0.1
57	W	UMi	1.7012	0.48±0.01	2.4±0.3	1.2±0.2
58	Z	Vul *	2.4549	0.43±0.05	5.4±0.4	2.3±0.2
59	RS	Vul *	4.4777	0.31±0.03	4.4±0.7	1.4±0.2
60	BE	Vul	1.5520	0.40±0.03	3.2	1.3

$$P_o = P/\exp(4.92(q^{1/2}-q_o^{1/2})+3.52(q^{-1/2}-q_o^{-1/2})-$$
$$-0.115(q^{-3/2}-q_o^{-3/2})+0.0028(q^{-5/2}-q_o^{-5/2}) \quad (7)$$

DISCUSSION

Our results may be profitably discussed if one bears in mind Popov's (1970) analysis. He found $1/3 \leq f \leq 2/3$ for the less massive binaries, $f \cong 2/3$, $g \cong 3$ for systems with $M \leq 6M_\odot$, $f=0.9$, $g \cong 1.5$ for systems with $6M_\odot \leq M \leq 12M_\odot$. The following conclusions can be drawn: (i) The conservative hypothesis can not account for the evolutionary scenario of all Algols; (ii) The PUP approach gives rise to serious problems for $M \leq 6M_\odot$ and too small g values with respect to Popov's, while for $M \geq 6M_\odot$ it is consistent with Popov's analysis; (iii) The DR approach can bring the $M \leq 6M_\odot$ systems into a consistent picture, but for $M \geq 6M_\odot$ the DR g value is greater than that given by Popov.

As regards the type of mass transfer, case A and B seem to predominate in systems with $M \geq 10M_\odot$ and $M \leq 6M_\odot$ respectively; both cases or case AB seem to be possible for $6M_\odot \leq M \leq 10M_\odot$.

REFERENCES

Batten,A.H.,Fletcher,J.M.,Mann,P.J.:1978,Publ.Dom.Astrophys.Obs.15,No.5.
Cester,B.,Giuricin,G.,Mardirossian,F.,Mezzetti,M.,Milano,L.:1979,Astron. Astrophys.Suppl.36, p.273.
Chambliss,C.R.:1976,Publ.Astron.Soc.Pacific 88, p.22.
Drobyshevski,E.M. and Reznikov,B.I.:1974,Acta Astron.24, p.29.
Mancuso,S.,Milano,L.,Russo,G.:1977,Astrophys.Space Sci.47, p.277.
Mardirossian,F.,Mezzetti,M.,Cester,B.,Giuricin,G.,Russo,G.:1979a,Astron. Astrophys.Suppl., in press.
Mardirossian,F.,Mezzetti,M.,Predolin,F.,Giuricin,G.:1979b,to be published.
Mezzetti,M.,Cester,B.,Giuricin,G.,Mardirossian,F.:1979a,Astron.Astrophys. Suppl., in press.
Mezzetti,M.,Cester,B.,Giuricin,G.,Mardirossian,F.:1979b,Astron.Astrophys. Suppl., in press.
Plavec,M.,Ulrich,R.K.,Polidan,R.S.:1973,Publ.Astron.Soc.Pacific 85,p.769.
Popov,M.V.:1970,Perem.Zvezdy 17, p.412.
Tomkin,J.:1978,Astrophys.J.221, p.608.
Tomkin,J. and Lambert,D.L.:1978,Astrophys.J.Letters 222, p.L119.

ACCRETION AND CHEMICAL STRUCTURE OF ALGOLS

Wim Packet
Astrophysical Institute, Vriji Universiteit Brussel
Brussels, Belgium

ABSTRACT

Evolutionary calculations show that during the last, slow stage of mass exchange in close binary systems, matter with gradually increasing molecular weight is transferred. If no mixing occurs the accreting star moves towards a position in the HR diagram hotter than the main-sequence band, in clear contradiction with observational evidence in Algols. So some efficient mixing mechanism must be present. Several mechanisms are examined, some of which can be definitely ruled out. A closer examination of some others is necessary.

1. INTRODUCTION

The purpose of this paper is to draw your attention to an important, but up to now neglected problem concerning the evolution of close binary systems. Consider a close binary undergoing mass exchange. Evolutionary calculations following the behavior of the mass-losing component tells us that it will end up with an atmospheric hydrogen abundance lower than the original zero-age abundance, due to the molecular weight gradient built up by central hydrogen burning. This phase with gradually decreasing surface H-abundance corresponds to the last, slow phase of mass transfer, during which the system resembles Algol-type binaries.

Whenever part of the lost matter is captured by the companion, this star should, in the absence of any mixing in the outer layers, reflect the surface hydrogen abundance of the mass-loser. In this way an inversed μ-gradient could be built up in the outer layers of the mass-gaining star. Direct measurements of the surface H-abundance are difficult, but a hydrogen-poor envelope of some extent can manifest itself indirectly by the following mechanism: lowering the hydrogen abundance causes a decrease of the opacity of the stellar material. This makes the envelope more compact, so that the surface temperature increases. In the HR diagram, the star thus shifts towards a position

M. J. Plavec, D. M. Popper and R. K. Ulrich (eds.), Close Binary Stars: Observations and Interpretation, 211–215.

to the left of the main sequence, provided the hydrogen-poor layer is deep enough. This of course clearly contradicts observation. It is well known that the mass-accreting star in Algols looks like a normal main-sequence star. This is confirmed by our Fig. 1, in which a number of Algol detached components are displayed. With one exception, all stars lie within the main-sequence band and their masses and luminosities correspond to theoretically computed main-sequence models.

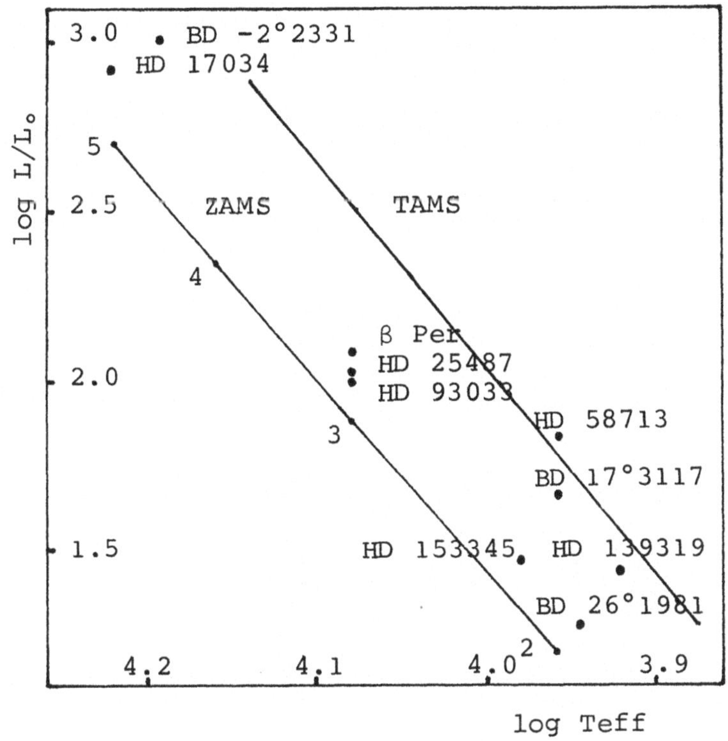

Fig. 1. Observational HR diagram for mass-gaining components in Algol systems. The systems are taken from the catalogue discussed by De Grève and Vanbeveren (1980). Only systems with good determinations of masses, luminosities and effective temperatures and with mass-loser masses between 0.6 and 1.0 $M_\odot$ (i.e. in the range of our computations) were retained.

2. COMPUTATIONS

In this section it will be shown that, in the absence of mixing, accreting material with increasing molecular weight μ does indeed shift towards too high effective temperatures to be compatible with observation.

For the mass-losing star, initial masses of $5M_\odot$ and $7M_\odot$ were taken. As 80% or more of all close binaries evolve through case B mass transfer

(De Grève and Vanbeveren 1980) only this case was considered. The final masses for both stars are 0.6M and 1.0M respectively. Unevolved binaries have a bimodal q-distribution, with peaks around 0.3 and 1.0 (Trimble 1974, 1978; De Grève and Vanbeveren 1979). From Table 1 it is obvious that initial q-values in the peak around 1.0 are excluded. We chose q_1 = 0.2 and 0.3 as representative values. Furthermore it is now established that close binary evolution in most cases will be non-conservative. De Grève and Vanbeveren (1979) find that in order to match observations more than 50% of the matter lost by the contact component must leave the system, with the best correspondence at $\beta = 0.2$, where β is the averaged fraction of the lost matter accreted by the other star. Additional support for this can be found in Table 1. In our calculations we used values $\beta = 0.1$ and 0.2. Actually we allowed β to vary during the mass transfer stage, with values 0.1, 0.2 and 0.5 during the slow phase of mass exchange. In total, 24 combinations of all parameters are possible. For each combination the evolution was followed from the zero-age main sequence through mass transfer up to core hydrogen exhaustion in the mass-gaining component

Table 1. Masses, maximum initial mass ratio and fraction of accreted material for the observed systems in Fig. 1

Star	M_1	M_2	q_1, max	$\beta(q_1=.2)$	$\beta(q_1=.3)$
HD17034	5.27	0.83	.86	.76	.64
HD25487	3.01	0.82	.49	.34	.22
HD58713	2.69	0.58	.55	.40	.28
HD93033	3.13	0.90	.48	.33	.21
HD139319	2.08	0.87	.33	.15	.03
HD153345	2.18	1.03	.30	.12	.01
BD-2°2331	2.30	1.00	.33	.15	.03
BD17°3117	2.44	0.90	.38	.20	.09
BD26°1981	1.91	0.63	.37	.19	.08
β Per	3.15	0.74	.55	.40	.29

From all these tracks the following general conclusions can be drawn:

As soon as transfer of matter with higher μ starts, the star moves to the left in the HR diagram and follows a track situated to the left of the main-sequence. Effective temperatures are too hot by 1500°K to 3000°K, depending on the amount of mass allowed to be accreted during the slow phase. After termination of mass transfer the stars evolve further, but 60% to 100% of their main-sequence lifetime is spent at the left side of the ZAMS. The luminosity is hardly affected by the anomalous composition of the envelope. Changing the values of q_1 or β (towards higher values) will not affect this picture very much, but will cause the masses of the gainers to become too large compared to observation.

This picture of Algol-evolution clearly does not fit observations. As the discrepancy is solely caused by the hydrogen-poor outer layers, it is clear that in Algols some mixing mechanism must be at work that is efficient enough to redistribute the helium excess across the star on a timescale shorter than the mass transfer timescale. The next section will deal with some possible mechanisms.

3. EXAMINATION OF MIXING MECHANISMS

3.1 For the sake of completeness we must mention the possibility that all matter leaves the system. In this case our problem clearly does not exist. This picture, however, is in contradiction with the theory of formation and structure of disks around close binary components (see e.g. Lynden-Bell and Pringle 1974; Lubow and Shu 1975).

3.2 A deep convective envelope is not present, as all our models have $\log T_{eff} > 3.9$.

3.3 Whem the inverse μ-gradient becomes steep enough, one expects a density inversion to occur, so that the envelope becomes Rayleigh-Taylor unstable. This, however, never occurred in our models, not even in those with an accreted fraction of 10%, which gives the steepest μ-gradient.

3.4 We included diffusion of hydrogen and helium in our evolutionary program, following the formalism developed by Noerdlinger (1977, 1978). The effect was marginal, because the diffusion timescale involved is much longer than the mass transfer timescale. One could argue that the underlying plasma physics are still uncertain, so that our diffusion rate is underestimated, but in order to have any effect the rate should be increasedby at least a factor 1000, which looks very improbable.

3.5 It can easily be seen that in case of accretion through a disk (which is observed in many Algols) the infalling matter has a specific angular momentum (per unit mass) that can easily be an order of magnitude larger than the specific angular momentum of the underlying stellar material. Shear turbulence induced by differential rotation could then act as a mixing agent. For the moment we are unable to describe this process in more detail, as it is quite complicated (in the case of accretion onto white dwarfs, a simplified model has been worked out by Kippenhahn and Thomas (1978)).

3.6 It might be that pulsational instability in the region with inverse μ-gradient causes some mixing. Stothers and Simon (1969) looked for pulsations in massive stars with a homogeneous helium-rich layer on top. This configuration, however, differs too much from ours for their results to be applicable for our purposes.

4. CONCLUSIONS

The results of our investigation can be summarized as follows:

unmixed models for the accreting component in close binaries during the last stage of mass transfer (associated with the Algol binaries) are found at a position hotter than the ZAMS;

convection, Rayleigh-Taylor instabilities, and diffusion are unable to mix in the accreted helium-rich material;

differential rotation and/or pulsational instability could do the job and should be looked at more closely. Three quantities determine the efficiency of the process: the vertical and horizontal extent of the mixed zone, and the timescale of the mixing process.

REFERENCES

De Grève, J.P. and Vanbeveren, D.: 1980, Astrophys. Space Sci., 68, 433.
Kippenhahn, R. and Thomas, H.-C.: 1978, Astron. Astrophys. 63, 265.
Lubow, S.H. and Shu, F.H.: 1975, Astrophys. J. 198, 383.
Lynden-Bell, D. and Pringle, J.E.: 1974, Monthly Notices Roy. Astron. Soc. 168, 603.
Noerdlinger, P.D.: 1977, Astron. Astrophys. 57, 407.
Noerdlinger, P.D.: 1978, Astrophys. J. Suppl. 36, 259.
Stothers, R. and Simon, N.R.: 1969, Astrophys. J. 157, 673.
Trimble, V.: 1974, Astron. J. 79, 967.
Trimble, V.: 1978, Observatory 98, 163.

COMMENTS FOLLOWING PACKET

Ziolkowski: I wonder why your models are so hot even after you stop accretion and permit them to readjust to thermal equilibrium. You probably do not deposit that much helium on them. Is it purely a chemical composition effect?

Budding: How reliably can your data establish the (often assumed) main sequence character of Algol-type primaries?

Packet: It is true that one sometimes assumes the more massive component to satisfy the zero-age mass-luminosity relation in order to establish absolute dimensions. So some of the data we used might indeed be biased in that way. However the discrepancy we find is essentially a discrepancy in effective temperature, and this parameter can be determined directly and rather accurately from spectroscopy.

Lacy (in response to a question by Dr. Budding): A double-line radial velocity curve has recently been determined for Algol by Joc Tomkin of the University of Texas at Austin and the absolute dimensions and masses are now known to within 5 percent or so. These new values do not differ significantly from estimates based on the assumption that the primary star is on the main sequence.

WHAT DO WE KNOW ABOUT ALGOL?

S. Söderhjelm
Lund Observatory

ABSTRACT

A comprehensive analysis of previously published observations of Algol has been submitted to Astronomy&Astrophysics under the title "Geometry and Dynamics of the Algol System". The main results can be summarized as follows:

The masses of the three components are m_A=3.6±0.1, m_B=0.79±0.01, m_C=1.6±0.1 solar masses.

The orbit of Algol C is co-planar with the AB-orbit within some 3°, but polarimetry of the eclipsing pair indicates that the senses of revolution are opposite.

The close orbit is circular, thus the 32 year period in the times of minima can not be explained by apsidal motion.

The photometric elements derived from a simultaneous analysis of 21 different light-curves show that the secondary fills its Roche lobe. The radii are R_A=2.89±0.04, R_B=3.53±0.04 solar radii, and the effective temperatures about T_A=12500 K, T_B=5000 K. Neither the gravity-darkening exponent for Algol B, nor the luminosity of Algol C can be determined from these analyses.

The photometric and spectroscopic observations of circumstellar matter can not be explained by a classical narrow stream between the components. Some magnetic interaction is probably involved, as can be inferred also from the radio synchrotron emission.

The alternate period-changes observed for nearly 200 years do not fit the standard Biermann-Hall model. At least 10^{-7} $m_\odot$ needs to be lost from the secondary each year. Part of it leaves the system, probably as a result of magnetic forces. A speculative proposal relates the 32 year periodicity in the times of minima to some stellar magnetic cycle.

M. J. Plavec, D. M. Popper and R. K. Ulrich (eds.), Close Binary Stars: Observations and Interpretation, 217–219.

COMMENTS ON SÖDERHJELM

Guinan

The observed erratic period decreases observed for Algol could arise from systemic mass loss rather than mass exchange. The radio observations indicate that the system may be enveloped by tenuous gas. Perhaps mass loss from the system supplies this gas.

Söderhjelm

I agree.

Linnell

The speaker deserves commendation for presenting light curve solution data in the form of magnitude residuals. It is always pleasing to see a theoretical light curve that appears to pass through the observations, but a plot of residuals is the best way to assess the quality of the fit.

Bolton

Did I understand correctly that you are estimating a mass transfer rate of $4 \times 10^{-7} M_{\odot}/yr$? My Hα profiles suggest that the rate does not exceed a few times $10^{-9} M_{\odot}/yr$.

Söderhjelm

Yes, it seems very hard to get the large period changes without that much mass involved.

Hall

In response to Bolton's comment about your large value of $\dot{M}$, let me comment that similar values of $\dot{M}$ were derived from observed period changes in two dozen other Algols by me and S.G. Neff (in Symposium No. 53). These rates, which correspond to the cooler star's thermal rate, are not steady-state values but rather averages over supposedly brief episodes of very high mass transfer. Let me also comment that a follow-up paper on the Biermann-Hall model (in Acta Astronomica 25, 1) emphasizes that the period changes require significant mass loss from the system as well as transfer.

Söderhjelm

I am aware of these works. Mainly, I disagree about the necessity for the mass-loss to come in large-scale bursts. I would rather have the path of a rather constant mass-flow modified by a variable magnetic field.

Brown

I would not want you to put binary polarimetry into disrepute by your statement that it gives the opposite sense of circulation of Algol to that from the other data. If you have taken the binary circulation to be the same as the circulation in the polarimetric Stokes parameters (Q, U) plane, then you may be incorrect. The (Q, U) plane circulation can be the same or opposite to the orbital circulation, depending on the quadrant of i (0 - 90° or 90° - 180°). Were you aware of that?

Söderhjelm

No, I was not. It would be very good if the methods did agree on the circulation.

Brown

It would be very disturbing if they did not!

Walter

Some months ago I also studied the problems of Algol. In my preceding report on the location of gas streams in Algol systems, I showed a diagram with spectroscopically derived values of the eccentricities, e, of the orbits and the positions of the periastrons, ω, for Algol systems with periods between 4 and 15 days, primary components of spectral types A and subgiant secondaries. It could be seen that the values of ω are situated in a distinct region, in this way making it very probable that the orbits are really circular and the eccentricities fictitious. Thus it seems to be very improbable that Algol has an elliptic orbit and that the thirty years period of the minima times may be explained by motion of periastron.

However, it offers another interpretation of this period. The large period of 600d in TV Cas could be explained by a precessional period of the rotational axis of the primary component, which is somewhat inclined to the orbital phase. Why should a similar effect not be working in the Algol system? I argue that the thirty years period is caused by precession of the rotational axis of the primary, and that the photometric effects occurring in connection with the gas stream distort the light curve. Especially in cases of partial eclipses it seems possible that minima times are derived which are systematically influenced. Moreover, the theoretical value of the precessional period of the Algol primary is in good agreement with the thirty years period.

Söderhjelm

The light-curve analysis shows that distortions large enough to influence the times of minima are unlikely.

ABUNDANCE OF SODIUM IN THE ATMOSPHERES OF ALGOL A AND B

Masayuki Kondo
Tokyo Astronomical Observatory, University of Tokyo,
Mitaka, Tokyo, 181 Japan
Akira Okazaki
Department of Astronomy, University of Tokyo,
Bunkyo-ku, Tokyo, 113 Japan

Abstract. The Na D1 and D2 lines of Algol A and B observed by Tomkin and Lambert(1978) were analyzed to obtain sodium abundance. A mild deficiency of sodium [Na/H]$\sim$-0.5 is found in Algol B, while a slight overabundance [Na/H]$\sim$+0.3 is found in Algol A.

Chemical composition of component stars in semi-detached binary systems of the Algol-type is one of the important factors for understanding of structure and evolution of those systems. Tomkin and Lambert (1978) detected the Na D lines of Algol B. We determined sodium abundance of Algol A and B from their line profiles.

We measured equivalent widths of the Na D lines of Algol A and B, which are less than those by Tomkin and Lambert(1978), as shown in the following table. The equivalent widths, belonging to both components, have to be converted to those with respect to the separated light of each component. Taking the reflection effect and the aspherical shape of Algol B into consideration and using model atmospheres, we obtain the light ratios at λ5893A to be L_A/L_B=5.63 and $L_C/(L_A+L_B+L_C)$=0.112 for the observed phases(0.306 and 0.716). These light ratios lead to the corrected equivalent widths from which we derive sodium abundance and microturbulence in the atmospheres of Algol A and B.

Fitting of the computed profiles based on our solution to the observed is fairly satisfactory for Algol A and B. Here we assumed the synchronized rotation for both components. About Algol C, however, we have little knowledge. If we adopt Teff=8400K, log g=3.8, solar abundance of sodium, and 3 km/s of microturbulence for Algol C, we obtain not small D lines of Algol C, that contaminate the profiles of Algol A in the observation. This situation suggests that the above obtained value of sodium abundance for Algol A should be regarded as the upper limit.

Different results have been reported among recent observational studies of metal abundance in the secondary component in semi-detached

M. J. Plavec, D. M. Popper and R. K. Ulrich (eds.), Close Binary Stars: Observations and Interpretation, 221–222.

binary systems of the Algol-type. We find that [Na/H]$\sim$-0.50 derived for for Algol B is comparable to the mild deficiency in the secondary of U Sge (Naftilan,1976). Our result [Na/H]$\sim$+0.30 for Algol A is consistent with Naftilan (1975), who found a slight overabundance of metals in the primaries of some Algol-type binary systems.

		Measured EW (A)	log ε	Teff (K)	log g	ξ_t (km/s)	vrot (km/s)
Algol A	D2	0.040	6.55	12000	4.0	1.5	54.4
	D1	0.029					
Algol B	D2	0.040	5.75	5900	3.3	2.5	57.0
	D1	0.032					

References.

Naftilan,S.A.:1975,Publ.Astron.Soc.Pacific,87,p.321.
Naftilan,S.A.:1976,Astrophys.J.,206,p.785.
Tomkin,J. and Lambert,D.L.:1978,Astrophys.J.(Letters),222,p.L119.

ROTATION OF THE PRIMARY COMPONENT OF ALGOL

S. M. Rucinski*
Warsaw University Observatory

1.8 Å segment of the Algol A spectrum centered at 1075.55 Å line of Ni II was scanned repeatedely with the 0.05 Å resolution from the Copernicus satellite. The numerical model is described in which 37 out of the total 55 scans obtained during two eclipses were combined in a least-squares solution, to determine the period of rotation and degree of non-solid rotation of Algol A. The period of rotation suggests full synchronism of rotational and orbital motions; the equatorial velocity is $V_e = 53 \pm 3$ km/s. The non-solid rotation parameter 's' measuring the distribution of angular velocity versus stellar latitude in $\omega = \omega_e (1 - s + s \cos^2 \vartheta)$ has been found equal $s = 0.07 \pm 0.25$ indicating the rotation law not far from solid body.

A full account of this work will appear in Acta Astronomica, Vol.29.

COMMENTS FOLLOWING RUCINSKI

Smith

Do you know how sensitive your result is to the assumed geometry?

Rucinski

The result is highly model-dependent since many assumptions about geometry were made (the axis of rotation parallel to that of revolution, Roche geometry for the secondary component, etc. etc.).

* Guest Investigator with the Princeton University telescope on the 'Copernicus' satellite, which is sponsored and operated by the National Aeronautics and Space Administration.

M. J. Plavec, D. M. Popper and R. K. Ulrich (eds.), Close Binary Stars: Observations and Interpretation, 223–224.

Popper

You show a limb-darkening coefficient in the line greater than unity. Can you comment on its significance?

Rucinski

The limb-darkening law is certainly non-linear; to have the smallest number of free parameters, the linear law was assumed. The accuracy of this determination is low, which might be related to this non-linearity.

Chen

I am glad to see Dr. Rucinski's result. The study of Mg II k line profiles during primary eclipses also indicates that the rotation of Algol A is synchronous with its orbital revolution.

Guinan

How did you determine your value of the limb-darkening coefficient?

Rucinski

The limb-darkening in the line was assumed to be identical to that in the continuum. This assumption might be obviously questionable but was made to keep the number of free parameters managable.

Hα EMISSION LINE PROFILES IN THE SPECTRUM OF ALGOL

C.T. Bolton and Donna J. Zubrod
David Dunlap Observatory, University of Toronto
P.O. Box 360, Richmond Hill, Ontario, CANADA L4C 4Y6

We have investigated variations in the strength and profile of the Hα emission line in the spectrum of Algol using 145 16 Å mm^{-1} spectrograms obtained during the period September, 1976 to December, 1977. The Hα emission line profile has been extracted by subtracting the absorption contributions of the three stars in the system. The resulting data set has been analyzed to look for variations related to orbital phase as well as shorter and longer term variations.

The emission profile is very similar to that usually seen in Be stars. It is not the type of profile usually associated with a disk. The V and R peaks are separated by 100 km s^{-1} during most of the orbit. The velocities of the maximum and minima on the profile repeat remarkably from orbit to orbit, even when there are large changes in the emission strength. All of the major features on the profile follow the velocity curve of the primary star with two exceptions noted below.

The emission profiles show a weak extended red wing in the quarter period centered approximately on primary eclipse and a similar extended blue wing around the time of secondary eclipse which may be due to a gas stream flowing from the secondary to the primary. The most marked orbital variation occurs in the V/R ratio and the velocities of the V and R peaks. Over most of the orbit $V/R \simeq 1$ and the V and R velocities follow the primary star velocity curve. Near $\phi = 0.22$ $V/R < 1$ and the R velocity increases by about 80 km s^{-1}. One-half period later $V/R > 1$ and the V velocity decreases by a similar amount. We believe these variations are due to aspect variations of a stream of gas moving tangentially to the surface of the primary star at an angle of 25° - 30° toward the trailing hemisphere from the substellar point. This is probably due to the reflection of the gas stream from the secondary star off a shock wave in the outer atmosphere of the primary star.

M. J. Plavec, D. M. Popper and R. K. Ulrich (eds.), Close Binary Stars: Observations and Interpretation, 225–227.

The central dip (self-absorption?) in the emission profile is most pronounced in the half of the orbit centered on primary eclipse, and most of the emission profile variability seems to occur in the interval from $0^p.75$ to $0^p.9$. The interpretation of these variations is not clear, but they may arise from interactions between the gas stream and the primary star.

The emission line strength implies that the gas density is at least 10^{10} cm^{-3}, assuming the entire volume of the Roche lobe surrounding the primary star is filled by the emitting material. If the disk/shell is highly flattened, the density may exceed 10^{12} cm^{-3}. The emission line strength can change by large amounts in less than one-half orbital period. If these changes are associated with expulsion of material from Roche lobe around the primary, the total energy released must be greater than 10^{37} ergs.

DISCUSSION FOLLOWING BOLTON AND ZUBROD

Budding

I have a few points to make. First I don't think the German data can be taken to unambiguously show that material around the primary extends to the Roche-limiting surface. The fit which I showed, for example, would require the disk radius to be about 30% greater than the photosphere radius. Second, I would like to know something about your formulary for the density calculations. What do you do about non-LTE effects, how are collisions accounted for and so on? Third, the paper of Lubow and Shu predicts certain definite features for the stream, like its thickness and the angle at which it departs from the LI region. Can you find any confirmation of these details in your observational results?

Bolton

First, if the disk does not fill up the Roche lobe in the plane then the densities I estimate would be increased--substantially increased if I adopt your number for the radius for the disk. The density calculation was done by assuming standard case B recombination theory. I assumed that the Hα emission originates from material uniformly distributed in the emitting volume.

I'm sure that we can derive some of the stream parameters once we begin work on the data obtained during eclipse.

Pringle

(i) May I congratulate you on being the first person today to present a mass transfer stream that actually obeys Newton's laws?

Iii) There are difficulties in associating the X-ray emission from β Per with the spot where the stream of transferred material strikes the primary star. One of these is that the X-ray emission is visible at phases when the spot is not.

Bolton

I agree, and that worries me. On the other hand the X-ray data presently available is spotty and not very good. Also, we don't know how large the X-ray emitting region produced by impact of the gas stream should be.

Rajamohan

Part of the Hα emission profile observed could arise in the hemisphere of the cool secondary that faces the primary star all the time, as most of these systems are synchronous rotators. The phase-dependent profile and velocity variation of this emission component which would blend with the Hα profile associated with the primary component should be taken into account before realistic models for gas streams in Algol systems are proposed. Spectroscopic reflection effects have not been given serious consideration so far in analysis of close binary systems.

Bolton

I have double-checked my profiles for this effect at the quadratures where emission from the secondary would be most readily distinguishable. There is no evidence for such emission at $V_{sec} \pm V_{rot}$ (synchr.).

ON THE LOCATION OF RADIO AND X-RAY EMISSION IN THE ALGOL SYSTEM

D. R. Florkowski
Department of Astronomy, University of Florida,
Gainesville, FL

The radio behaviour of Algol at centimeter wavelengths has been reviewed by Hjellming (1976, 1977), and by Gibson (1976). The observed radio emission can be roughly divided into two types: a quiescent type and a flare-like or outburst type. The quiescent emission is characterized by a low flux density, usually between 20 - 50 mJy, and a nearly flat spectral index. The latter means that the value of the flux density is nearly independent of wavelength. The variations in flux density, when present, are gradual and have a time scale on the order of days. The flare-like behaviour shows rapid and large changes in flux density. The amplitude of an outburst is wavelength dependent, the shorter wavelengths having larger amplitudes. Woodsworth and Hughes (1976, hereafter WH) attributed the two types of behaviour to two physically distinct sources. The quiescent type of emission is produced by a very large, thermally emitting cloud which surrounds the eclipsing system. The flare-like variations are due to a synchrotron source that is associated with mass loss. Their model is inconsistent with optical and X-ray information concerning the Algol system. However, a model with two radio components can be modified to agree with other types of data.

Guinan _et al_. (1976) detected a disk of material surrounding Algol A. The radius of the disk was estimated to extend to the Roche lobe of Algol A. Material from the disk could be lost from the binary system via the outer Lagrange point, L_3. The amount of material that is ejected from the system is not known. However, it is likely that some material does escape, and could form a cloud of ionized hydrogen that surrounds the binary. WH tried models for this cloud that have density distributions of the form $n_e \propto r^m$. The models were constrained by the condition that the cloud must be optically thin, and that the maximum radius, deduced from the time scale of variability, must be roughly one light-day. Two solutions were found: m = -1 with $T_e \geq 10^5$K, and m = -1.5 with $T_e > 10^6$K. These solutions were rejected by WH because a stellar wind was considered to be the only possible mode of mass ejection, and a wind would have a m = 2 distribution. However, a cloud formed by mass loss from the L_3 point is not required to have this distribution. Thus either of the above solutions is possible. The cooler model of uniform density

M. J. Plavec, D. M. Popper and R. K. Ulrich (eds.), Close Binary Stars: Observations and Interpretation, 229–232.

that was adopted by WH is harder to accept. Also, the suggestion that the variations in the free-free emission from the cloud are caused by changes in the ionizing flux from the stars seems very unlikely. No large intrinsic variations in the ultraviolet have been observed by Eaton (1975), or by Chen et al. (1977). A possible mechanism, related to the flare activity, that explains the radio variations and the high temperature of the cloud will be described below.

Pooley and Ryle (1975), and Gibson (1976) found no correlation between the orbital phase and the occurrence of a flare. Gibson (1976) found a correlation between the maximum flux of a flare and the spectral index. Also, he noted that 6 of the 9 strongest flares occurred near secondary eclipse. Assuming that all flares are caused by the same physical mechanism, then differences in flare amplitude would be due to differing numbers of particles involved, and by possible eclipse or other geometrical effects. Bielicki et al. (1974) found for a binary similar to Algol, that the disk material is loosely bound. An increase of about ten percent in the kinetic energy of a particle in the disk will eject the particle. The mass loss rate from Algol B could be irregular. Short term random increases in the mass loss could transfer enough kinetic energy to eject a sizeable fraction of the disk material. The ejected material would hit the inner regions of the cloud, and a compressional shock would form. In the impact region the material becomes optically thick, and some acceleration process supplies energy to the electrons. Gyrosynchrotron emission could occur and form a flare. The turbulence of the material in the impact zone causes the magnetic field to be randomly oriented, and the flare would have no net polarization. This is in agreement with the observations of Gibson et al. (1975). An observer is looking directly at the impact area between phases 0.6 - 0.7. This region is eclipsed by Algol A and B during phases 0.04 - 0.15. The impact area is close to the location calculated by Bielicki et al. (1974). If the flare source expands during an outburst, then the source will have a density gradient. This implies that the size of the source is wavelength dependent. The longer wavelength radiation will be emitted from the less dense, outer layers, and the shorter wavelength radiation will be emitted from the denser, inner layers. An eclipse of the flare source will be deeper at shorter wavelengths because the smaller size implies that a larger percentage of the flux at that wavelength will be blocked from view. This could explain, at least partially, the peak flux density and the spectral index of a flare. An eclipse would explain the very rapid decline and the inverted spectral index of the flare that was observed at phase 0.04 on July 11, 1972 by Pooley and Ryle (1973) and by Hjellming et al. (1972). Also, an eclipse could also explain the large differences in the expansion velocities, obtained by VLBI observations, of two flares (B. Clark et al., 1975; T. Clark et al., 1976). The first flare listed in the table below was observed during an eclipse. In their analysis both groups assumed that changes in the observed visibilities were only caused by changes in the brightness distribution due to expansion. However, this is not true for the first flare because an eclipse also changes the brightness distribution. Thus, for the first flare the value determined for the expansion velocity is spurious.

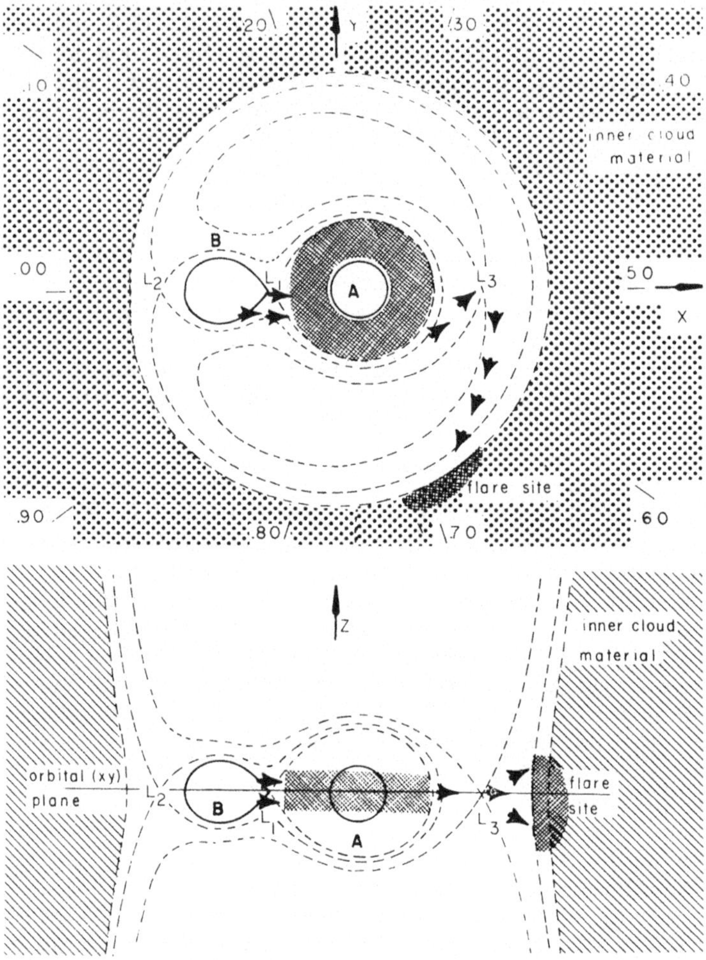

Figure 1. (Top)Orbital plane view of Algol. The indicated flare site is approximate. The enhanced mass flow is shown schematically by arrows. (Bottom) View of Algol perpendicular to the orbital plane. An arbitrary disk thickness is shown.

Table 1
Very Long Baseline Interferometry of Algol at 3.8 cm

Date	Angular dia.* (milli-arc sec)	S_{max} (Jy)	Velocity Expansion (km/sec)	Orbital Phase	Notes
May 4-5, 1974	3.6 - 4.6	0.6	500 - 1000	.02 - .15	1
Jan 15-16, 1975	1.2	1.	<100	.42 - .53	2

Notes: 1. B. Clark et al. (1975)
2. T. Clark et al. (1976)

*Gaussian brightness distribution assumed.

The decay of a flare could release energetic (faster than thermal) electrons which interact with the cloud material. This would heat the cloud, and produce nonthermal bremstrahlung radiation at radio (the observed quiescent emission) and at X-ray wavelengths. The variability is caused by variations in the numbers of the "fast" electrons released from different flares. The energy distribution of the X-ray emission (Harnden et al. 1977) indicates temperature of (3 - 6) x 10^6 K which is consistent with the values needed for the quiescent emission. Hence, both of them could originate in the cloud. Takakura (1975), for example, discusses physical mechanisms to produce both solar and X-ray flares. Analogous mechanisms, operating over much larger scales, may be at work in the circumstellar material in the Algol system.

More observations are needed to further out understanding of Algol. In particular radio observations should be made to check the various correlations mentioned above. Multi-wavelength observations of eclipses of the flare source would be very valuable. Since the depth of an eclipse depends on wavelength, viewing angle, and the surface brightness of the flare source (the latter two being independent functions of time), and since flares occur randomly, observed eclipses will be rare. The nonspherical expansion of the flare source should be detectable by means of VLBI. Optical observations of dilution sensitive lines should be made. Lines formed in densities $<10^{11}$ cm^{-3}, such as HeI 10830 A, would be produced in the same region as the radio emission. Also, coordinated observations in different spectral regions may reveal new insights.

REFERENCES

Bielicki, M., Piotrowski, S., and Ziolkowski, K.: 1974, Astrophys. Space Sci., 26, p. 173.
Chen, K-Y., Merrill, J. E., and Richardson, W. W.: 1977, Astron. J., 82, p. 67.
Clark, B. G., Kellermann, K. I., and Shaffer, D.: 1975, Astrophys. J. Letters, 198, L123.
Clark, T., et al.: 1976, Astrophys. J. Letters, 206, L107.
Eaton, J. A.: 1975, Pub. Astron. Soc. Pacific, 87, p. 745.
Gibson, D. M.: 1976, Dissertation, University of Virginia.
Gibson, D. M., Viner, M. R., and Peterson, S. D.: 1975, Astrophys. J. Letters, 200, L143.
Guinan, E. F., McCook, G. P., Bachmann, P. J., and Bistline, W. G.: 1976, Astron. J., 81, p. 57.
Harnden, F. R., Jr., Fabricant, D., Topka, K., Flannery, B. P., Tucker, W. H., and Gorenstein, P.: 1977, Astrophys. J. 214, p. 418.
Hjellming, R. M.: 1976, in "The Physics of Non-Thermal Radio Sources" ed. G. Setti (Dordrecht:Reidel) p. 203.
Hjellming, R. M.: 1977, Veroff. Remeis-Stern. Bamberg, 11, No. 121, p. 41
Hjellming, R. M., Webster, E., and Balick, B.: 1972, Astrophys. J. Letters, 178, L139.
Pooley, G. G., and Ryle, M.: 1973, Nature, 244, 270.
Takakura, T.: 1975, in "Solar Gamma-, X-, and EUV Radiation" ed. S. R. Kane, IAU Symp. 68, (Dordrecht: Reidel) p. 299.
Woodsworth, A. W., and Hughes, V. A.: 1976, Mon. Not. R. Astron. Soc., 175, p. 177 (WH).

TIME-RESOLVED SPECTROSCOPY OF RW TAU AND X PER

R. H. Kaitchuck, R. K. Honeycutt, S. L. Mufson
Astronomy Department, Indiana University
Bloomington, Indiana, U.S.A.

Abstract. Time-resolved spectroscopy of the gaseous ring of RW Tau has revealed characteristics of its structure. Observations of H-alpha emission during one eclipse show a partial eclipse of the ring which is brighter at second contact than at third contact. The radial velocities became nearly constant after mideclipse. During two other eclipses the emission continued to strengthen after second contact. X Per has shown variability on a time scale of minutes. On one occasion H-alpha emission decreased in strength by a factor of two in 90 seconds.

RW Tau often shows emission lines during the total phase of eclipse whose radial velocity behavior suggest the presence of a gaseous ring rotating about the primary star (Joy, 1942). Previous studies have been unable to follow the rapid variations of these lines as the ring undergoes eclipse. We have begun a study of this system at Goethe Link Observatory using the 0.91-meter telescope and a SIT vidicon spectrometer. This equipment can record a spectrum every five minutes during the 80 minute totality. The Doppler components from each side of the ring are not separated with the 8 A resolution used for most of this study. The table below lists the eclipses observed thus far and the average equivalent width of H-alpha during totality. On three nights the emission was

Date (UT)	Eq. Width	Date (UT)	Eq. Width
Feb. 7, 1978	7.8 A	Nov. 8	1.2 A
Sept. 8	3.8	Nov. 19	1.4
Oct. 25	0.7	Dec. 28	7.3
Oct. 28	< 0.1	Feb. 2, 1979	4

strong enough to be measured on individual spectra. On the remaining nights the spectra were co-added to detect the emission. The observations of October show that the emission can disappear in one orbit.

M. J. Plavec, D. M. Popper and R. K. Ulrich (eds.), Close Binary Stars: Observations and Interpretation, 233–236.

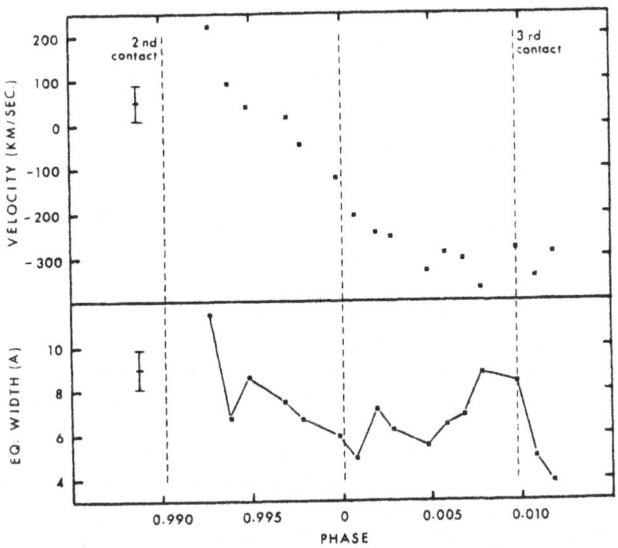

Fig. 1. The radial velocity and equivalent width of H-alpha emission during the December 28, 1978 eclipse of RW Tau.

The December eclipse represents a major outburst of the system. The start of totality was lost due to an equipment problem. Figure 1 shows the equivalent width of H-alpha during totality. The ring is partially eclipsed and brightest at the contact points when the maximum area is visible. Data points outside totality are low due to the bright continuum of the primary star. The ring is asymmetric, being brighter at second contact than at third contact. Apparently the material on the receding side of the primary is brighter compared to the advancing side. This appears consistent with a hydrodynamical calculation for the similar system U Cep by Prendergast and Taam (1974). These models predict a different distribution of density on the leading and trailing sides of the primary star. Assuming that the brightening which begins at phase 0.005 marks third contact of the ring eclipse and that the ring is roughly circular, the ring has a radius of 1.4 times the primary star. Figure 1 also shows the radial velocity of H-alpha during eclipse obtained by fitting a Gaussian to the wide (19 A FWHM) profiles. In addition to the error bar shown, a systematic error of ± 46 km/sec is possible. Even taking this into account, the rest wavelength is reached well before mideclipse. The velocity varies as expected until phase 0.005 when it becomes approximately constant and the line begins to strengthen. In the models of Prendergast and Taam, this is approximately the point where the majority of gas on the receding side of the primary is eclipsed. Furthermore, the models imply that the dense inner regions emerging from eclipse should not show a large variation in velocity. Olson (1979) obtained photometry of this same eclipse up to second contact. There was an ultraviolet excess of over

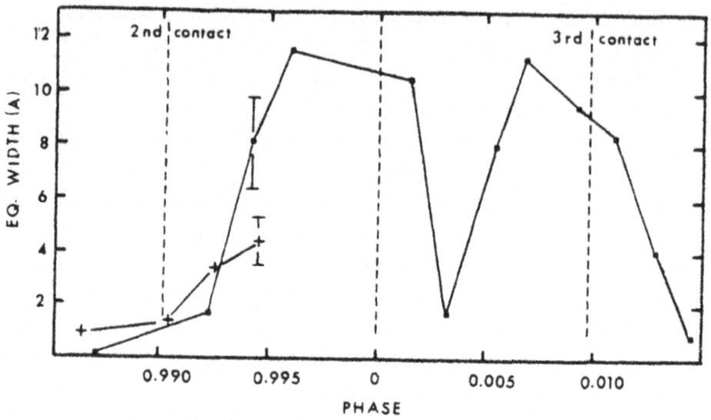

Fig. 2. The equivalent width of H-alpha emission during the February 7, 1978 (dots) and the February 2, 1979 (crosses) eclipse of RW Tau.

one magnitude and smaller excesses at longer wavelengths. This data, together with the emission line strengths, are consistent with hydrogen free-free and free-bound emission from a region optically thick in the continuum at 15,000 K and $N_e = 2 \times 10^{12}$ cm^{-3} .

Figure 2 shows the data for February 1978. Unlike the December eclipse the line strength continues to increase after second contact. This same behavior was apparent in the February 1979 eclipse until clouds ended observing. Simultaneous photometry by Burkhead (1979) clearly shows that the contact point occurred as indicated in Figure 2. This brightening is still apparent after the points prior to second contact are corrected for the light of the primary star. This behavior implies material is uncovered at the advancing side of the primary. However, it is then difficult to explain the dimming at phase 0.003. It is interesting that the average equivalent width for December and February 1978 are almost the same; yet the ring appears to have a different structure, at least on the receding side of the primary.

X Per is an unusual Be star. It is an intrinsically weak x-ray source. White et al. (1976) has found 13.9 minute and 22.4 hour periodicities in the x-ray flux. The current model for the system is a Be star with a neutron star companion. On August 23, 1978 X Per was observed simultaneously by Goethe Link Observatory and HEAO-A1. At this time the x-ray data is not available; however the optical spectroscopy recorded a remarkable event. The upper part of Figure 3 shows the equivalent width of H-alpha versus time. A resolution of 2 A and 90 second integration times were used. Between 9:28 and 9:31 UT the equivalent width dropped from 7.3 to 3.3 A. This was followed by a more extended event of 540 seconds duration in which the equivalent width dropped from 7.7 to 2.6 A. Standard stars were observed and the event does not appear to be instrumental in origin. The line widths became smaller and the central wavelength increased during the event. This strongly suggests a weakening of the unresolved, blueshifted component. This type of event is not uncommon for X Per. Murdin et al. (1976) reported

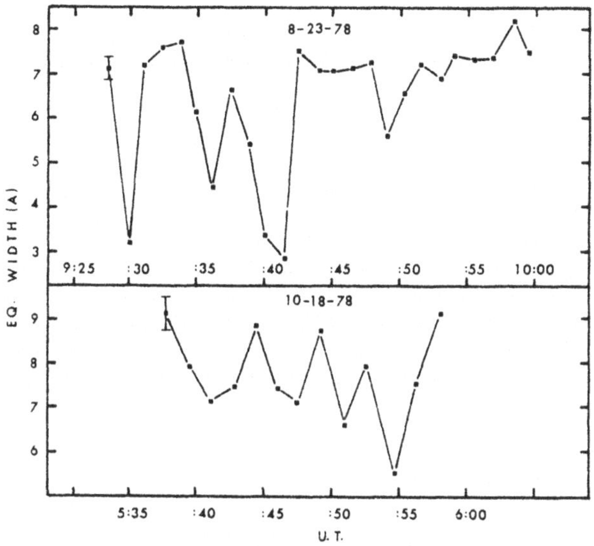

Fig. 3. The equivalent width of H-alpha emission from X Per on August 23 and October 18, 1978.

a similar event from H-alpha photometry with a duration of 600 seconds. Campisi and Treves (1976) reported dips in the ultraviolet continuum. One large event had a 600 second duration. It is interesting that these events are always characterized by a drop in flux followed by a return to the previous level. The strongest events observed have time scales of 500 to 600 seconds. On September 25, 1978 the equivalent width was higher (10.2 A) and showed only a hint of variation over a three hour interval. The bottom part of Figure 3 shows the observations of October 18, 1978. The equivalent width appears to oscillate between 7 and 9 A. Unlike the August data, the line width increases as the equivalent width decreases. There appears to be no central wavelength variation.

References

Burkhead, M. S. : 1979 (private communication).
Campisi, I. E. and Treves, A. : 1976, Monthly Notices Roy. Astron.Soc. 176, pp. 225-231.
Joy, A. H. : 1942, Publ. Astron. Soc. Pacific 54, pp. 35-37.
Murdin, P., Penston, M. V., and Penny, A. J. : 1976, Monthly Notices Roy. Astron. Soc. 176, pp. 233-236.
Olson, E. C. : 1979 (private communication).
Prendergast, K. H. and Taam, R. E. : 1974, Astrophys. J. 189, pp.125-136.
White, N. E., Mason, K. O., and Sanford, P. W. : 1976, Monthly Notices Roy. Astron. Soc. 176, pp. 201-215.

IUE ULTRAVIOLET SPECTRA OF THE INTERACTING BINARY U CEPHEI

Yoji Kondo
University of Pennsylvania, Philadelphia, Pennsylvania, USA
(Permanent Address: Goddard Space Flight Center)
George E. McCluskey
Lehigh University, Bethlehem, Pennsylvania, USA
Robert E. Stencel
Goddard Space Flight Center, Greenbelt, Maryland, USA

1. INTRODUCTION

The eclipsing binary U Cephei has proven to be of great interest in the study of stellar evolution in close binary systems. Batten (1974), Hall and Walter (1974), Rhombs and Fix (1976), Markworth (1977), and Olson (1978), among others, have recently reported on their intensive ground based studies of U Cephei. Kondo, McCluskey and Wu (1978) have investigated the ultraviolet light curves of U Cephei obtained with Astronomical Netherlands Satellite (ANS). Kondo, McCluskey and Stencel (1979) have discussed the International Ultraviolet Explorer (IUE) spectra of U Cephei. This paper discusses results incorporating additional IUE high resolution spectra of U Cephei obtained in both far-ultraviolet and mid-ultraviolet spectral regions.

According to Batten (1974), the system consists of a B7 V primary of 4.2 $M_\odot$ and 2.9 $R_\odot$ and a G8 III-IV secondary with 2.8 $M_\odot$ and 4.7 $R_\odot$. Its orbital period is about 2.5 days. The G-star is evolved and losing mass, some of which is accreted by the B-star, some leaves the system and some may return to the G-star (Figure 1).

2. OBSERVATIONS

The IUE instrumentation has been discussed in detail by Boggess et al. (1978). All of the observations were obtained in the high resolution modes: three exposures in the far-ultraviolet (1200-1900 A at $\lambda/\Delta\lambda \sim 10^4$) and sixteen exposures in the mid-ultraviolet (1900-3200 A at $\lambda/\Delta\lambda \sim 10^4$). The sixteen mid-ultraviolet spectra, taken over a period of about one year, cover most phases of interest with the exception of primary eclipse which is too deep in the ultraviolet for high resolution observations to be obtained. The three far-ultraviolet spectra were obtained at phases 0.184, 0.680 and 0.903 over a period of about ten months.

M. J. Plavec, D. M. Popper and R. K. Ulrich (eds.), Close Binary Stars: Observations and Interpretation, 237–241.

3. DISCUSSIONS AND CONCLUSIONS

The lines showing the effect of the gas streaming most distinctly are the resonance line of Fe II at 2599.395 A and the resonance doublet of Mg II at 2795.523 and 2802.698 A. The temperature and the density of the gas stream is such that it is seen primarily in terms of phase-dependent, Doppler-shifted excess absorption in those resonance lines. The signature of the gas stream is rather subtle; hence, it can be unambiguously observed only in the strong resonance lines of the ions of these abundant elements. For actual spectral profiles and their variations, the interested reader is referred to Kondo et al. (1979, 1980). The spectral variations cannot be accounted for in terms of orbital Doppler-shifting of the photospheric absorption in the B-star; the gas stream can be seen clearly as shifting excess absorption indicating the directions of the gas stream.

We note that several mid-ultraviolet spectra obtained months apart at nearly the same phase are quite similar. This indicates that U Cephei was relatively stable over this period of time. This is useful when comparing the far-ultraviolet spectra with one another since they were obtained at wide intervals and are few in number. Olson also informs us that U Cephei was in its quiescent state during the period of the IUE observations judging from his photometric data.

Lines of Si II, Si III, Si IV, C II, C IV, Fe III, Al II and Al III have been identified in the far-ultraviolet. Most of these absorption lines are shallow and broad, indicative of the rapid rotation (~300 km s^{-1}) of the B-star and hence of their at least partial association with the photosphere of the B-star. The phase variation of the radial velocities also shows this partial association with the B-star. However, significant total absorption variations, perhaps due to hot/cool spots on the B-star and gas streaming effects, are observed.

To recapitulate, the effects of mass streaming are most clearly seen in the resonance lines of Fe II and Mg II in the mid-ultraviolet spectrum of U Cephei; the far-ultraviolet absorption lines of Si II, Si III, Si IV, C II, C IV, Fe III, Al II and Al III are probably also affected.

A model consistent with these observations is shown in Figure 1. The gas stream leaves the G-star from the hemisphere facing the B-star. It circles around the B-star with some of the gas probably falling onto the forward and trailing hemispheres of this star. The rest escapes from the system after orbiting about 270° around the B-star. Some of this gas may fall back onto the trailing hemisphere of the G-star rather than escaping from the system. Observations near phase 0.15 obtained on June 2, September 7, September 12, 1978 and May 9, 1979 indicate a relatively stable stream pattern during this eleven month period.

The radial velocity of the edges of the absorption features are as high as 400-500 km s^{-1} along the projected line of sight. With the exception of locations near the B-star, which has an escape velocity of about 740 km s^{-1}, this velocity is sufficient to allow the gas to escape from the system.

It is possible that one source of the kinetic energy of the gas stream is the same mechanism that may drive stellar winds. Wolff and Kondo (1978) noted that the non-linear mode (g-mode) of oscillations in one of the components in a binary system (e.g., HZ Her = Her X-1) may lead to a surge of mass from the star. Modisette and Kondo (sumbitted for publication) indicate that this mass surge could be a continuous phenomenon if the beat periods of the harmonics involved are sufficiently close to each other. Another source of energy for the gas stream is heating by the B-star of the G-star's facing hemisphere; it could also contribute or give rise to the observed mass ejection (cf. Modisette and Kondo).

The gas in the stream is hot enough to keep most Fe, Si and Mg atoms in ionized states; it is well above 10000°K. The density of the gas is low enough so that practically all atoms are in the ground state; it must be below about 10^{15} particles cm^{-3}. Batten's estimate of 10^{13} electrons cm^{-3} is consistent with our data. The absence of emission probably indicates that the total volume occupied by the gas stream having significant densities is not much larger than the volume of the B-star.

With the advent of high capacity computers, it has become possible to treat the gas flow problem hydrodynamically. Some pioneering efforts are already being made in this direction. However, in order to construct a viable model, knowledge of the physical parameters involved (e.g., the direction of the gas flow, temperatures and densities of the gas, and boundary conditions) is essential. At the temperatures and densities of the plasma involved, the signature of the gas stream of the plasma can be investigated effectively by analyzing the resonance lines of ionized abundant elements in the ultraviolet. Extensive ultraviolet observations of selected binary systems portend to yield this much needed information.

REFERENCES

Batten, A. H.: 1974, Pub. Dom. Ap. Obs., 14, pp. 191-269.
Boggess, A. et al.: 1978, Nature, 275, pp. 387-392.
Hall, D. S., and Walter, K.: 1974, Astr. Ap., 37, pp. 263-273.
Kondo, Y., McCluskey, G.E., and Wu, C.-C: 1978, Ap. J., 222, pp. 635-640.
Kondo, Y., McCluskey, G.E., and Stencel, R.E.: 1979, Ap. J., (to appear).
Kondo, Y., McCluskey, G.E., and Stencel, R.E.: 1980, Ap. J.,(submitted).
Markworth, N.L.: 1977, Ph.D. Thesis, Univ. Florida.
Modisette, J.L., and Kondo, Y.: 1979, Ap. J., (submitted).
Olson, E. C.: 1978, Ap. J., 220, pp. 251-260.
Rhombs, C. G., and Fix, J. D.: 1976, Ap. J., 209, pp. 821-828.
Struve, O.: 1963, Pub.A. S. P., 75, pp. 207-211.
Wolff, C. L., and Kondo, Y.: 1978, Ap. J., 219, pp. 605-612.

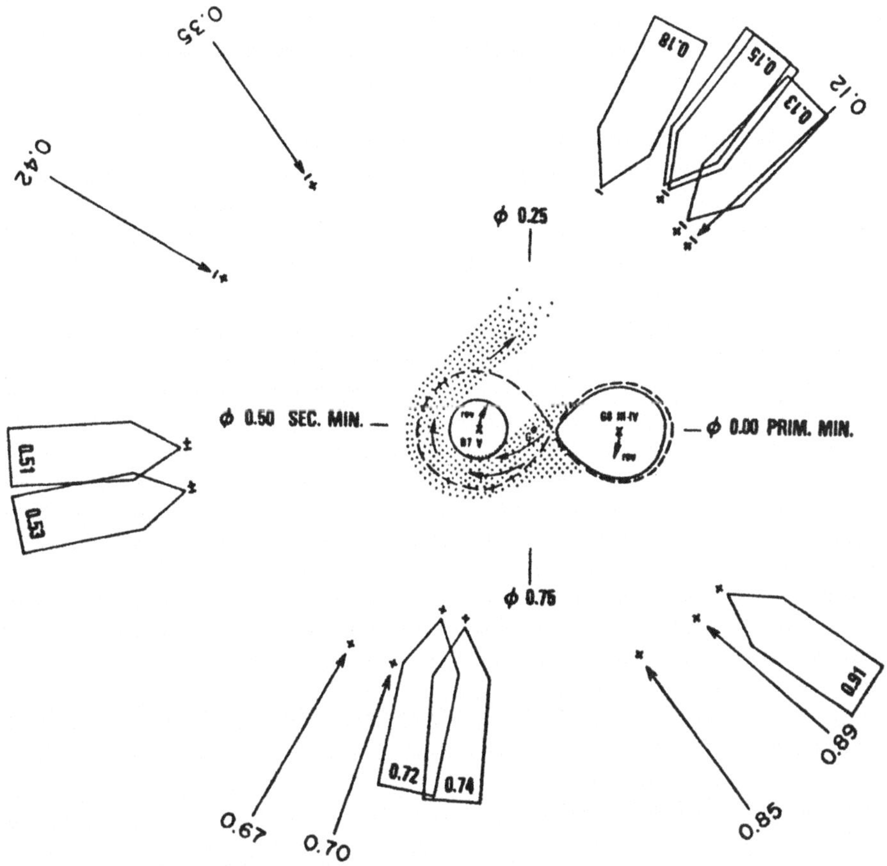

Figure 1. Schematic representation of the gas stream as observed in the Fe II and Mg II resonance lines. The broad arrows are the data from 1978 June and September and the thin arrows are the data from 1979 March and June: the phases of observation are also given. The + sign indicates that the radial velocity of the gas stream projected against the B-star is away from the observer, the - sign that the radial velocity is toward the observer, the $\pm$ sign both away and toward the observer, and the $\mp$ sign mostly toward, but with a minor component away from, the observer. The + sign at phases between 0.67 and 0.91 is probably due to the combination of the gas streaming around the B-star and the gas falling onto that star creating a hot spot which faces the observer shortly after phase 0.75. The $\mp$ sign at phases between 0.12 and 0.15 is probably due to the partial occultation of the B-star by the matter leaving the G-star.

COMMENTS FOLLOWING KONDO, MC CLUSKEY AND STENCEL

Stencel: These data do not rule out the possibility of a disk around the B star. Lines of singly ionized Fe and Mg are lower ionization. We do not yet have sufficient phase coverage in the farther UV where higher ionization lines might reveal the 12000°-13000°K disk discussed here by Olson.

Peters: Have you attempted to obtain an estimate of the column density in the stream from the observed strengths of the Fe II absorption?

Kondo: We have attempted to do so using various approaches but have thus far been unable to separate the photospheric components in the observed absorption feature. The problem is also complicated by the possible presence of hot (or cool) spots on the B star; this might give rise to varying photospheric component.

Smak: Do you exclude the possibility of contamination by emission components?

Kondo: We have been unable to detect evidence of any emission lines. The possibility of contamination by emission components could not be ruled out simply from the current data.

Nariai: I wonder if you can accelerate the flow by the gravitational accelerator.

Kondo: Simple conversion of gravitational potential energy to kinetic energy would not give rise to the observed gas stream which is leaving the binary system. We do need an additional source of energy. However, we do not claim that the local heating due possibly to the convergence of the non-linear (g-mode) oscillations is the only conceivable mechanism causing the mass ejection.

Shu: I wish to disagree somewhat with Nariai's comment. In the case of Roche lobe overflow, with matter leaving L_1 at $\sim$ sonic speeds, the matter is effectively trapped inside the Roche surface of the detached component (unless there are non-gravitational forces which intervene).

Kondo: (Please see my reply to Dr. Nariai's query.)

TRANSIENT DISKS AND MASS TRANSFER IN U CEPHEI*

EDWARD C. OLSON
University of Illinois, Urbana, Illinois

Conspicuous, variable photometric contamination was observed during primary eclipses of the totally eclipsing Algol-like system U Cep from late-1974 to late-1977. Excess light was particularly prominent between second and third contacts, and originated mainly in optically thick accretion disks surrounding the B-type hot component. Observations of such disks were obtained with uvby (I) photometry during 10 eclipses, using the 1.0 M and 0.4 M reflectors at Prairie and Kitt Peak National Observatories. Figure 1 shows ultraviolet observations between internal contacts for active (October, 1975) and quiescent eclipses. The light excess over quiescent levels was converted to flux distributions L_λ^c relative to the quiescent visual flux of the G-star.

Some light variation was present even at mid-eclipse. This light declined in brightness during October, 1975, and its spectral distribution was that of the G-type cool component. Most of the extra light evident closer to internal contacts came from accretion disks. Ingress disk light was always stronger than egress light, and varied significantly on a time scale as short as 10 days. It is possible that the brightness variations of the outer hemisphere of the G-star were related to the erratic mass flows that produced the accretion disks.

Incremental fluxes $\Delta L_\lambda^c(\phi_1,\phi_2)$ between orbital phases ϕ_1 and ϕ_2, being independent of G-star variations from eclipse to eclipse, were used to describe the disk light. Phases 0.995, 0.990, 0.985, and 0.9815 ($\approx$ second contact), and egress counterparts, were convenient in calculating ΔL_λ^c. Similar progressive changes in flux distribution with phase were noted in 16 ingress/egress disks. Light from disk edges showed Balmer jumps in emission, and was characteristic of near-10,000 K plasma of finite optical thickness. Most disk radiation was truly optically thick; the Balmer jump, in absorption, grew slightly but consistently stronger moving in phase away from the disk "limb". A satisfactory model (Figure 2) explains this radiation as near-limb emergent intensity from a slight extended stellar atmosphere. That is,

*A more detailed paper has been submitted to the Astrophysical Journal.

M. J. Plavec, D. M. Popper and R. K. Ulrich (eds.), Close Binary Stars: Observations and Interpretation, 243–247.

ΔL_λ^c(calc) $\sim I_\lambda(o,\mu)$ ΔA(disk), where ΔA(disk) is the radiating cross section. Four parameters enter these models: T(disk), log g, μ, and A(disk). Disk properties were fairly consistent on all nights. T(disk) was usually between 12000 and 13000 K; log g, as expected, was lower than that of the B-star; μ was $\approx$ 0.05 to 0.2, implying near-grazing light emergence. H/R(B), disk semi-thickness normal to the orbital plane and relative to the B-star radius, varied from 0.2 to 0.6. Outer disk radii were (1.3 to 1.5) R(B). Thus, at peak disk size, U Cep resembled peculiar close binaries such as W Ser and RX Cas, recently discussed by Plavec and Koch (1979).

Such geometrically thick disks were not in hydrostatic equilibrium since the sound travel time from disk center to top was about three times the orbital period of a typical disk particle. To be optically thick at all observed wavelengths, minimum required disk masses were $\approx 2 \times 10^{-10}$ solar masses. An order-of-magnitude estimate of the rate of mass transfer follows by assuming that much of the transferred mass went into the disk, and that accretion from the disk supplied the disk luminosity. If the disk was continuous around the outer hemisphere of the B-star, and had mean properties of the disk segments seen at ingress or egress phases, the luminosity increase with disk was $\Delta L \approx$ (0.1 to 0.6) L(B), where L(B) is the normal B-star luminosity. Since most of the orbital kinetic energy is unavailable for accretional heating because of the rapid B-star rotation, let L(B) $\Delta L \approx [0.3\ G\ M(B)/R(B)]\dot{M}$, where M(B) is the B-star mass and $\dot{M}$ the mass accretion rate. This gives $\dot{M} \approx (1 \text{ to } 6) \times 10^{-6}$ solar masses yr^{-1}. Significant disk light was present from late-1974 to late-1977 about 0.3 of the time, suggesting a mean transfer rate $\approx (0.3 \text{ to } 2) \times 10^{-6}$ solar masses yr^{-1}. The period was essentially constant in this same interval (Crawford and Olson 1979). It therefore seems likely that this time of active disk formation and collapse was one of normal or below-normal mass transfer. A large period decrease, implying loss of orbital angular momentum occurred in 1972 (Hall and Keel 1977). If the historical period trend of U Cep is to continue, a significant period increase must occur in the near future.

The author thanks the Director of Kitt Peak National Observatory and the National Science Foundation for support, and Richard Crawford for numerous helpful discussions.

References

Crawford, R. C., and Olson, E. C.: 1979, Publ. Astron. Soc. Pacific, in press.
Hall, D. S., and Keel, W. C.: 1977, Acta Astron. 27, pp. 167-177.
Plavec, M., and Koch, R. H.: 1979, preprint.

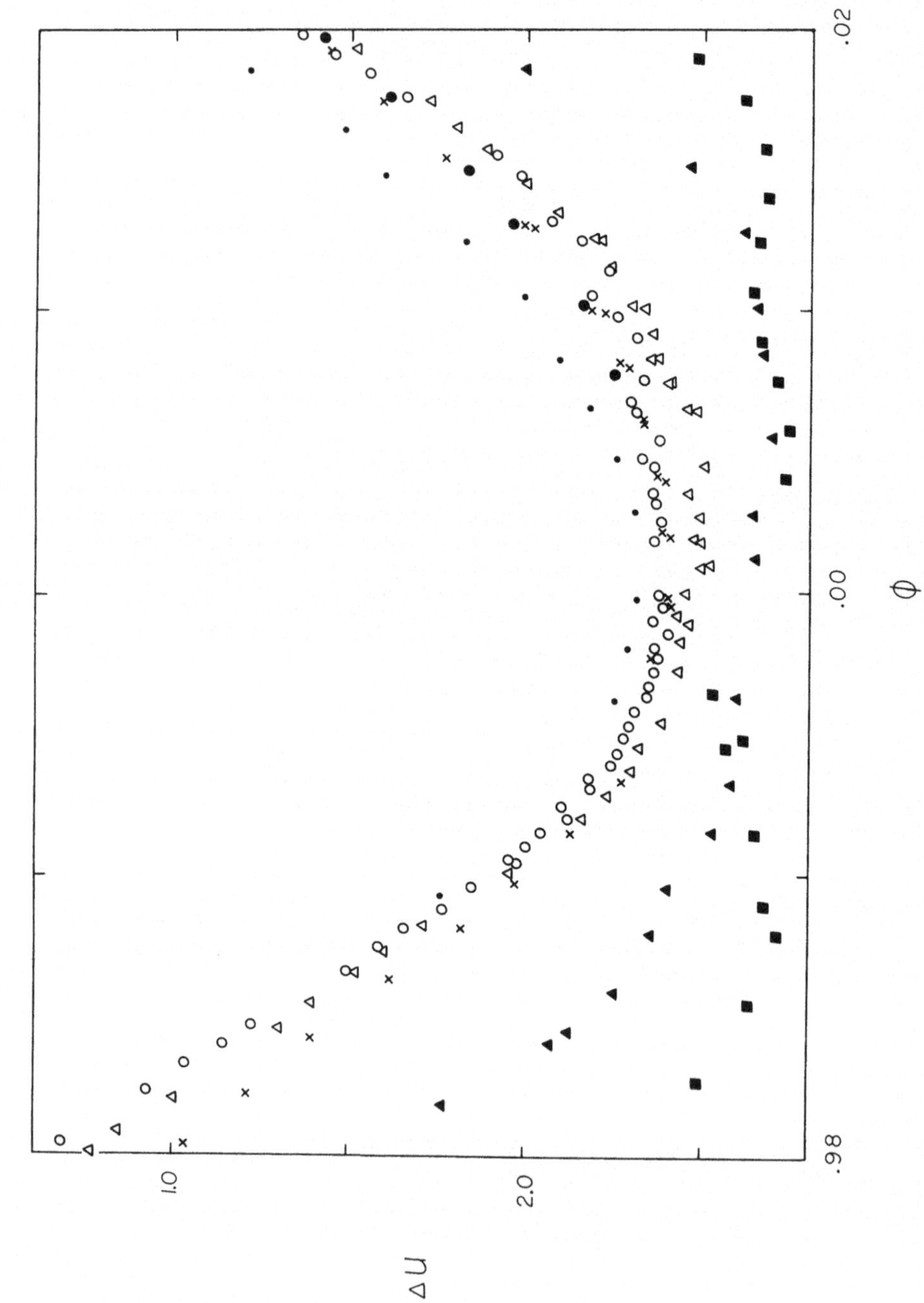

Fig. 1. - Ultraviolet observations of U Cep near mid-primary eclipse. Small filled circles, UT 1975 October 6; crosses, October 11; large filled circles, October 16; open circles, October 26; open triangles, October 31; filled triangles, November 20, filled squares, 1975 December 20.

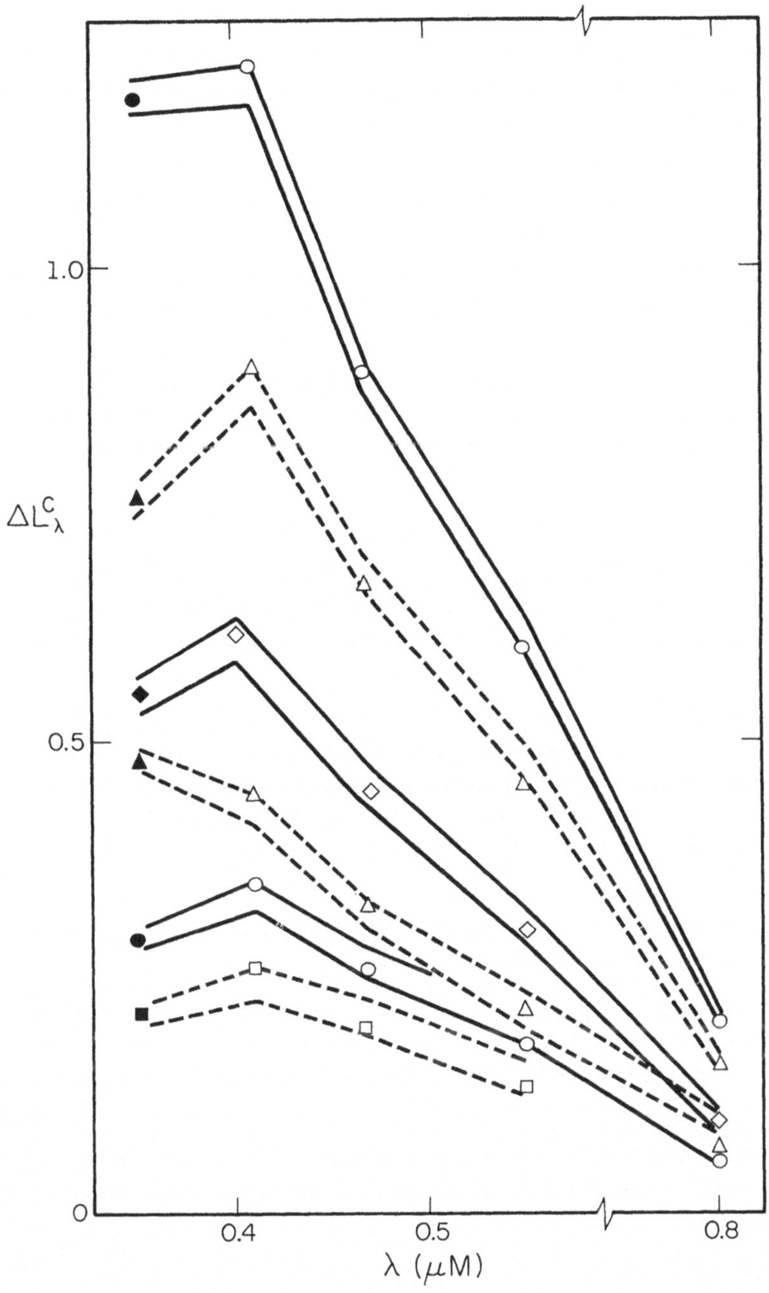

Fig. 2. - Flux distributions from various portions of ingress and egress disks. Solid and broken lines show observations with estimated errors. Symbols are model calculations, and represent (L to R) u, v, b, y, and I wavelengths.

COMMENTS FOLLOWING OLSON

Zuiderwijk: Do you have an estimate of the total mass in this disk-like structure? If so, can you build it, or destroy it, on the timescales you observe?

Olson: Minimum disk masses, estimated by supplying enough mass to produce optically thick disks, $\sim 2 \times 10^{-10}$ solar masses. Rapid disk collapse is not via viscosity effects. Disk particles may simply carry insufficient angular momentum to complete orbits around the B-star. Therefore, mass accretion from the disk may supply the disk luminosity.

Kondo: The CIV ($\lambda\lambda$1548 and 1550) and SiIV ($\lambda\lambda$1393 and 1402) resonance lines showed variations at different phases of U Cep. This may be related to the presence or absence of the hot or cool spots.

Olson: Yes.

Smak: A comment and a question: It would seem to me that the nearly identical values of O-C you obtain from "distorted" light curves may indicate that the source of this distortion is the hot spot. And how did you determine the moments of minima when distortions were present?

Olson:

The disk light asymmetry and extra light from the hot spot both contribute to delays in apparent times of minima. Distorted light curves also may show light losses on egress branches, which also contribute to the delays. Kwee-van Woerden method was used to determine t(min), but only undisturbed eclipses were used to determine the true "undisturbed ephemeris"(see Crawford and Olson, P.A.S.P. in press).

Hall: Concerning the brightness changes you ascribe to the cooler star and the brightness changes you ascribe to the hotter star and its temporary disk, were they related in any way? For example, were they coincident in time or did the first precede the second?

Olson: In both the fall 1974 and fall 1975 series of distorted eclipses, peak brightness of the outer hemisphere of the G-star preceded disk maxima by many orbital cycles. G-star brightness fell monotonically during each fall. There is a suggestion that brightening of the G-star precedes mass-transfer bursts that produce transient disks.

Smak (after Hall): As a follow up to Doug Hall's question: Did you get any IR observations for the secondary eclipse?

Olson: I have some observations of secondary eclipse in late October, 1975, but no detailed time sequence that might show progressive changes in the G-star.

POLARIZATION OBSERVATIONS OF U CEPHEI

V. Piirola
Observatory and Astrophysics Laboratory
University of Helsinki
Tähtitorninmäki, SF-00130 Helsinki 13, Finland

ABSTRACT

Polarimetric and photometric observations of U Cephei during the 1974-75 outburst are discussed. An optically thin envelope or disk cannot explain the observations. The polarization too small by a factor of 10, when compared with the amount of additional light during the totality of the eclipse of the B-star. This suggests that most of the additional light came from optically thick matter around the B-star, i.e. the effective radius of the photosphere of the B-star, r_B, was increased by 15-20 % during the period of high mass transfer. The polarization observations could then be explained by a simple model containing an optically thin spherical envelope of outer radius 1.1 r_B and an equatorial disk of outer radius 1.7 r_B surrounding the expanded B-star. The mean electron density is about 1.4×10^{12} e/cm^3 in the spherical envelope and 1.6×10^{12} e/cm^3 in the preceding and 0.9×10^{12} e/cm^3 in the trailing side of the disk, the thickness of which is assumed to be 0.1 r_B normal to the orbital plane. The variations in the position angle can be explained if the inclination of the orbit is about 83°. The details will be published elsewhere.

COMMENTS FOLLOWING PIIROLA

Poeckert: I feel it is an important feature that there is very little polarization structure at times when no mass transfer is taking place. The fact that the photosphere of the B star is not strongly polarized is important. This fact suggests that high polarization is a consequence of the disk alone, not the stellar photosphere or any extended spherical photosphere.

Piirola: It is simply impossible to produce the observed polarization curves with a model containing an equatorial disk only. Especially in the case of the parameter P_y, which shows changes of opposite sign before and after mid-eclipse, the spherical envelope-type component and i ≈ 83° give a good fit to the observations.

M. J. Plavec, D. M. Popper and R. K. Ulrich (eds.), Close Binary Stars: Observations and Interpretation, 249–250.

Mitrofanov: I think that the observed polarization may also be explained if the possible nonsphericity of the B-star envelope is taken into account.

Piirola: Yes, I think that an ellipsoidal envelope would give practically the same results

Budding: Your figures appear to show higher densities towards the preceding hemisphere than the trailing hemisphere of the primary, yet the usual picture is that the action of the Coriolis force is to accumulate matter toward the following hemisphere. Have you any remarks about this?

Piirola: I remember computations of Prendergast and Taam which indicate that the circumstellar matter tends to have greater extension and density in the preceding side of the primary, just as we observe.

Plavec: I notice that the electron densities you give for the "envelope" and for the "disk" are actually the same--so is it necessary to distinguish two structures?

Piirola: The use of spherical envelope- and equatorial disk-components is for computing purposes only. One could try also with an ellipsoidal envelope.

Rucinski: You encountered problems in that the optically thin disc would give too much polarization; have you attempted to use the optically (and perhaps geometrically) thick disc to model your observations?

Piirola: Not yet. Calculations involving multiple scattering in optically thick envelopes will make the modeling rather complicated.

Smak: Why do you put most of the blame on the B star rather than on the disk?

Piirola: The model only assumes that most of the additional light comes from optically thick matter around the B star, i.e. the effective radius of the photosphere of the star is increased. This can be caused either by accumulation of matter around the star, expansion of the star itself or both. The crucial point is that more or less spherical structure around the B star seems necessary to explain the polarization curves.

IUE OBSERVATIONS OF LONG PERIOD ECLIPSING BINARIES: A STUDY OF ACCRETION ONTO NON-DEGENERATE STARS

Mirek J. Plavec
Department of Astronomy, University of California, Los Angeles

ABSTRACT

It has long been thought that β Lyrae is a unique system, by virtue of its UV spectrum and its nature. I will argue that a whole class of interacting long-period binaries exists, similar to β Lyrae. According to IUE observations made in 1978-79 by myself and R. H. Koch, this group comprises: RX Cas, SX Cas, V 367 Cyg, W Cru, β Lyr, and W Ser. AR Pav is a transition case linking them with the symbiotics. I also suggest that HD 218393 (KX And), HD 72754, and HD 51480 are their non-eclipsing counterparts. I will call the whole group the W Serpentis stars.

These systems are mass-transfering binaries (case B) in which the mass transfer rate is relatively high, probably on the order 10^{-6} to 10^{-4} $M_{\odot}$/year. They display an ultraviolet continuum with a color temperature definitely higher than the one observed in the optical region. Even more characteristical is the presence of strong emission lines of N V, C IV, Si IV, Fe III, Al III, and lower ions of C and Si. I will attempt to discuss these phenomena on the assumption that they are due to accretion onto non-degenerate stars.

INTRODUCTION

We seem to understand reasonably well how a star reacts to mass loss. Evolutionary sequences describing the evolution of the losers in mass-transfering binary systems were calculated already more than ten years ago. We understand very poorly how the on-flowing mass is accreted by the gainer. Several authors (to name only some: Flannery and Ulrich, 1977; Packet and de Grève, 1979) argue that the gainer will swell up rapidly soon after the onset of mass transfer and will also fill its critical Roche lobe. With less certainty, these authors suggest that rapid mass loss from the system will ensue, causing the system to shrink and thereby aggravate its instability against mass loss. One is tempted to wonder how all the observed Algols could have survived this ordeal!

M. J. Plavec, D. M. Popper and R. K. Ulrich (eds.), Close Binary Stars: Observations and Interpretation, 251–261.

This paper is a preliminary report on a systematic effort aimed at studying accretion observationally. Every Algol system offers us some information, and not all of it has been extracted yet. For example, the chemical composition and atmospheric structure of the gainers have been almost neglected, although they should provide important clues on the problem mass transfer vs. mass loss from the system. This is why J. Dobias and myself at UCLA have embarked on a program studying not only the composition of the losers (a controversial topic, see Plavec and Polidan, 1976), but also of the gainers. Spectrograms to be used in spectrum synthesis are being supplemented by scanner data on the energy distribution in the continuum, and by IUE observations in the far UV. The IUE spectra I have obtained (U CrB, RZ Sct, TX UMa, even the more anomalous AU Mon and V356 Sgr), and those obtained by Koch (U Cep, TT Hya, RW Per) appear normal and agree with the spectral classification in the optical region. This is not surprising, since the rate of mass transfer in ordinary Algols is probably quite low. The Algols are simply ladies with a most interesting past, but are (contrary to what is fashionable nowadays) extremely uncommunicative about that past.

DISCOVERY OF THE W SERPENTIS CLASS OF BINARIES

In order to study cases with higher rates of mass transfer, I began to study certain Algol-like systems of longer period, such as W Ser, V 367 Cyg, W Cru, and of course β Lyr. Fortunately, R. H. Koch had other similar stars on his (originally Sobieski's) program: SX Cas, RX Cas, and AR Pav (which can probably be classified both as an Algol and a symbiotic star). These systems have one important feature in common (apart from a relatively long period, 13 - 605 days): They display emission lines (H, He I, Fe II as the case may be) which, if the energy supply is radiative, imply the presence of a much hotter source in the system than the observed optical spectra indicate. For example, at Lick, Plavec, Polidan and Peters discovered emission lines of He I in W Ser (Fig. 1), whose optical spectrum is usually classified as F5 II, and which certainly shows no traces of anything hotter than early A. Because of these emission lines, the UV spectra promised important answers. What we found is exciting, in both senses of this word, but poses more problems than answers (for a preliminary report, see Plavec and Koch, 1978).

The seven systems enumerated above show two prominent features in the far ultraviolet: (1) A continuum with a color temperature higher than the one inferred from the optical spectrum, but apparently not hot enough to excite He I (about 12,000 K). (2) A very rich and prominent emission line spectrum with the N V doublet at 1240 A, C IV doublet at 1550 A, and lines of Si II, Si III, Si IV, C II, Fe III, Al III, etc. These are mostly zero-volt lines, that is transitions to the ground state level of the ion, and are no doubt excited collisionally. The source of energy maintaining the relatively high level of ionization is not so clear. If C IV and N V are ionized collisionally under conditions similar to the transition zone between the chromosphere and the corona in the Sun, then they require electron temperatures up to and above 200,000 K. The He II recombination lines, such as λ 1640 A, are absent (except in AR Pav).

Figure 2 shows the UV spectrum of SX Cas with its numerous strong emissions and deep absorptions, as well as its optical spectrum observed with the Lick scanner. The Figure shows that the whole energy distribution cannot be fitted by the Kurucz model atmosphere for a giant with T_{eff} = 8,500 K, which appears quite appropriate for the optical spectrum. Figure 3 shows the UV spectrum of W Serpentis and demonstrates that at primary mid-eclipse, which is partial, the continuum flux is diminished but the emission lines are almost unafffected. Observations of a total eclipse of SX Cas in February 1979 (Fig. 4) give an even more definite picture: When the optical spectrum of the gainer, usually classified as A6 III, disappears, the hotter UV continuum disappears as well, but the emissions stay and are not markedly weakened. A more complete discussion of our observations will be published elsewhere, jointly with R. H. Koch. Here I would like to present my own thoughts on them.

ACCRETION ON NON-DEGENERATE STARS

Obviously, both the optical continua (A6 III in SX Cas, F5 II in W Ser) and the hotter (perhaps B8) UV continua are associated with the gainers; therefore the gainers cannot be ordinary stars, and must have a more complex structure. Perhaps a more complete run through the total eclipse of SX Cas will reveal a phase shift in the eclipses of the two continua. On the other hand, the emission lines cannot originate anywhere near the gainer, since they are not eclipsed together with it. The emission lines are not too different from those typically observed in giants like Capella. In view of the close similarity between Capella and the G5 III loser in SX Cas, it is tempting to explain the emissions as the loser's chromosphere-corona signature. However, there are serious objections: (1) The energy radiated in these lines in SX Cas and other W Ser stars is several orders of magnitude greater than in Capella or even the RS CVn stars. (2) The N V line competes in strength with C IV, or is stronger, while in giant chromospheres and in all regions of the Sun, N V is always considerably weaker (Doschek et al., 1978). (3) In SX Cas and RX Cas a suitable G-type giant is available, but there is none in β Lyr (where the loser is B8 II). (4) In β Lyr, most of the emissions do not participate in the orbital motion of either component, and those few that do indicate trend to move with the gainer (Hack et al. 1976, 1977; Plavec and Dobias, 1979).

I believe that both the hot continua and the very hot plasmas whose presence has been detected by means of the IUE spectra, are related to a large rate of mass transfer in the W Ser stars. β Lyr, W Ser, SX Cas, RX Cas, and V 367 Cyg are known to display unusually large period variations; the two long-period systems, W Cru and AR Pav, are of course less well observed. All are case B systems, where a higher mass transfer rate must be anticipated at all times except at the very end of the process. Large and often erratic fluctuations of light are known to exist in all of them. The optical absorption spectra of the gainers in SX Cas, RX Cas, V 367 Cyg, and W Ser are shell spectra, indicating a dense circumstellar envelope, which is also manifested by the presence of emission lines even in the optical spectra.

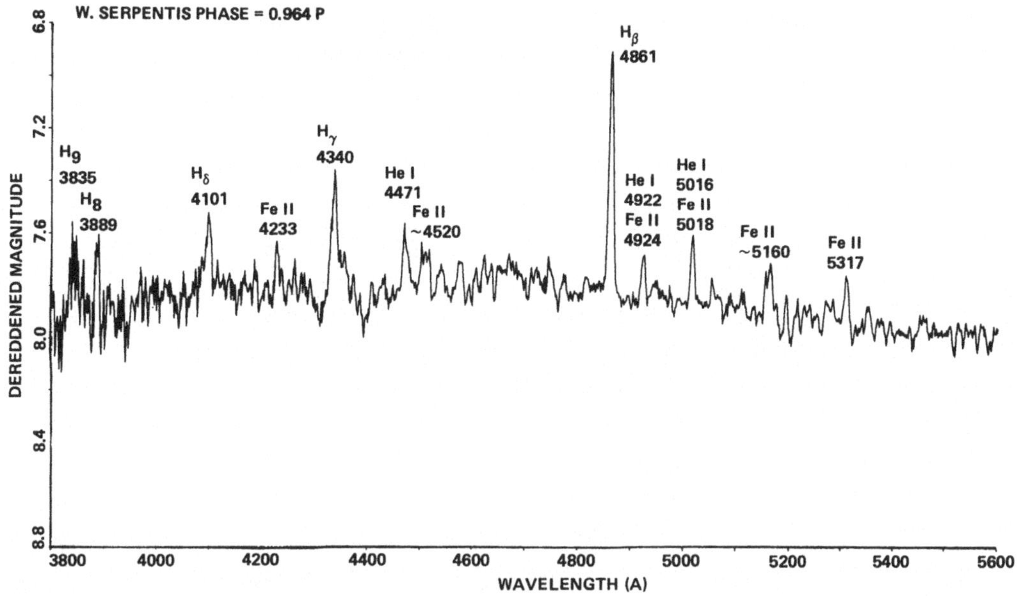

Fig. 1: Optical spectrum of W Serpentis at primary eclipse.

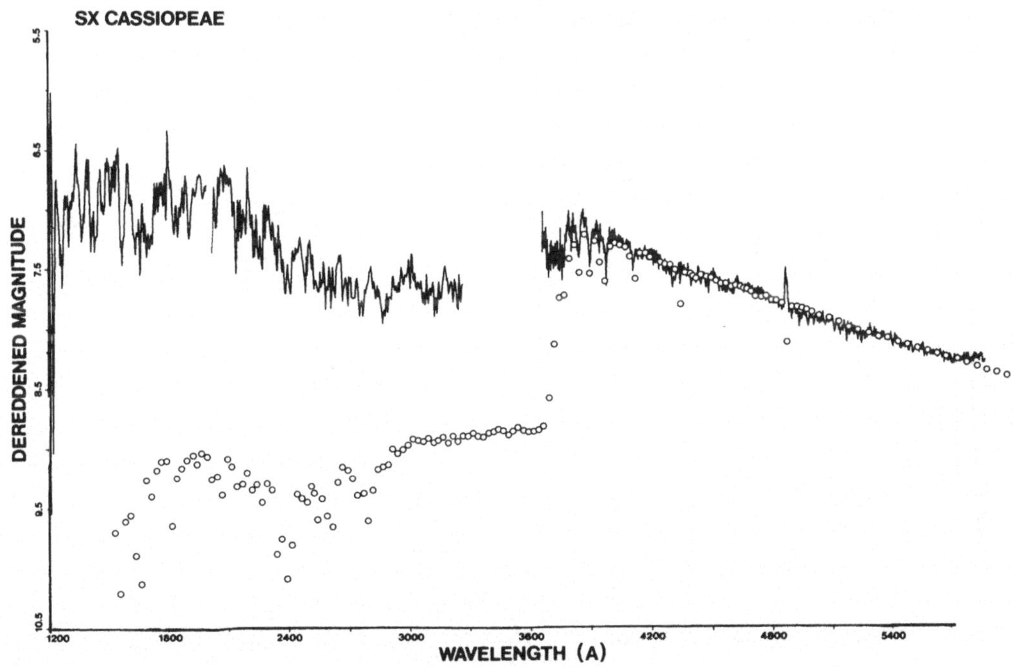

Fig.2: SX Cassiopeae: Energy distribution outside eclipse.

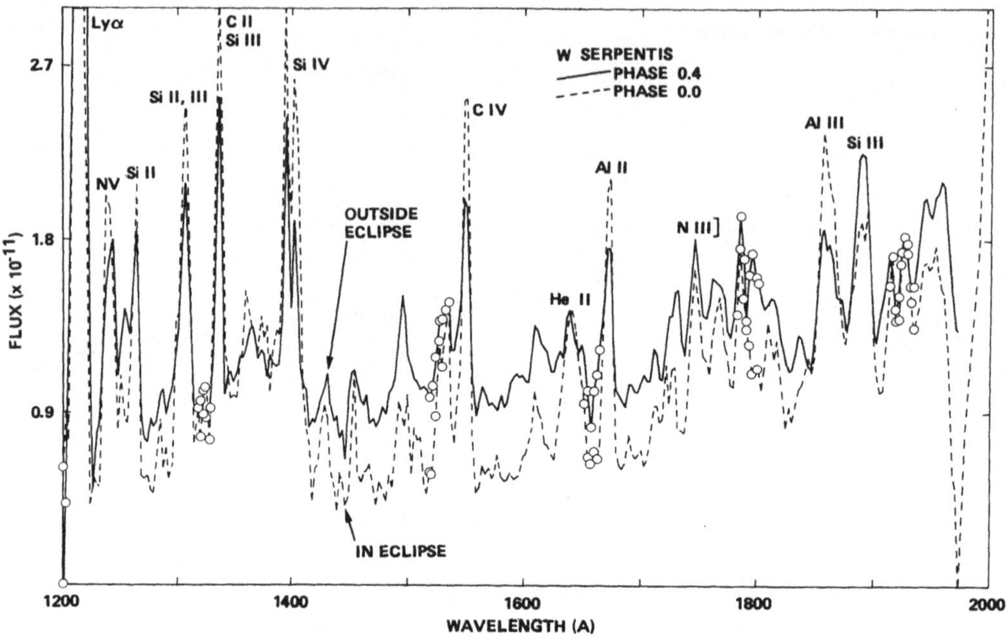

Fig. 3: Ultraviolet spectra of W Serpentis.

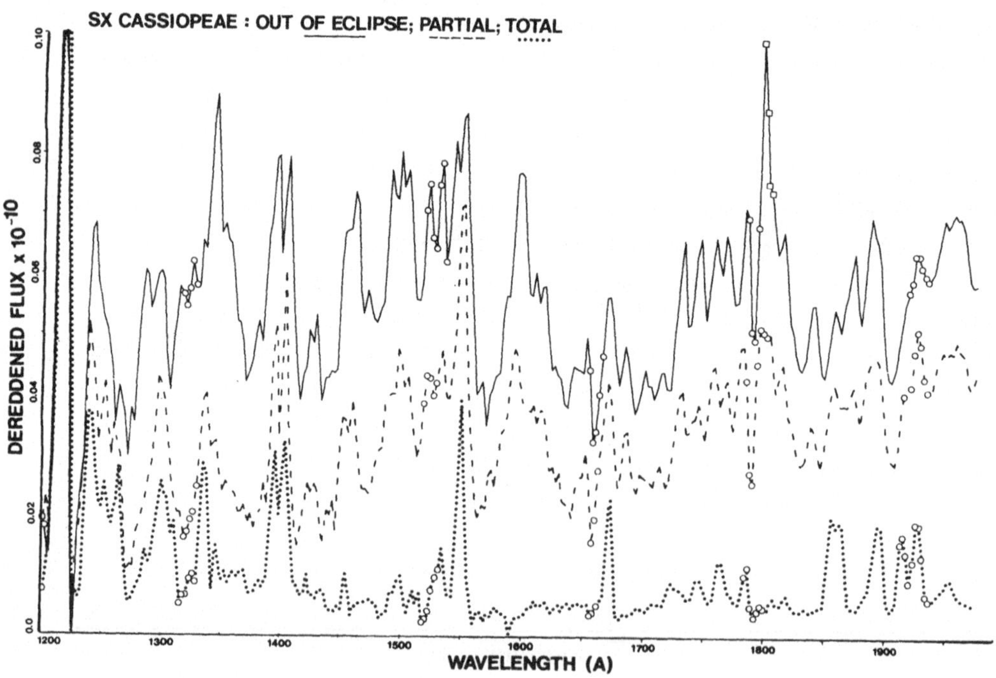

Fig. 4: Ultraviolet spectra of SX Cassiopeae.

All these facts suggest that we should study accretion processes in these stars, and apply the concepts developed for cataclysmic variables to accretion onto non-degenerate stars. We do not have a fully convincing evidence that the gainers in the W Ser systems are (or, better, initially were) main-sequence stars, but the systems appear to be a natural continuation of the Algols to longer periods. For simplicity, I will assume that the gainers are MS stars of moderate mass. My estimate of 3.6 $M_\odot$ for the gainer in SX Cas may be typical, or may be a bit below average). For some order-of-magnitude estimates it is useful to assume that the radii of the MS stars scale approximately as $R \sim M^{0.7}$ and intrinsic luminosities as $L \sim M^{3.5}$.

If mass is transferred ultimately to the surface of the gainer, the total accretion energy available is $L_a = 3.1\times10^7\ \dot{M}\ M/R$, in solar luminosities. (I will consistently express M, R, L in solar units, and the mass transfer rate $\dot{M}$ in solar masses per year). Comparing the accretion luminosity L_a with the star's intrinsic luminosity L_* , we find for MS stars

$$L_a/L_* \simeq 3.1\times10^7\ \dot{M}\ M^{-3.2}\ . \qquad (1)$$

For low- and moderate mass stars, it is fairly easy to produce more energy by accretion than by nuclear generation. If we assume a 3.6 $M_\odot$ main-sequence gainer in SX Cas, both luminosities will be equal for $\dot{M} = 2\times10^{-6}\ M_\odot$/year, which may very well be the correct value for SX Cas. β Lyrae, with its much more massive gainer ($\sim$ 15 $M_\odot$) requires $2\times10^{-4}\ M_\odot$ per year for $L_a = L_*$, but this value, too, is quite possible. Thus we must realize that in some case B systems, the true nature of the accreting star may be greatly obscured by the accretion process.

An upper limit to $\dot{M}$ for the gainer is set by the Eddington luminosity, which (for electron-scattering opacity) is $L_c = 3.2\times10^4$ M, and is reached for $\dot{M}_c = 1.1\times10^{-3}$ R. For most W Ser systems, where we can anticipate $\dot{M} \sim 10^{-5}\ M_\odot$/year, we are comfortably in the subcritical domain. However, even for MS gainers supercritical accretion is not ruled out, and certain observations of the emission lines in βLyr suggest that that system may not be far from critical accretion.

With MS gainers, it may happen that the characteristic accretion temperature will fall below 10^5 K, in which case photoionization tends to become more important than electron scattering. Buff and McCray (1974) showed that for photoionization as the main opacity source, the Eddington limit may fall below the above-given value by a factor of 10 if not 100. The effect is difficult to evaluate, since Buff and McCray assumed a simple continuous spectrum with a cutoff, while actually the EUV spectrum at these temperature is dominated by individual emission lines (Stern et al., 1978).

On the other hand, the peak rate of mass outflow from losers with radiative envelopes is $\dot{M} = 3.2\times10^{-8}$ R L / M , where all the values (in solar units) refer to the loser at the onset of mass transfer (Paczynski, 1971). For a 7 $M_\odot$ starting to lose mass just beyond the Main Sequence, this value is $\dot{M} = 10^{-4}$. Supercritical rate of mass loss will induce an optically thick wind blowing the material away from the gainer.

This effect should probably be taken into account when we discuss if and how the gainer survives the peak rate of mass transfer. Apparently, mass outflow from deep convective envelopes may easily lead to supercritical conditions, and may be responsible for the outbursts of the recurrent giant nova T CrB (Plavec, Ulrich, and Polidan, 1974; Webbink, 1976).

We often say that systems like β Lyr or SX Cas are in the rapid phase of mass transfer, but actually they are already in its slower, advanced stage, since in both of them, the masses are already reversed in favor of the gainer. It would be most exciting to catch a system at a peak mass transfer rate, even in the radiative case B.

Typically, in the W Ser systems, the size of the gainer is so small compared to the dimensions of the system that accretion occurs via an accretion disk. It is possible to scale up the disk models used for dwarf novae. If we do so, we find that a truly extended, optically thick disk is unlikely to form unless the mass transfer rate is quite high. Consider again SX Cas, which is a better determined case than the rest: It is totally eclipsing, has a reasonably good photometric solution (Shao, 1967; Koch, 1972), and Andersen (1973) published a fairly reliable radial velocity amplitude for the G5 III loser. Assuming that the giant fills its critical Roche lobe, the mass ratio is found to be ~ 8 in favor of the gainer, the separation of the components is $A \simeq 75\ R_\odot$, and the radius of the gainer's Roche lobe is $\sim 0.56\ A \simeq 42\ R_\odot$. The radius of the gainer itself is only $\sim 0.08\ A \simeq 6\ R_\odot$, i.e. the star appears to be located somewhat above the Main Sequence if its mass is really $3.6\ M_\odot$. A fully developed accretion disk could probably extend as far as some $30\ R_\odot$. Perhaps the double-peaked Balmer emission lines found by Struve (1944) indicate an edge of the disk, with their circular velocities of ~ 150 km/s. But for moderate rates of mass transfer (10^{-6} to $10^{-5}\ M_\odot$/year), the temperatures of the scaled-up disks may fall below 10^4 K. In the disk models (e.g. Shakura and Sunyaev, 1973), the peak disk temperature comes at $r = 1.36\ R \sim 8\ R_\odot$, and is for our model of SX Cas $T_{max} = 10^5\ \dot{M}^{\frac{1}{4}}$. At such low temperatures, the gas opacity is low, and a truly optically thick disk will probably not develop. Eclipse observations do not signal any presence of a very extended, optically thick disk. However, even an optically thick disk would be geometrically thin perpendicularly to the orbital plane, with a maximum thickness of perhaps 1/10 of the star's radius in SX Cas, and seen edge-on, it would hardly cause more than a slight perturbation in the eclipse light curve.

If an extended, optically thick disk existed in SX Cas, we would expect to see a hot spot where the on-flowing gas stream hits the disk. Because the potential well of the outer edge of the disk is quite shallow, the hot spot would emit a small fraction of the total accretion energy. The ratio L_{hs}/L_a scales approximately as $R/R_d \sim 0.2$ in SX Cas. A weak hot spot may exist in W Ser, where it is suggested by observations, but not in SX Cas, where there is no evidence for it. Part of the released accretion energy is radiated in the emission lines from a thin disk. An optically thick (but geometrically rather thin) disk may exist in SX Cas much closer to the surface of the gainer.

Most of the accretion energy is probably released at the interface between the disk and the accreting star. In the scaled-up disk model, accretion occurs along a narrow equatorial belt of the gainer. Adopting the expressions derived by Pringle (1977), the boundary layer temperature can be estimated to be no higher than

$$T = 2.3\times10^{6} \ \dot{M}^{\alpha} \ M^{\beta} \ R^{-\gamma} \ , \qquad (2)$$

where $\alpha = 6/19$, $\beta = 8/19$, and $\gamma = 18/19$. The color temperature of the "hot" UV continuum observed by IUE in SX Cas and W Ser is near 12,500 K , which indicates a mass transfer rate $\dot{M} \simeq 2\times10^{-6}\ M_{\odot}$/year, a very plausible result. When the corrected IUE 1978 spectra become available, it will be possible to be more definite about the relative luminosities of the transition zone and the underlying star, and estimate the thickness of the accretion belt; from my preliminary results, the semithickness is $H/R \simeq 0.06$.

A tentative picture that emerges is that in SX Cas, the UV continuum is due to accretion in the boundary layer, while the optical region is dominated by the radiation of the star itself. It should be stressed, however, that the observed absorption spectrum of the gainer is a shell spectrum, so that some kind of gaseous envelope -- optically thick at least in the absorption lines of Fe II, Ti II, etc. -- surrounds most of the gainer, not only a thin equatorial belt.

THE EMISSION LINES

The high-ionization emission UV lines in SX Cas radiate a total power of several solar luminosities. All the emission spectra of the "W Serpentis stars" are very similar to each other. Similarity with β Lyrae is very important, since for that system it was possible to obtain high-dispersion spectra. All the emission lines in β Lyr have P Cygni profiles indicating an outflow at a velocity of about 150 km/s. Moreover, high dispersion spectra permit to decide with confidence that among the expected intercombination lines, only Si III] λ 1892 A is present. Its comparison with the permitted Si III line at 1298 A, as calibrated by Nicolas (1978), and the absence of the other intercombination lines indicate that the electron density is surprisingly high, $N_e \simeq 3\times10^{12}\ cm^{-3}$. If the chromospheric model is applied, the resulting emission measures indicate that the lines are formed in a very thin slab. This is hard to reconcile with the P Cyg character. The emission lines do not participate in the orbital motion of either component, with a very few exceptions (C III λ 1175 A seems to be emitted by a region close to the gainer, and is moving with it and expanding at the same time: Hack et al., 1977). Moreover, the strength of the emission lines does not appear affected by the eclipses.

It can be concluded that most probably the line-emitting plasma surrounds the whole system. Quantitatively, this leads to a very high rate of mass outflow from β Lyr, if we interpret the observed outflow velocity as the local escape velocity, and at the same time accept the high value of electron density mentioned above. The separation of the components in β Lyrae is ∿ 50 $R_\odot$, and the radius of the emitting region should be several times as large. Adopting for it D = 200 $R_\odot$, we get $\dot{M} \sim 3\times10^{-3}$ $M_\odot$/year or more, i.e. we are essentially in the domain of the critical accretion. The required kinetic energy itself is close to 10^4 $L_\odot$. Often there exists an approximate equipartition between the energy radiated and the energy locked in mass motions. It is therefore puzzling that the UV continuum of SX Cas is rather similar to that of β Lyr. It is true that the accretion disk in β Lyr is certainly optically thick, since it behaves like a star; thus a much higher $\dot{M}$ is to be anticipated for β Lyrae. My estimates based on the rate of the period variation (Hack et al., 1975) gave for it only 5×10^{-5} $M_\odot$/year. Perhaps we should think in terms of a very anisotropic distribution of both the mass motions and the radiation.

It appears that the origin of the emission lines remains a puzzle. They are no doubt also connected with the accretion process, and fed from the accretion energy.

CONCLUDING REMARKS

It is very fortunate that we have a whole group of systems with similar properties. Each of the systems is so complicated that it is hardly possible to extract all the necessary information from it. It is hoped that a "group treatment" will be more succesful. β Lyrae is of course the brightest and best known member of the group. It may not be the simplest case, however, because the gainer is at all wavelengths accessible to us now fainter than the loser (B8 II). The most complex behavior we have observed is that of W Ser during primary eclipses. This may be due to the fact that -- unless our model is quite wrong -- the loser in W Ser acts more or less like a dark screen, thereby permitting us to see all the complexities of the structure of the gainer. It is because of this promise that I think the systems should be labeled the W Serpentis stars.

The research reported here has been supported by grants from NSF and NASA. Assistance of the IUE Observatory staff is acknowledged.

REFERENCES

Andersen, J.: 1973, Publ. Astron. Soc. Pacific 85, 191.

Buff, J. and McCray, R.: 1974, Astrophys. J. 189, 147.

Doschek, G.A., Feldman, U., Mariska, J.T., and Linsky, J.L.: 1978, Astrophys. J. Lett. 226, L 35.

Flannery, B.F. and Ulrich, R.K.: 1977, Astrophys. J. 212,533.

Hack, M., Hutchings, J.B., Kondo, Y., McCluskey, G.E., Plavec, M., and Polidan, R.S.: 1975, Astrophys. J. 198, 453.

Hack, M., Hutchings, J.B., Kondo, Y., and McCluskey, G.E. : 1977, Astrophys. J. Suppl. 34, 565.

Koch, R.H.: 1972, Astron. J. 77, 500.

Nicolas, K.R.: 1977, PhD thesis, University of Maryland
Paczynski, B.:1971, Ann. Rev. Astron. Astrophys. 9, 183.
Packet, W. and De Grève, J.P.: 1979, Astron. Astrophys. 75, 255.
Plavec, M. and Polidan, R.S.: 1976, in "Structure and Evolution of Close Binary Stars" (ed. P. Eggleton, S. Mitton, and J. Whelan: Reidel) p. 289.
Plavec, M. and Koch, R.H.: 1978, Inf. Bull. Var. Stars No. 1482.
Plavec, M., Ulrich, R.K., and Polidan, R.S.: 1974, Publ. Astron. Soc. Pacific 85, 769.
Pringle, J.E.: 1977, Mon. Not. R.A.S. 178, 195.
Shakura, N.I. and Sunyaev, R.A.: 1973, Astron. Astrophys. 24, 337.
Shao, C.Y.: 1967, Astron. J. 72, 480.
Stern, R., Wang, E., and Bowyer, S.: 1978, Astrophys. J. Suppl. 37, 195.
Struve, O.: 1944, Astrophys. J. 99, 89.
Webbink, R.F.: 1976, Nature Phys. Sci. 262, 271.

DISCUSSION FOLLOWING PLAVEC

Andersen: A few years ago, I spent quite some time measuring Alan Batten's plates of SX Cas (PASP 85, 191, 1973), and it left me with the conviction -- very much in agreement with your model -- that I had never seen the spectrum of the gainer itself, only the shell. The radial velocities of the gainer also show very large scatter, as you know, while the few measurements of the loser seemed more reasonable. Now, in your model, the loser fills its Roche lobe. Don't you think that the mass ratio implied by this assumption is extreme?

Plavec: You are quite right. In view of the probably large rate of mass transfer, I am ready to accept it. If we assume mass loss from the loser via stellar wind only, we may reduce the mass ratio to about 3, but hardly below this value, since otherwise both masses will become unrealistic. It would be necessary, however, to explain why the G5 III star, i.e. a star of relatively low luminosity, should have such a strong stellar wind.

Leung: From my preliminary analysis of the light curves of V 367 Cygni, I found that this system is extremely close. It has either a semidetached or contact configuration. I welcome your suggestion that the temperature of the gainer should be higher. The previously published spectral type leads to unacceptable underluminosity of this star.

Plavec: The gainer is probably a complex structure. The system is rather similar to β Lyrae, and may possess a thick accretion disk simulating a contact component.

Friedjung: A similar situation seems to exist for the symbiotic star Z Andromedae about which I shall talk on Friday. The high ionization emission lines and the hot continuum observed by IUE do not seem to come from the same region.

Plavec: R.H. Koch and myself found that the star AR Pavonis can be classified both as a symbiotic object and as a member of our W Ser class. There indeed seems to exist a profound similarity.

Collins: A definitive consequence of your model is that the observations of the hot star's equatorial region should reveal a very large polarization (of the order of 10%) in the far ultraviolet. Thus polarization observations in the UV should provide significant observational check on mass transfer models where high temperatures are anticipated.

Plavec: I would also recommend attempts to detect flickering, in particular in W Ser.

Scarfe: Would you include RZ Ophiuchi among your W Serpentis-type binaries?

Plavec: Yes, I believe so. Other suspects are RS Cep and RX Gem. Douglas Hall suggested some ten years ago that these two latter stars have accretion disks.

Wilson: What is the prospect for finding the rotational velocities of the higher mass components in any of your systems?

Plavec: The observed optical spectrum of SX Cas is a shell spectrum. Typically in shell stars, we observe an underlying photospheric spectrum due to a fairly rapidly rotating central star. The same may be the case in SX Cas, but so far I have not been able to detect the broadened lines. I read the abstract of your and Twigg's talk on rotationally unstable gainers (to use my terminology), and I think that your ideas should be seriously pursued. We know that U Cephei rotates fast, no doubt because of accretion. That star is of course not shielded by a large disk, since it is itself large relatively to the size of the system.

ON THE EXISTENCE OF DOUBLE-CONTACT BINARIES

R. E. Wilson and L. W. Twigg
Department of Astronomy
University of Florida
Gainesville, Florida 32611

We present observational evidence relevant to the existence of binaries of the morphological type proposed by Wilson (1979), called double-contact binaries. Consider normal evolution with Roche lobe overflow, in the rapid phase of mass exchange. It is well known that such mass-exchange spins-up the accreting star (most of whose mass is transferred material by the end of the rapid phase). Many examples are known (Table 1, for example) of Algol-types with fast-spinning primaries (original secondaries). Suppose the accreting star has been spun-up so much that it cannot accept further high-angular momentum material. It will then be filling its limiting rotational lobe (but not its Roche lobe). The secondary (original primary) will still be filling its Roche lobe. The overall result is that both stars fill their limiting lobes accurately, but are not in contact with one another. Since it is common to speak of "the contact component" of a semi-detached system - meaning the one in contact with its limiting lobe - it seems natural to call members of the new morphological type "double-contact" systems. Of course, there must be a repository for the mass lost by the Roche lobe filling and rotational lobe filling components. This would naturally be the circumstellar disks around the primary components of such systems as β Lyr and V356 Sgr (Wilson and Caldwell, 1978). The end of the double-contact phase would occur when the rate of transfer of rotational angular momentum back into the orbit, by tidal braking, begins to exceed the rate of transfer of angular momentum the opposite way by mass exchange.

Table 1 summarizes information on a number of possible double-contact systems we have investigated so far. One should direct attention toward Algol-types with rapidly rotating primaries. Here we have used the rotational disturbance (Rossiter effect) to estimate the ratio, F, of rotation rate to the synchronous rotation rate. Figures 1 to 7 show fits to the rotational eclipse disturbance for 6 binaries. Figure 8 is an example of a light curve fit. Column 5 of Table 1 gives quantitative estimates of F for U Cephei, RZ Scuti, AQ Peg, SW Cygni, U Sagittae, and RY Persei. By analysing light curves of the same binaries,

M. J. Plavec, D. M. Popper and R. K. Ulrich (eds.), Close Binary Stars: Observations and Interpretation, 263–268.

we have estimated also the percent of the rotational lobe which the primary fills and the percent of the Roche lobe which it fills. If the former number is about 100%, we have a strong candidate for a double-contact binary. As the table shows, U Cep is the only strong candidate at present, but others may emerge when allowance is made for several effects which reduce the apparent rotational disturbance. All of the binaries listed in Table 1 show peculiarities, such as variable emission lines near eclipse and shoulder distortions in the light curves.

This work was done under N.S.F. grant AST-09188.

Table 1

Summary of Results

Name	P(days)	M_2/M_1	SP_1	$\frac{V}{V_{syn.}}$	% Lobe Filled	% Lobe If Syn.	V_{eq} (km/sec)
U Cep	2.49	0.644	B7 V	10	100	32	380
RZ Sct	15.19	0.209	B2 II	10	61	19	268
U Sge	3.38	0.371	B8.5 V	5-6	72	37	260-312
AQ Peg	5.55	0.390	A2e	9	70	22	158
RY Per	6.86	0.260	B3 V	9	48	19	198
SW Cyg	4.57	0.301	A2e	5	46	23	125

REFERENCES

Gaposchkin, S. 1952, Harvard Ann. 113, p. 2 (AQ Peg, RY Per).
Hansen, H. K. and McNamara, D. H. 1959, Ap. J. 130, p. 789 (RZ Sct).
Hardie, R. H. 1950, Ap. J. 112, p. 542 (U Cep).
Hiltner, W. A. 1946, Ap. J. 104, p. 396 (RY Per).
Markworth, N. 1977, Ph.D. Dissertation, Univ. of Florida (U Cep).
McNamara, D. H. 1951, Ap. J. 114, p. 513 (U Sge).
McNamara, D. H. and Feltz, K. A. 1976, Pub. A.S.P. 88, p. 688 (U Sge).
Neubauer, F. J. and Struve, O. 1945, Ap. J. 101, p. 240 (RZ Sct).
Struve, O. 1946a, Ap. J. 103, p. 76 (AQ Peg).
Struve, O. 1946b, Ap. J. 104, p. 253 (SW Cyg).
Struve, O. 1949, Monthly Notices R.A.S. 109, p. 492 (U Sge).
Walter, K. 1971, Astr. and Ap. 13, p. 249 (SW Cyg).
Wilson, R. E. 1979, Ap. J. (December 15, in press).
Wilson, R. E. and Caldwell, C. N. 1978 Ap. J. 221, p. 917.

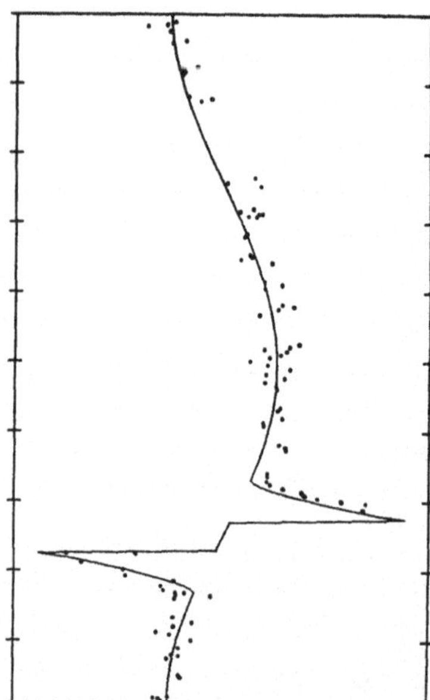

Figure 1. Observed and theoretical radial velocity curves for U Cep.

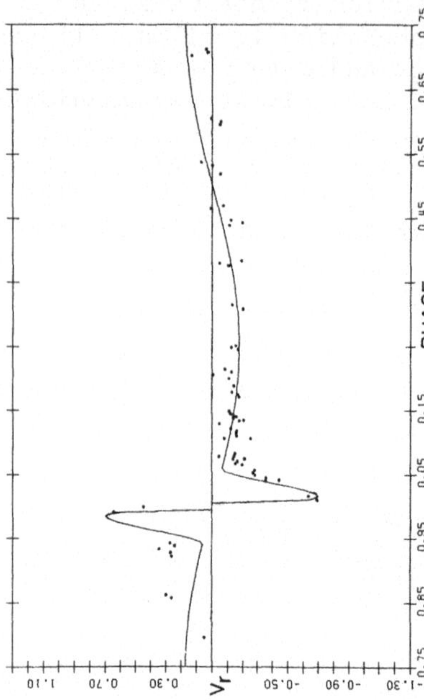

Figure 2. Radial velocity curves for RZ Sct (Hansen and McNamara observations).

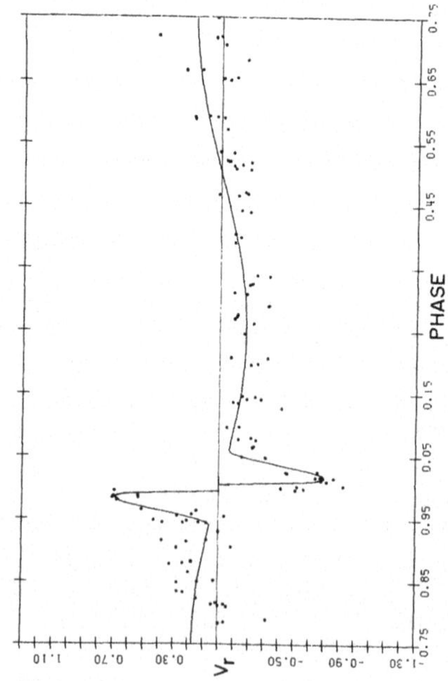

Figure 3. Radial velocity curves for RZ Sct (Neubauer and Struve observations).

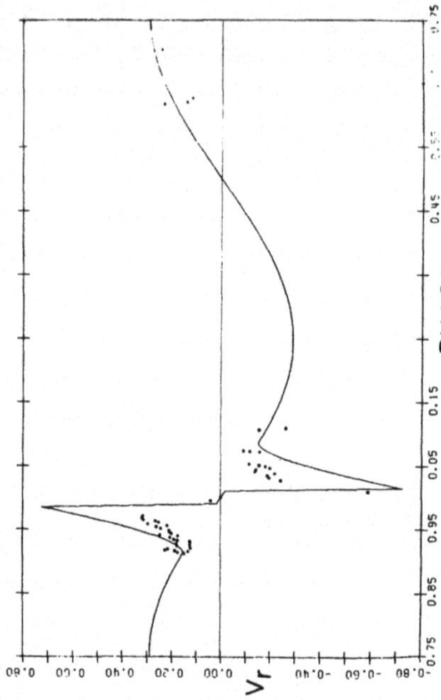

Figure 4. Radial velocity curves for U Sge.

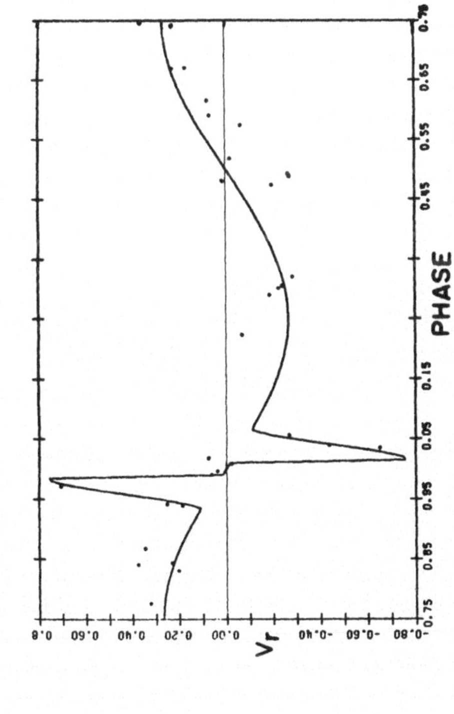

Figure 5. Radial velocity curves for AQ Peg.

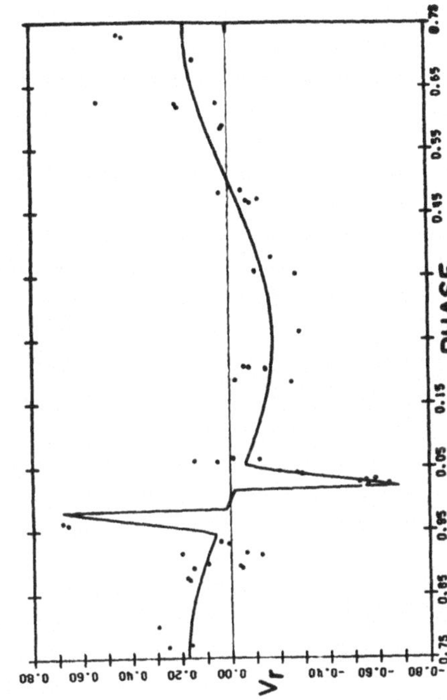

Figure 6. Radial velocity curves for RY Per.

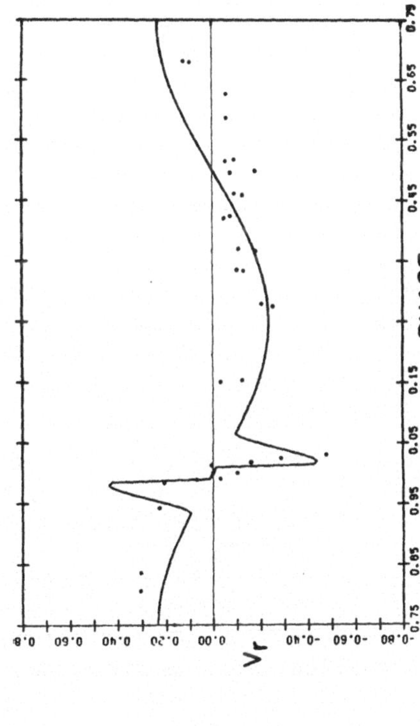

Figure 7. Radial velocity curves for SW Cyg.

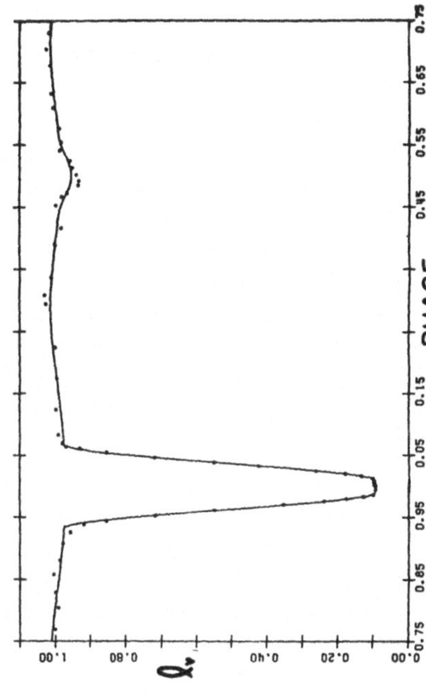

Figure 8. V light curve of SW Cyg and differential corrections solution curve.

COMMENTS FOLLOWING WILSON AND TWIGG

Smak: I am worrying about very serious systematical errors which are likely to affect your basic conclusion. It is rather obvious that the velocity of rotation can be obtained quite accurately from the Rossiter effect. They have to be compared with theoretical predictions involving several uncertain parameters, which come from the radial velocity and light curves. As for the radial velocities, all the systems you mentioned have been single-line objects, with radial velocity curves severely distorted by the presence of emission and absorption component (most dramatic examples being RZ Sct and U Sge); we could see, in many cases, how poor was the fit of your theoretical curves to the observed points outside of eclipse. To summarize, the K_1 values are very uncertain. As for the light curves, you need them to determine the radii and--indirectly--the mass-ratio. Again, these light curves are known to contain severe distortions due to disks, etc. and yet you are adding extra free parameters appropriate for a rapidly rotating star (flattening, gravity darkening). The basic quantity that enters your comparison is $GM_1/R_1 \sim \frac{K_1^2}{r_1}(1 + \frac{M_1}{M_2})\frac{M_1}{M_2}$ and one can see how sensitive it is to the uncertainties in k_1, r_1, and $g = g(r_2)$.

Wilson: I should emphasize that the photometric solutions were done from good photoelectric light curves for most of the systems. They were done by differential corrections, with the cooler star constrained to fill the Roche lobe. The geometry is rather well determined. We proceed tentatively with the figure of the hot star correctly computed from the rotation rate derived from the rotational disturbance. Thus there are not very many free parameters at all. I do not believe the masses could be wrong enough to undermine the overall result, which is that there do exist Algols (with spectroscopically visible hot components) which are fairly close to the double-contact configuration. By inference, the slightly faster rotating (and spectroscopically invisible) actual double-contact systems are likely to exist also.

Kuzma: You might run into trouble using the Rossiter effect to obtain rotational velocities. Gravity darkening and limb darkening lead to non concentric isophotal lines. This is also strongly dependent on the wavelength used. This can introduce a large ambiguity into the radial velocity obtained.

Wilson: These effects are all accounted for rather rigorously in our computations. The only possibly significant effect we do not treat is the variation of line strength over the surface.

Polidan: Later in this session I will discuss two non-eclipsing semi-detached systems that support your suggestion that there is a limit to the accretion rate for a rapidly rotating star. In these systems the "mass accreting" star is very rapidly rotating. Both stars display UV resonance line profiles indicative of strong stellar winds. It appears that the majority of the material being lost by the secondary is not accreted by the rotating star, rather it is ejected from the binary.

Pringle: I wish to dispel the idea that the accreting star can spin so fast that it cannot accrete any more mass from an accretion disc. An accretion disc is very efficient at redistributing angular momentum and can enable the star to accrete without problem.

Wilson: It seems to me that the time scale for tidal braking of the outer stellar envelope could be sufficiently long to allow a thick disk to be built up. As I understand the present situation, the amount of dissipation in such an envelope is at least somewhat controversial.

Rucinski: The shape of the component at "rotational contact" depends on the distribution of angular momentum versus radius; what sort of a distribution do you assume in your calculations?

Wilson: We assume uniform rotation (constant angular velocity). However differential rotation should affect the figure significantly only if the core rotates faster than the envelope. Here the core would rotate slower than the envelope, if anything.

Collins: If you actually believe that the star is rotating at near critical velocity then you must use great care in producing the line profile. For instance, the use of the Stoeckley-Mihalas atmosphere is inappropriate for velocities corresponding to a $W > 0.5$. These problems can perhaps be best demonstrated by noting that the radius which results from the light curve analysis cannot be unambiguously associated with any radius on the star. Since in the case of critical rotation whose radius varies by 50%, this is not a negligible problem.

Wilson: Since we compute the rotational and tidal distortion (and gravity darkening, etc.) in detail, there is not just one radius which characterizes the photometric solution. I agree that the refinement you propose could be significant, but it could hardly change our basic conclusions.

Hall: The question of the reliability of Wilson's absolute dimensions can be checked by seeing if they have the cooler star filling its Roche lobe and also have the hotter star obeying the mass luminosity relation. Such a test was used in Acta Astronomica 24, 215 to check absolute dimensions of many such systems, which have incomplete or unreliable radial velocity curve amplitudes.

Wilson: The solutions were done with the cooler star constrained to fill its Roche lobe. I maintain that the solution radii are less uncertain than is the mass-luminosity relation.

STRUCTURE OF THE GASEOUS DISK IN RX CAS

S. Kříž
Department of Astronomy and Astrophysics, Charles University
Švédská 8, Prague, Czechoslovakia

RX Cas appers to be a long-period Algol-like binary with very fast mass exchange (Struve 1944). In the period 1975-1977, this star was observed photometrically by Arsenijevič, Grygar, Harmanec, Horn, Koubský, Kříž and Pavlovski at Hvar Observatory (Yugoslavia) and by Zverko at Skalnaté Pleso Observatory (Czechoslovakia). The resulting UBV light curves seem to be rather peculiar, for the following reasons:

1) Combining our observations with the older ones, we obtain the following formula for the times of minima:

$$\text{Min} = \text{J.D. } \underset{\pm .053}{2416250.953} + \underset{\pm .00025}{32.31195}\,E + \underset{\pm .026}{1.030} \cdot 10^{-5}\,E^2.$$

The period increase is surprisingly high.

2) The scatter of observed points is considerable (about 0.2 mag.) and cannot be caused by observational errors. It corresponds to short period light variations on the time scale of several days.

3) After J.D. 24422660, the brightness of the whole system decreased in all colours. This decrease is about 0.2 mag. in V colour and 0.5 mag. in U colour. Only after J.D. 2442900 the brightness increased again. Such a behaviour confirms the reality of old observations by Gaposchkin (1944) and Martynov (1950) who found a long-term light variation with a period of approximately 500 days.

4) The part of the light curve around the phase 0.75 is definitely much lower than the corresponding part around the phase 0.25. This effect is most pronounced in the ultraviolet; the depression around the phase 0.75 is much deeper than the secondary minimum. This behaviour is opposite to the behaviour of dwarf novae. Instead of a bright shoulder connected with a bright spot, we observe some darkening.

We propose the following working model of RX Cas: The late-type component overflows the corresponding Roche lobe. The outflowing

M. J. Plavec, D. M. Popper and R. K. Ulrich (eds.), Close Binary Stars: Observations and Interpretation, 269–270.

material is transferred to the detached mass-gaining component through an accretion disk. Contrary to the dwarf novae, the accretion disk almost fills its Roche lobe. Therefore, the transferred material cannot accelerate and form a hot spot. This cold material is mixed with the material of the disk in the region visible around the phase 0.75, covers the hot central parts of the disk and causes the depression of the light curve near this phase. A considerable part of the light arises from the disk and from the transition region between the disk and the mass-gaining component (especially in the ultraviolet). The long-period light variations are connected either with the variable mass transfer predicted by Bath (1972) or with some instability of the disk itself which may be similar to the instability proposed by Osaki (1974) and Paczynski (1978) for explanation of the outbursts of dwarf novae.

A detailed paper dealing with the photometry of RX Cas will be published in Bull. Astron. Inst. Czechosl.

REFERENCES

Bath, G.T.: 1972, Astrophys. J. 173, pp. 121-134.
Gaposchkin, S.: 1944, Astrophys. J. 100, pp. 230-241.
Martynov, D.J.: 1950, Bull. Astron. Obs. Englehardt 27, pp. 3-47.
Osaki, Y.: 1974, Publ. Astron. Soc. Japan 26, pp. 429-436.
Paczynski, B.: 1978, in Nonstationary Evolution of Close Binaries, ed. A.N. Zytkow (Warsaw: Polish Scientific Publishers), pp. 89-98.

DISCUSSION FOLLOWING KŘÍŽ

Pringle: Is the disc you are talking about two-dimensional, or does it fill the whole three-dimensional Roche lobe?

Kříž: It is essentially flat, i.e. two-dimensional. I have presented only a phenomenological model. I computed a beautiful theoretical model, but it does not fit my observations.

K. Walter: The fluctuations in the light curves shown here seem to be of a similar type to those I found for the Algol system TV Cas. I observed that orbital system, with a period of 1.8 days, during six years and found clear fluctuations with a basic period of 600 days. This period could be interpreted as the precessional period of the rotational axis of the primary component, if it is assumed that the this axis is somewhat inclined to the orbital plane. My paper will appear in the Astronomy and Astrophysics Supplement. For RX Cas, with its orbital period of 32 days, a very long precessional period is to be expected.

Plavec: In long-period systems such as RX Cas, circumstellar disks and envelopes play a dominant role, namely in the optical region, and their instabilities may easily cause the fluctuations.

IUE OBSERVATIONS OF THE TWO PECULIAR BINARY SYSTEMS BETA LYRAE AND UPSILON SAGITTARII+

Margherita Hack, Umberto Flora, Paolo Santin
Astronomical Observatory, Trieste, Italy

The common peculiarities of these two systems are: a) the companion is a massive object (probably $m_2 \gtrsim 10$) whose spectrum is not observable; b) both systems show evidence, though in different degrees, of mass-transfer and mass-loss; c) both present, in different degrees, hydrogen deficiency; d) ultraviolet observations have shown, in both cases, the presence of lines of highly ionized elements like N V, C IV, Si IV, probably formed in an extended envelope because they do not show orbital radial velocity shifts, and cannot be explained by the effective temperature of the star whose spectrum we observe. The latter property seems to be common to several close binaries, as shown by the ultraviolet observations with IUE by Plavec and Koch (1979); e) both systems present infrared excess, suggesting the presence of an extended envelope (Gehrz et al. 1974; Lee and Nariai, 1967; Humphreys and Ney, 1974; Treffers et al. 1976).

The main characteristics of the two systems are summarized in Table 1.

The IUE observations and their purpose.

Although Beta Lyrae had been observed very extensively with Copernicus (Hack et al. 1975, 1976, 1977) a few problems were left, because the spectral region 1500-2000 A, where the sensitivity of the spectrometer was extremely low, had not been observed. In this region are located two lines, λ 1640 He II and λ 1908 C III] , which are important for deriving the physical conditions of the envelope. During the first run of observations of Beta Lyrae with Copernicus, it was a surprise to observe the strong emission of the resonance lines of N V and

+ Based on observations by the International Ultraviolet Observer (IUE) collected at the Villafranca Satellite Tracking Station of the European Space Agency.

M. J. Plavec, D. M. Popper and R. K. Ulrich (eds.), Close Binary Stars: Observations and Interpretation, 271–286.

the apparent absence of λ1085 He II (blended with the resonance lines of N II). A confirmation of the weakness or absence of He II was given by low resolution spectra obtained with an objective prism spectrograph on Skylab by Kondo et al. (1976), and its complete absence was confirmed by IUE observations. This result suggested that the far UV emission lines are not radiatively but collisionally excited.

Table 1 - Properties of the systems

	β Lyr			ν Sgr		
P (d)	12.9349			137.9567		
f(m) ($\odot$)	8.73			1.677		
K (Km s^{-1})	184.0			49.12		
γ (Km s^{-1})	-17.8			-13.29		
a sini (Km)	32.7×10^6			9.3×10^7		
i (°)	$\geq 75°$			47°		
Sp.Prim.	B8-B9			B8,F0 Ia		
H/He	1/2			1/40		
m_1/m_2 (i=90°)	1/1.6	m_1=14	m_2=21.4			
" "	1/6	1.6	11.8			
m_1/m_2 (i=47°)				1/2	m_1=4.8	m_2=9.64
" "				1	17.1	17.1
" "				2	38.6	19.3
Probable type sec.	B0-B5 III-IV or Black Hole			09 V		

The existence of strong emission in the region 1900-2100 A was shown by the low resolution observations obtained by Hack (1974) with experiment S2/68 on TD-1 and by Kondo et al. (1976) with Skylab. Hack attributed these emissions mainly to the strong Fe III lines present in this region. However, a possible contribution could be due to λ1908 C III] whose emission is strong in several Wolf-Rayet stars, symbiotic stars, and chromospheric spectra of late-type stars. This line requires a gas density of 10^{10} cm^{-3} or less (Osterbrok, 1970). Now the IUE observations have excluded its presence, putting the lower limit for the density of the envelope above 10^{10} cm^{-3}.

Upsilon Sgr was also observed with Copernicus, but only in two limited spectral regions (Duvignau et al. 1979) and with very low S/N. The present observations cover the epochs of the two minima and the two quadratures in the spectral range 1170-3200 A. Table 2 gives the list of the observations. All the exposures were obtained with the 3" aperture.

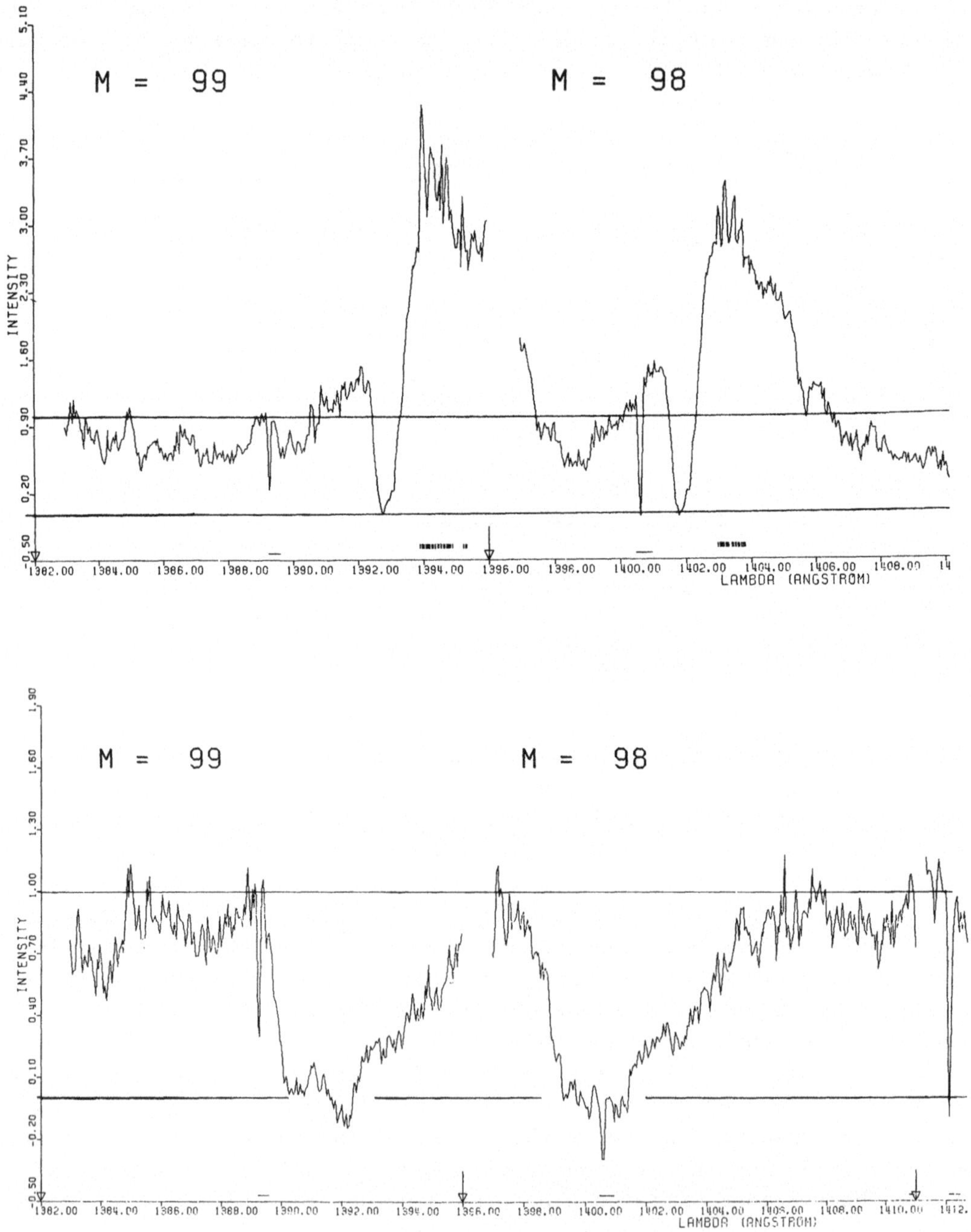

Figure 1. The profiles of the Si IV resonance lines. Top: Beta Lyr (phase 0.51); bottom: Upsilon Sgr (phase 0.02).

Table 2. Observations
(High Resolution, 3" Aperture)

β Lyr Image	J.D	Phase	Exposure
	2443		
3.1375	615.87	0.512P	8 min.
2.1328	.91	0.515	4
3.1386	17.90	0.669	2
3.1426	25.78	0.279	2
2.1397	.81	0.281	2
υ Sgr			
	2443		
3.1398	620.72	0.02 P	60 min.
2.1361	.72	0.02	10
2.1528	49.59	0.23	10
3.1592	.63	0.23	60
3.1856	85.50	0.49	60
2.1909	718.47	0.73	10
3.2136	.50	0.73	60
2.1910	.60	0.73	20
3.2137	.62	0.73	75

The line-spectra.

The most conspicuous characteristics of the spectra of the two systems are the following: the far ultraviolet spectrum is dominated by strong ground level or metastable lines of multi-ionized atoms like C II, C III, C IV, N V, Si II, Si III, Si IV, Al II, Al III, and low excitation lines of Fe III. Copernicus observations of Beta Lyr (Hack et al. 1975) in the region at $\lambda < 1150$ A have shown also the presence of N II, S III, S IV, ground level lines of Fe III, and the absence of the lines of O VI.

The most striking difference between the spectra of the two systems is the shape of the line profiles. The spectrum of Beta Lyr is characterized by the presence of strong emissions cut by sharp absorption cores shifted shortward by about 150 km/s; the same lines, in the spectrum of Upsilon Sgr, present very broad, asymmetric profiles in absorption, with a sharp edge on the side of the short wavelengths, indicating a terminal velocity of about 700 km/s. (Fig. 1)

These lines do not show appreciable shifts due to orbital motion.

The regions where there are no resonance lines of abundant ions are very similar in the two systems. The Fe II lines, which dominate the region 1600-1700 A, present radial velocity shifts indicating that they are formed in the atmosphere of the primary. In the spectrum of Beta Lyr λ1640 He II is absent. In Upsilon Sgr there is a sharp absorption at λ1640.5; moreover there is a broad depression (missing in Beta Lyr) centered at λ 1640 and extending from λ 1639 to λ 1643. This depression may be due to a broad He II absorption. (Fig. 2)

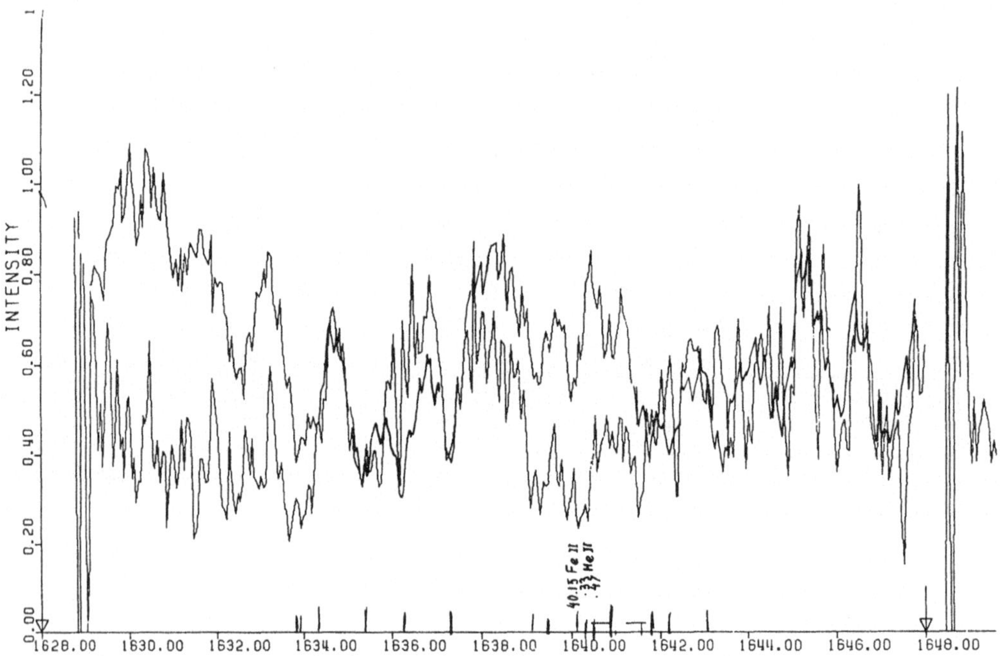

Figure 2. The region of λ 1640.5 He II: top Beta Lyr (phase 0.51), bottom Upsilon Sgr (phase 0.49).

The region 1850-2200 A, where many strong lines of Fe III are located, differs in the two systems, because in Beta Lyr the strongest lines of Fe III present P Cygni profiles with rather strong emission lines, while in Upsilon Sgr they are in absorption and fainter than in Beta Lyr. The region 2200-3100 A is dominated by lines of once-ionized metallic atoms (mainly Fe II). The radial velocity shifts clearly indicate that these lines are formed in the atmospheres of the primary stars. The absorption lines of the once-ionized metals are generally stronger in Upsilon Sgr than in Beta Lyr, a fact which can be explained by the lower temperature of Upsilon Sgr, which in the visual range has a metallic-line spectrum very similar to that of Alpha Cyg (A2 Ia), or Epsilon Aur (F0 Ia) (Hack and Pasinetti, 1963).

The Mg II resonance lines in Beta Lyr present a P Cygni contour with an absorption core shifted shortward by 1.4 A relatively to the interstellar absorption line, and a strong emission wing. Upsilon Sgr presents no emission wing but only a strong absorption core in the broad depression in the continuum due to the photospheric lines, which extend from about 2788 to 2810 A. The shape of the depression is very similar to that observed in Epsilon Aur (Hack and Selvelli, 1979) and indicates the existence of a central emission which fills up and distorts the gaussian profile of the photospheric core. The sharp absorption core may be of interstellar or circumstellar origin. To decide which, we have compared the equivalent widths of other interstellar lines in Beta Lyr and Upsilon Sgr. The K line of Ca II and the Na I lines measured at phase 0.22 in Beta Lyr and at phase 0.71 in Upsilon Sgr, when the separation between stellar and interstellar lines is largest, have about the same value. We expect that also the Mg II interstellar components of the resonance multiplet have about the same intensity in the two stars. Since the equivalent widths of the cores in Upsilon Sgr are about two times stronger than in Beta Lyr, we conclude that probably the sharp cores in Upsilon Sgr are a blend of interstellar and circumstellar lines.

A large discrepancy between the equivalent widths of the Mg II cores measured with Copernicus (Duvignau et al. 1979) and with IUE has been found; it is due to the different central depth, which is about 1 in the IUE spectra and only 0.3 in those of Copernicus. This difference depends on the background correction. Since the particle background in the Copernicus data is large in this spectral region, due to the South Atlantic Anomaly, and since the IUE spectra in the Mg II region permit an easy, direct determination of the background, we think the latter to be more reliable.

Radial velocities.

Only relative determinations of radial velocity are possible with IUE. We have assumed rest velocity for the several interstellar lines easily detectable in our spectra, and have determined the stellar radial velocity relative to them. About 20 interstellar lines are well detectable and measurable in the short-wave range, and about another 10 in the long-wave range. They allow us to establish with sufficient accuracy (0.1 A) the wavelength scale over the whole spectral range.

By comparing the plots for different phases one can easily distinguish the stellar lines which present the orbital radial velocity shifts, from the almost stationary lines of the envelope. The results for the photospheric lines are given in Table 3, and compared with the results obtained in the visual range.

Table 3. Radial velocity shifts of the photospheric lines.

$\Delta\varphi$	β Lyrae IUE		σ	Sahade et al. 1959	Flora, Hack 1975
0.67- 0.51	+ 142	SW	11	+ 120	+ 150
0.28- 0.51	- 213	SW	17	- 215	- 185
	- 194	LW	25		
0.67- 0.28	+ 346	SW	18	+ 335	+ 345
	υ Sgr IUE			Wilson 1914	
0.23- 0.73	+ 99	SW	12	+ 95	
	+ 75	LW	9		

Table 4 gives the radial velocity of the absorption cores of the envelope lines for Beta Lyr. Their behavior is strictly similar to that observed with Copernicus in 1973 and 1974 by Hack et al. (1975, 1976) for the lines included in the spectral range 1030-1400 A.

Table 4. Mean radial velocity of the Beta Lyr envelope.

Ion	IP (ev)	EP (ev)	RV (km s^{-1})
C II	11.20	0.0	-163
C III	24.28	6.5	-128
C IV	47.67	0.00	-151
N V	77.09	0.00	-163
Mg II	7.61	0.00	-144
Al II	5.96	0.00	-156
Al III	18.75	0.00	-171
Si II	8.11	0.0	-146
Si III	16.27	0.00	-174
Si III	"	6.5	- 98
Si IV	33.32	0.00	-182
Fe III	16.16	3.71	-175
Fe III	"	5.06	-143
Fe III	"	7.8	-128
Fe III	"	8.2	
Fe III	"	8.6	
Fe III	"	8.7	
Fe III	"	9.5	-129
Ni III	18.07	6.8	-144
Ni III	"	7.7	

Table 5 gives the terminal velocity (measured at half intensity relative to the nearby continuum) for the ground-level lines of various ions at different phases for Upsilon Sgr.

Table 5. Terminal velocity (km/s) (Upsilon Sgr)

line	phase							
	0.02		0.23		0.49		0.73	
	v	σ	v	σ	v	σ	v	σ
1334 C II	-434	20	–		–		–	
1548 C IV	-713	19	-636	19	-791	19	-694	19
1238 N V	-726	25	-702	25	-822	:	-678	25
1242 N V	-677	:	-652	:	–		–	
1206 Si III	-821	20	–		–		-746	16
1393 Si IV	-829	21	-678	21	-829	21	-742	21
1402 Si IV	-828	21	-742	21	–		-700	21

We point out one characteristic of the spectrum of Upsilon Sgr shown by past observations in the visual range: sometimes, but not always, at phase 0.5 a violet-shifted component of Hα indicating a velocity of about -300 km/s has been observed; it has been interpreted as evidence of a jet of gas moving from the primary toward the secondary and projecting itself on the stellar disk of the primary. Another red-shifted component, at about +150 km/s, has occasionally been observed (Hack, 1960) and interpreted as due to a jet of gas from the secondary toward the primary. A search for evidence of such jets in the ultraviolet spectrum was made in the short wave range (we had no time to observe the long-wave range at phase 0.5). We have observed: a) a possible component of N V line λ 1238 shifted shortward by about 240 km/s relatively to the terminal velocity of the main component. Nothing can be said about the other line λ1243 which is blended with λ1238.
b) the two resonance lines of Al III show two features longward of the main line, which are absent at the other phases; they could indicate the presence of absorbing gas moving at + 113 km/s.

The continuous spectra.

The continuum was determined by passing through the highest points of the tracings, in positions where no emission contribution is expected. By comparing these well defined "windows on the continuum" on the spectra of Beta Lyr and Upsilon Sgr, it appears that the latter does not present detectable emission lines.

The flux (in erg cm^{-2} s^{-1} A^{-1}) is obtained by the relation $F_\lambda = S_\lambda^{-1}$ N/t, where S_λ is the chromatic sensitivity for the high resolution mode (Cassatella and Selvelli, 1979), N the IUE flux number, t the exposure time in seconds. Because of the unknown fraction of radiation occulted by the 3" aperture, the values of the flux are reduced by a factor of 2 or larger (Bohlin and Snijders, 1978) and no information can be obtained about the light variation with the phase.

The continuous energy distribution observed at the various phases is compared with the theoretical models by Kurucz (1979) and with the observations from S2/68 (Jamar et al. 1976). No apparent variation with the phase in the energy distribution curve has been found either for Beta Lyr or for Upsilon Sgr.

Beta Lyr: The best fit of the observed spectrum with the theoretical ones, after normalization at λ 2500, is found for T_e = 13000 K; the dependence on the gravity is negligible. However, the observed spectrum presents a flux higher than the theoretical one in the region 1800-2200 A where many strong and moderately strong lines of Fe III are situated.(Fig. 3)

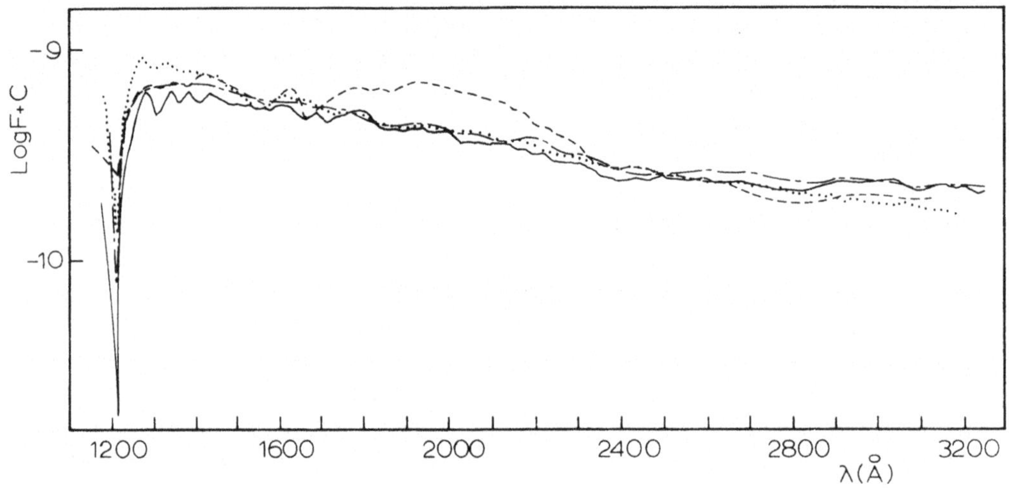

Figure 3. The logarithm of the flux (in erg cm^{-2} s^{-1} A^{-1}) for Beta Lyr (mean of three far UV and two near UV images) versus the wavelength (dashed line) and comparison with Kurucz models (—— T_e = 11000, log g = 3;— - -T_e = 13000, log g = 2.5; ••••• T_e = 15000 log g = 3). The curves have been arbitrarily normalized for 2500 Å.

Probably the continuum we have traced is too high, passing through several faint emission wings. This is confirmed by the fact that the line blocking appears larger in this region than in the contiguous region

2300-2400 A. After having corrected this first approximation continuum, the fit with the models for T_e = 13000 K is very good from λ1200 to λ 3100. This result agrees with the data obtained with Copernicus (Hack et al. 1975), where the observations fit an energy distribution intermediate between B6 and B8 in the range 1200-1400 A; however it was found that at λ<1150 the spectrum better fits a distribution curve corresponding to B5.

A determination of the interstellar extinction for normal spectra can be made by using the relation given by Thompson et al. (1978) for B-type stars: $E_{B-V} = 0.40\ (m_{1565}-m_{2740}) - 0.60\ (m_{1565}-m_{2365})$. Since the continuous spectrum of Beta Lyr in the range 1565-2740 can be considered normal, being fitted very well by models with T_e = 13000 K, we have applied this relation and found E_{B-V} = +0.02.

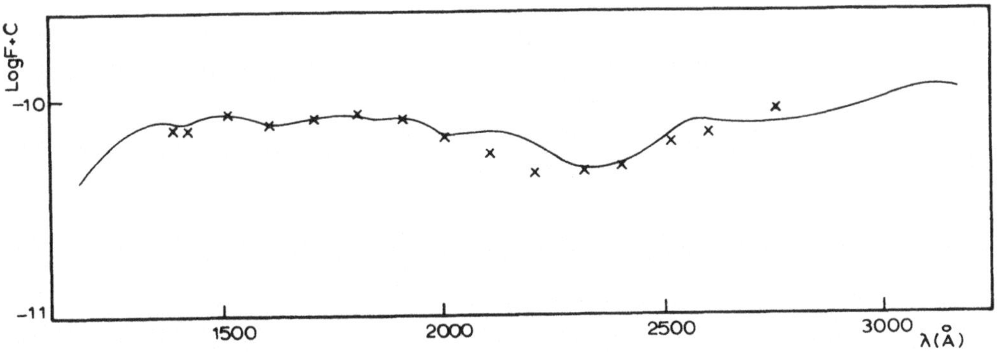

Figure 4. The logarithm of the flux (in erg cm^{-2} s^{-1} A^{-1}) for Upsilon Sgr (mean of four far UV and three near UV images) versus the wavelength) (continuous line). The crosses represent the logarithm of the flux observed for Zeta Oph reduced by a factor 35 plus the flux observed for Alpha Cyg and reddened by E_{B-V} = 0.31 (S2/68 observations)

Upsilon Sgr: The average spectrum is compared with several theoretical models. No one of them fits the observations, and it is clear that two different models, one for $T_e \sim 8000$ K and another for $T_e \gtrsim 11000$ K are necessary to fit the whole spectrum. The spectrum presents a slight depression at 2100-2300 A, giving evidence of appreciable interstellar extinction. (Fig. 4)

Discussion.

A model for Beta Lyr has been extensively discussed on the basis of the data obtained with Copernicus (Hack et al. 1977) and the present observations generally confirm it. The lower limit for the envelope density, which was above 10^8 cm^{-3}, because of the absence of forbidden lines in the spectrum, is now raised above 10^{10} because of the ascertained absence of λ 1908 C III] .

A point which needs further discussion is the stratification of the ions in the envelope. By considering all the observations of the radial velocity and rotational broadening of the lines formed in the envelope in the visual and ultraviolet range, we can infer that the expansional velocity is not regularly increasing with the distance from the photosphere of the primary star.

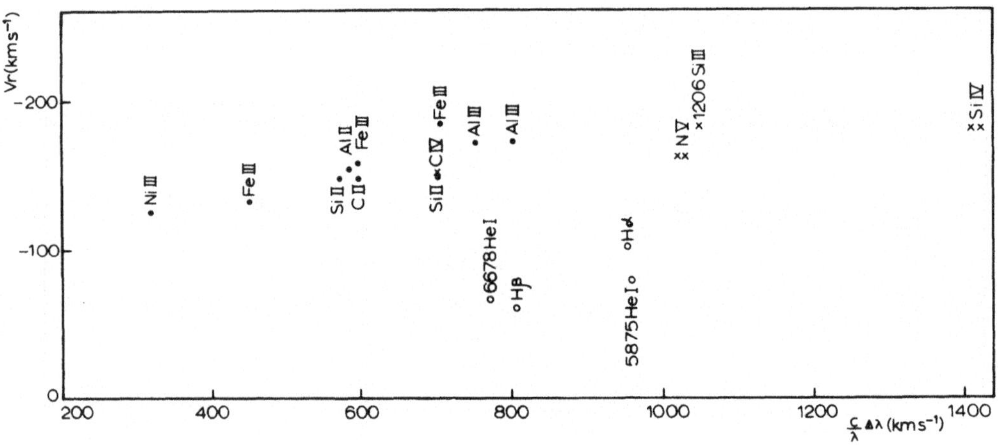

Figure 5.Lines formed in the Beta Lyr envelope: expansional radial velocity versus the rotational broadening. Dots and open circles: low ("chromospheric") ionization lines in the UV and visual range respectively; crosses: high ("coronal") ionization lines.

In fact, if we plot the expansional velocity of the absorption cores versus the width of the whole line (emission wings and absorption core) measured on the continuum (in km/s) we find the curve of Figure 5. We observe that the lines of lower ionization are grouped in the region of lower rotational velocities, while the lines of multi-ionized atoms have generally higher rotational velocities. This suggests the existence of two regions, one at $T \sim 10^4$ K, $N_e \simeq 10^{12}$ cm^{-3} (as indicated by the strength of the metastable lines of He I), that we will call "chromosphere", and another at $T \sim 10^5$ K, $N_e \sim 10^{10}$ cm^{-3} that we will call "corona". The lines of Ni III, Fe III, Al II, Si II, C II, Si III, Al III, He I and H I are formed in the "chromosphere"; the widths increase from Ni III to Hα and 5875 He I. When several lines of the same ion, originating from different low excitation levels, are measurable (as is the case for Fe III) the expansional velocity decreases with increasing excitation potential.

Since we can reasonably expect that the largest optical depth in this region is at Hα and 5875 He I, we can deduce that the linear rotational velocity increases with the distance from the primary, and therefore that the expansional velocity at first increases with the distance, reaches a maximum and then decreases again.

In the "corona" the C IV lines have lower rotational velocities than the N V lines. It seems improbable that the optical depth in the corona is lower in the C IV lines than in the N V lines, because of the larger abundance of carbon, its lower ionization potential and the comparable gf values. It is reasonable to think that the rotational velocity in the extended envelope follows a keplerian law, decreasing outward; in this case the expansional velocity is also slowly decreasing outward, from -180 (Si IV, Si III) to -160 (N V) and -150 km/s (C IV).

The radial velocities in the "chromosphere" and "corona" at the epoch of the second quadrature are generally less negative by 20 to 30 km/s than at the epoch of the first quadrature (a fact already observed in the visual range by Flora and Hack, 1975, and with Copernicus by Hack et al. 1976, 1977), probably because of the presence of streams or condensations in the vicinity of the primary.

The resonance lines of Ca II and Na I do not present the slight radial velocity variations shown by the other envelope lines, but indicate a constant expansional velocity of -80 km/s relatively to the primary, suggesting that they are formed in a cooler outward circumstellar envelope.

A different behavior is that of 1175 C III, which probably has a component associated with the invisible companion, and was discussed by Hack et al. (1977).

While the envelope profiles of Beta Lyrae are explained by a shell rotating at about 300 to 400 km/s and expanding at about 150 km/s, those of Upsilon Sgr are not easily explained. In order to understand, at least qualitatively, the broad asymmetric absorption profiles of the lines due to multi-ionized atoms in the far ultraviolet spectrum of Upsilon Sgr, we have compared them with the theoretical P Cyg profiles computed by Castor and Lamers (1979). These profiles are computed for spherical envelopes surrounding a single star, and neglecting collisional excitation. Hence we can expect only a very rough indication of the properties of the envelope in our particular case.

One striking characteristic of our profiles is that the intensity left at $v = v_{term}$ is almost zero, and this indicates a high value for the total optical depth of the envelope. But the higher the total optical depthis, the higher the expected intensity of the emission wing; but this is missing on our spectra. The effect of collisional excitation is to increase still more the intensity of the emission. The only way out is to admit that a strong absorption stellar line is present. The theoretical profile computed on this hypothesis explains our observations. But what are the implications of this assumption? The photospheric lines neutralizing the envelope emissions cannot be formed in the primary, whose temperature is lower than 10^4 K. Only supergiants earlier than B5 or main sequence stars earlier than B1 present absorption profiles of N V (Snow and Jenkins 1977). Hence the photospheric lines of N V, C IV and Si IV must be formed in the hot companion. Its spectral type can be inferred from the distribution of the continuum, taking into account the need to correct for interstellar extinction. Hence we have attempted to fit the observed continuum by combining spectra of early type stars with those of A-F supergiants observed with S2/68 (Jamar et al. 1976). A very good agreement with the observed spectrum (average of spectra at different phases) is obtained by adding the flux of Alpha Cyg (A2 Ia) reddened for E_{B-V} = +0.31 and the flux of Zeta Oph (O9.5 V, E_{B-V} = +0.31), reduced by a factor of 35. Hence if the primary is an A-type supergiant, M_V = -7 or -8 (as indicated by the visual spectrum and by the width of the central emission in the Mg II lines), the companion should be an O9 dwarf which at λ1500 has about the same luminosity as the primary, but in the visual is about 100 times less luminous than the primary.

The absorption lines of the highly ionized elements are deeper at phase 0.5, when the secondary is in front, than at the other phases, confirming that the contribution of the spectrum of the secondary is more important at this phase.

The lines formed in the photosphere of the primary star have on the average lower central depths in the far UV than in the near UV; this fact can be explained by the presence of the continuum of the hot companion which fills up the lines of the primary more efficiently at the short than at the long wavelengths.

The terminal velocity of the envelope lines is systematically lower at the quadratures than at the epoch of the minima, suggesting a flow of matter through the lagrangian points L2 and L3.

A model for Upsilon Sgr was discussed by Duvignau et al. (1979). The present data give for the primary M_V -7 and M_{bol} -7.3; for the secondary M_V -2 and M_{bol} -5; R_1 and R_2 are equal to about 120 $R_\odot$ and 3 $R_\odot$ respectively. Since the semi-axis of the orbit of the primary is about 170 $R_\odot$ the system is almost in contact, $(R_1 + R_2) / a$ being about 0.7. The hydrogen deficiency of the primary indicates that it has lost a large amount of mass, and this has been transferred to the companion, producing the thick envelope indicated by the profiles of

C IV, N V, SI IV. If this envelope is associated with the secondary, we should observe the orbital radial velocity shifts in the lines. However, if the two stars have comparable masses, or if the secondary is the more massive one, the shift is too small ($<$0.4 to 0.7 A) to be observable on such broad lines, with such a complex structure.

As discussed by Duvignau et al. (1979), it seems probable that Upsilon Sgr is an example of case C evolution, i.e. the primary reaches the Roche lobe at the end of central He burning.

Acknowledgements.

We gratefully thank Dr. P.L. Selvelli for obtaining part of the spectra of Upsilon Sgr, and the whole staff of Villafranca Observatory for their assistance in observing and reducing the data.

This work was supported by a CNR-SAS contract.

The data processing has been made on the DEC PDP 11/45 computer system of Trieste Observatory, with programs prepared by one of us (P.S.).

References.

Bohlin, R.C. and Snijders, M.A.J., 1978 IUE NASA Newsletter No. 2, November 1978

Cassatella, A. and Selvelli, P.L., 1979, Communication to the users, March 7, 1979

Castor, J.I. and Lamers, H.J.G.L.M., 1979 Astroph. J. Suppl. 39, 481

Duvignau, H., Friedjung, M. and Hack, M., 1979 Astron. Astroph. 71, 310

Flora, U. and Hack, M., 1975 Astron. Astroph. Suppl. 19, 57

Gehrz, R.D., Hackwell, J.A. and Jones, T.W., 1974 Astroph. J. 191, 675

Hack, M., 1960 Contr. Obs. Milano-Merate No. 152

Hack, M., 1974 Astron. Astroph. 36, 321

Hack, M., Hutchings, J.B., Kondo, Y., McCluskey, G.E., Plavec, M. and Polidan, R.S., 1975 Astroph. J. 198, 453

Hack, M., Hutchings, J.B., Kondo, Y., McCluskey, G.E. and Tulloch, M.K., 1976 Astroph. J. 206, 777

Hack, M., Hutchings, J.B., Kondo, Y. and McCluskey, G.E., 1977 Astroph. J. Suppl. 34, 565

Hack, M. and Pasinetti, L., 1963 Contr. Obs. Milano-Merate No. 215

Hack, M. and Selvelli, P.L., 1979 Astron. Astroph. 75, 316

Humphreys, R.M. and Ney, E.P., 1974 Astroph. J. 190, 339

Jamar, C., Macau-Hercot, D., Monfils, A., Thompson, G.I., Houziaux, L. and Wilson, R., 1976 Ultraviolet Bright Star Spectrophotometric Catalogue, European Space Agency, Paris

Kondo, Y., Parsons, S.B., Henize, K.G., Wray, J.D., Benedict, G.F. and McCluskey, G.E., 1976 Astroph. J. 208, 468

Kurucz, R.L., 1979 Astroph. J. Suppl. 40, 1

Lee, T.A. and Nariai, K., 1967 Astroph. J. 149, L 93

Osterbrok, D.E., 1970 Astroph. J. 160, 25

Plavec, M.J. and Koch, R.H., 1979 UCLA Preprint No. 60

Snow, T.P. and Jenkins, E.B., 1977 Astroph. J. Suppl. 33, 269

Thompson, G.I., Nandy, K., Jamar, C., Monfils, A., Houziaux, L., Carnochan, D.J. and Wilson, R., 1978 Catalogue of Stellar Ultraviolet fluxes, The Science Research Council

Treffers, R.R., Woolf, N.J., Fink, U. and Larson, H.P., 1976 Astroph. J. 207, 680

COMMENTS FOLLOWING HACK, FLORA AND SANTIN

de Loore: I understand that you invoke case C evolution in order to explain the large helium-hydrogen ratio. However, in my opinion it is possible to explain the evolutionary system by a case B of mass transfer as well, (starting with initial mass of $40M_\odot$ and $10M_\odot$ or masses of $27M_\odot$ and $7\ M_\odot$ with separation of $250R_\odot$ and $500R_\odot$ respectively). Such binaries, considered now at the end of the Roche lobe overflow or having just finished, can explain the circumstellar material and the large He-abundance. Since the probability of case B mass transfer and exchange is so large, one should also think about a case B solution for this case.

Hack: Maybe yes--I was thinking of case C because of the large H-deficiency and the rather long period, ∿133 days.

Guinan: During May through November, 1978 Hα wide- and narrow-band and uvby photometry of υ Sgr was done on about 90 nights. The observations were obtained at Biruni Obs., Shiraz Iran by J.D. Donnen, I. Siah, and myself. I hope that all this data will be reduced by September.

Did you obtain data during the light minimum? Were there any changes in the line strengths?

Hack: Yes, I obtained data at the two minima and the two quadratures. The envelope lines are all deeper at phase 0.5. The primary photospheric lines do not show appreciable change events, but just RV-shifts.

Nariai: υ Sgr shows violet-shifted Hα line which corresponds to the velocity of about -200 km/sec for certain phase. What is the relation between this flow and the high-speed flow you have just shown with the far UV spectra?

Hack: The high speed flow ($\sim$800 km/s) is probably wind from the hot 09 companion (and is present at all phases), while the violet-shifted Hα component which is sometimes observed at and near phase 0.5 only is due to a gaseous stream from the primary toward the secondary, projecting itself on the primary disk.

Hα OBSERVATIONS OF ALGOL-TYPE INTERACTING BINARY SYSTEMS

Geraldine J. Peters
Department of Astronomy
University of Southern California
Los Angeles, CA 90007 USA

ABSTRACT

An observational study of the Hα feature in Algol type binary systems has revealed the presence of emission in nearly all systems with periods greater than 6 days. The disks which produce this emission are generally thin and can extend out to 90% of the Roche surface of the primary star. If emission is found in the systems of shorter period, it tends to be transient, lasting only a few orbital cycles. In several systems, there is evidence of mass loss from the vicinity of the primary at phase 0.5.

1. INTRODUCTION

In this paper, initial results are presented from a study of the Hα feature in interacting binaries of the Algol type. Algol systems displaying a wide range in period are being observed at Kitt Peak National Observatory in order 1) to look for the presence of emission in Hα which implies the existence of a circumstellar disk of relatively low density ($N_e \simeq 10^{10} cm^{-3}$) and 2) to look for and interpret phase dependent profile variations in Hα which provide information on the dimensions of the Hα emitting disks, the extent of the asymmetry in the disks, and whether there is evidence of mass outflow in the systems. Since circumstellar disks radiate most strongly at Hα, this feature can reveal more information about the geometry of and physical conditions in these disks than the other emission features which the pioneering researchers on Algol disks (i.e. A. H. Joy and O. Struve) analyzed.

The data base for this investigation include 1) Varo image tube spectrograms (11 - 33 A mm^{-1}) obtained at Lick Observatory during 1973 - 1976 in collaboration with Drs. M. Plavec and R. S. Polidan and 2) Carnegie image tube plates recently obtained at Kitt Peak National Observatory with the 0.9 m Coudé Feed telescope. To date nearly 50 systems have been observed at least once and we have good phase coverage for about 10 systems. About half of the sample are "short" period systems ($P \simeq 3^d$) while the other half have periods > 5^d.

M. J. Plavec, D. M. Popper and R. K. Ulrich (eds.), Close Binary Stars: Observations and Interpretation, 287–292.

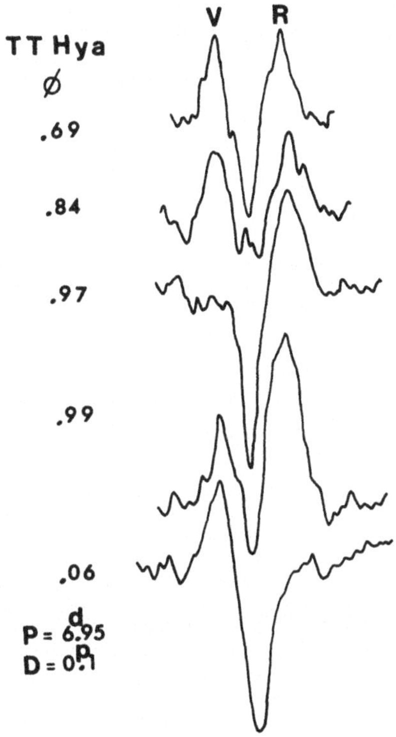

Fig. 1. Phase dependent Hα profile variations in TT Hya.

2. RESULTS

This investigation has revealed that Hα emission is common in systems with periods greater than 6 days. Furthermore, this emission persists for a large number of orbital cycles ($\gtrsim$100). However, there appears to be no correlation between the intensity of the Hα emission and the period of the system. In fact, one system, V889 Aql (P = $11^{d}.1$) shows no trace of Hα emission, even during eclipse. Alternatively, Hα emission is rare in systems with periods less than 6 days. If emission is found, it tends to be transient, lasting only a few orbital cycles or less.

Let us now direct our attention toward the long period systems. Shown in Figure 1 is a series of Hα profiles observed in TT Hya, a system which displays strong ($I \simeq 1.3 I_c$) Hα emission. These profiles illustrate the type of phase dependent variations frequently seen in long period Algols. Near quadrature, we observe a two component feature which is formed in a disk which co-rotates with the primary. In TT Hya, the width of the emission feature implies that the Hα emitting region is located $7.5R_{\odot}$ ($3.75R_p$) from the primary. As the disks of Algol primaries undergo eclipse, we typically observe the type of Hα profile variation shown in the latter three profiles in Figure 1. The fact that one observes a complete occultation of the V and R lobes in systems such as

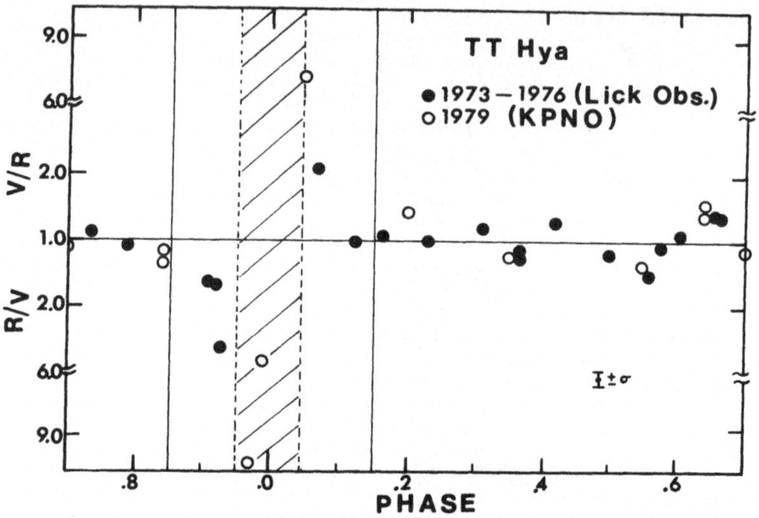

Fig. 2. V/R in Hα versus phase for TT Hya. The thin vertical lines between $0.85 < \phi < 0.15$ delineate the domain where the secondary occults the Roche surface of the primary. The cross-hatched region indicates the duration of primary eclipse.

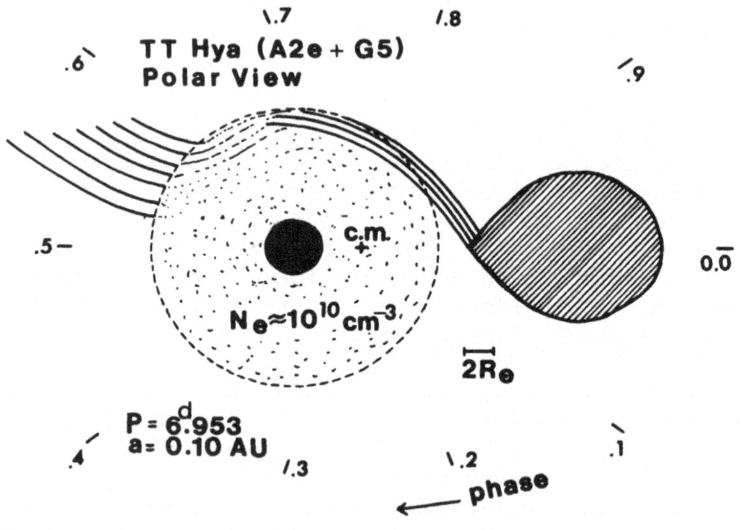

Fig. 3. A model suggested for TT Hya based upon the Hα observations discussed in this paper. The system is drawn to scale using the masses and radii given by Popper (1979).

TT Hya and RX Gem which barely display a total eclipse, implies that the disks are thin, $\simeq 1R_p$, in vertical extent. The emission observed at central eclipse is from the periphery of the disk. If the particles are in stable Keplerian orbits about the primary, the width of the Hα feature implies that the disk extends out to $10.4R_\odot$. As the radius of the Roche surface of the primary in TT Hya is $10.5R_\odot$, the observations suggest that the disk extends out to this limit. The latter result is supported by the general phase dependent behavior of the Hα profile. In Figure 2, we see that the V lobe of Hα begins to weaken at $\phi = 0.85$, the phase at which one would expect occultation to commence if the disk extended out to the Roche surface. Also notice that the receding part of the disk appears to be less extended than the approaching side. Outside of primary eclipse, $V/R \simeq 1.0$, except for the region around $\phi = 0.5$ where $V<R$. Observations of both TT Hya and RX Gem reveal the presence of deep, blue-shifted cores ($v \simeq -30$ km s^{-1}) near $\phi = 0.5$. In TT Hya, the strength of the core suggests a mass outflow from the vicinity of the primary of $2.5 \times 10^{-8} M_\odot$ yr^{-1}. Finally, it can also be seen from Figure 2 that the cyclic profile variations have persisted for at least six years. Variations recently observed at KPNO are identical to ones seen in the earlier Lick Observatory data.

A model has been developed for the TT Hya system based solely on the Hα observations. This model is shown, drawn to scale, in Figure 3. An Hα emitting disk of density 10^{10}cm^{-3} extends to the Roche surface of the primary; it is less than $1R_p$ in vertical extent. The density was obtained from the equivalent width of Hα with the aid of a formula developed by Wellman (quoted in Pagel 1960). We assume the secondary fills its critical Roche surface. The nature of the gas stream should be regarded as speculative. Its thickness and angle are based upon the assumption that it is similar to the stream observed in HR 2142 (Peters 1976). The streamlines near $\phi = 0.5$ represent the observed mass loss discussed above.

All long period Algol systems observed to date with sufficient phase coverage display the same type of profile variations observed in TT Hya. Some minimum sizes deduced for disks (in terms of the Roche radius) are: RX Gem (0.9), AU Mon (0.6), V393 Sco (0.3), AD Her (0.5), and AW Peg (0.4). Recent theoretical studies by Lubow and Shu (1975) suggest that the disks should be small (<0.3) while Paczynski (1977) predicts that the disks can extend out to 90% of the Roche surface of the primary. The Hα observations discussed in this paper are more compatible with the latter.

REFERENCES

Lubow, S. H. and Shu, F. H.: 1975, Astrophys.J. 198, pp.383-405.
Paczynski, B.: 1977, Astrophys.J. 216, pp.822-826.
Pagel, B.E.J.: 1960, Vistas in Astronomy 3, 203-257.
Peters, G.J.: 1976, in A. Slettebak(ed.), "Be and Shell Stars, IAU Symp.70", pp.417-428.
Popper, D.M.: 1979, personal communication.

COMMENTS FOLLOWING PETERS

Shu: A number of developments of the past few years have convinced me that accretion disks are often larger than what Lubow and I had proposed in 1975. Nevertheless, I would like to make the following cautionary remark. The densities you and other speakers this morning have deduced refer to relatively low density material. This material might well contribute to most of the emission lines and polarization, but it may tell us little about what most of the mass is doing.

Peters: I agree. Perhaps the detailed analysis of the sharp absorption features will make my analysis more complete.

Hall: Since Hα emission must come from optically thin regions, your observations probably have outlined the maximum extent of the thin, outer parts of the disk. It is possible that the disk may be optically thick closer in, in which case it would appear as part of the "radius" in a standard solution of UBV light curve, for example. That prompts me to point out that radii of the mass-gaining star in most long-period Algols (listed in Acta Astronomica 24, 215) fall clearly above the main sequence in an M vs. R diagram. I suggest that a fuller picture will be given if the wide-band and narrow bands are looked at together.

Peters: Such a photometric project would be a nice supplement to a spectroscopic one such as the one I described.

Smak: Could you estimate the inner radii of disks?

Peters: The density in the disk most assuredly increases uniformly toward the primary star. It is the gas which has a (low) density of 10^{10} cm^{-3} which produces most of the Hα emission we observe. I intend to analyze the detailed Hα profile (emission plus sharp absorption) as a function of phase throughout eclipse and perhaps this analysis will provide information on the density and extent of the denser inner disk.

Whelan: Could you directly determine the electron density, N_e, from the Balmer decrement, rather than using estimates of the volume of the disc?

Peters: Yes, upon making a number of assumptions about the gas in the disk (e.g. radiative transfer in the functional form of the density gradient in the disk, whether LTE prevails, etc.); currently my data base does not include the higher Balmer lines but I intend to supplement my Hα data with "blue" spectrograms and ultimately use Balmer decrements as well as the strengths of other Balmer lines to obtain a better estimate for N_e in the disk.

Bolton: What was the velocity separation of the Hα emission peaks in the spectra you showed us?

Peters: It is 500 km s^{-1} outside of eclipse and 400 km s^{-1} during central eclipse when the material with a higher velocity of rotation is occulted by the secondary star. Incidentally, the velocity separation during eclipse is compatible with the disk extending out to the Roche surface of the primary if it is assumed that the emitting H atoms are in Keplerian orbits.

Bolton: The emission profiles are lying on top of a rapidly varying continuum--the absorption profiles of the underlying stars. If this continuum can be rectified by subtracting the absorption profiles, you can extract considerable kinematical information even outside eclipse. My experience with Algol and other stars indicates this is a worthwhile procedure.

Peters: I intend to model these systems by analyzing the complex Hα feature as a function of phase.

ULTRAVIOLET OBSERVATIONS OF CLOSE BINARY STARS

R. S. Polidan
Princeton University Observatory
Princeton, NJ 08540 USA

G. J. Peters
Department of Astronomy
University of Southern California
Los Angeles, CA 90007 USA

Abstract. In this paper, we present far ultraviolet observations of three interacting binary stars (HR 2142, HR 7084, and λ Tauri). All stars exhibit evidence of a gas stream and extensive mass outflow from the system. Physical parameters are derived for the gas stream in HR 2142.

The physical conditions in the gas streams in close binary systems produce the strongest spectral features in the far ultraviolet. This has made direct study of the mass flow difficult until the advent of satellite astronomy. Observations of three active systems (Table 1) were obtained with the *Copernicus* satellite during the past two years.

Table 1

Observations

Star	Components	P(Days)	M_2/M_1	Dates of Observation
HR 2142	B1Ve+gK:	80.86	$\sim$0.1	Dec. 76 - Feb. 79
HR 7084	B3Ve+F-K	6.70	$\sim$0.2	Aug. 78 - May 79
λ Tau	B3IV+gA	3.95	0.26	Feb. 78

Selected spectral features between 1000 A and 1400 A were investigated at high resolution (0.05 A) in HR 2142 and λ Tauri and at low (0.20 A) resolution in HR 7084. Details of the equipment used and technique of observation can be found in Rogerson *et al.* (1973).

M. J. Plavec, D. M. Popper and R. K. Ulrich (eds.), Close Binary Stars: Observations and Interpretation, 293–297.

HR 2142

HR 2142 was suggested to be a mass transfer binary by Peters and Polidan (1973) on the basis of ground based observations. The binary model was refined (Peters 1976) with more observations suggesting a binary with a permanent accretion disk, a thick gas stream, and a counter-stream. The ultraviolet observations were obtained at twelve phase points over seven orbital cycles.

Strong, variable absorption lines, consistent with the existence of a gas stream, were seen in C II, N II, Si II, Si III, S II, S III and Fe III. The gas stream was not detected in lines of C I, C III, N I, N III, O I, and Fe II. Figure 1 shows the phase dependence of the gas stream line strengths normalized to the strength at phase 0.95. The points connected by the solid line was obtained from ground based observations of the hydrogen Balmer lines (Peters 1976). This curve was found to fit the strength variation of the lower ionization (Si II and S II) UV lines. The points with error bars represent the strength variation of the higher ionization species (C II, N II, Si III, S III, and Fe III). Figure 2 shows one of these lines, Fe III 1130.404, at three phases: 0.7 (no stream), 0.95 (peak stream), and 0.02 (counter-stream).

The radial velocity variation of the gas stream lines is shown in Figure 3. Again, the points connected by the solid line were obtained from analysis of ground based data and the low ionization UV lines. The points with error bars represent the velocities of the higher ionization species in the gas stream.

Analysis of the ultraviolet line strengths, velocities, and shapes has yielded three principal results. The gas stream is quite complex. At least two major, repeatable from cycle to cycle, components exist: one producing the higher ionization species and one responsible for the lower ionization species. The slower variation of the higher ionization lines suggests that they are formed in a more extended region than the low ionization lines. Both components appear to be composed of many (>3)

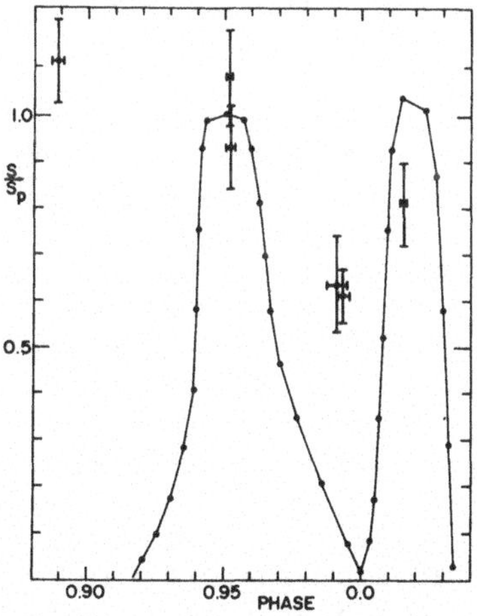

Figure 1. Phase dependence of the gas stream lines in HR 2142.

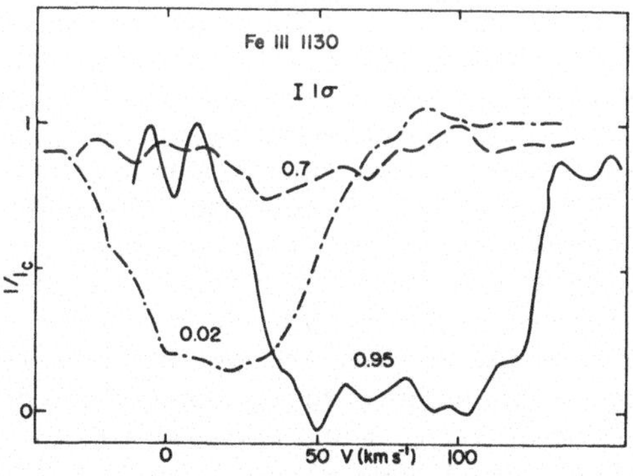

Figure 2. Fe III 1130.404 at three phases in HR 2142.

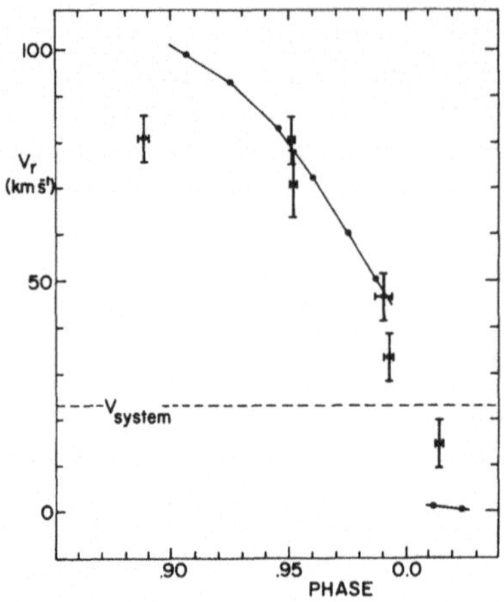

Figure 3. Radial velocity versus phase for the gas stream lines in HR 2142.

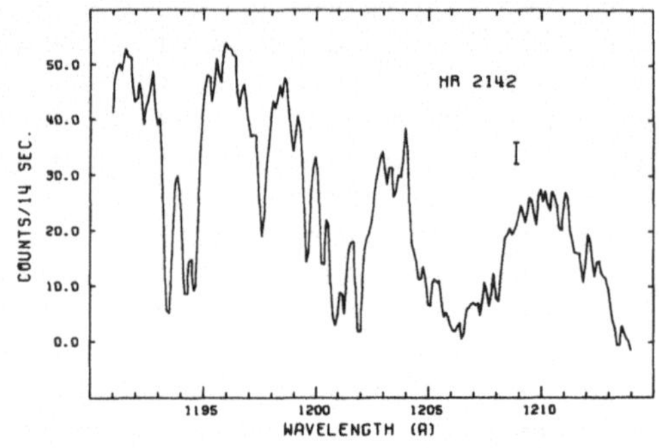

Figure 4. The Si III 1206 region in HR 2142.

discrete "clouds" that do not repeat from cycle to cycle. Analysis of the principal "cloud" in each component suggests that in the higher ionization zone Te $\sim$ 14000 K and $N_e \sim 10^{12}$ cm^{-3} and in the lower ionization zone $T_e \sim$ 11000 K and $N_e \sim 10^{13}$ cm^{-3}. These conditions suggest that the gas stream has a high density "core" and a hot, lower density "halo". The physical conditions discussed above and the model presented earlier (Peters 1976) imply a minimum mass transfer rate of 10^{-7} $M_{\odot}$ yr^{-1} in this system. The final piece of information obtained from the analysis regards the angle at which the gas stream crosses the system. The observed velocity variation (Figure 3) argues in favor of a much greater angle with respect to the line of centers than is predicted by theory (Lubow and Shu 1975). The velocity field is, however, complex. Further analysis and more observation will be required before this deviation can be confirmed.

Observations of HR 2142 at phases when the gas stream is not visible suggest that most of the material that leaves the secondary is ejected from the system and not accreted. In particular, near phase 0.5 negatively displaced (V $\sim$ 100 km s^{-1}) absorption components, comparable in strength to the strongest gas stream lines are observed. Also, despite the impinging gas stream the rapidly rotating primary star displays a reasonably strong stellar wind at Si III 1206 (Figure 4). Evidence of mass ejection from the system can also be seen in C II 1335 where moderately strong, sharp absorption components not associated with the interstellar medium are seen. The most probable origin of these components is an extensive circumsystem disk or cloud.

Summarizing the results for HR 2142, the physical parameters derived for the gas stream are within the range predicted from theoretical calculations (cf. Lubow and Shu 1975) with the possible exception of the angle the stream makes with the line of centers. Extensive mass outflow from the binary is observed, suggesting that at this stage of evolution little material is being accreted by the primary star in HR 2142.

HR 7084

This Be star has been shown to be an interacting binary star by Koubsky (1978). The ultraviolet observations of this system are less extensive and of lower resolution than those of HR 2142. The general behavior of HR 7084 in the ultraviolet is quite similar to that of HR 2142. A prominent gas stream is seen between phases 0.8 and conjunction and an extensive outflow of material is seen near phase 0.5. The physical conditions in the stream are similar to those found for HR 2142. A strong stellar wind is also seen in the Si III 1206 line of the primary star.

λ Tauri

The eclipsing binary star λ Tauri has been briefly surveyed at high resolution. A gas stream was dectected but at a substantially reduced strength compared to HR 2142 and HR 7084. Similarly, only weak evidence was found for the existence of mass outflow from the system. However, numerous (3 to 6) weak, non-interstellar absorption components were found in C II, N I, and N II, suggesting the presence of multiple circumsystem disks or clouds.

This work was supported by NASA grants NAS 5-23576 (RSP) and NSG 5356 (GJP).

REFERENCES

Koubsky, P.: 1978, Bull. Astron. Inst. Czechosl., 29, pp. 288.
Lubow, S. and Shu, F.: 1975, Astroph. J., 198, pp. 383.
Peters, G. J.: 1976, in Be and Shell Stars (Ed. A. Slettebak), IAU Symp. No. 70, pp. 417.
Peters, G. J. and Polidan, R. S.: 1973, in Extended Atmospheres and Circumstellar Matter in Spectroscopic Binary Systems, IAU Symp. No. 51, pp. 174.
Rogerson, J. B., Spitzer, L., Drake, J. F., Dressler, K., Jenkins, E., Morton, D. C., and York, D. G.: 1973, Astroph. J., 181, pp. L97.

DISCUSSION FOLLOWING POLIDAN AND PETERS

Shore: In light of the wind blowing from the B star, have you looked for X-ray emission from the inflowing matter slamming into the stellar wind? It could be a soft source.

Polidan: On the Copernicus satellite there is a co-aligned X-ray proportional counter operated by the University College, London. Throughout the ultraviolet observations of these systems X-ray data were accumulated. In no case was a significant X-ray flux detected.

The X-ray detector is sensitive only in the 3-8 keV range, however, so the presence of soft X-rays cannot be ruled out.

Shore: What is the velocity of the wind?

Polidan: The wind velocity is typical of that found in single rapidly rotating stars--a few hundred kms^{-1}.

Hall: Two questions about HR 2142: First, you said that no mass is assimilated, that all left the system. One hundred per cent loss would have far-reaching implications for the theory of mass loss but $\sim$50% loss, which would be easier to explain would involve only a factor 2 in your derivation of $\dot{M}$ from the observations. Can the observational determination of $\dot{M}$ really be so precise? Second, would it be worthwhile to monitor it photometrically, say, in UBV for several years.

Polidan: At this time we can say with a fair degree of confidence that more than 50%, possibly as much as 90%, of the transferred material is lost from the environment of the components of the binary. With the completion of our detailed analysis the fraction of the transferred material that is lost will be better established. The weaker unsaturated lines in the UV allow a very accurate determination of the relative column densities of the stream and outflow. We do believe that the observations are of sufficient quality that as little as 10% accretion could be detected.

It would be very worthwhile to photometrically monitor HR 2142. The star is almost certainly variable in light. The existence of light variations, periodic or non-periodic, would help in understanding the mass transfer in this system. Similarly, I would like to encourage polarimetric observations of this system over its 81 day period. Also, the other system that I mentioned, HR 7084, would benefit greatly from photometric and polarimetric observations.

NARROW BAND PHOTOMETRY AND OBSERVABLE EFFECTS IN ALGOL TYPE SYSTEMS

E. Budding
Department of Astronomy, University of Manchester, England

The application of narrow band photometry to Algol type systems is considered and advantageous circumstances pointed out. Observable effects can be divided into two kinds: classical (associated essentially with differential centre to limb variation) and interactive (where an emission component may be present in the line). Observations of U Sge and β Per are discussed from this basis and deductions made about the size and shape of an emitting region. Conclusions are compared with theoretical ideas about such regions and their expected stability, both for these and other similar systems.

1. Motivation

The circumstance of eclipsing binarity can make highly significant addition to the effectiveness of narrow band photometry as an astrophysical tool. This can be particularly effective in Algol type systems, where one star, though of comparable size to the other, may be of very different luminosity. The eclipsing star then acts as a screen, allowing differences between the centre to limb variation of a line and the surrounding continuum to be monitored. Differential effects originating in the atmosphere of the brighter star or some other surrounding material, may, in this way, be capable of detailed investigation. The technique has already been exploited by various authors, notably by Guinan et al. (1976). Special points relevant to the present work are given in more detail by Budding and Marngus (1980).

2. Technicalities

A convenient approach to narrow band photometry is made with the aid of interference filters. The basic principle is one of comparison between flux in the vicinity of some spectral feature and the surrounding continuum. The Hβ line appears to have received most attention in this way, probably because of its strength and position in relation to the spectral sensitivity of photocathodes. Some preliminary observations of U Sge, to be discussed later, were made with Hβ filters and the notation

M. J. Plavec, D. M. Popper and R. K. Ulrich (eds.), Close Binary Stars: Observations and Interpretation, 299–303.

in what follows will correspond to this, though the formulae are, in principle, quite general. Ideally, the narrow filter should be tuned to the line centre and just wide enough to comfortably accommodate broad wings. The wide filter could also be an interference filter, though of lower finesse, and a transmission minimum at the line centre would be advantageous.

3. Formulary

A formula which relates a quantity of theoretical specification (equivalent width) to something observable, and which can be shown to be fair for suitably selected filter pairs is of the form

$$w_\beta = a + b\beta \ , \tag{1}$$

(cf. Crawford, 1958); i.e., a linear transformation. Such a transformation applies to standardize actual β index values observed with different apparatus, i.e.

$$\beta_{\mathrm{standard}} = c + d\beta_{\mathrm{local}} \ , \ (a,b,c,d \text{ are constants}). \tag{2}$$

A general formula relevant to the situation obtaining in eclipsing binaries where circumstellar gaseous material may be present is:

$$\overline{w}_\beta = \frac{L_{c_1} w_{\beta_1}\left(1 - (p\alpha_c - (p-1)\alpha_\beta)\right) + L_{c_2} w_{\beta_2} - i_\beta(1-\alpha_{em})}{L_{c_1}(1-\alpha_c) + L_{c_2} + L_{c_3}} \tag{3}$$

where L_{c_1}, L_{c_2} and L_{c_3} represent fractional continuum luminosities (their sum = 1) of primary, secondary and circumstellar continuum sources respectively; w_{β_1}, w_{β_2} and $\overline{w}_\beta$ are fairly self evident, and α_c, α_β and α_{em} represent light loss functions. The first two have well known forms, though α_β requires us to consider a quantity $\overline{u}_\beta$ which represents a linear limb darkening coefficient for a line region of effective width p. A suitable definition of p is $L_{c_1}/(L_{c_1}-L_{\beta_1})$ where L_{β_1}/L_{c_1} is the fractional depth of the line centre. The third component of the numerator in (3) represents an emission contribution of equivalent width i in terms of the composite spectrum, associated with circumstellar plasma. Attention is then directed to the form of the light loss function α_{em} for this source. Explicit expressions for this have been evaluated for some simplified, optically thin shell and disk models, though if asymmetry, stratification or self-absorption effects require consideration, numerical integration may be called for. In any case, some parameter h, indicatint the relative extent of the emission région above the primary photosphere in terms of its radius, in addition to the quantities i_β and $\overline{u}_\beta$ mentioned already, will be of some significance in representing the observed effects. For various reasons disk models seem to offer the most appropriate approach to observations of the Algol type systems mentioned in the first section. For such a case we have (for central "occultation" eclipses)

$$
k \geq 1+h \quad
\begin{cases}
d \geq 1+h+k \; ; & \alpha_{em} = 0 \\
1+h+k > d \geq 1+k \; ; & \alpha_{em} = \alpha(s)/c \\
1+k > d \geq k-1 \; ; & \alpha_{em} = \left(\alpha(s)+\alpha(s_1)-\alpha(s')/(1+h)^2\right)/2c \\
k-1 > d \geq k-1-h \; ; & \alpha_{em} = 1 - (1-\alpha(s))/c \\
k-1-h \geq d \; ; & \alpha_{em} = 1
\end{cases}
\qquad (4)
$$

where the "straight edge" α-function $\alpha(s)$ is given by $\alpha(s) = (\cos^{-1}(s) - s\sqrt{1-s^2})/\pi$ and the various other quantities are as follows: $d = (\sin^2\theta \sin^2 i + \cos^2 i)^{\frac{1}{2}}/r_1$, $k = r_2/r_1$, $s' = d-k$, $s_1 = 1/(1+h)$, $s = s's_1$, $c = \alpha(s_1) + \frac{1}{2}(h^2+2h)s_1^2$; the basic quantities being, in the usual notation, primary and secondary relative radii r_1, r_2, and orbital inclination and phase i and θ. If i differs from 90° the foregoing formula still approximately holds good, but k is replaced by $k' = (r_2^2-\cos^2 i)^{\frac{1}{2}}/r_1$. The same formula also describes the "transit" eclipse, but to account for the reappearing part of the disk we introduce $d' = d-2k'$ and then

$$\alpha_{em,tr} = \alpha_{em,oc}(d) - \alpha_{em,oc}(d') \qquad (5)$$

4. Observations

Observations have been made on a few nights in July 1978 and 1979 at Kottamia in Egypt. The region of primary minimum of U Sge was concentrated upon. Targets were to assess the extent of limb darkening (cf. Redman, 1936) and the possible role of any circumstellar emitting material (McNamara, 1951). Simultaneous comparison of the two fluxes was effected by means of a two beam system equipped with an achromatic beam-splitter. DC recordings were measured and processed with the aid of a large screen digitizer at the University of Manchester's Regional Computer Centre.

5. Results

An encouraging result appears from the fitting of the simple disk model of section 3 to the Hα observations of Guinan et al (1976)(Fig.1). In

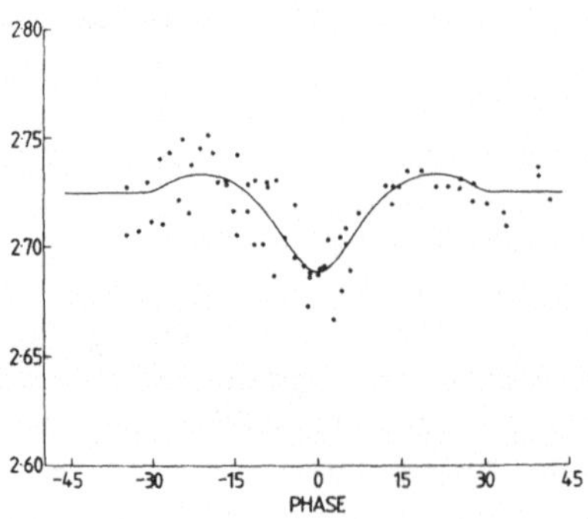

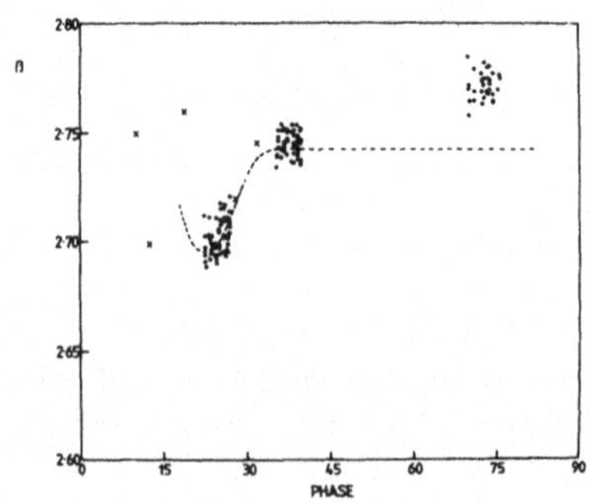

x = Redman's observations,
• = Author's observations

the model zero limb darkening in the line with a representative value of u_c = 0.45 for the continuum have been assumed. The relative height of the disk in terms of the primary radius is 0.3, suggesting a relatively large structure but not extending so far as the Roche lobe itself. The apparent scatter at ingress may reflect a real photometric variation at different observation periods.

In the case of the preliminary observations of U Sge (Fig.2) the position is less clear. The drop in equivalent width around fourth contact is in the opposite sense to the expected. It could be matched by some choice of parameters but an unacceptable level of emission is required, or some asymmetric distribution is implied.

6. Ideas

The idea of an overflow of gaseous material from the contact component of a "semi-detached" close binary system has become widely discussed in the literature (for a review see Sahade and Wood's book, 1978). Some appreciation of the situation may be afforded by reference to Fig.3.

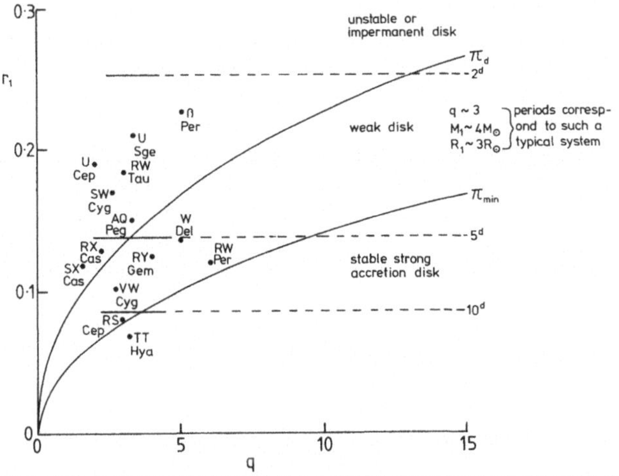

Fig.3. THE DIFFERENT ZONES OF DISK FORMATION SHOWN IN THE DIAGRAM RELECT THE STRINGENT FIXING OF THE π_d, π_{min} CURVES, WHICH DO NOT TAKE INTO ACCOUNT THE ROLE OF DISK VISCOSITY OR RADIATION PRESSURE. OBSERVED POINTS (•) ARE RATHER UNCERTAIN IN SOME CASES.

The curves of closest approach $_{min}$ and stable quasi-circular particle orbit distance π_d have been taken from the work of Lubow and Shu (1975). Those systems for which the primary radius is less than π_{min} may be expected to form accretion disks in the manner proposed by Lubow and Shu, however for systems like β Per and U Sge the outer boundary of the star would be interposed before the particles could take up steady circular orbits. It may be expected, therefore, that while there could still be mass transfer and a circulation effect around the primary in such "semi-detached" systems, such a transfer would not lead to a thick accretion disk, but should be characterized by low densities and unsteadiness. Perhaps the asymmetric distribution suggested by the U Sge results could be understood in terms of such unsteadiness as well as the fluctuations observed in Guinan's data for Algol.

REFERENCES

Budding, E., Marngus, N.: 1980, Astrophys. and Space Sci., 67, p.477.
Crawford, D. L.: 1958, Astrophys. J., 128, p.190.
Guinan, E., McCook, G., Bachmann, P., Bistline, W.: 1976, A.J.,81,p.57.
Lubow, S.H., Shu, F.H.: 1975, Ap.J., 198,p.383.
McNamara, D.H.: 1951, Publ. Astron. Soc. Pac., 63, p.38.
Redman, R.O.: 1936, Mon.Not.R.A.S., 96, p.488.
Sahade, J., Wood, F.B.: 1978, "Interacting Binary Stars",Pergamon Press.

DISCUSSION FOLLOWING BUDDING

Andersen: In your last diagram, I see SX Cas at a mass ratio q = 1 - 2. Was this based on any new data, or how was it derived? If we assume that the secondary fills its Roche lobe, we get q = 8 - 10.

Budding: I will check this point. I think I have taken these stars from an onld paper by Struve or maybe some other old catalogue.

Plavec: I have studied SX Cas very thoroughly in recent months. Andersen is quite right. Struve's original radial velocity curves published in 1944 are completely unreliable since they reflect the complex motions in the envelope of the hotter component, and nothing more. If there exists any reliable radial velocity determination at all for this system, it is Andersen's estimate for the cool component. In order to get the mass ratio, we must make additional assumptions; an indeed, if we postulate that the cool star fills its critical Roche lobe, the mass ratio must be at least 8. There is growing suspicion, though, that considerable mass loss may occur even for late-type giants not filling their respective Roche lobes. Even if we relax this condition, the mass ratio can hardly be smaller than 3, otherwise we get quite unrealistic masses of both stars.

ON THE LOCATION OF GAS STREAMS IN ALGOL SYSTEMS

K. Walter
Astronomical Institute of the University of Tübingen
Federal Republic of Germany

Seven years ago, at the Parksville Symposium, I reported on first results obtained from light curves of typical Algol systems which were discussed according to principles taking into account the presence of gas streams. According to this modified method the reflection parameter is derived from parts of the light curve outside eclipses which are least influenced by gas stream effects. Light curves rectified in this manner, or expressed more exactly, roughly released from reflection and ellipticity effects, only in parts display a horizontal course, in the surplus in intensity after primary eclipse and by part also before it. This indicates an additional light source visible during about half the orbital period. On the preceding side of the primary eclipse a loss in intensity is frequently also observed, which is caused by absorption effects of the gas stream, and often overlaps the surplus intensity.

Figure 1 gives an example of a conventionally rectified light curve with distortions unexplained by the author (Knipe 1974). Here, for XZ Sgr, the reflection parameter is not zero, as Knipe derived, but reflection has its normal course, as it is seen from the second quarter of the light curve, and the surplus in intensity after primary eclipse is clearly present.

Also, the light curves of the primary eclipses are distorted by gas stream effects. For this reason the geometric parameters of the system, the relative radii of the components, and the inclination of the orbit should be derived first of all in accordance with the observed geometric data, especially phases of contacts. Then among the remaining confined possibilities one has to search for that combination of parameters which gives an efficient representation of the observations. Of course, a smooth adaptation to the light curve outside the eclipses must be included.

Figure 2 gives an example found to be typical in carrying out this method. The data are taken from working documents used for the photometric study of U Cep by Hall and Walter (1974). The figure represents the differences between observed and computed intensities for U Cep in total eclipse and the following phases for a sequence of approximate

M. J. Plavec, D. M. Popper and R. K. Ulrich (eds.), Close Binary Stars: Observations and Interpretation, 305–310.

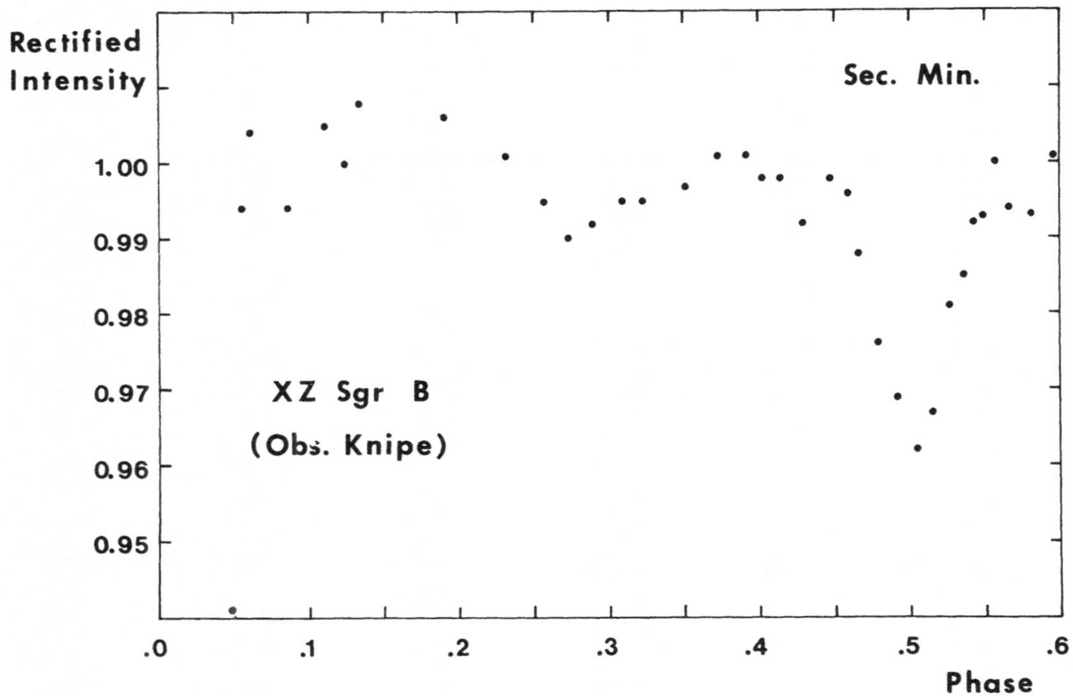

Figure 1. Rectified intensities of normal points for the first half of the light curve outside primary eclipse, given by Knipe (1974) from his observations of XZ Sgr in B. Here, additional light from the hot spot is seen until phase 0.25; the intensity variation between phases 0.25–0.4 is caused by reflection. The figure is taken from the article on XZ Sgr by Kappelmann and Walter, Astron. Astrophys. Suppl., in press.

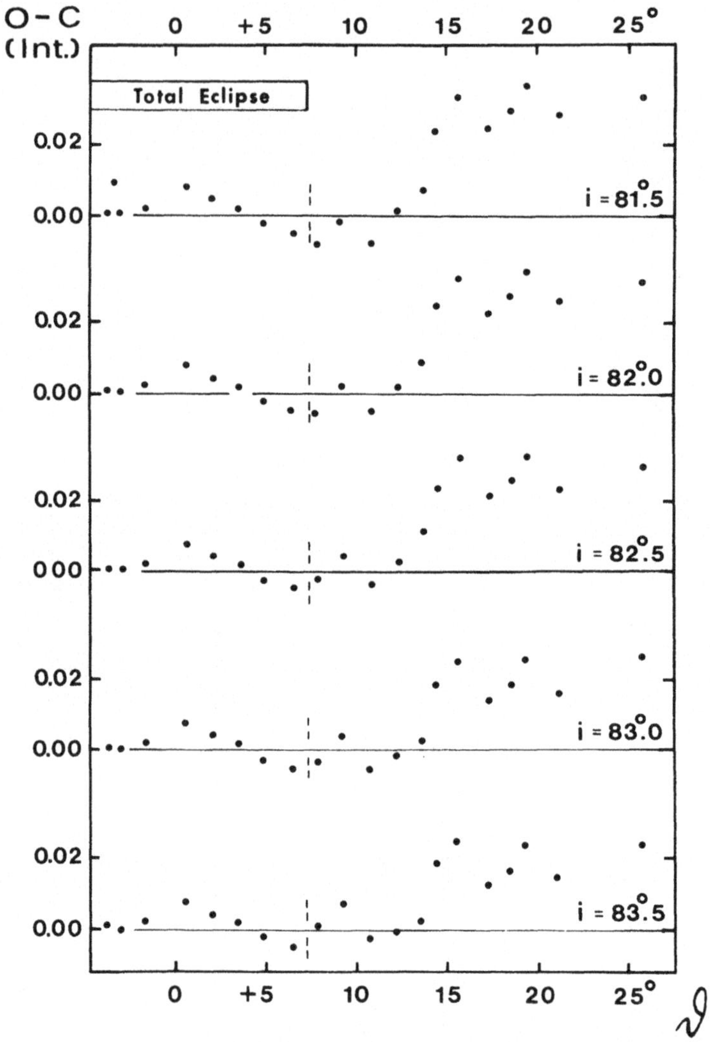

Figure 2. Representation of the light curve of U Cep (observer Tschudovitchev) for a part of the primary eclipse with use of geometric parameters r_1, r_2, and i derived from contact phases. See text. Ordinates are the differences between observed and computed intensities.

solutions, where the inclination is chosen as the free parameter, while the radii are determined from contact conditions. It is seen that after a short phase interval following the third contact there is a rather sudden increase in intensity of about 2-3% of the intensity of the uneclipsed system. This is a typical effect. We ascribe it to the reappearance of a hot region on the primary component. Similar losses in intensity are also observed on the descending branches. From the phases of disappearance and reappearance the hot regions can be located on the disks of the primaries. In most cases they were found to be situated near the polar regions.

According to the modified method to date the following systems have been investigated: SW Cyg (Walter 1971), TW And (Ammann and Walter 1973), Y Psc (Walter 1973), U Cep (Hall and Walter 1974), RX Gem (Hall and Walter 1975), RW Ara (Walter 1976), X Gru (Schulz and Walter 1976), TW Dra (Walter 1977), and XZ Sgr (Kappelmann and Walter 1979). In all systems polar hot spots on the primary components were found, although, as in the case of Y Psc, not at each epoch of observation. In U Cep and RX Gem equatorial spots were also found, and in RX Gem a ring around the primary.

The presence of hot spots near polar regions suggests the existence of magnetic forces which cause ionized matter to leave the orbital plane. Probably the primary component works like a coaxial magnetic dipole, and its main body rotates synchronously with the subgiant in its orbit. Then that part of the gas stream which reaches the polar region has its steady site relative to the components, and the eclipse of the polar spot will coincide with the primary eclipse, as is observed.

However, a branch of the gas stream which flows from the subgiant into a high latitude region of the primary does not flow on the shortest way, but is first shifted away by Coriolis forces alongside the trailing hemisphere of this component and reaches the polar region along magnetic lines on a curved path. It is interesting to note that Batten (1974) observed a component of the CaII K absorption line (as well as Mg+-lines) in the spectrum of U Cep, which "appears with a velocity of about -200 km/s at around phase $0\overset{p}{.}57$ and then rapidly decreases in velocity, merging itself in the central component of the line by $0\overset{p}{.}63$". These observations were made at a normal state of the system. Hall and Walter (1974) suggested that this line originates from particles of the polar branch of the gas stream which cause the polar spot, and which are forced to change the direction of their velocities within a short phase interval in the described manner.

Let us now ask what happens in times of extremely large mass transfer. It seems obvious that near the orbital plane the magnetized layers around the primary are pushed aside by the overflowing matter and that the equatorial spot on the trailing hemisphere and ring-like structures are strengthened. However, what occurs later on, when the largest outbreaks of matter from the subgiant are passed over and the outbreaks become weaker? One may expect that in this state the magnetic field is going to restore its former structure. It begins again to approach near the equatorial regions. Because the transferring matter is diminished, this matter is no longer able to push aside the plasma of the magnetic field as nearly completely as before, but can only disturb it and cause a wide and deep field of turbulent motions on the trailing side of the primary component. Perturbations of this kind can then appear and disappear within intervals of several days. Presumably they can effect large absorptions of the light of the primary, especially at phases which are situated near those, where at normal states of mass transfer, the large variations of radial velocities of the mentioned lines are observed; that is, around phase 0.6. Indeed, in late 1975 extremely strong variations in the light curve of U Cep were observed by Olson (1978) and Kondo et al. (1978), the largest ones near phase 0.6. To explain this, Olson (1978) suggested the formation of a short-lived large cool spot covering about 80% of the projected area of the B star around this phase.

This report may be finished by a short comment on results about elliptical spectroscopic orbits of typical Algol systems with periods between 4 and 15 d and main sequence primaries of spectral type A. In a forthcoming paper (Walter 1979) I show that these systems have circular orbits. The apparent ellipticity of the orbits and the position of the periastron are artifacts connected with the different strengths of the polar and equatorial branches of the gas stream in these Algol systems.

REFERENCES

Ammann, M., Walter, K.: 1973, Astron. Astrophys. 24, 131.

Batten, A.H.: 1974, Publ. Dominion Astrophys. Obs. 14, No. 10.

Hall, D.S., Walter, K.: 1974, Astron. Astrophys. 37, 263.

Hall, D.S., Walter, K.: 1975, Astron. Astrophys. 38, 225.

Kappelmann, N., Walter, K.: 1979, Astron. Astrophys. Suppl., in press.

Knipe, G.: 1974, Monthly Not. 167, 369.

Kondo, Y., McCluskey, G.E., Wu, C.: 1978, Astrophys. J. 222, 635.

Olson, E.C.: 1978, Astrophys. J. 220, 251.

Schulz, E. and Walter, K.: 1970, Astron. Astrophys. Suppl. 29, 51.

Walter, K.: 1971, Astron. Astrophys. 13, 249.

Walter, K.: 1973, Astrophys. Space Sci. 24, 189.

Walter, K.: 1976, Astron. Astrophys. Suppl. 26, 227.

Walter, K.: 1977, Astron. Astrophys. Suppl. 32, 57.

Walter, K.: 1979, in preparation.

COMMENTS FOLLOWING WALTER

de Loore

Can you give an order of magnitude estimate of these magnetic fields? If they are sufficiently large (say some 1000 gauss) they could be measured by Zeeman analyzers. Do such measurements exist?

Walter

As far as I know such measurements do not exist. Roughly estimated magnetic fields of some hundred gauss could be sufficient.

Smak

I want to use this opportunity to put a commercial on XZ Sgr: It belongs to that group of extreme Algols, exemplified by AS Eri and DN Ori, which have very low mass, highly overluminous secondaries. It certainly deserves more photometric and spectroscopic attention.

Walter

I observed this system in UBV at La Silla in 1972-74. The observations and the results from their evaluation are in print [Kappelmann and Walter, Astron. Astrophys. Suppl.].

VI. X-RAY BINARIES

X-RAY BINARY SYSTEMS

David Crampton
Dominion Astrophysical Observatory, Victoria, B.C.

ABSTRACT

X-ray emission has now been detected from a large number and variety of binary systems. A brief survey of the properties of these binary systems is given, primarily from an optical standpoint.

I. INTRODUCTION

In the past few years, rapid progress has been made in the observations of X-ray binary systems and in the theoretical interpretation and understanding of the processes occurring in close binary systems and of their evolution. In this paper, I shall treat the X-ray binary systems primarily from an optical observational viewpoint, and try to relate these systems to conventional binary systems starting first with the massive binary systems and working down to the systems of lowest mass.

Although the initial impetus for studying X-ray binaries was to discover and to determine parameters of degenerate objects, we are learning considerably more about close binary systems as a whole, because the degenerate object essentially behaves as a test particle for which and from which we can determine precise orbital parameters, at least in the case of the pulsing, eclipsing systems. The complications which we encounter in trying to determine equally precise parameters from optical observations indicate areas or problems that need to be carefully scrutinized in the study of conventional systems.

II. BLACK HOLE CANDIDATES

Naturally, there is keen interest in establishing the existence of, and determining the parameters for the mythical black holes. Since Bolton is reviewing the object that is still the prime candidate, I will be brief. I would like to note, however, that HDE 226868 =

M. J. Plavec, D. M. Popper and R. K. Ulrich (eds.), Close Binary Stars: Observations and Interpretation, 313–321.

Cyg X-1 is not particularly unusual optically, the He II λ 4686 A emission that has become the standard signature of the optical counterparts of X-ray sources being very weak, and the spectrum otherwise (except for broad Hα emission) being that of a normal B0 supergiant. The hypothesis that the unseen star is a black hole is still the most plausible interpretation of the observations, although by taking extreme values of all the parameters the mass of the degenerate object can be nearly reduced to that of a massive neutron star.

Two other X-ray sources exhibit very similar X-ray characteristics to Cyg X-1 and hence may contain black holes; Cir X-1 and GX 339-4. Samimi et al. (1979) have reported observations of the latter source and have summarized the X-ray observations of the prime black hole candidates. These objects all exhibit X-ray variability on timescales of milliseconds, have high and low states and have similar X-ray spectra. Although faint and reddened, we can look forward to optical observations of the optical counterparts of Cir X-1 and GX 339-4 to see whether they too have large mass functions.

III. THE MASSIVE SYSTEMS

It is the pulsing, eclipsing X-ray binaries from which we have learned the most, particularly those in the Magellanic Clouds since then the distances are known as well. Orbital elements for those binaries with relatively well-determined parameters are given in Table 1, with those of Cyg X-1 for comparison. For references, see Hutchings (1979).

TABLE 1. ORBITAL ELEMENTS

Name	P days	V_o km s^{-1}	K_o km s^{-1}	K_{HeII} km s^{-1}	K_x km s^{-1}	e_o	e_x	i
HD77581	9.0	0	22	-	273	.14	.09	>75°
SMC X-1	3.9	180	19	245	299	.36±.09	<.0007	∿65°
1538-52	3.7	-172	33	302:	323	.14:		70°
Cen X-3	2.1	+ 39	24		415		.0008	83°
LMC X-4	1.4	271	50	498	-	0:		75°
HD153919	3.4	- 75	19	-	-	.2, .05		87°
Cyg X-1	5.6	0	72			.06		∿60°

The following aspects of the quantities in the table are noteworthy:

- Periods are typical.
- Systemic velocities are not peculiar apart from that of

4U1538-52. A large stellar wind is probably responsible for the negative velocity of HD 153919.

- $K_{HeII} \sim 0.8\ K_x$, indicating that the He II emitting region is closer to the centre of mass. Actually, in several systems He II emission arises both from regions near the X-ray source and near the heated face of the companion.

- As revealed by SMC X-1, the spectroscopic eccentricity may be spurious due to a variety of reasons (Hutchings 1979). The eccentricity of HD 77581 is well determined and limits have been placed on the apsidal motion by Rappaport, Joss and Stothers (1979).

- $i > 60°$ for all systems. X-ray beaming or obscuration must be responsible for this anisotropy.

Some of the physical parameters which have been deduced for the massive binary systems are listed in Table 2. These systems are confined to a very restricted range in luminosity and effective temperature, and, despite their unusual evolutionary status (see, for example, van den Heuvel 1976), their spectra are completely normal.

TABLE 2. PHYSICAL PROPERTIES

Name	P_{pulse} s	Sp	M_V mag	M_{bol} mag	q	$M_o/M_\odot$	$M_x/M_\odot$	L_o/L_x
HD77581	283	B0.5Ia	-7.2	-10.0	12	23	1.9	1300
SMC X-1	0.7	B0 Ib	-6.6(-6.1)	- 9.0	16	16	1	4
1538-52	529	B0 I	-6.5	- 9.1	10	20	2	500
Cen X-3	4.8	07 III	-5.9	- 9.3	17	17	1	100
LMC X-4		07 III-V	-5.2(-5.0)	- 8.3	10:	25	2.6	30
HD153919		06.5f	-6.5	-10.2	20:	27	1.3	5000
Cyg X-1		09.7 Iab	-5.6	- 9.1	3	18	6	100

The intrinsic luminosities of the optical primaries derived from spectral classification and Hγ equivalent widths are also apparently normal, since the agreement between these and the values derived from the distance to the Magellanic clouds (absolute magnitudes given in parenthesis in Table 2) is very good. It is very clear now that the masses of the optical primaries are all low by a factor of two or more. This fact is dramatically displayed if these systems are plotted on an HR diagram with the stellar evolution tracks of stars of various mass (see Hutchings et al 1979 Fig. 3). Rappaport (1979) has shown that the masses of the neutron stars in the pulsing binary systems may all be very similar and equal to $\sim 1.4\ M_\odot$. Perhaps we should adopt this value in difficult cases and derive the other parameters on this basis. Although all of these systems have an X-ray luminosity $L_x \sim 10^{38}$ ergs sec^{-1}, the ratios of the optical to X-ray luminosities are very different from each other. No particular correlation with any of the system parameters except perhaps with the pulse period is evident.

There is one additional massive system for which we do not yet know the orbital parameters, 4U1223-62 or WRAY 977. From pulse timing measurements it is obvious that the orbital period is long, but the exact value is still unknown. The spectral type of WRAY 977 is B2Iae with strong P Cyg profiles in the lines of most of the elements. It is apparently the most evolved of the massive X-ray binaries and hence its parameters will be of considerable interest.

IV. Be SYSTEMS

The X-ray sources associated with Be stars have a lower X-ray luminosity, $L_x \sim 10^{34}$ ergs sec^{-1} ($L_o/L_x \sim 10^5$), harder spectra, transient behaviour and in general longer pulse periods ($\sim$ 300s) and orbital periods ($\sim$ 25d?). None of these systems have both optical and X-ray orbits. The X-ray orbit (Rappaport et al. 1978) for 4U0115 + 63 indicates that the orbit is quite eccentric (e = 0.3), a factor which is probably related to the transient behaviour of the X-rays. It is not clear what is responsible for the observed 580-day velocity variations (Hutchings, Crampton and Redman 1975), in X Per because if assumed to be orbital, a massive black hole is implied by the deduced mass function. The discovery of a pulse period points to a slowly rotating neutron star rather than a black hole however. It may be relevant that Poeckert (1979) has recently discovered weak He II emission in the spectrum of the Be star ϕ Per which apparently is associated with an unseen companion that is probably a neutron star.

As for the supergiant systems, the optical counterparts to these X-ray sources are restricted to a very narrow range of spectral type and luminosity class (O9-B1, III - Ve). The SMC sources (Crampton, Hutchings, and Cowley 1978) again indicate that the luminosities are normal for their spectral classification.

V.1. LOW MASS SYSTEMS I: NON-DEGENERATE STAR VISIBLE

Two sources in this category, Her X-1 and Cyg X-2, now have reasonably well determined parameters. Thorstensen et al (1979) have recently shown that a probable third member, 4U2129 + 47, has a large 1.5 mag. photometric variation with P = 5.2 hours.

In the well studied Her X-1 = HZ Her system (e.g. Boynton 1978), a very large X-ray heating effect is primarily responsible for the large spectral type (B-F) and light (1.5 mag.) variation, and produces complicated radial velocity variations (Crampton and Hutchings 1974). It is still not clear where the weak variable emission lines are formed. Extensive observations by Boynton and his collaborators show that the 35-day ON-OFF X-ray variability is due to a large precessing disk. The masses of the components are $M_O \sim 2\ M_\odot$, $M_X \sim 1\ M_\odot$.

Recent observations of Cyg X-2 (Cowley, Crampton and Hutchings 1979) indicate that instead of being a short period system similar to the cataclysmic variables as once thought, the period is 9.8 days and the optical primary is an F giant located about 8 kpc from the sun with $z \sim 1.5$ kpc and a systemic velocity of -222 km s^{-1}. In this case, most of the He II emission probably originates from the heated face of the F star, although a variable component arising from a region near the X-ray source is also present. The light curve is very complex, being composed of an ellipsoidal variation, a single-peaked variation most prominent in U that is presumably due to X-ray heating, and erratic outbursts. A combination of the restrictions on q and i determined from the ellipsoidal variation together with the spectroscopic observations yields masses of $M_O \sim 0.7\ M_\odot$ and $M_X \sim 1.5\ M_\odot$.

The X-ray luminosity of these sources is $\sim 10^{37.5}$ erg s^{-1} and the ratio L_X/L_O is $\sim$ 10-100. Both Cyg X-2 and Her X-1 are situated far from the galactic plane and have high systemic velocities. Either these systems were imparted large velocities as a result of a supernova explosion or they are population II objects.

V.2. LOW MASS SYSTEMS II: SCO X-1 TYPE

There are now a large number of X-ray sources which are identified with objects having blue continuous spectra with He II emission, weak C III - N III emission, and, sometimes, H emission superposed as in the prototype, Sco X-1. A list of these objects which includes some of the burst sources is given by Cowley (1979).

While all the optical light of Sco X-1 appears to come from reprocessed X-rays originating from a luminous disk around the X-ray source, a neutron star, and the companion is not visible, it seems probable that it is a low-mass, slightly evolved, non-degenerate star which is losing mass by Roche Lobe overflow (Cowley and Crampton 1975). Analysis of the radial velocity variations of the He II emission combined with the light variations yields masses $M_O \sim 1\ M_\odot$, $M_X \sim 1\ M_\odot$ and $i \sim 25°$. The low amplitude of the light curve suggests a very low inclination, $i < 10°$, but this would lead to unreasonable masses and so it is probable that beaming or obscuration of the X-rays prevents the expected large X-ray heating of the primary.

If Sco X-1 is typical, the X-ray luminosity of this class of binaries is $L_X \sim 10^{37.5}$ ergs s^{-1} as in the case of Cyg X-2 or Her X-1, but L_X/L_O is $\gtrsim 10^3$. Cowley (1979) shows that the galactic distribution of the Sco X-1 systems is concentrated about the galactic centre.

V.3. LOW MASS SYSTEMS III: AM HER TYPE

According to Cowley (1979) there are now four X-ray sources identified which have optical counterparts characterized by strong emission-line spectra, short periods ($\sim$ few hours) and high polarization indicating large magnetic fields. These systems are related to the cataclysmic variables although the transitions from minimum to maximum light is not cataclysmic and is only a few magnitudes. The X-ray luminosity of these systems is $L_x \sim 10^{34}$ ergs sec^{-1} and the optical luminosity is comparable. It is generally assumed that these systems contain a low mass dwarf which fills its Roche Lobe and a magnetic white dwarf. The emission lines in AM Her, at least, originate primarily in a stream of material between the two stars and so they yield no direct information about the masses of the stars. Estimates of $M_0 \sim 0.4\ M_\odot$ and $M_x \sim 1\ M_\odot$ have been given by analogy to the cataclysmic variables (Cowley and Crampton 1977).

V.4. LOW MASS SYSTEMS IV: X-RAY NOVAE

In some respects (e.g. a fast rise of $\sim$ 7 mag. to maximum light) the prototype of this class, Nova Mon 1975 or AO 0620-003 resembles the classical novae. At maximum light, however, the spectrum was nearly continuous with only weak He II and H emission and no substantial mass outflow was detected. The optical and X-ray outburst was accompanied by a non-thermal radio outburst. At minimum light Oke (1977) observed the spectrum of a late type K dwarf which is presumably the companion to a degenerate star in the system, but no convincing orbital periodicity has yet been observed.

Cen X-4 has recently undergone a similar outburst (Kaluzienski and Holt 1979) from which we may expect to determine further parameters for these systems.

VI. SOFT X-RAY BINARY STARS

There are two other classes of X-ray systems which emit predominately soft X-rays, the SS Cyg and RS CVn systems. These are both classes of binary systems in which accretion of material onto a white dwarf is responsible for the soft X-ray emission, although hard X-rays are apparently produced by coronal processes in a few of the RS CVn systems. The SS Cyg stars were discussed in detail at IAU Colloquium No. 53 in Rochester and the RS CVn stars at several workshops (e.g. Hall 1978).

VII. 4U1626-67

The optical counterpart to this pulsing source has colours similar to those of the Sco X-1 type sources. The 7 s pulsation has been observed in visible light (Ilovaisky, Motch and Chevalier 1979) as well as in the X-rays, but pulse timing observations have only yielded an upper limit to the orbital period of P < 1 hour (Rappaport 1979). The X-ray variations exhibit a characteristic time scale of $\sim$ 1000 secs and this has led to the suggestion that this system is a very close binary system containing a neutron star and white dwarf with P $\sim$ 1000 sec. (Rappaport 1979).

VIII. SS433

Recent spectroscopic observations obtained at Victoria indicate that this remarkable source (e.g. Margon et al. 1979) is a binary system with P $\sim$ 13 days, although it is not yet clear how it fits into the classes discussed above. The radial velocity variations of the Hβ and Hα emission peaks of the "stationary" system of lines are periodic with P = 13.0 ± 0.1 days, a semi-amplitude, K, of $\sim$ 75 km s^{-1}, and with time of maximum positive velocity occurring on JD 2444058.5 ± 0.2. An eccentricity of 0.2 ± 0.1 is deduced from the observations. The very broad emission lines are typical of those observed to originate in accretion disks, and so it is assumed that the observed velocity variation represents orbital motion of the X-ray emitting object. If this object is further assumed to be a neutron star with $M_x \sim 1.4\ M_\odot$, then the companion would have $M_0 \sim 1.8\ M_\odot$ for i $\sim$ 90°. The period and mass function are very similar to those of Cyg X-2 although the optical observations are of the non-degenerate star in that case. The X-ray luminosity, $L_x \sim 10^{35.5}$ ergs s^{-1}, for an assumed distance of 3.5 kpc (Margon et al. 1979) is lower than that of Cyg X-2 but higher than that of AM Her.

The He I "stationary" emission lines also probably exhibit the same 13 day velocity variation although poorer phase coverage and larger observational error preclude the detection of small differences. He I absorption lines which strengthened considerably in July 1979 probably also exhibit a variable velocity of comparable amplitude although their systemic velocity is $\sim$ 350 km s^{-1} more negative. The observations of the He II λ 4686 A emission are not adequate to determine whether they display a similar variation. No 13 day periodic variation of the intensities of any of the lines has been detected, although the profile of Hβ progressively changed during the period of observation, JD 2443987-2444085.

Current models deduced from the "moving" lines (Milgrom 1979, Margon and Abell 1979) ascribe the 164 day, 81,000 km s^{-1} variation to motion arising in two colinear rotating beams. The source of the energetics involved is still a mystery however, since our recent observations of a 13 day periodicity indicate that otherwise this system is similar to other low mass systems.

CONCLUSION

Accretion of material supplied by a normal star onto a compact object is apparently responsible for the observed X-ray emission in most, if not all, of the binary systems considered here. In many cases there is evidence of considerable interaction between the components, gas streaming, accretion disks etc. Further studies of these systems should help us to understand these and other phenomena encountered in close binary systems, as well as revealing the nature of the compact objects.

REFERENCES

Boynton, P.E.: 1978, Proc. Int. School of Phys. "Enrico Fermi", R. Giacconi and R. Ruffini (eds.), North Holland, Amsterdam, pp. 111.
Cowley, A.P.: 1979, in Proc. NATO Advanced Study Institute, Cape Sounion, Greece. Cambridge U. Press.
Cowley, A.P., and Crampton, D.: 1975, Astrophys. J. Letters, 201, L65.
Crampton, D., and Cowley, A.P.: 1977, Publ. Astron. Soc. Pacific, 89, pp. 374.
Crampton, D., and Hutchings, J.B.: 1974, Astrophys. J., 191, pp. 483.
Crampton, D., Hutchings, J.B., and Cowley, A.P.: 1978, Astrophys. J. Letters, 223, L79.
Hall, D.S.: 1978, Astron. J., 83, pp. 1469.
Hutchings, J.B.: 1979, in Proc. NATO Advanced Study Institute, Cape Sounion, Greece. Cambridge U. Press.
Hutchings, J.B., Crampton, D., and Redman, R.O.: 1975, Monthly Notices Roy. Astron. Soc., 170, pp. 313.
Hutchings, J.B., Cowley, A.P., Crampton, D., van Paradijs, J., and White, N.E.: 1979, Astrophys. J., 229, pp. 1079.
Ilovaisky, S.A., Motch, C., and Chevalier, C.: 1978, Astron. Astrophys., 70, L19.
Kaluzienski, L.J., and Holt, S.S.: 1979, IAU Circ. 3360, 3362.
Margon, B., and Abell, G.O.: 1979, Nature 279, pp. 701.
Margon, B. et al.: 1979, Astrophys. J. Letters, 230, L41.
Milgrom, M.: 1979, Astron. Astrophys. in press.
Oke, J.B.: 1977, Astrophys. J., 217, pp. 181.
Poeckert, R.: 1979, Astrophys. J., in press.
Rappaport, S.: 1979, in Proc. NATO Advanced Study Institute, Cape Sounion, Greece. Cambridge U. Press.
Rappaport, S., Clark, G.W., Cominsky, L., Joss, P.C., and Li, F.: 1978, Astrophys. J. Letters, 224, L1.
Rappaport, S., Joss, P.C., and Stothers, R.: 1979, submitted to Astrophys. J.
Samimi, J. et al.: 1979, Nature, 278, pp. 434.
Thorstensen, J. et al.: 1979, submitted to Astrophys. J.
van den Heuvel, E.P.J.: 1976, in IAU Symposium No. 73, pp. 35, P. Eggleton, S. Mitton, J. Whelan (eds.), Reidel, Dordrecht, Holland.

COMMENTS FOLLOWING CRAMPTON

Ziolkowski: I would like to make a comment on Cyg X-2. We are working now with Bohdan Paczynski on a model of this system. We believe that the apparent lack of the X-ray heating effect tells us something not only about Cyg X-2 but also about other low mass X-ray binaries. It appears, for example, that, in order to explain the smallness of the X-ray heating effect in Sco X-1, it is enough to assume that X-rays are coming from a flat, limb darkened disc. However this would not work for Cyg X-2. It seems unavoidable that there is some screen between the optical component and the X-ray source. If this screen is a thick disc, then, in order to avoid observable heating of the outer parts of the disc, it has to be thick quite close to its center (at a distance of less than 5 percent of the separation between the components of the binary system). I will present preliminary results of our work on Cyg X-2 during the meeting of Commission 42 at Montreal.

Mitrofanov: Have you estimated the lifetime of a binary system containing neutron star and degenerate dwarf? It seems it may be very small due to the generation of gravitational waves.

Vanbeveren: The overluminosity with respect to the mass of the optical companions of massive X-ray binaries suggests that they have lost a large amount of mass due to stellar wind (see e.g. the paper by Vanbeveren and de Loore, this symposium). If these stars should be at the end of core hydrogen burning one therefore would expect a lowering of the hydrogen abundance in the surface by at least a factor of 2 or 3. Is it possible to observe an effect like that?

Crampton: As I indicated, the spectra which we have appear to be completely normal as far as abundances are concerned. However, I suspect that we would not detect a deficiency of H of a factor of 2-3.

Mayo: What is the latest news on the X-ray emission from V861-Sco? You have told us it is now not considered to be the source of the 38 sec pulsing. There was a Copernicus report of an X-ray eclipse in the IAU circulars (in June/July 77). Are these X-ray eclipses confirmed? Indeed, is V861-Sco now thought to be an X-ray source at all? And finally, in the light of your answers, what type of binary system do you now think V861-Sco is?

NON-CONSERVATIVE MASS EXCHANGE AND ORIGIN OF X-RAY CLOSE BINARIES

Daiichiro Sugimoto and Shigeki Miyaji
Department of Earth Science and Astronomy, College of General Education, University of Tokyo, Tokyo 153, Japan

1. TWO TYPES OF X-RAY BINARIES

Recently, masses of component stars have been determined for many X-ray close binaries (XCBS). For relatively well determined sources the masses of X-ray components are plotted against the masses of their optical components in Figure 1 and their orbital periods are summarized in Table 1 (Cowley 1977; Bradt, Doxsey, and Jernigan 1979; Wheeler 1978; and references quoted therein). Cowley (1977) prepared a table and noticed that there are two distinct types of XCBS. The Type I XCBS consists of an X-ray star and an early type star more massive than about 12 $M_\odot$. On the contrary, the Type II XCBS consists of an X-ray star and a star less massive than about 2 $M_\odot$. As seen in Figure 1, there is not any distinct intermediate type for which the mass of the optical component lies in the range of about 2-12 $M_\odot$. The aim of the present paper lies in interpreting the origin of these types of XCBS on the bases of the conditions for the formation of a neutron star and of mass exchange in close binary systems.

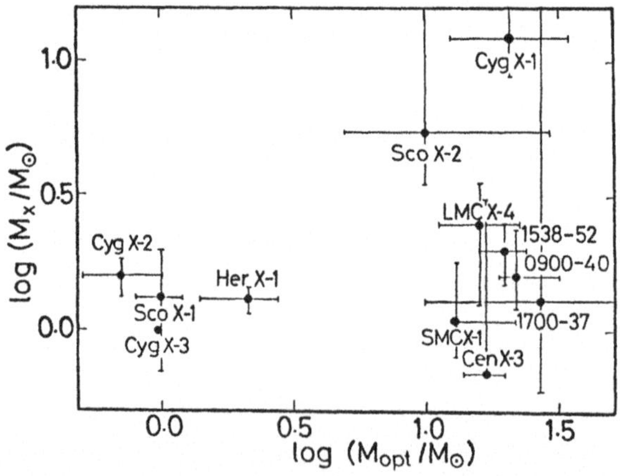

Fig.1. — Masses of X-ray components are Plotted against their optical components.

TABLE 1.— ORBITAL PERIODS

Source	P_{orb} (days)
Type I	
SMC X-1	3.89
LMC X-1	1.41
0900-40	8.97
Cen X-3	2.09
1538-52	3.73
Sco X-2	7.85
1700-37	3.41
Cyg X-1	5.60
Type II	
Sco X-1	0.79
Her X-1	1.70
Cyg X-3	0.20(16.9)
Cyg X-2	11.17(0.86)

M. J. Plavec, D. M. Popper and R. K. Ulrich (eds.), Close Binary Stars: Observations and Interpretation, 323–327.

2. CONSERVATIVE MASS EXCHANGE AND TYPE I XCBS

In a separate paper presented to this symposium (Sugimoto and Miyaji 1979), which will be referred to as SM, it was discussed that a Type I XCBS is produced from an early-type contact binary as a result of conservative mass exchange. However, the resultant XCBS was shown to have a relatively long period. Such discussion is generalized to include different initial conditions. We denote the mass of the component stars A and B by M_A and M_B ($M_{A0} > M_{B0}$), respectively. The total mass of the system is $M_t \equiv M_A + M_B$ which is conserved. Here and hereafter the subscript 0 denotes its initial value. As the mass is transferred from Star A to Star B, the separation changes. It reaches the smallest value when $M_A = M_B = M_t/2$. The stage just before the supernova explosion of Star A will be denoted by the subscript 1. At this stage the hydrogen-rich envelope of Star A has been almost stripped off and we approximate $M_{A1} = M_{A0}/4$.

Even at the stage of the smallest separation, Star B should be accommodated within its Roche lobe. This condition is transformed into the condition for the period at the pre supernova stage as where

$$(P_1/\text{day}) > \frac{11.2}{(3 + x)^3 \ (1 - x)^3} \ f^{3/2}(M_t/M_\odot) \ , \qquad (1)$$

$x \equiv M_{B0}/M_t$ is the initial mass fraction of Star B and $\underline{f}$ is the ratio of radius to mass in solar units for main-sequence stars. For resonable ranges in $x = 0.2 - 0.5$ and $M_t/M_\odot > (12 - 20)$, condition (1) yields $P_1 > (4.7 - 12)$ days, respectively. After the supernova explosion the period becomes somewhat longer. As far as Type I XCBS's in Table 1 are concerned, this condition is not satisfied though it is marginal for 0900-40. Therefore we have to seek another interpretation for the origin of such XCBS.

3. NON-CONSERVATIVE MASS EXCHANGE AND ORIGIN OF TYPE I XCBS

What happens if the initial separation is much wider than those discussed in the preceding section? In such case the mass transfer becomes very rapid and non-conservative as was discussed in SM. Then the component stars of early-type contact binary come so close each other that they will dissipate or coalesce.

Here, we will consider how large initial separation is required in order to avoid the coalescence. We will consider the case of Cen X-3 as a typical example. Its history is illustrated in Figure 2. The mass of its optical component Star B stays almost constant, which we take $M_B = 17\ M_\odot$ (Cowley 1977). In order to avoid the excessive closing up of the stars, we assume that the initial mass of Star A was close to but a little larger than M_B, i.e., $M_{A0} = (17 + \varepsilon) M_\odot$. After non-conservative mass exchange, the core of Star A is left whose mass is $M_{A1} = M_{A0}/4 = 4.25\ M_\odot$. From equation (2) of SM with $\ell = 1.7$ we obtain

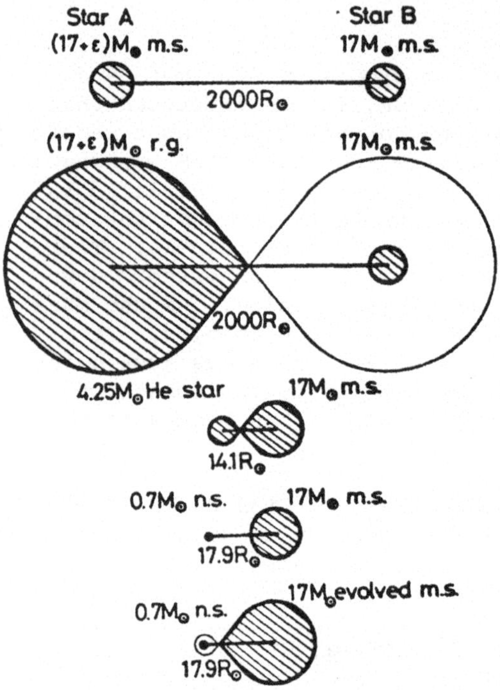

Fig.2. — A model of Cen X-3.

$a_1/a_0 = 0.0070$. Then Star A makes supernova explosion and a neutron star of mass $M_A = 0.7\ M_\odot$ is left as observed in Cen X-3. If the explosion is instantaneous and spherical, the orbit becomes elliptical with the eccentricity $e = 0.215$, and the separation becomes $a = (1-e)^{-1}a_1 = 1.27\ a_1 = 0.0089\ a_0$. If we assign $a = 17.9\ R_\odot$ for the present value of the separation corresponding to the orbital period of 2.087 days, we obtain the initial separation to have been $a_0 = 2000\ R_\odot$ for which the radius of the critical Roche lobe was $R_{cr,A0} = 760\ R_\odot$.

How large stellar radius did Star A reach in its course of evolution? The stellar luminosity should be fainter than the critical luminosity, which is $L_{cr} = 1.6 \times 10^5\ L_\odot$ for $M_{A1} = 4.25\ M_\odot$. If we assume the luminosity of a half of L_{cr} and the effective temperature of $T_{eff} = 3500$ K, we obtain the stellar radius to have been $R = 770\ R_\odot$. As this radius is not smaller than $R_{cr,A0}$, the star expanded large enough to initiate the mass transfer. In other words we reach a consistent model with some margin. For other Type I XCBS's the situation is essentially unaltered, i.e., we can construct consistent models: The mass of their optical component M_B is comparable with the mass escaping from the system $(M_{A0} - M_{A1}) \simeq M_B$ so that the exponential term in equation (2;SM) does not become excessively small.

4. ORIGIN OF TYPE II XCBS

For Type II XCBS, on the other hand, a serious difficulty arises if we imagine a scenario similar to those for Type I. It comes from the smallness of M_B, which results in an excessive closing up of the component star. Let us consider Her X-1 as a typical example, for which we take $M_B = 2.2\ M_\odot$ (Wheeler 1979). Because Star A is now a neutron star, its initial mass should have been more massive than the upper mass limit to the carbon deflagration supernova, for which we employ the lowest estimate of $M_{A0} = 5\ M_\odot$ (Wheeler 1978). The helium core mass for this star will be close to the Chandrasekhar limit so that we assume $M_{A1} = 1.5\ M_\odot$. Then we obtain $a_1/a_0 = 4.26 \times 10^{-4}$ by means of equation (2;SM) with $\ell = 1.7$. Because the mass of the neutron star is $1.3\ M_\odot$, we obtain further $a/a_1 = 1.06$. The observed orbital period of 1.70 days corresponds to $a = 9.11\ R_\odot$. Therefore the initial separation should have been as large as $a_0 = 2490\ a = 2.3 \times 10^4\ R_\odot$, which is too large to initiate the mass exchange. As seen from equation (2;SM), such results are common among the systems with $(M_{A0} - M_{A1}) > M_B$, i.e., among Type II XCBS's.

However, there exist such XCBS's in nature. How can we remove such difficulty? The fact that M_B is small indicates in itself that the mass exchange has been non-conservative. Therefore, only one possibility is to reduce the mean value of angular momentum which is carried away with unit mass of escaping matter. When Star A is a red giant, there may be a self-excited mass ejection, in which the envelope of Star A will be expelled by Star A itself as in the case of mass loss from a single star. In such case the gas particle has a velocity higher than the thermal speed when it leaves from the binary system. Therefore the effective value of ℓ will be appreciably smaller than 1.7. In the limiting case of high speed mass ejection, for example, only the angular momentum which was associated originally with the escaping gas, is carried away. It is about $\ell \simeq 0.25$ in the case of $M_A = M_B$, for example. Thus the result of the self-excited mass ejection will be the reduction of M_A and a gradual closing up of Star A to Star B. When the component stars close enough, the non-conservative mass exchange commences and the component stars become closer and closer according to equation (2;SM) with ℓ = 1.7. However, an excessive closing up will be avoided, if Star A has lost most of its hydrogen-rich envelope by the preceding self-excited mass ejection.

One may ask why such self-excited mass ejection does not take place in the precursor of Type I XCBS. In this connection it is interesting to note the following difference in stellar evolution. The stars more massive than 12 $M_\odot$ stay around yellow supergiant in and near the helium burning phase (see e.g., Lamb, Iben and Howard 1976). On the other hand the stars less massive than 12 $M_\odot$ evolve to red supergiant (see e.g., Barbaro, Chiosi and Nobili 1972), where appreciable self-excited mass ejection will take place. In the later phases of evolution the stars in both mass ranges become red supergiants but their lifetimes are too short to allow appreciable amount of mass ejection.

This research was supported in part by the Scientific Research Fund of the Ministry of Education, Science and Culture (464080).

REFERENCES

Barbaro, G., Chiosi, C., and Nobili, L.: 1972, Astron. Astrophys., 18, pp. 186-197.
Bradt, H.V., Doxsey, R.E., and Jernigan, J.G.: 1979, to be published in Advances in Space Exploration 3.
Cowley, A.P.: 1977, Ann. New York Acad. Sci., 302, 1.
Lamb, S.A., Iben, I.Jr., and Howard, W.M.: 1976, Astrophys. J., 207, pp. 209-232.
Sugimoto, D., and Miyaji, S.: 1979, paper presented to this symposium.
Wheeler, J.C.: 1978, to be published in the Proceedings of the Workshop on Supernovae and Supernova Remnants held in Erice, Sicily, May 1978.

DISCUSSION FOLLOWING SUGIMOTO AND MIYAJI

Shu: I would like to comment that the use of the Roche model becomes somewhat dubious when dealing with separations as large as a few thousand solar radii. The assumption of synchronism must almost certainly break down for such widely separated systems.

Sugimoto: The important parameter is not the geometry of the Roche model itself but the angular momentum carried away together with the gas outflowing from the system. The outflowing gas is, in any case, not in the synchronism. Such effects were taken into account when the particle trajectories were computed. For the size of the critical Roche lobe, the important thing is the size within which the gravitational pull from the companion can be neglected in determining stellar structure. In this sense such size is not so much different even when the synchronism breaks down.

Meyer-Hofmeister: The mass ratio is important for the question whether the mass transfer is conservative or not. For mass ratios far from 1 the thermal timescale of the secondary is much longer than the thermal timescale of the primary which forces loss of mass from the system.

Sugimoto: It depends not only on the mass ratio but also more critically on the evolutionary stages of the star, because the thermal timescale of the envelope depends strongly on the stellar radius. Therefore it depends on the initial separation between the components. When the mass exchange commences as in the cases late A, B, or C, it is well non-conservative, as has been shown in my first paper presented at this symposium.

THERMONUCLEAR BURNING IN THE ENVELOPE OF AN ACCRETING NEUTRON STAR

E. Ergma, T. Tutukov

ABSTRACT

A simple analytical method has been used to investigate thermonuclear flashes in a degenerate envelope of an accreting neutron star. The heating source is the compression of envelope and the cooling is due to the radiative cooling of the envelope. The role of hydrogen burning is discussed. A possible evolutionary scenario for formation of preburster binaries consisting of a neutron star and a low mass main sequence star is proposed.

The most promising model for explanation of bursters to date is the thermonuclear one proposed originally by Woosley and Taam (1976), Maraschi and Cavaliere (1977). A thermonuclear model has been developed in a number of papers by Joss (1977, 1978a, b) and Ergma and Tutukov (1979a) and Tutukov and Ergma (1979b).

In this paper we present results of the investigation of the thermal evolution of a shell accreted onto the surface of a neutron star.

Thermal evolution of an accreted shell strongly depends on several factors: 1) the accretion rate $\dot{M}$, 2) thermal history of the neutron star, 3) the fraction of the rest which goes into heating, 4) the chemical composition of accreted matter, 5) the opacity, and 6) the mass of the neutron star.

Let us give a brief description of our analytical method (for more details we refer to our previous papers: Tutukov and Ergma (1979) and Ergma and Tutukov (1979).

We consider a neutron star undergoing spherically symmetric accretion of the nuclear fuel from a close second component. The size of the latter and the distance between the components is of the order of or somewhat smaller than the solar radius according to the three second delay of the optical burst (Grindlay et al., 1978).

M. J. Plavec, D. M. Popper and R. K. Ulrich (eds.), Close Binary Stars: Observations and Interpretation, 329–334.

For the flat thin envelope the condition of thermal equilibrium

$$\frac{16\pi\sigma R^2 T_{sh}^4}{3\kappa\rho_{sh}\Delta R} = \alpha c^2 \dot{M}' \qquad (1)$$

allows an estimation of the shell temperature T_{sh}. Here σ is the Boltzman constant, R is the radius of the neutron star, κ is the opacity, ρ_{sh} and ΔR are average density and thickness of the shell, c is the light velocity, and $\dot{M}'$ is the accretion rate in g/s.

α is the main parameter determining the variation of the values T and ρ. Nuclear energy gives $\alpha \approx 0.007$. The selfcompression of the shell due to the mass accumulation leads to $\alpha = GM\Delta R/R^2 = \gamma P_{sh}/(\gamma-1)/\rho_{sh}/c^2$, where P_{sh} is the pressure at the bottom of the shell and γ is the adiabatic index. The thermal evolution of the accreting shell for $\alpha = 0.01$ was investigated in our previous paper (Ergma, Tutukov, 1979). Now we shall discuss the thermal evolution of the shell with heating due to the selfcompression only. The main equation (1) in this case transforms to:

$$T_{sh} = \begin{cases} 10^{9.64} \left(\frac{\kappa\dot{M}}{M}\right)^{1/2} \frac{\rho_{sh}^{1/2}}{\mu} & \text{for nondegenerate gas with the average molecular weight } \mu \\ 10^{7.36} \left(\frac{\kappa\dot{M}}{M}\right)^{1/4} \frac{\rho_{sh}^{7/12}}{\mu_e^{5/6}} & \text{for degenerate electron gas} \\ 10^{8.5} \left(\frac{\kappa\dot{M}}{M}\right)^{1/4} \frac{\rho_{sh}^{5/12}}{\mu_e^{2/3}} & \text{for relativistic degenerate electron gas} \end{cases}$$

$\dot{M}$ is the accretion rate in solar mass per year. The main results are given in Fig. 1 and Table 1. The ignition lines $\varepsilon_{CP} = \varepsilon_{DIF}$, $\varepsilon_{HE} = \varepsilon_{DIF}$, $\varepsilon_C = \varepsilon_{DIF}$ are shown in Fig. 1, where ε_{CP}, ε_{HE} and ε_C are the energy production rate in $C^{12}(P,\gamma)\,N^{13}$, $3\alpha \rightarrow C^{12}$ and $C^{12} + C^{12}$ reactions, respectively, and ε_{DIF} is the average radiative cooling rate (see equation (1)) of the envelope. The degeneracy border is marked by the dashed line. The dotted line marks places of stationary burning of hydrogen with $\varepsilon_{DIF} = {}^6 10^{15}$ erg/s/g for $N_{CNO}/N_P\ 10^{-3}$, where N_{CNO} and N_P are numbers of CNO and H atoms per gram. The shell temperature in this case may be determined by

$$T_{sh} = 10^{4.14} \kappa^{1/4} \rho^{5/6}$$

If the hydrogen burning is stationary, $\alpha = 0.007$. Then equation (2)

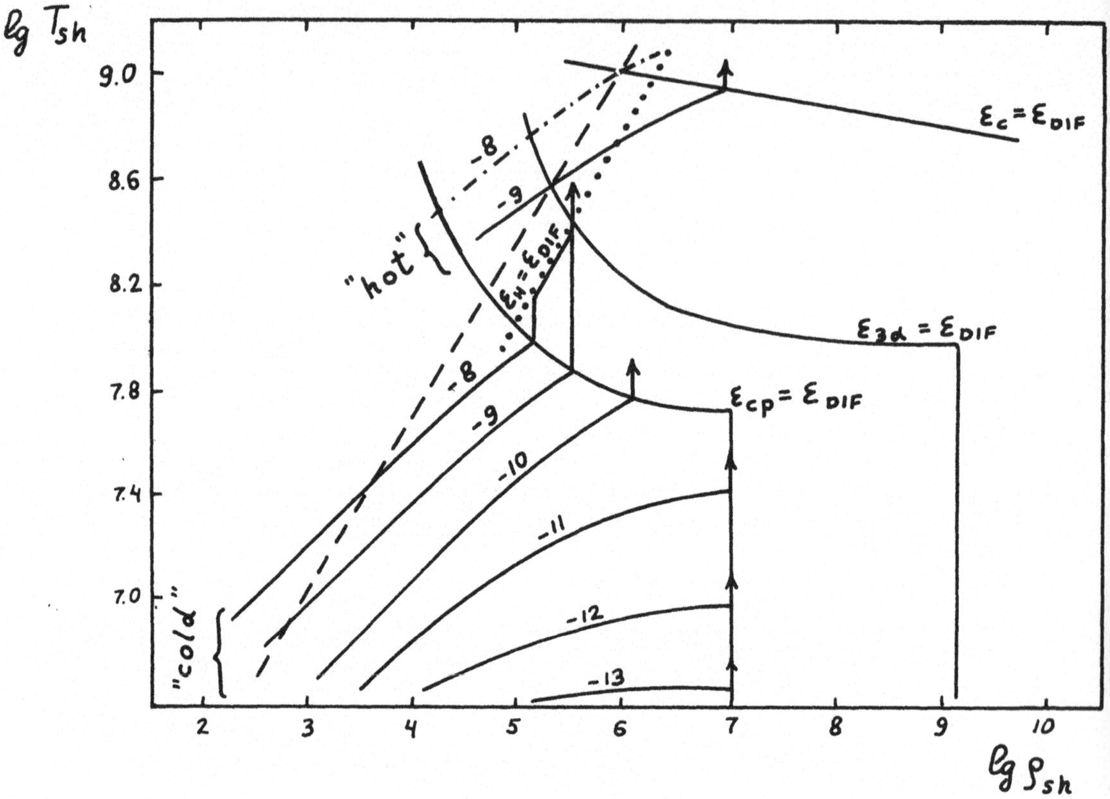

Fig. 1. Variation of temperature and density of the envelope of the accreting neutron star. The mass of the neutron star is 1.41 $M_\odot$ and the radius is 6.57 km. The numbers near tracks are the logarithm of the accretion rate in solar mass per year.

transforms to:

$$T_{sh} = \begin{cases} 10^{10.0}\left(\frac{\kappa\dot{M}}{M}\right)^{1/3}\left(\frac{\rho_{sh}}{\mu}\right)^{1/3} & \text{for nondegenerate gas} \\ 10^{8.75}\left(\frac{\kappa\dot{M}}{M}\right)^{1/4}\left(\frac{\rho_{sh}}{\mu_e}\right)^{5/12} & \text{for degenerate electron gas} \\ 10^{9.25}\left(\frac{\kappa\dot{M}}{M}\right)^{1/4}\left(\frac{\rho_{sh}}{\mu_e}\right)^{1/3} & \text{for relativistic degenerate electron gas} \end{cases} \tag{3}$$

These tracks are marked in Fig. 1 as "hot".

Equations (2) and (3) have been obtained under the assumption that

Table 1. Main parameters of nuclear bursts in the degenerate envelope of the neutron star $M = 1.41\ M_{\odot}$ $R = 6.6\,10^{15}$cm

lg $\dot{M}$	lg E_{nuc} ergs	lg E_{He} ergs	lg T sec	lg τ sec	lg M_{sh} (g)	comment
- 9	40.0	-	5.5	2.0	22.3	hot
- 8	39.0	38.2	2.4	0.2	20.2	cold
- 9	39.0	38.2	3.4	0.2	20.2	cold
-10	40.0	39.2	5.4	1.2	21.2	cold
-11	41.2	40.4	7.6	2.4	22.4	cold
-12	41.2	40.4	8.6	2.4	22.4	cold
-13	41.2	40.4	9.6	2.4	22.4	cold
-14	41.2	40.4	10.6	2.4	22.4	cold
-15	43.4	43.4	14.6	5.4	25.4	cold

T is the interval between bursts, $\tau = \frac{E}{L_{ed}}$ is the duration of bursts, L_{ed} is Eddington luminosity, M_{sh} is the mass of the shell.

that the neutron star is a cold incompressible sphere with the low cooling energy flux. This assumption is consistent with the hypothesis that "bursters" are old neutron stars without a magnetic field. According to Tsuruta (1974) a hot neutron star cools very quickly and for $t \gtrsim 10^7$ years its temperature is less than 10^7°K.

Let us sketch the picture of thermonuclear burning in the shell for $\dot{M} \lesssim 10^{-9}\ M_{\odot}$/yr (see Fig. 1). The accretion leads to heating of the shell up to the "ignition" temperature of the reaction $C^{12}(P,\gamma)N^{13}$. That reaction heats the shell to 10^8°K (if $N_{CNO}/N_P = 10^{-3}$). Stationary hydrogen burning becomes established 100 seconds later and the helium flash with a total energy release of more than 10^{38} ergs occurs at that moment if $\dot{M} \lesssim 10^{-9}\ M_{\odot}$/yr. The time interval between bursts exceeds one hour and the burst duration is equal to several seconds.

If the accretion rate exceeds $\sim 10^{-9}\ M_{\odot}$/yr, then the "hot" solution leading to accumulation of a carbon envelope becomes possible and bursts disappear. Thus, our simple model accounts for the main properties of observed X-ray bursters, although to get more information the numerical calculations are needed (Ergma, Kudryashov, 1979).

Let us now discuss briefly the problem of the origin of a close binary consisting of a neutron star in pair with a detached low mass main sequence star ($M \lesssim M_{\odot}$) with the distance between components of the order of $\sim 10^{11}$ cm. Tutukov and Yungelson (1979) found that such a system contracting due to gravitational wave radiation has a lifetime of $\sim 3\,10^9$ years and a mass exchange rate of $\sim 10^{-9} M_{\odot}$/yr which is responsible for the steady X-ray luminosity of bursters. But gravitational wave radiation becomes effective for a binary with solar mass components only if a $\lesssim 10^{11.3}$ cm. It is possible that the cold component with the strong magnetic stellar wind can lose its orbital angular momentum effectively if a $\lesssim 10^{12}$ cm (Eggleton, 1976, Kraitcheva et al., 1978), but we know that too poorly as yet. Taking into account that the galactic distribution of bursters is similar to the globular clusters' distribution, it seems probable that most of the prebursters were formed in globular clusters. After the tidal disruption of the latter, prebursters populated the bulge of our galaxy (Schakura, 1978) hence their galactic distribution.

We can point out two types of the recapture mechanism which can work in dense globular clusters according to Hills (1975). Hills found that if a binary collides with a star whose space velocity is smaller than the orbital velocity and the mass exceeds the mass of any component, then exchange capture is practically inevitable. The massive star forms a binary with the more massive component and the lightest star leaves the system. Supposing that half of all globular cluster stars are double we could estimate the number of close preburster systems forming in this way.

The total number of stars N_c of a cluster with the mass M_c in solar units and the Salpeter mass spectrum $dN \sim dM/M^{2.5}$ is $N_c = M_c/3\bar{m}$, where m is the minimal mass of stars in solar mass. The number of stars with $M \gtrsim 10\ M_\odot$ N_s which become neutron stars through the supernova explosion is $N_s \approx 0.01\ m^{0.5}\ M_c$. Now on the grounds of simple statistics the number of close collisions which is supposed to be equal to the number of close captured binaries with semiaxes $a < a_{min}$ is:

$$n = \frac{v\tau N_s N_c a_{min}^2}{2R_c^3} \tag{4}$$

where v is the average space velocity, τ is the lifetime and R_c is the cluster radius.

The number of binaries with $a_{min} \lesssim 10^{11.3}$ consisting of a neutron star and a low mass star formed during 10^{10} years in a cluster with $M_c = 10^6\ M_\odot$, $m \approx 0.1\ M_\odot$ and $R_c \approx 3$ pc is $n_1 \approx 0.3$. Such a close binary can leave the parent cluster due to the impulse conservation law. If the newly formed binary is wider, but a $\lesssim 100\ R_\odot$ and $M_{sec} \approx M_\odot$, then the secondary fills its Roche lobe in the course of its evolution. After the loss of part of the mass during the common envelope state of evolution a hydrogen model with a helium core will be established on a nuclear timescale. The mass loss rate will depend then on the initial period and values $\sim 10^{-9}\ M_\odot$/yr seem possible.

The second possibility is the formation of binaries with $a_{min} \lesssim 10^{13.6}$ cm consisting of a star with $M \gtrsim 10\ M_\odot$ and a low mass star. After expansion of the evolved massive component, the common envelope state (Tutukov, Yungelson, 1978) and the supernova explosion the system can become close enough for gravitational wave radiation (Paczynski, 1976). As $\tau = 10^7$ years in this case, the number of systems formed in the same cluster is $n_1 \approx 12$. Taking into account the distribution of binaries over a we suppose that only part of them will turn out to be close enough. As the lifetime of the normal component filling the Roche lobe is $\sim 10^9$ years and the number of bursters ~ 30, the total number of prebursters previously formed in our galaxy is $\sim 3 \cdot 10^2$. Our estimations display a possibility of preburster close binary formation due to exchange capture, but uncertainties prevent improving these numerical estimations.

Let us note that the number of preburster binaries formed in the very center of our galaxy with $R \lesssim 100$ pc and $M \approx 10^9 M_\odot$, according to equation (4), three times exceeds this number for a rich globular cluster. So we can expect to find several burster-like systems in the very center.

Four sources with $L_x \approx 10^3 L_\odot$ in the center were found by Cruddace et al., 1978. Low luminosity of the optical counterpart is then due to its low mass and, possibly, due to the anisotrophy of disk X-ray radiation (Milgrom, 1978).

REFERENCES

Cruddace R.G., Fritz G., Shulman S., Friedman H., McKee J., Johnson M., 1978, Astrophys. J.L., 222, L95.
Eggleton P.P., 1976, Proceed. IAU Symp. No. 73, Eds. P. Eggleton, S. Mitton, J. Whelan, p. 209.
Ergma E.V., Kudryashov A., 1979 (in preparation).
Ergma E.V., Tutukov A.V., 1979, Astron. Astrophys. (in press).
Grindlay J.E., McClintock J.E., Canizares C.R., van Paradijs J., Cominsky L., Li F.K., Lewin W.H.G., 1978, Nature, 274, 567.
Hills J.G., 1975, Astron J., 80, 809.
Joss P.C., 1977, Nature, 270, 310.
Joss P.C., 1978a, Preprint CSR-P-78-56.
Joss P.C., 1978b, Astrophys. J.L., 225, L123.
Joss P.C., 1979, Invited talk at the Ninth Texas Symposium Preprint MIT
Kraicheva Z.T., Popova E.I., Tutukov A.V., Yungelson L.R., 1978, Astron. Zh., 55, 1176.
Milgrom M., 1978, Astron. Astrophys., 67, L25.
Maraschi, Cavaliere A., 1977, Highlights of Astronomy, 4, Reidel, Dordrecht, Part 1, p. 126.
Paczynski B., 1976, Proceed. of IAU Symp. No. 73, Eds. P. Eggleton, S. Mitton, I. Whelan, p.
Schakura N.S., 1978, Private communication.
Tutukov A.V., Ergma E.V., 1979, Pisma Astron. Zh., 5, 34.
Tutukov A.V., Yungelson L.R., 1979, Acta Astron. (in press).
Tsuruta , 1974.

ARE MOST OF X-RAY BINARIES RECURRENT TRANSIENTS?

Janusz Ziółkowski
N. Copernicus Astronomical Center
Polish Academy of Sciences
ul. Bartycka 18, 00-716 Warsaw, Poland

INTRODUCTION

The aim of this paper is to argue that some day in the not-too-distant future one of the well-known X-ray pulsars, say, Cen X-3 (although better examples would be probably GX 301-2 or GX 1+4), will disappear from the X-ray sky. It will reappear again after some, say, tens of years and its pulse period will be then longer than it is now.

The basis for making such a prediction comes from an apparent conflict between the lifetimes of X-ray pulsars as deduced from the observed number of these objects and their association with massive X-ray binaries (Ziółkowski 1977, 1978) and the time scales of the observed spin-ups of these pulsars (Rappaport and Joss 1977, Schreier 1977, Sanford 1977). The latter time scales are found to be by 2 to 3 orders of magnitude shorter than the lifetimes. To remove this conflict, I suggested (Ziółkowski 1978) that X-ray pulsars are, in fact, recurrent transient sources with intermittent phases of: a) X-ray emission and the associated spin-up due to accretion and b) no accretion, no X-ray emission and spin-down due to magnetic braking (sometimes called "Kundt's mechanism"). In this picture, the increase of the pulsar period during phase b) roughly cancels out the decrease during phase a) and the average period remains equal to certain equilibrium value P_{eq} (period at which the outer edge of the magnetosphere rotates with the Keplerian velocity [Davidson and Ostriker 1973]). This "saw-tooth" model is, of course, a very crude one, but for our purposes in this paper we do not need to go into more sophisticated modelling (which is available) of either phase a) or b). I want only to demonstrate that this simple model recently found observational support in the behaviour of the transient X-ray pulsar A 0535-26 (Li et al. 1979). Furthermore, some quantitative estimates made using the above simple model do agree with the observed characteristics of A 0535-26.

LIFETIMES AND SPIN-UP TIME SCALES OF X-RAY PULSARS

Most of the X-ray pulsars are associated with massive X-ray binaries. Out of about 17 pulsars known so far, 13 are confirmed or

M. J. Plavec, D. M. Popper and R. K. Ulrich (eds.), Close Binary Stars: Observations and Interpretation, 335–341.

suspected members of massive binaries and only two are known to be members of low-mass systems. For this reason, we may assume that the lifetimes of X-ray pulsars are comparable with the lifetimes of massive X-ray binaries. These latter lifetimes may be estimated by the following simple considerations. Present masses of the optical O-B components of massive X-ray binaries are $\gtrsim 15\ M_\odot$ (Conti 1978). Their original masses were probably $\gtrsim 30\ M_\odot$ (Ziółkowski 1977, 1978). The surface density of massive early-type stars in the disc of the galaxy is 9.4 kpc^{-2} for stars more massive than 15 $M_\odot$ and 2.4 kpc^{-2} for stars more massive than 30 $M_\odot$ (Ostriker et al. 1974). Roughly one-third of them are probably members of close spectroscopic binaries (Blaauw and van Albada 1974, Bohannan and Garmany 1978). Let us assume that in half of these binaries the companion is a compact object at a distance small enough to become a strong X-ray source during certain phase of binary evolution. Then we have in the galaxy no more than, respectively, 1000 ($M \gtrsim 15\ M_\odot$) or 250 ($M \gtrsim 30\ M_\odot$) potential massive X-ray binaries. These objects are now seen only as massive O-B stars, but are expected to become bright X-ray sources some day. Let us note that in the above considerations the history of the binary systems prior to the formation of the compact component is not important; what is important is the fact that the number of potential massive X-ray binaries cannot be larger than certain fraction of all massive O-B stars in the galaxy. Now, if we assume that about 50 out of $\gtrsim$ 150 galactic X-ray sources are massive X-ray binaries, then we obtain, respectively, 20 and 5 for the ratio of main sequence lifetime of O-B component to the lifetime of X-ray emission phase of binary evolution. Recalling that main sequence lifetimes are 10^7 years for a 15 $M_\odot$ star and 5×10^6 years for 30 $M_\odot$ star (Ziółkowski 1972), we get finally 5×10^5 to 10^6 years for the typical lifetime of a massive X-ray binary. This implies that the typical lifetime of an X-ray pulsar should be also of this order.

On the other hand, the observed spin-up time scales of X-ray pulsars are much shorter, typically in the range 10^2 to 10^4 years (Rappaport and Joss 1977, Schreier 1977, Sanford 1977). The difference is 2 to 3 orders of magnitude. In the "saw-tooth" model, the slope of the small sections (phase a)) corresponds to the time scale of the spin-up, while the slope of the line $P = P_{eq}$ is comparable to the lifetime of the X-ray pulsar.

I should mention that the difference between the two time scales could be somewhat decreased by arguing that counts of massive stars are incomplete and that there is a concentration of these stars towards galactic center; e.g., van den Heuvel rather prefers 10^5 years as a typical lifetime of a massive X-ray binary. This is still much longer than a typical spin-up time scale. On the other hand (as we shall see in the next section), phase b) of the "saw-tooth" is typically longer than the phase a), which means that we observe less than half of the relevant X-ray sources (and perhaps only as little as one-tenth of them). This, of course, works in the opposite direction, i.e., makes the conflict of time scales more severe.

SPIN-UP AND SPIN-DOWN MECHANISMS

The change of the rotational period $P = 2\pi/\Omega$ of an X-ray pulsar can be described by the equation (Lamb et al. 1973, Rappaport and Joss 1977):

$$\frac{\dot{P}}{P} = \frac{\dot{M}}{I}\frac{dI}{dM} - \frac{\dot{M}}{I\Omega} h_a + \frac{T}{I\Omega}, \tag{1}$$

where I and M are the moment of inertia and the mass of the neutron star, h_a, is the component of the specific angular moment parallel to $\vec{\Omega}$ of the matter crossing the Alfvén surface, and T is the external (magnetic and viscous) torque. The first term on the right hand side of equation (1) is in most cases much smaller than either the second term (accelerating accretion torque) or the third term (braking magnetic torque). If we have the rotational period close to the equilibrium value ($P \approx P_{eq}$), then depending on whether the instantaneous value of P is slightly higher or lower than P_{eq}, we might encounter two different situations:

1) If $P \gtrsim P_{eq}$, then accretion is possible and the spinning-up torque is present. If we assume for simplicity that the external torque is negligible in such a situation, then the characteristic time scale of the spin-up is:

$$t_{su} = -\frac{P}{\dot{P}} = \frac{I\Omega}{\dot{M}}\frac{1}{R_A^2\,\Omega} = \frac{I}{\dot{M}\,R_A^2}, \tag{2}$$

where R_A is the radius of the magnetosphere, or (Rappaport and Joss 1977):

$$t_{su} = 1.4 \times 10^4\, f_1\, P^{-1}\, L_{37}^{-6/7} \tag{3}$$

$$f_1 = (\xi\, v_{ff}/v_r)^{-1/7}\, I_{45}\, \mu_{30}^{-2/7}\, R_6^{-6/7} \times \times (M/M_\odot)^{3/7} \tag{3A}$$

where P, L_{37}, I_{45}, μ_{30} and R_6 are (respectively): period, X-ray luminosity, moment of inertia, magnetic moment and radius of the neutron star in the units of (respectively): sec, 10^{37} erg sec^{-1}, 10^{45} g cm^2, 10^{30} Gs cm^3 and 10^6 cm. The quantity ξ is the fractional solid angle subtended at the neutron star by the infalling matter at the Alfvén surface and v_{ff}/v_r is the ratio of the free-fall velocity to radial drift velocity at the Alfvén surface. Let us note that in the case of $P = P_{eq}$, the magnetic pressure $B^2/8\pi$ should be balanced against $\rho\, v_r^2$ and this leads to a different dependence of f_1 on the ratio v_{ff}/v_r than found by Lamb et al. (1973) or Rappaport and Joss (1977). This remains true as long as $|P/P_{eq} - 1| \lesssim v_r/v_{ff}$.

The comparison with observational data demonstrated (Rappaport and Joss 1977) that the relation (3) is obeyed reasonably well by pulsars for which spin-up rates are known from the observations. This agreement strongly supports the view that: 1) X-ray pulsars are indeed rotating neutron stars accreting from the discs (Rappaport and Joss 1977) and that 2) their rotational periods are close to equilibrium periods (Lamb 1977).

2) If $P \lesssim P_{eq}$, then accretion is inhibited and the braking magnetic torque is present (Kundt's mechanism). This torque T can be roughly estimated (Kundt 1976, van den Heuvel 1977)* as equal to $\rho v_r^2 R_A \times \xi 4\pi R_A^2 \approx \dot{M} v_r R_A$. Equation (1) then yields the characteristic spin-down time scale:

$$t_{sd} = \frac{P}{\dot{P}} \approx \frac{I\Omega}{\dot{M} v_r R_A} = \frac{I}{\dot{M} R_A^2} \frac{v_K}{v_r}, \tag{4}$$

where v_K is the Keplerian velocity at the Alfvén surface. From equations (2) and (4) we have:

$$\frac{t_{sd}}{t_{su}} \approx \frac{v_K}{v_r} \tag{5}$$

If roughly constant period is to be maintained, this means that phase b) of the "saw-tooth" is longer than phase a) by a factor of the order of v_K/v_r (the theory of the accretion discs guess is that $v_K/v_r \sim$ 10 to 100). We should remember, however, that if the braking torque during phase a) is not negligible, then the time scale for the spin-up might be significantly longer and we have $t_{sd}/t_{su} < v_K/v_r$. The observations indicate that, indeed, in some cases, e.g. Her X-1 (Rappaport and Joss 1977), the external torques are not negligible during the accretion phase.

OBSERVATIONAL EVIDENCE

A 0535-26 is a recurrent transient X-ray pulsar, which was observed to undergo five outbursts since its discovery in April 1975 (Li et al. 1979). The interval between the outbursts was in the range 4 to 18 months and the typical active state (i.e., X-ray emission phase) was roughly by one order of magnitude shorter than the typical interval between the outbursts. During the active phases the pulsar was observed

*Davis et al. (1979) argued recently that Kundt's approach leads in general to serious overestimate of the braking torque, but their criticism does not apply to the case of $P \approx P_{eq}$.

to spin-up on the time scales ~ 70 to 160 years, but the average pulse period appeared to be constant on much longer time scale (Li et al. 1979). This indicates that during the inactive phases, slow-downs on the time scale ~ 10^3 years must occur that roughly cancel the spin-ups of the active phases. It is clear that the behaviour of this pulsar is quite consistent with the "saw-tooth" model. Now, let us make some quantitative estimates. The average pulse period of A 0535-26 is 103.8 sec and the typical X-ray luminosity is ~ 0.5 I_{Crab} (Li et al. 1979) which with the distance 2.8 kpc (Giangrande et al. 1978) gives L ~ 10^{37} erg sec^{-1}. The equation (3) gives then the theoretical spin-up time scale t_{su} ~ 140 years, in excellent agreement with the observations. The lack of the precise correlation between the observed spin-up time scales and the observed X-ray luminosities during the different outbursts (Li et al. 1979) might indicate that the braking torques and/or orbital effects are not quite negligible during the active states. The observed ratio of spin-down and spin-up time scales is ~ 10, again in good agreement with equation (5).

Another example of a pulsar which exhibits similar "saw-tooth" behaviour is perhaps Her X-1, which is known to be a transient source with active and non-active states lasting 1 to 15 years (Jones et al. 1973). Unfortunately, nothing is known about the possible spin-downs during the non-active states. We will have to wait perhaps 10 years or more before this information is available. If the "saw-tooth" model is correct in this case, then the observed t_{sd}/t_{su} is ~ 1, and the observed t_{su} (~ 3 x 10^5 years) is by an order of magnitude longer than the theoretical t_{su} calculated from the equation (3). This fact is probably due to significant braking torques acting during the active phase (Rappaport and Joss 1977) or due to the fact that a transition region between the disc and the magnetosphere is possibly not thin (Ghosh and Lamb 1978).

One might hope that more observational data relevant for the "saw-tooth" model will be available after the next outbursts of the transient X-ray pulsars A 1118-61 and 4 U 0115+63 are observed. These data might come sooner than the data from Her X-1.

Finally, I would like to comment on the fact that the theoretical equilibrium periods calculated for different pulsars often fail to agree with the observed periods. The theoretical formula for equilibrium period may be written as follows (van den Heuvel 1977, Lamb 1977):

$$P_{eq} = 2.7\ f_2\ L_{37}^{-3/7}\ \text{sec} \tag{6}$$

$$f_2 = (\xi\ v_{ff}/v_r)^{3/7}\ \mu_{30}^{6/7}\ R_6^{-3/7}\ (M/M_\odot)^{-2/7} \tag{6A}$$

It is clear that factor f_2 is much more sensitive to the intrinsic pulsar parameters than factor f_1 in the equation (3). Therefore, it is not surprising that the observational data fit the equation (3) (derived under the assumption $P \approx P_{eq}$) much better than equation (6) ($P = P_{eq}$) if both f_1 and f_2 are assumed to be constants.

ACKNOWLEDGEMENTS

I am grateful to B. Paczyński, F. Li and S. Rappaport for helpful discussions.

REFERENCES

Blaauw, A. and van Albada, T.S.: 1974, preprint.
Bohannan, B. and Garmany, C.D.: 1978, Astrophys. J. 223, p. 908.
Conti, P.S.: 1978, Astron. Astrophys. 63, p. 225.
Davidson, K. and Ostriker, J.P.: 1973, Astrophys. J. 179, p. 585.
Davies, R.E., Fabian, A.C. and Pringle, J.E.: 1979, Monthly Notices Roy. Astron. Soc. 186, p. 779.
Ghosh, P. and Lamb, F.K.: 1978, Astrophys. J. 223, p. L83.
Giangrande, A., Giovanelli, F., Bartolini, C., Guarnieri, A. and Piccioni, A.: 1978, preprint.
Jones, C.A., Forman, W. and Liller, W.: 1973, Astrophys. J. 182, p. L109.
Kundt, W.: 1976, Phys. Letters 57A, p. 195.
Lamb, F.K.: 1977, in Eight Texas Symposium on Relativistic Astrophysics, Annals New York Academy Sciences 302, p. 482.
Lamb, F.K., Pethick, C.J. and Pines, D.: 1973, Astrophys. J. 184, p. 271.
Li, F., Rappaport, S., Clark, G.W. and Jernigan, J.G.: 1979, Astrophys. J. 228, p. 893.
Ostriker, J.P., Richstone, D.O. and Thuan, T.X.: 1974, Astrophys. J. 188, p. L87.
Rappaport, S. and Joss, P.C.: 1977, Nature 266, p. 683.
Sanford, P.W.: 1977, in Eight Texas Symposium on Relativistic Astrophysics, Annals New York Academy Sciences 302, p. 386.
Schreier, E.J.: 1977, in Eight Texas Symposium on Relativistic Astrophysics, Annals New York Academy Sciences 302, p. 445.
van den Heuvel, E.P.J.: 1977, in Eight Texas Symposium on Relativistic Astrophysics, Annals New York Academy Sciences 302, p. 14.
Ziółkowski, J.: 1972, Acta Astron. 22, p. 327.
Ziółkowski, J.: 1977, in Eight Texas Symposium on Relativistic Astrophysics, Annals New York Academy Sciences 302, p. 47.
Ziółkowski, J.: 1978, in Nonstationary Evolution of Close Binaries, ed. A.N. Zytkow (Warsaw: Polish Scientific Publishers), pp. 29-54.

DISCUSSION FOLLOWING ZIOLKOWSKI and ERGMA AND TUTUKOV

Shu: I would like to make a mathematical comment. It might be useful to regard your $\dot{P}/P$ equation as a differential equation for a relaxation oscillator. Have you thought about what possible dependence $\dot{M}$ might have on P so that your equation has solutions ('sawtooth') characteristic of a relaxation oscillator?

Ziolkowski: No, I have not. Perhpas I should try. I know that different groups of people were discussing physical reasons which might make an equilibrium solution unstable, but, according to my knowledge, no one used the approach you suggest.

Mitrofanov: Matter accreted by a neutron star increases its angular momentum. On the other hand, its interaction with the rotating magnetosphere near the Alfven surface results in decreasing the angular momentum either because of outflow (the Illarionov-Sunyaev mechanism) or because of heating (Mitrofanov). So, the equilibrium between these two processes seems to occur - rather than the alternating spin-up and spin-down phases.

Sugimoto (to Eergma and Tutukov): According to our generalized theory of shell flashes, the pressure of the burning shell remains almost constant to a good approximation. Therefore the corresponding arrow in your Figure should bend somewhat to the left. I would like to say, however, that your picture and treatment appear to be very good.

Tutukov: Our simple model is sufficient to study main properties of thermonuclear bursts in the nuclear-fuel-rich envelope of an accreting neutron star. The model can be further developed to take into account, within the same degree of approximation, effects such as rotation, magnetic field, or the change in pressure at the bottom of the nuclear burning shell.

A CRITIQUE OF THE POLARIMETRIC EVIDENCE ON THE NATURE OF CBS

J.F.L. Simmons, C. Aspin and J.C. Brown
Dept. of Astronomy, University of Glasgow

ABSTRACT. Determination of the values and confidence intervals for physical parameters and the orbital inclination of binaries from polarimetry is critically discussed. A new method based on χ^2 best fitting and incorporating constraints imposed by the canonical model is developed and applied to polarimetric data for Cygnus X-1. Whereas the χ^2 best fitting procedure yields an inclination of $|i| = 79°$ for Cygnus X-1, the level of noise, and the model dependency of the parameter determination give rise to a very broad confidence interval around this value. Our method also indicates the accuracy of the polarimetric measurements necessary for more precise parameter determinations.

It can be shown (eg. Brown et al 1978) that polarisation of light caused by Thomson scattering in the envelope of a binary system in which (the canonical model) (i) the geometry of the system is time independent in a frame uniformly rotating with the binary period, (ii) no eclipsing or variable absorption effects take place, (iii) variations of the direct starlight intensity are small will behave according to

$$Q_t(\lambda) = p_o + q_1 \cos \lambda + p_2 \cos 2 \lambda + q_1 \sin \lambda + q_2 \sin 2 \lambda$$
$$U_t(\lambda) = u_o + u_1 \cos \lambda + u_2 \cos 2 \lambda + v_1 \sin \lambda + v_2 \sin 2 \lambda$$

(Q_t and U_t are the (theoretical) Stokes parameters, $\lambda/2\pi$ is the phase). The harmonic coefficients are functions of inclination and weighted integrals of the electron density distribution. In a natural frame in which the z-axis lies along the line of sight and the binary axis lies in the x,z plane

$$p_1/v_1 = -q_1/u_1 = \cos i, \quad p_2/v_2 = -q_2/u_2 = (1+\cos^2 i)/2 \cos i.$$

If the density distribution is symmetric about the orbital plane only second harmonic variations arise. Observations are made in a frame rotated by an angle ϕ(unknown a priori) relative to this natural frame. In principle fourier analysis of the observed $Q(\lambda)$ and $U(\lambda)$ curves would yield the p_k, q_k, u_k, v_k and hence determines, indeed overdetermines, (via the constraints above) both i and ϕ. Noise in

M. J. Plavec, D. M. Popper and R. K. Ulrich (eds.), Close Binary Stars: Observations and Interpretation, 343–345.

the data, however, forces one to adopt a statistical procedure of 'best fitting' model and data.

Usually a simple minded formal linearized error approach is adopted when a model is used to extract parameter values from given data. Comparison of the model M {p} , where p represents the model parameters, and the data values, X_r, yields optimal values for p that depend only on X_r, i.e. $p_{opt} = P(X_r)$. If the model agrees well, and if σ_{X_r} are small then under the linear approximation

$$\sigma^2_{p_{opt}} \simeq \sum_r \left(\frac{\partial P}{\partial X_r}\right)^2_{\bar{X}_r} \sigma^2_{X_r}$$

where $\bar{X}_r$ are the expected, i.e. theoretical, values for X_r. There are a number of sources of confusion here. Firstly, even if the model conflicts with data for all p values this approach will yield formal errors in **p**, and perhaps very small ones, but say nothing of this incompatibility. Also linearity assumptions can breakdown, depending on the size of σ_{X_r} and the function form of p.

We adopt the more rigorous approach of determining the domain of parameter values for which the data and the model agree at a given significance level, and hence simultaneously test the acceptability of the canonical model which has been challenged by Kemp et al (1979) and Milgrom (1978) in the case of Cyg X-1.

Extensive polarimetric data for Cygnus X-1 have been provided by Kemp (Kemp et al 1979), and takes the form of N equally spaced phase bin averaged Stokes parameters, $Q_{ob,r}$, $U_{ob,r}$, which have been folded on a 5.600^d period. The statistic

$$\chi^2_{2N} = \sum_{r=0}^{N-1} \frac{(Q_{ob,r} - Q_{t,r})^2}{\sigma^2_r} + \frac{(U_{ob,r} - U_{t,r})^2}{\sigma^2_r}$$

measures the agreement between model and data, and has a chi-squared distribution with 2N degrees of freedom if $Q_{ob,r}$ and $U_{ob,r}$ are normally distributed. $Q_{t,r}$, $U_{t,r}$ are the model predicted values; and for given data χ^2_{2N} will be a function of the 8 independent model parameters, which we refer to as i,p. A 90% confidence region will be that domain of parameters for which the model must be accepted at 10% significance, i.e. $\chi^2_{2N}(i,p) < \chi^2_{2N,10\%}$. Of course, to determine this entire 8 dimensional parameter domain is computationally very difficult. Since we are interested primarily in a confidence interval for inclination, i, we can adopt a simplifying method that admits an analytical solution. Fixing i, we determine by the method of Lagrange's multipliers the optimal values of the remaining 7 independent parameters for this i and the corresponding value of χ^2_{2N}, which we call $\inf \chi^2_{2N}(i)$. The 90% confidence interval for i can then be obtained by finding that range of i for which $\inf \chi^2_{2N}(i) < \chi^2_{2N,10\%}$.

This analysis yields for the asymmetric and symmetric cases respectively 90% confidence intervals of $85^\circ \lesssim i \lesssim 150^\circ$ and $90^\circ \lesssim i \lesssim 125^\circ$, with optimal value $i \simeq 102^\circ$ ($|i| \simeq 78^\circ$) in both cases, the confidence interval on i greatly exceeding the formal error estimates of $\pm 7^\circ$ obtained as above. Optimal values for the other parameters give a lower limit on the mass of scattering material of $\sim 10^{24}$ gm, which is comparable to the mass transfer per orbital period needed to produce the X-ray luminosity. The optimal polarimetric value of $|i| \simeq 78^\circ$ differs greatly from $|i| = 28^\circ$ of the light curve analysis (Hutchings 1979). However the polarimetric confidence interval we obtain is so broad that polarimetric and photometric results are not in contradiction statistically. Furthermore the true confidence interval on i from spectrophotometric modelling may also be much larger than the formal range $28^\circ \pm 2^\circ$ usually quoted, for similar reasons (cf. Bolton, this meeting). We plan to extend our error analysis technique to the spectrophotometric case.

Though the polarimetric results for Cyg X-1 may seem disappointing, Kemp's data are very noisy, and his phase binning technique may give rise to large bin errors because of systematic changes (eg. of the system's geometry). Use of fewer orbital periods could give better parameter estimations with an adequate photon count. We are working on determining the accuracy required for useful model testing and parameter estimation.

A fuller discussion of this work can be found in Simmons et al (1979).

REFERENCES

Brown, J.C., McLean, I.S. and Emslie, A.G.: 1978, Astron. & Astrophys. 68, 415.
Hutchings, J.B.: 1974, Ap.J. 188, 341.
Kemp, J.C., Barbour, M.S., Parker, T.E. and Herman, L.C.: 1979, Ap.J. Letters 228, L.23.
Milgrom, M.: 1978, Astron. & Astrophys. 65, L.1.
Simmons, J.F.L., Aspin, C. and B rown, J.C.: 1979, preprint.

DISCUSSION FOLLOWING SIMMONS, ASPIN AND BROWN

Guinan: Have you tested your procedure of determining the orbital inclination from polarization measures against systems with i known through an eclipse analysis?

Simmons: One of our assumptions is that no eclipsing takes place. The analysis could be extended to cover this case, and we intend to do this, but the technical problems in the mathematics will be more difficult. But yes, it is a good idea to test the method in this way. Of course, it would be necessary to have good polarimetric data for an eclipsing system which could otherwise be considered to reasonably comply with the assumptions made in the canonical model.

RADIO FLARES IN THE CIRCINUS X-1 SYSTEM

G.D. Nicolson
National Institute for Telecommunications Research,
CSIR, Johannesburg.

Abstract

The 16.6 day periodic radio flares in Circinus X-1 have been observed at 6 cm wavelength from February 1978 to May 1979. Two strong flares occurred in February 1978 followed by a long period of quiescence until flaring recurred in January 1979. The onset of the flares was however 18 hours earlier in the 16.6 day cycle than previous flares. Similar behaviour has been reported for the X-ray emission and it is concluded that both the X-ray and radio emission are caused by mass transfer at peristron in an eccentric binary with a 16.6 day period. Each radio flare generally shows a multipeaked structure with the first maximum being the strongest. There is also a systematic trend for the subsequent peaks to become weaker with each cycle and this suggests that the compact object precesses in relation to an accretion disc.

DISCUSSION FOLLOWING NICOLSON

Zuiderwijk: I was rather surprised by the accuracy you give for your periods. Can you be sure that the onset of a flare has a fixed timing in the orbital phase?

Nicolson: The time of onset of a flare can be determined to within 1 hour. For 1979 the flares apparently all begin at a well-defined phase. Assuming that this is always so then errors quoted are correct but refer to mean period over many orbits.

M. J. Plavec, D. M. Popper and R. K. Ulrich (eds.), Close Binary Stars: Observations and Interpretation, 347.

OPTICAL AND ULTRAVIOLET[1)] OBSERVATIONS OF HZ HER

R. Kippenhahn, H. Ritter, H.U. Schmidt, H.-C. Thomas
Max-Planck-Institut für Physik und Astrophysik
Forschungsgelände, D-8046 Garching

ABSTRACT

Results of five nights of photoelectric UBV photometry of HZ Her and of simultaneous ultraviolet spectroscopy with IUE are presented.

1 INTRODUCTION

In a recent paper, Kippenhahn and Thomas (1979) have discussed a model according to which energy is transported by circulation from the area of HZ Her which is illuminated by Her X-1 to the area in the shadow. Although there exists already a wealth of observational data about HZ Her (see e.g. Gerend and Boynton, 1976; Czerny-Schwarzenberg, 1978; Kilyachkov and Shevchenko, 1978), these data are not well suited for studying the observational consequences of the Kippenhahn-Thomas model. The authors found it therefore worthwhile to start their own observations which in particular address the phase of the X-ray eclipse. Results of the first of these observations, on the one hand photoelectric UBV photometry and on the other hand simultaneous ultraviolet spectroscopy with the IUE are reported here. A detailed presentation of these data will be given elsewhere (Kippenhahn et al. 1979).

2 THE OPTICAL OBSERVATIONS

Photoelectric observations in the UBV system have been made with the 123 cm reflector of the Calar Alto observatory near Almeria, Spain, on five successive nights, starting on May 20th, 1979. The journal of these observations is listed in Table 1. The orbital phases $\varphi_{1.7}$ are based on data of Giacconi et al. (1973), neglecting period changes, whereas for the X-ray phase φ_{35} the zero point has been taken from Giacconi et al.

1) Based on observations by the International Ultraviolet Explorer at the Villafranca Satellite Tracking Station of the European Space Agency.

M. J. Plavec, D. M. Popper and R. K. Ulrich (eds.), Close Binary Stars: Observations and Interpretation, 349–353.

Table 1: Journal of the UBV observations.

Night No.	Start of observations JD 2440000+	$\varphi_{1.7}$	φ_{35}	End of observations JD 2440000+	$\varphi_{1.7}$	φ_{35}
1	4014.396	0.153	0.932	4014.655	0.306	0.939
2	4015.375	0.729	0.960	4015.648	0.890	0.967
3	4016.369	0.314	0.988	4016.483	0.381	0.991
4	4017.383	0.910	0.017	4017.645	0.064	0.024
5	4018.360	0.485	0.045	4018.563	0.604	0.051

(1973) and the average period of $34\overset{d}{.}95$ from Davison and Fabian (1977). Accordingly, turn on of cycle No. 77 should have occurred on May 23rd, just before the observations of night No. 4. However the prediction of turn on can only be made with an uncertainty of about ±2 days.

The transformation from the instrumental to the standard UBV system has been done in the usual way, for details see Kippenhahn et al. (1979). The resulting light curves are shown in Fig. 1. Each data point corresponds to one integration, the length of the vertical bar indicates the rms error of the observation. The eclipse of the X-ray source is assumed to occur between the phases $\varphi_{1.7} = 0.93$ and $\varphi_{1.7} = 0.07$ (dashed lines in Fig. 1), corresponding to a half angle of $25\overset{\circ}{.}2$. The overall shape of the light curves agrees rather well with previous observations (Gerend and Boynton, 1976). But the totality of the eclipse and the asymmetric shape near primary minimum are features which have not been identified by Gerend and Boynton (1976) in their data analysis and are not reproduced by their "tilted disk" model. Details of these observations will be discussed by Kippenhahn et al. (1979).

3 THE ULTRAVIOLET OBSERVATIONS

Simultaneously to the optical observations of the minimum of HZ Her on night No. 4 (see Table 1), two low resolution spectra in the long wavelength range have been taken with the IUE. The journal of these observations is given in Table 2.

Table 2: Journal of the IUE observations.

Camera Image	Dispersion	Exposure (min)	Exposure midpoint JD 2440000. +	Phase midpoint $\varphi_{1.7}$	φ_{35}
LWR 4583	Low	150	4017.610	0.0435	0.0235
LWR 4584	Low	207	4017.751	0.1268	0.0275

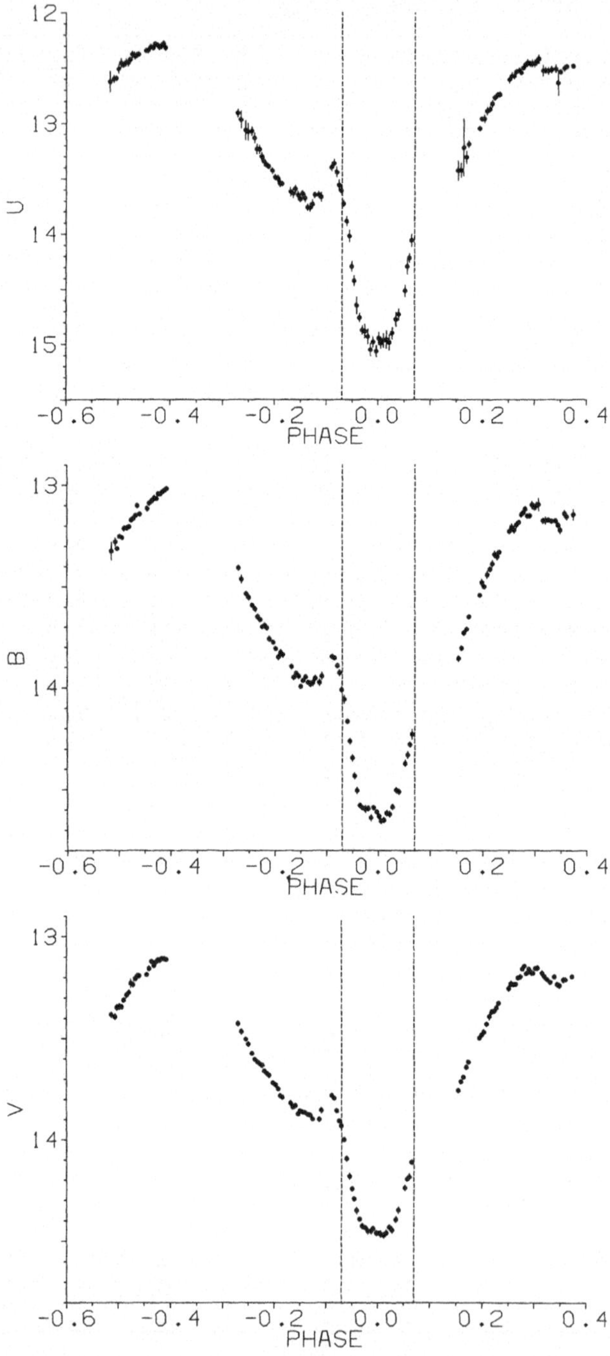

Fig. 1: The observed light curves of HZ Her in U, B and V (from top to bottom) versus the orbital phase. For details see text.

In spite of the relative long exposure the spectrum which has been taken during the eclipse of the Her X-1 is underexposed by about a factor of two, resulting in a low signal to noise ratio. The mean UV flux in the usable wavelength range $\lambda\lambda$2400-3000 Å is found to be about $5\cdot10^{-15}$ erg cm^{-2} s^{-1} Å^{-1}. This is about half the estimated flux in the short wavelength range (Dupree et al., 1978). Due to the low signal to noise ratio, neither emission nor absorption lines could be identified. The shape of the continuum is consistent with black body temperatures in the range 7000-9000 K.

The second image, taken outside the X-ray eclipse, is well exposed in the ranges $\lambda\lambda$ 2000-2400 Å and $\lambda\lambda$ 2900-3100 Å, but saturated in between due to the long exposure. Again neither emission nor absorption lines could be identified with certainty. The UV fluxes at $\lambda\lambda$ 2000 Å and 3000 Å are about $4\cdot10^{-14}$ erg cm^{-2} s^{-1} Å^{-1} and $2.5\cdot10^{-14}$ erg cm^{-2} s^{-1} Å^{-1} respectively.

REFERENCES

Czerny-Schwarzenberg, A.: 1978, Inf. Bull. Var. Stars No. 1491.

Davison, P.J.N., Fabian, A.C.: 1977, Monthly Notices Roy. Astron. Soc. 178, pp 1P-8P.

Dupree, A.K., Davis, R.J., Gursky, H., Hartmann, L.W., Raymond, J.C., Boggess, A., Holm, A., Kondo, Y., Wu, C.-C., Macchetto, F., Sanford, M.C.W., Willis, A.J., Wilson, R., Ciatti, F., Hutchings, J.B., Johnson, H.M., Jugkau, J., Morton, D.C., Treves, A., van den Heuvel, E.J.P.: 1978, Nature 275, pp. 400-403.

Gerend, D., Boynton, P.E.: 1976, Astrophys. J. 209, pp. 562-573.

Giacconi, R., Gursky, H., Kellogg, E., Levinson, R., Schreier, E., Tananbaum, H.: 1973, Astrophys. J. 184, pp. 227-236.

Kilyachkov, N.N., Shevchenko, V.S.: 1978, Sov. Astron. Lett. 4, pp. 191-194.

Kippenhahn, R., Thomas, H.-C.: 1979, Astron. Astrophys. 75, pp. 281-290.

Kippenhahn, R., Schmidt, H.-U., Thomas, H.-C.: 1979, preprint.

DISCUSSION FOLLOWING KIPPENHAHN, RITTER, SCHMIDT AND THOMAS

Mazeh: Two Russian observers (Kilyachkov and Shevchenko) published recently UBV light curve of HZ Her near eclipse. Are their results similar to the light curves you have presented?

Ritter: I don't know because we have not been aware of the Russian observations.

Budding: How sensitive are the derived temperatures to what is assumed about interstellar reddening?

Ritter: The spectra that I have presented have not been corrected for interstellar reddening. This certainly must be done, and until it has been done I cannot say how sensitive the derived temperatures depend on the reddening. Thus these temperatures should be considered as a rough estimate.

RE-EVALUATION OF THE ORBITAL ELEMENTS OF HDE 226868 = CYG X-1

C.T. Bolton and Douglas R. Gies
David Dunlap Observatory, University of Toronto,
P.O. Box 360, Richmond Hill, Ontario, CANADA L4C 4Y6

We have scanned 75 high dispersion spectrograms of HDE 226868 = Cyg X-1 with the PDS microdensitometer of the David Dunlap Observatory in preparation for a detailed spectrophotometric study of the system. The plate material is identical to that described by Bolton (1972a,b, 1975) and includes those plates. We have determined velocities for all of the spectrograms by fitting a parabola to the bottom of each line profile. The lines were divided into three groups, hydrogen lines, He I lines, and high excitation lines (He II, Si III and IV, N III), according to excitation, and separate mean velocities were determined for each group on each plate. We have computed the orbital elements of the system separately for each line group in order to look for differences caused by the effects of mass transfer or stellar wind.

The elements determined from the three groups of lines are in excellent agreement. The only significant difference among the elements is that the H lines give a K approximately 5 km s^{-1} larger than that of the other two groups. Individual plates show weak evidence for a small Balmer progression. We attribute these effects to variable blending of absorption and emission lines. The three groups of lines have essentially the same V_0 velocity, and the few plates taken during x-ray high states fit the velocity curve well. Thus any stellar wind or x-ray heating effects are apparently undetectably small.

The orbital elements obtained from the He I and high excitation line groups are very similar to those derived by Bolton (1975) with two exceptions. First, the eccentricity in the new solutions is not statistically significant. Second, the K is slightly larger. The origin of these differences is not yet clear.

REFERENCES

Bolton, C.T.: 1972a, *Nature*, 235, 271.
Bolton, C.T.: 1972b, *Nature Phys. Sci.*, 240, 124.
Bolton, C.T.: 1975, *Ap.J.*, 200, 269.

M. J. Plavec, D. M. Popper and R. K. Ulrich (eds.), Close Binary Stars: Observations and Interpretation, 355–356.

DISCUSSION FOLLOWING BOLTON AND GIES

Zuiderwijk: Which spectral lines are varying?

Bolton: I think that everything varies some. The most variable lines are the He I singlets and the Si III and O II lines.

Simmons: The inclination determined previously by Hutchings from light curves for Cyg X-1 was $28^{\circ} \pm 1^{\circ}$. You are now giving, or quoting, a value more like 60°. How do you think the error analysis involved in the $28^{\circ} \pm 1^{\circ}$ result could give such a small error?

Bolton: I don't remember Hutchings quoting such a small error. I certainly never did. You might get such a small error by forcing the visible star to fill to Roche lobe, but that would be an assumption. There is no observational requirement for this.

van Paradijs: How large are the random variations in the spectrum, expressed in spectral and luminosity classes?

Bolton: So far we have found no variation in the luminosity class. There are a number of line ratios defining spectral type. These all show variations from O9.5-B0 with O9.5-O9.7 being more common than O9.7-B0. However, I am not yet certain that all line ratios on the same plate give consistent results.

Whelan: Do you assume that CYG X-1 is a binary system, or do your articles of faith permit the existence of a trinity?

Bolton: We are searching for evidence of a third body. We have not yet found anything, but we haven't ruled out too many of the possibilities either. The new enlarged data set has not been exploited in this regard yet, but in a few weeks ... .

OPTICAL OBSERVATIONS OF X-RAY SOURCES IN THE LMC

M. Pakull
European Southern Observatory, Geneva, Switzerland

ABSTRACT*

Photoelectric and spectroscopic observations of optical counterparts of X-ray sources in the LMC are described.

Image-dissector scanner spectra of the recently discovered low-mass system LMC X-2 (Pakull, IAUC 3313; Pakull and Swings, IAUC 3318) are presented and compared with those of faint galactic bulge sources. Their similar appearance, namely strong UV excess and the presence of He II 4686 A, Hα , and C III + N III emission at 4640 - 4650 A, indicate that their optical emission is mainly a result of the intense X-ray heating of the stellar atmosphere and/or matter sourrounding the system.

Photometry of the star "19" and of R 148, the optical candidates for the recurrent LMC transient A0538-66 and for LMC X-1, respectively, reveals variability of about 0.05 mag. The data for the B5 supergiant R 148 suggest a 20-day double sinusoidal variation.

The ellipsoidal light curve of the Sanduleak-Philip star = LMC X-4 is discussed by means of light curve synthesis. The large amplitude conflicts with the spectroscopically determined mass ratio. However, two well separated brightness states have been found in each of the two maxima. The resulting smaller amplitudes are in agreement with the standard picture of light curve analysis. Several mechanisms to explain the third light phenomenon are discussed.

*As submitted before the meeting; no update has been received.(Editors).

DISCUSSION FOLLOWING PAKULL

Wilson: There are at least two additional parameters on which your two empirical measures depend, namely R/Roche, the size of the star with respect to its lobe, and (L_x/L_{opt}) bolometric. Did you assume that the

M. J. Plavec, D. M. Popper and R. K. Ulrich (eds.), Close Binary Stars: Observations and Interpretation, 357–358.

star fills its lobe and that the bolometric X-ray luminosity is zero (i.e. that there is negligible X-ray heating)? If either of these assumptions is significantly wrong, it would affect your results systematically.

Pakull: I assume that the star fills its Roche lobe. The fact that the minimum at phase $\phi_X = 0.5$ is deper than at the X-ray eclipse tells us that the X-ray heating is negligible. Some of the assumptions might be wrong, I agree. However, the aim of the simple model presented was to give one possible selfconsistent explanation of the system.

VARIATIONS IN THE PROFILES OF Hβ AND HE II 4686 A in VELA X-1

Ed. J. Zuiderwijk
European Southern Observatory, Geneva, Switzerland

ABSTRACT*

The first results are presented of an extensive spectroscopic study of HD 77581 = Vela X-1, based on several averaged Coude spectra. The profile of Hβ is phase-dependent and consists of two absorption components superimposed on one another. The profile of He II 4686 A is of the P-Cygni type around phase 0.7, whereas the line is in absorption with variable strength between phase 0.9 and phase 0.5 (with respect to the X-ray eclipse).

These observations indicate that a gaseous stream is present in the system, as well as an asymmetrically expanding envelope of HD 77581. The phase dependence of Hβ further supports an eccentric orbit of Vela X-1.

*As submitted before the meeting; no update has been received. (Editors).

COMMENT ON THE PAPER BY ZUIDERWIJK

Crampton: HD 77581 = Vela X-1 is interesting in that it does have a well-established eccentricity. Recently Rappaport, Joss, and Stothers (preprint) showed that there is no apsidal motion evident in their excellent X-ray observations. We can expect to have much better limits on the apsidal motion -- and hence on the internal mass distribution -- in the near future.

M. J. Plavec, D. M. Popper and R. K. Ulrich (eds.), Close Binary Stars: Observations and Interpretation, 359.

EVIDENCE FOR A 6 YEAR PERIODICITY IN X PERSEI

J.D. Dorren
Biruni Observatory, Shiraz University, Shiraz, Iran.
E.F. Guinan
Villanova University, Villanova, Pennsylvania, 19085, USA

Intermediate and narrow-band photoelectric observations obtained at Biruni and Villanova Observatories from 1977 to the present, when combined with previous data, strongly suggest that light increases of about 0.5 mag occur at 6 year intervals. Possible explanations are considered.

I. INTRODUCTION

The X-ray pulsator X Persei (4U 0352+30) is an intriguing object at both X-ray and optical wavelengths. From the X-ray data a period of 13.9 minutes now appears well-established, but early suggestions of a 22.4 hour periodicity remain unconfirmed (White et al.,1976, Becker et al., 1979). The nature of the X-ray emission is unusual; the source is about 3 orders of magnitude weaker than other X-ray binary pulsars, and the spectrum contains a strong high-energy X-ray component (Mushotzky et al., 1977). Long known as an irregular variable, X Per has been classified as a Be star with an O9.5 (III-IV)e spectral type and showing broad hydrogen emission and absorption features (Brucato and Kristian 1972, Hutchings et al., 1975). A 581-day period is indicated by the radial velocity study of Hutchings et al., (1975). This has been variously interpreted as an orbital period (about a black hole) or an apsidal period, the models having in common accretion on to a neutron star as the source of the pulsed X-rays. In 1977 an increase in brightness of the star was observed by Kalv (1977) and Guinan et al., (1978), and by April 1979 we had accumulated sufficient evidence to suggest that X Per undergoes periodic ($\sim$ 6-year) outbursts (Dorren et al., 1979a).

2. PRESENT OBSERVATIONS

Photoelectric observations of X Per commenced at Biruni Observatory in August, 1977 using intermediate and narrow-band Hα filters and a Stromgren u filter (λ3500) and continued up to the present (September 1979) both there and at Villanova University (using an Hα filter set matching that at Biruni and a pair of intermediate and narrow-band filters centered on the OI λ7774 feature). Descriptions of the filters and

M. J. Plavec, D. M. Popper and R. K. Ulrich (eds.), Close Binary Stars: Observations and Interpretation, 361–365.

instrumentation at Biruni and Villanova are to be found in Guinan et al., (1979) and Guinan and McCook, (1974) respectively. A detailed discussion of this very extensive material (a total of 110 nights in Hα) will be published elsewhere (Dorren et al., 1979b).

To trace the history of X Per we extract from the compilation of Mook et al., (1974) the photoelectric V and B-V observations from 1964 to 1973. A comparison of our data with the V band may be made by determining the wavelength dependence of the light variation from the 3 bandpasses λ6685, λ7790 and λ3500 and then scaling the λ6585 data to the effective V wavelength (λ5500). The V observations from 1964 to the present, Kalv's data, and the interpolated Biruni-Villanova V measures are displayed in the figure, together with the B-V index where observed. Also included are V and B-V observations obtained during January 1975 by Margon et al. (1976). To avoid overcrowding, only a portion (about one third) of the Biruni-Villanova data was plotted. From the data obtained from 1974 onward, it is clear that there was a period of relatively constant light, perhaps slowly decreasing up to the spring of 1977 when V=+6.71 mag. Thereafter there was a fairly sharp increase in light by August 1977, indicated both by Kalv's data and our own. Maximum brightness of around V=+6.38 mag occurred in February 1978 and was followed by a decline of similar slope with a pause about 120 days long in mid-1978. The decline continued to a brightness of V=+6.79 mag, the faintest the star has been since 1903, and the most recent data (September 1979) indicates that the decline has slowed or possibly halted. This may be the start of another interval of roughly constant light similar to that which occurred between 1973 and early 1977. An inspection of the figure indicates two other maxima occurring in mid-1966 and early 1972 which appear to be brighter than the most recent one, and the three occur at intervals of approximately 6 years. Although there is no data from 1968 to 1972, the whole figure is suggestive of a repetitive pattern of outbursts lasting roughly 2 years separated by intervals of nearly constant light of about 4 years duration. It is interesting to note that, at times of constant low light, the B-V index is bluest at B-V≈+0.16 while near the maxima, B-V=+0.25 to +0.30. Additional UBV data has been published by Harris (1956) where V=6.13, B-V=+0.29 and by Hiltner (1956) where V=6.07, B-V=+0.31. The dates of observation, though uncertain, are probably 1954 or 1955 (Harris) and 1954 (Hiltner). These colors and magnitudes suggest that the star is near maximum light, which is consistent with the 6-year cycle indicated by the later data. The only other reliable material is the visual photometry of Müller and Kempf (1907) from 1888-1902 which shows cyclic behavior in which the maxima occur near times expected with the 6-year period, in 1894 and 1900.

3. DISCUSSION

The evidence for a 6-year periodic behavior is sufficiently compelling to warrant a discussion of possible explanations of the phenomenon. The two most likely mechanisms would appear to be (a) the ejection of an envelope by the O9 star and (b) accretion on to a compact object (probably a neutron star) leading to nuclear burning. The first

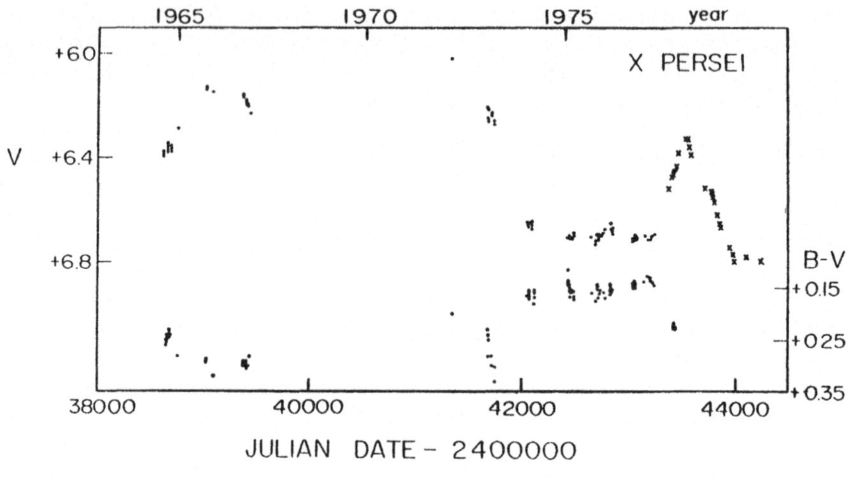

Fig. 1. V-band photometry of X Persei. The crosses are the interpolated Biruni-Villanova data points.

possibility is supported by the fact that X Per has long been considered to be a Be star with broad hydrogen emission and absorption features. The increase in light would then be associated with the ejection of an envelope, initially optically thick, but which, as it expands outwards, becomes less dense and less luminous. However no spectroscopic evidence for velocity changes corresponding to shell ejection have been reported, and such periodic behavior or range of light fluctuation is not known to be characteristic of other Be stars (Underhill, 1960).

There are similarities between the photometric behavior of X Per and that of recurrent novae and symbiotic stars. Periodic outbursts followed by longer intervals of quiescence on time scales of months to tens of years are characteristic of the recurrent novae and, to a lesser extent, of some symbiotic stars. Both of these types are believed to be binaries with orbital periods of hours for the former and typically from 100 to 1000 days for the latter. The outbursts are believed to be produced by accretion on to a more compact companion (a white dwarf) leading to repetitive thermonuclear shell flashes. The light changes in the novae are typically 5 mag or more, in symbiotic stars 2-3 mag, while for X Per we observed a change of about 0.5 mag. Hence a possibility which should be investigated further is a model for X Per in which there is an O9 star with a neutron star companion where the light changes arise from the burning of material accreted on to the neutron star. Changes in brightness of the neutron star of several orders of magnitude would result in a relatively small overall change in light due to the presence of the highly luminous O-star which dominates the total light of the system, a circumstance which does not occur in the novae. This model is in accord with others in which 13.9 min. X-ray period is associated with the rotation of a neutron star, and the spin-up rate and low X-ray luminosity point to a relatively small accretion rate (Becker et al., 1978 and refs. therein). There is no explanation for a 22.4 hour modulation in this model but evidence for this period is rather weak. Continuing observations at all wavelengths are clearly desirable. An examination of the existing X-ray and spectroscopic data for flux changes over a 6-year period would be particularly interesting.

REFERENCES

Becker, R.H., Boldt, E.A., Holt, S.S., Pravdo, S.H., Robinson-Saba, J., Serlemitsos, P.J., and Swank, J.H.: 1979, Astrophys. J. Letters 227, pp 21-24.

Brucato, R.J., and Kristian, J.: 1972, Astrophys. J. Letters 173, pp. 105-107.

Dorren, J.D., Guinan, E.F., and McCook, G.P.: 1979a, I.A.U. Circ. No. 3352.

Dorren, J.D., Guinan, E.F., McCook, G.P. and Siah, M.J.: 1979b, in preparation.

Guinan, E.F., Dorren, J.D., Siah, M.J. and Koch, R.H.: 1979, Astrophys. J. 229, pp. 296-303.

Guinan, E.F. and McCook, G.P.: 1974, Pub. Astron. Soc. Pac. 86, pp. 947-951.

Guinan, E.F., McCook, G.P. and Dorren, J.D.: 1978, I.A.U. Circ. No. 3182.

Harris, D.L.: 1956, Astrophys. J. 123, pp. 371-372.

Hiltner, W.A., 1956, Astrophys. J. Suppl. 2, pp. 389-462.

Hutchings, J.B., Cowley, A.P., Crampton, D., Redman, R.O.: 1974, Astrophys. J. Letters 191, pp. 101-104.

Hutchings, J.B., Crampton, D., and Redman, R.O.: 1975, Monthly Notices Roy. Astron. Soc. 170, pp. 313-324.

Kalv, P.: 1977, Inf. Bull. Var. Stars. No. 1359.

Margon, B., Bowyer, S., and Penegor, G.: 1976, Monthly Notices Roy. Astron. Soc. 176, pp. 217-223.

Mook, D.E., Boley, F.I., Foltz, C.B., and Westpfahl, D.: 1974, Pub. Astron. Soc. Pac. 86, pp. 894-898.

Müller, G., and Kempf, P.: 1907, Astron. Nachr. 175, pp. 162-166.

Mushotzky, R.F., Roberts, D.H., Baity,W.A., and Peterson, L.E.: 1977, Astrophys. J. 211, pp. 129-133.

Underhill, A.: 1960, in Stellar Atmospheres, ed. J.L. Greenstein, Univ. of Chicago Press.

White, N.E., Mason, K.O., Sanford, P.W. and Murdin, P.: 1976, Monthly Notices Roy. Astron. Soc. 176, pp. 201-215.

COMMENTS FOLLOWING DORREN AND GUINAN

Henrichs: The stability of the 13.9 min. X-ray pulse period indicates a neutron star rather than a white dwarf (Mason 1977; Rappaport and Joss 1977; Rappaport 1979). This is perhaps the most firm requirement for any model.

Guinan: I agree with you that the X-ray source has characteristics of a neutron star rather than a magnetic white dwarf.

Kaitchuck: It is perhaps worth mentioning that the rapid Hα variations I reported on this morning all occurred during the last brightening of X Per. The observations made after it faded have shown a constant line strength.

Chambliss: Is there a reliable radial velocity curve or an estimate of f(M) for X Persei?

Guinan: Yes. A radial velocity study of X Per has been published by Hutchings *et al.* in 1974.

Mazeh: Can you suggest an observational test to verify your model against the other models?

Guinan: Yes. The 6-year light variation apparently found should have an X-ray counterpart.

PERIODICAL VARIATIONS IN THE SPECTRUM OF X PERSEI. DISCOVERY OF A 22-HR PERIOD IN THE VARIATIONS OF THE RADIAL VELOCITIES OF Hα.

T. S. Galkina
Crimean Astrophysical Observatory
Crimea, U.S.S.R.

The bright B0e star X Persei has been observed for over 50 years as an emission-line object and as a variable star. Throughout this time, the star displayed typical behavior for such an object. The discovery that it lies very close to the position of a weak X-ray source renewed interest in the star, which is now under close scrutiny. The probable association of X Persei with 3U 0352+30 has been subject to several discussions (Hutchings, Crampton and Redman, 1975; Margon, Bowyer, and Penegor, 1976; Hutchings and Walker 1976; Persi, Viotti and Ferrari-Toniolo 1977, 1978).

Hutchings et al. (1974) found evidence for a period of 581 days and an amplitude of 66 km/s in the radial velocities of broad Balmer absorption lines, which may indicate the presence of a massive companion. The mass function was found to be $f(M) = 18.2 \pm 3.5\ M_{\odot}$. Models based on these and on low-energy X-ray observations range from one involving a black hole (Hutchings et al. 1975) to another postulating an accreting white dwarf (Garavoglia and Treves 1976). Henrichs and van den Heuvel (1977) suggested a pulsating neutron star with a pulsation period of 13.9 minutes and an orbital period of 22.4 hours. The period of 581 days in then intepreted as due to apsidal motion.

Until now, observations could not definitely exclude any of the proposed models. No clear optical evidence existed for optical flux modulations with periods of 13.9 minutes and 22.4 hours (Hutchings 1977) corresponding to the variations in the X-ray range between 0.6 and 7.5 kev (White et al. 1976). Spectroscopic observations were therefore carried at the Crimean Observatory since 1974, covering about two long periods by about 100 spectrograms with dispersions 30 and 36 A/mm at Hα and in the region λλ 4950 - 3650 A.

Figure 1 shows that the broad Balmer absorptions vary in V_r in the long period found by Hutchings, with an amplitude of 50 km/s. The V emission component of Hα has an amplitude of 30 - 35 km/s. The He I lines have a smaller amplitude (22 km/s), large scatter, and a phase shift of about 0.25 P. A large scatter of radial velocities is observed

M. J. Plavec, D. M. Popper and R. K. Ulrich (eds.), Close Binary Stars: Observations and Interpretation, 367–373.

at phases 0.45 - 0.70 P. There may be several contributing factors, hard to separate: (1) Rapid rotation of the star (V sin i = 270 km/s), (2) Gas motions, (3) Possible apsidal motion. The rotational velocity of the absorption component of the surrounding envelope was estimated at 62 km/s.

We searched for the 22.4 hr period in the radial velocities of the V and R emission edges of Hα. Three well-covered phase intervals were studied: (1) Phases 0.73 - 0.86, observed between November 1976 and January 1977; (2) 0.27 - 0.32 P, Sept.-Oct. 1977; (3) 0.41 - 0.49 P, Dec. 1977 - Jan. 1978. Fig. 2 shows these variations with phase of the 22.4 hr period (top curve). The zero point was taken to be the time of the minimum of the ratio of intensities (I_V/I_R) of the two components. The phase dependence of this intensity ratio is shown in the bottom curve. We find a good agreement for the three data groups: the radial velocities vary in a period of 22.4 hours with an amplitude of 15-17 km/s. We believe that Hutchings failed to detect this period because he studied too broad phase intervals.

We believe that our observations reveal an optical periodicity analogous to the known 22.4-hr X-ray modulation of 3U 0352+30. If the radius of the B0e star is 4.9 R_O and V sin i = 270 km/s, its period of rotation would be 22.5 hours. We think that the long period of 581 days is caused by the motion of the binary around a remote third body. Apsidal motion of a close eccentric orbit is not impossible either.

Now a few remarks on the gaseous envelope surrounding the primary component of X Persei. Great variations occurred in the emission spectrum of X Persei between 1962 and 1974. In 1974, the first three Balmer lines and the He I lines were much weaker than before. (Fig. 3). The activity continues to be weaker than 10 years ago: recently, emission was observed only at Hα , Hβ , and weakly at Hγ an He I 5876 A. We believe that this means weakening of matter ejection from the star, and decrease in the outflow velocity. Sometimes sharp components of He II 4686 A are observed in emission. Hα is the strongest emission line. Its variations have been observed as continuously as possible between November 1974 and March 1978. Marked changes in the intensity ratio V/R occur on a scale of 3-4 months. Between 11/5/76 - 12/30/76 we observed V/R < 1, then near Jan. 19, 1977 an inversion occurred. A new set of observations started in September 1977. Again, there was V/R < 1, then a reversal occurred, and from 11/30/77 to 3/15/78 there was V/R >1.

There is no definite explanation of the phenomenon of the V/R variations. Struve's model (1931) of a rotating gaseous ring was subsequently elaborated upon by Huang (1972, 1973, 1974). Kriz (1976) believes that an elliptical ring provides an adequate model. Huang (1977) invoked interaction between the existing ring and the newly ejected matter from the star. Boyarchuk (1957) showed that a stationary outflow cannot originate from the equatorial region of the Be star, and that active processes occur on the star's surface. Our observations

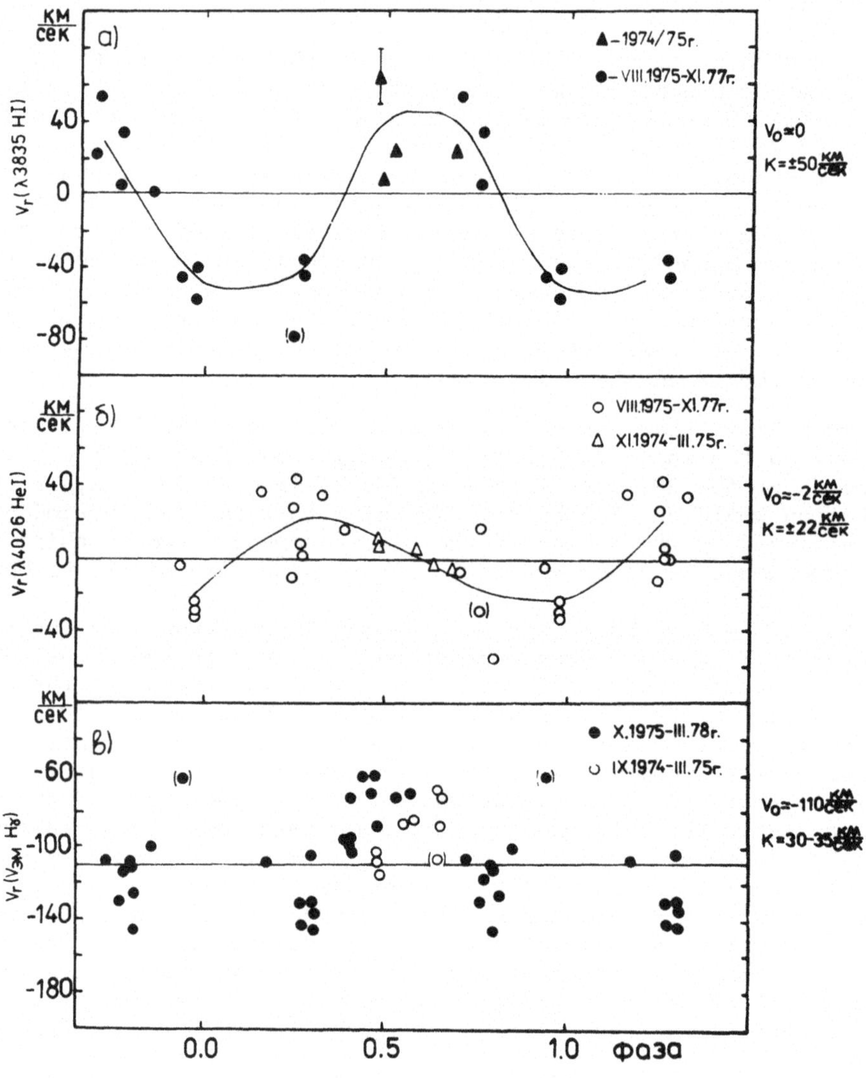

Figure 1. Radial velocity variations of X Persei in a period of 581 d.

confirm Boyarchuk's idea. Boyarchuk remarked that the variation in the ratio V/R might be a measure of the expansion of the Be star envelope: when V<R the envelope is extended, when V>R the envelope is shrunken. However, the envelope as a whole cannot pulsate.

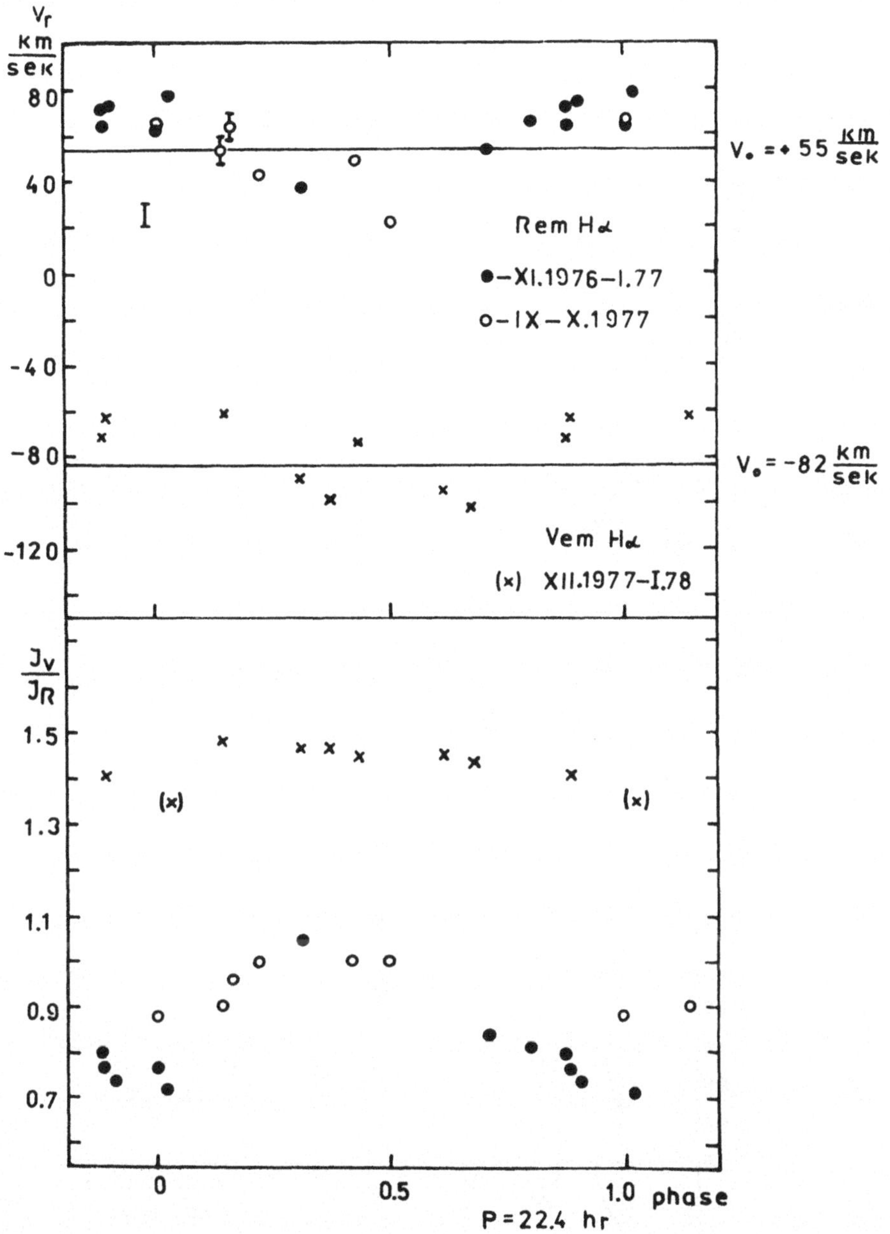

Figure 2.: The 22.4-hour periodicity in X Persei.

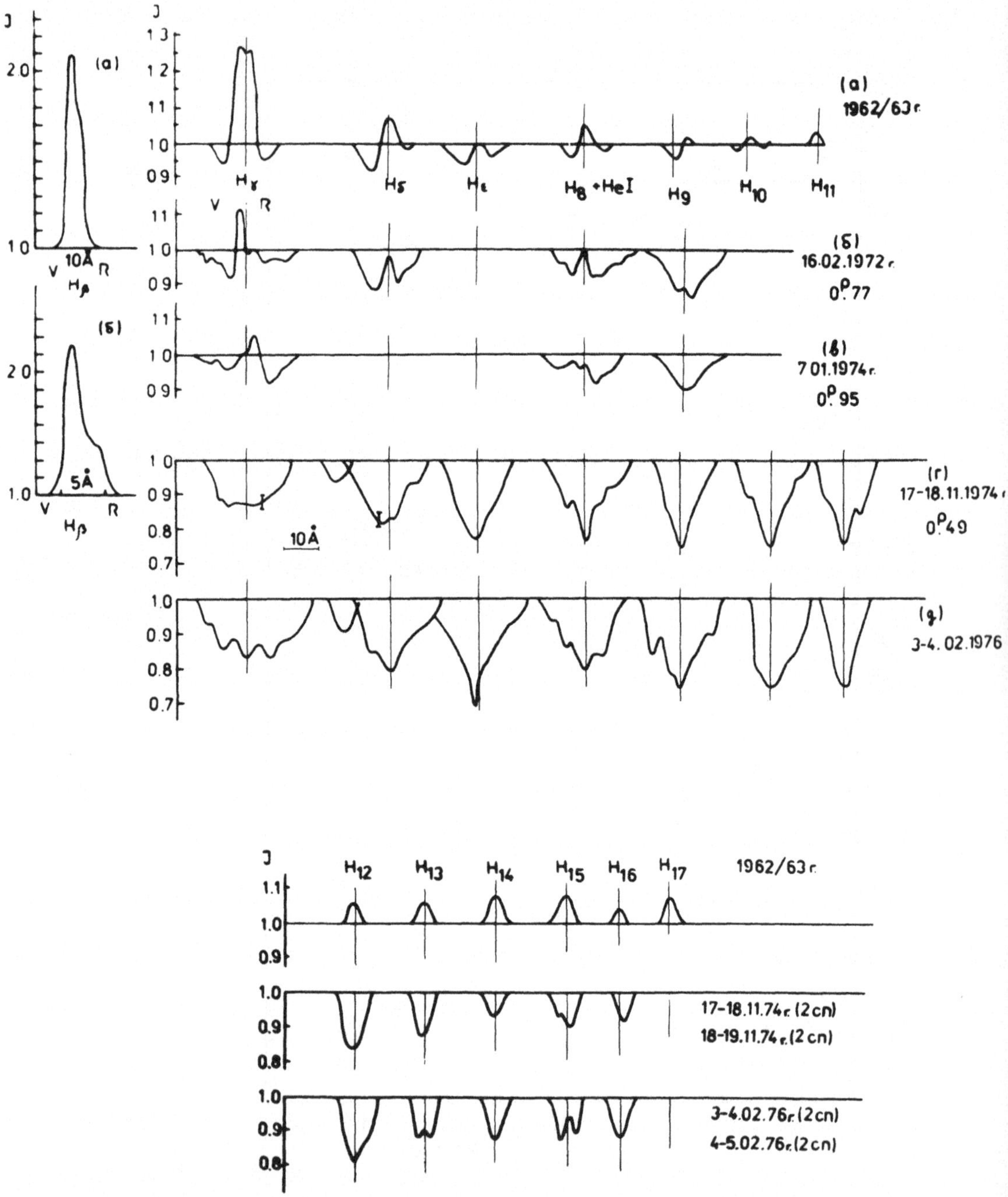

Figure 3. Profiles of the Balmer lines in X Persei.

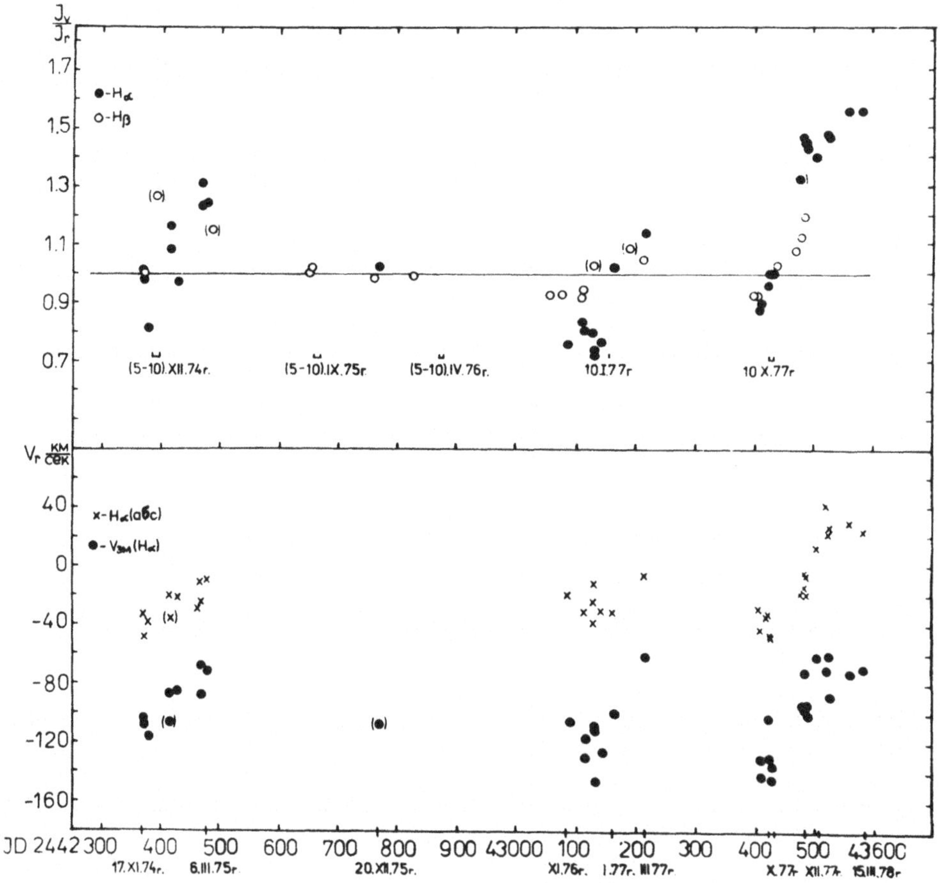

Fig. 4. V/R variations and radial velocity changes in X Persei.

Figure 4 shows the variation of V/R with time between November 1974 and March 1978 (top), and the corresponding variations of the radial velocities of the V emission edge and of the absorption core of Hα (bottom). Three V/R reversals were observed. Interval between them may be about 9 months, i.e. one half of the 581-day period. The phenomena seem to be correlated. Unfortunately, the descending branch of the V/R variation curve cannot be observed from the Crimea. The equality V=R lasts only 5-10 days, while the growt of the V/R ratio takes 4-5 months. The reversal of this trend indicates the start of expansion. We have not been able to determine the phase of this reversal. The amplitude of the V/R variations seems to differ from cycle to cycle. This behavior of V/R may be due to irregular ejections and interactions in the close eccentric orbit of the neutron star, and the wide orbit about the remote companion. Highly ionized material ejected from the B0e star probably recedes to some definite distance, depending on the power of ejection and the velocity of the flow. The ejected material is not dissipated and may be returned to the star's surface, as suggested by Boyarchuk. Interaction of the ejecta with the existing envelope/ring may lead to the observed contractions (V>R) and expansions (V<R) of the envelope.

Persi *et al.*(1977, 1978) found, from optical and infrared observations, a correlation between the star's brightness and the extent of the circumstellar envelope. They also explain the brightness variations as due to contraction and expansion of the envelope. Our observations overlapped with those by Persi. During the faint stage of X Persei (1974-76), the ratio V/R is >1, which corresponds to contraction, while at the phase of higher brightness (September 1977) the reversed ratio indicated expansion.

We urge further observations, in particular between March and September, in order to determine the phase of the V/R reversal.

REFERENCES

Boyarchuk, A.A.: 1957, Astr. Zhur. USSR, 34, 201.
Garavoglia, M. and Treves, A.: 1976, Astron. Astrophys. 49, 235.
Huang, S.-S.: 1973, Astrophys. J. 183, 541.
Huang, S.-S.: 1977, Astrophys. J. 212, 123.
Hutchings, J.B.: 1977, Mon. Not. R.A.S. 181, 619.
Hutchings, J.B. and Walker, G.A.H.: 1976, Pub. Astr. Soc. Pacif. 88, 754.
Hutchings, J.B., Cowley, A.P., Crampton, D., and Redman, R.O.: 1977, Astrophys. J. 191, Lett. 101.
Hutchings, J.B., Crampton, D., and Redman, R.O.: 1975, Mon. Not. 170, 313.
Kříž, S.: 1976, Bull. Astr. Inst. Czech. 27, 321.
Persi, P., Viotti, R. and Ferrari-Toniolo, M.: 1977, Mon. Not. 181, 685.
Persi, P., Viotti, R. and Ferrari-Toniolo, M.: 1978, Mon. Not. 185, 841.
White, N.E. *et al.*: 1976, Mon. Not. R.A.S. 176, 201.

SS 433 AS A MAGNETIC X-RAY BINARY

G. W. Collins, II, G. H. Newsom and R. N. Boyd
The Ohio State University

ABSTRACT

In this paper we expand upon a proposed model for SS 433. Specific attention is given to the manner in which the light travel time within the system places constraints upon the geometrical parameters defining the system. Consideration is also given to the physical situation which exists in the high Doppler shift emitting region. Implications of recent observations of "a 13-day" period spectral and light variability upon the central source are briefly discussed.

Since the announcement by Margon et al. (1979a) that the anomalous lines in the spectrum of SS 433 could be interpreted as highly Doppler shifted Balmer and Neutral Hydrogen lines, this object has received a remarkable amount of attention. Previous to this announcement, it had been listed by Stephenson and Sanduleak (1977), and had been noted by Seaquist et al. (1978) as a radio source, Marshall et al. (1978) as an X-ray source and Clark and Murdin (1978) as positionally coincident with the supernova remnant W50. Margon et al. (1979b) have placed the object at a distance in excess of 3.5 kpc while Liebert et al. (1979) have measured the absolute fluxes of various components of the spectra. Abell and Margon (1979) have fit the periodic behavior of the highly Doppler-shifted line spectrum with a kinematic model having a period of about 164d, an orbital inclination of either 17° or 78° and possessing radial beams inclined to the rotation axis of the orbit by either 78° or 17°. Collins and Newsom (1979) have shown that all these data are consistent with the picture of a magnetically accelerated stellar wind arising in the primary of an X-ray binary. Their model suggests that the beams arise when the magnetically accelerated particles pass through the region of cool gas and perhaps dust located in the orbital plane of the system beyond the Stromgren sphere of the X-ray binary (see fig. 1). In this paper we shall note some of the further implications of the model.

Since the orientation of the beam with respect to the radius vector will depend on the complicated hydromagnetic interaction of the beam with the cool gas, we have let the orientation β (see fig. 2) be arbitrary within the range $-\pi/2 < \beta < \pi/2$. The radial

M. J. Plavec, D. M. Popper and R. K. Ulrich (eds.), Close Binary Stars: Observations and Interpretation, 375–379.

velocity then becomes

$$v_r = v[\sin(\alpha+\beta)(\cos\theta \cos\phi \sin i + \sin\theta \cos i) + \cos(\alpha+\beta) \sin\phi \sin i] \quad (1)$$

where ϕ = o occurs when the intersection of the magnetic and orbital

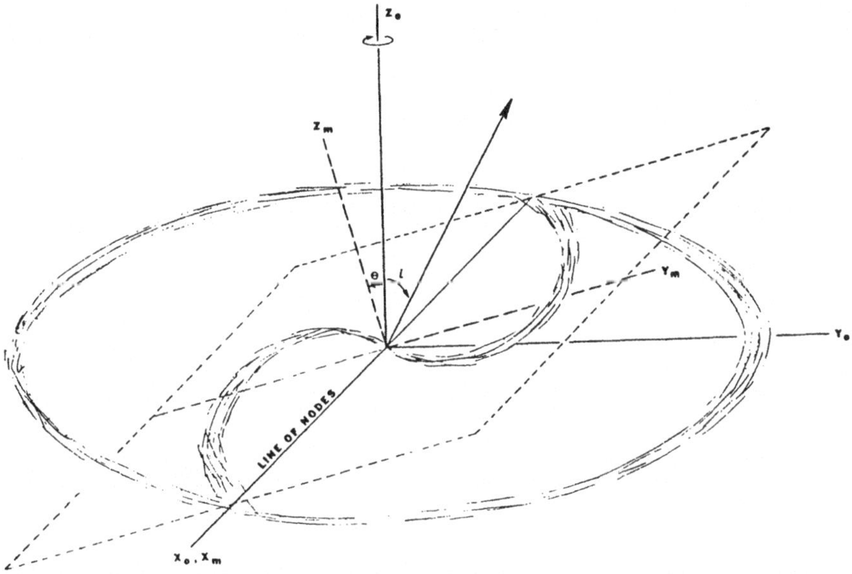

Fig. 1. Particle Beams Accelerated From X-Ray Binary Encounter Gas Disk at the Intersection of Magnetic and Orbital Planes.

planes is in the plane of the sky. In the disk model the beams will be produced at the nodes and hence $\alpha = 0$. It is clear from eq. 1 that in the Abell-Margon model ($\beta = 0$, $\alpha = \pi/2$) an ambiguity in i and θ exists as equation 1 is invariant under the transformation $[i = \pi/2 - \theta,\ \theta = \pi/2 - i]$ (Note: our θ is the complement of Abell and Margon's angle). For a wide range of β limited by possible values of θ (i.e. $12° < |\beta| < 90°$, $\alpha = 0$), only $i = 17°$ will preserve the observed amplitude and phase difference between the maximum and first crossing point. Indeed only a combination of $|\alpha+\beta| = \pi/2$ produces the ambiguity in θ and i.

Because the beams occur at a large distance from the central binary in this model, a substantial light travel time effect exists. This causes the beams to display equal radial velocities at a time slightly later than when their motion is across the line of sight. Thus a net positive redshift with respect to the average of the extreme should occur at the crossing point. In the case where $\alpha = 0$, this addition shift is given by

$$\Delta Z = -\frac{4\pi}{P}\left(\frac{v\gamma}{Rc^2}\right) \sin^2 i \sin\phi_c \left[\cos\beta \cos\phi_c - \sin\beta \cos\theta \sin\phi_c\right] \quad (2)$$

where ϕ_c is the value of the phase at the first crossing after maximum, P is the beam period, $\gamma = (1-v^2/c^2)^{-1/2}$ and R is the radial distance from the binary to the point of beam emission. Since the crossing points are symmetrically placed about $\phi=\pi$, the measured values of the redshift increments ΔZ at the two crossing points provide sufficient information to specify β and R. Thus measurement of these values is of the utmost importance, although the day to day "flickering" in radial velocity will require that a mean curve be derived from several cycles. The specification of these parameters places significant constraints on the physical situation in the beam-radiating region.

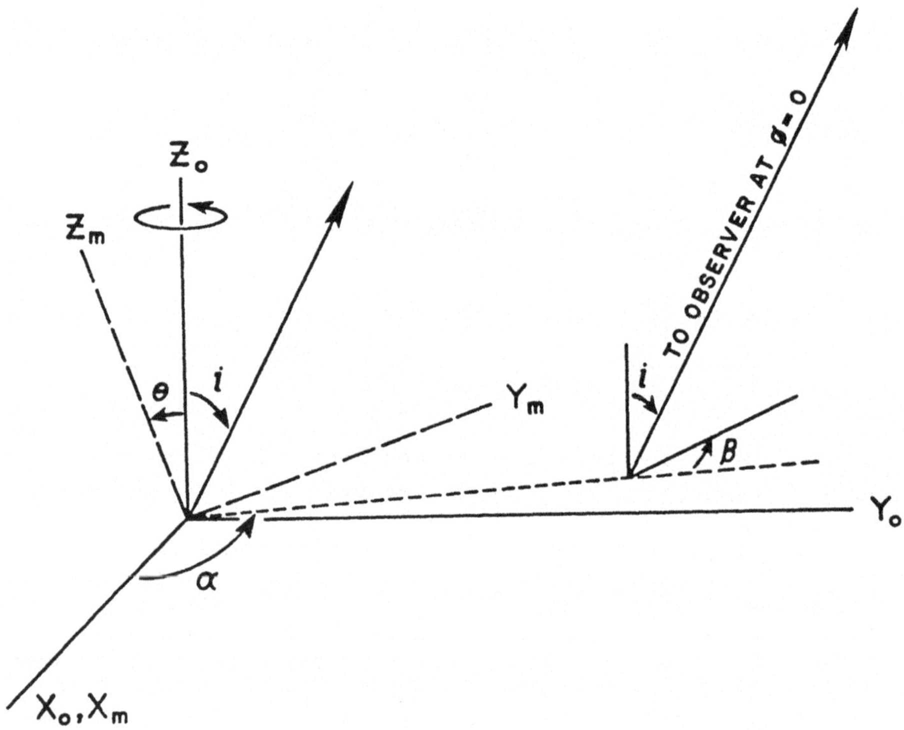

Fig. 2. Coordinate Frame for X-Ray Binary Model

Before passing to considerations of the nature of the central binary source itself, it is worth describing conditions in the beam. Several authors (e.g. Liebert et al. 1979) have concluded that the atoms radiating in the beam have a velocity of about 80,000 km/sec. This, combined with the daily variation and implied maximum recombination time, present a picture of a 33 Mev proton beam incident on a cool gas with a density of at least 10^8 particles/cc. Although collisionally produced electrons will pass through a speed resonance with the beam protons producing a recombination spectrum in the moving frame, the bulk of the line spectrum appears to be formed via a charge exchange reaction with atomic hydrogen. Each incoming 33 Mev proton undergoes about 10^4–10^5 radiative captures during its lifetime in the disk. Thus with plausible values for the mass outflow (i.e. about $10^{-8} M_{\odot}/yr$) it is possible to match the observed line intensities. A more detailed treatment of the beam physics is presently being completed.

Lastly we must make certain that the physical situation required by the beam does not present an impossible situation for the central binary. Plausible beam densities (Collins et al. 1980) and the known particle energies imply a magnetic field of about a gauss at the point where the beam reaches its maximal energy. Collins and Newsom (1979) show that this places minimal constraints upon the stellar mass of the primary required to contain the field. We originally suggested that the beam motion was driven by the orbital motion of the binary system itself but were rather concerned that the 160 day period was vastly longer than that of any known X-ray binary. Recent observations by Crampton et al. (1979) of a 13 day spectroscopic variation of the unshifted Hα component and apparent confirmation of a similar photometric light variation (Kemp and Arbabi 1979) would seem to rule out orbital motion as the direct driving force. Using an inclination of 17° and the mass function of Crampton et al. (1979), a very approximate calculation indicates that a magnetically distorted sychronously rotating primary would precess with a period of about 1/2 year. In this case the orbital motion would still be the ultimate source of the beam energy, but it would be supplied via the precession. Further work is in progress to strengthen the quantitative aspects of this argument.

REFERENCES

Abell, G. O., and Margon, B. (1979), Nature 279, pp. 701-703.

Clark, D. H., and Murdin, P. (1978), Nature 276, pp. 44-45.

Collins, G. W., II and Newsom, G. H. (1979), Nature 280, pp. 474-475.

Collins, G. W., II, Newsom, G. H., and Boyd, R. N. (1980), Ap. J., in preparation.

Crampton, D., Cowley, A. P., and Hutchings, J. B. (1979), IAU Circular 3388.

Kemp, J. C., and Arbabi, M. (1979), IAU Circular 3391.

Liebert, J., Angel, J. R. P., Hege, E. K., Marton, P. G., and Blair, W. P. (1979), Nature 279, pp. 384-387.

Margon, B. et al. (1979a), IAU Circular 3345.

Margon, B., Ford, H. C., Katz, J. I., Kwitter, K. B., Ulrich, R. K., Stone, R. P. S., and Klemola, A. (1979b), Ap. J. Lett. 230, pp. L41-L45.

Marshall, F. E., Mushotzky, R. F., Boldt, E. A.,Holt, S. S., and Serlemitsos, P. J. (1978), IAU Circular 3314.

Seaquist, E. R., Gregory, P. C., and Crane, P. C. (1978), IAU Circular 3256.

Stephenson, C. B., and Sanduleak, N. (1977), Ap. J. Suppl. 33, pp. 459-469.

DISCUSSION FOLLOWING COLLINS, NEWSOM AND BOYD

Massey: How do you reconcile your model with the wealth of other observed properties--the unshiften emission lines, for example?

Collins: Only time has constrained me from dealing with the additional observed properties, all of which to date do not appear to be in conflict with this picture. For instance, much of the unshifted emission spectrum arises from the quiscent material in the disk which has been ionized by the beam. However we do not rule out the possibility than a component may arise within the central binary system inself since such a spectrum is not uncommon in other X-ray binaries.

VII. THE RS CANUM VENATICORUM STARS

PERIOD CHANGES AND MASS LOSS RATES IN 34 RS CVn-TYPE BINARIES

Douglas S. Hall
Copernicus Astronomical Center, Polish Academy of Sciences

Jerzy M. Kreiner
Institute of Physics, Silesian University, Katowice

Steven N. Shore
Astronomy Department, Columbia University

We have collected all available times of minimum, over 1650, for 23 regular, 7 short-period, and 4 long-period eclipsing RS CVn binaries using the card catalogues at the Sternberg Astronomical Institute and the Krakow Astronomical Observatory, as well as unpublished data. We examined them critically, discarded those grossly in error, and assigned weights. For AD Cap, RV Lib, and ϵ UMi there are virtually no data. For the remaining 31 we determined quadratic ephemerides by weighted least squares. Values of dlnP/dt are given in the Table with their errors. Roughly 2/3 are variable at greater than the 2 σ level. Significant decreases outnumber significant increases by almost 2:1. We show that light curve asymmetry produced by the distortion wave probably accounts for the relatively small, rapid fluctuations in the O-C curve but not the long-term period changes.

In our model one star, usually the cooler and more evolved one, has a convectively driven wind which corotates with the star (and with the system, since rotation is synchronous) out to the Alfven radius R_A. Because the well-defined long-lived photometric wave in most RS CVn systems indicates strong longitudinally asymmetric spot activity, we allow for possible anisotropy in the wind, stronger from the unspotted hemisphere, which should be a large coronal hole. The equation is

$$\frac{d\ln P}{dt} = 3\left[\frac{U}{V} f \sin\theta(\min) - 2\frac{M_c}{M} + 3\frac{M}{M_h}\left(\frac{R_A}{a}\right)^2\right]\frac{d\ln M_c}{dt}, \quad (1)$$

where M is mass, subscript c refers to the cooler mass-losing star, U is wind velocity, f is a geometrical factor about $4/\pi^2$, $\theta(\min)$ is phase of wave minimum, V is relative orbital velocity, and a is orbi-

M. J. Plavec, D. M. Popper and R. K. Ulrich (eds.), Close Binary Stars: Observations and Interpretation, 383–387.

Table. Derived Rates of Mass Loss From the Active Star.

Name	dlnP/dt (yr^{-1})	minimum B (Gauss)		$\dot{M}$ ($M_\odot$ yr^{-1})		$\dot{M}_{th}$ ($M_\odot$ yr^{-1})
		n=3	n=2	minimum	maximum	
RT And	-1.17±0.06(-7)	526	243	2.1(-9)	2.6(-9)	3.1(-8)
CQ Aur	+2.45±0.42(-6)					
SS Boo	+2.16±1.38(-7)	0	0	8.2(-8)	3.7(-7)	5.5(-8)
SS Cam	+2.07±1.72(-7)	0	0	4.1(-7)	4.5(-6)	2.1(-6)
SV Cam	+1.58±0.64(-8)	0	0	1.3(-8)	3.1(-8)	1.0(-8)
RU Cnc	-2.09±0.80(-7)	111	40	4.8(-11)	2.4(-10)	2.1(-8)
RZ Cnc	-2.39±2.06(-7)	8	5	3.2(-13)	6.0(-11)	9.2(-6)
WY Cnc	-1.74±2.50(-8)	0	0	0	1.7(-8)	4.5(-8)
RS CVn	-3.17±0.48(-7)	156	68	4.3(-10)	7.8(-10)	3.4(-7)
UX Com	-2.06±0.96(-6)	579	220	1.4(-8)	1.1(-7)	1.1(-7)
RT CrB	+5.61±3.85(-7)					
CG Cyg	-6.53±1.49(-8)	1118	325	1.4(-9)	3.7(-9)	1.9(-8)
WW Dra	-1.70±0.55(-7)	106	48	9.2(-11)	3.6(-10)	3.2(-7)
RZ Eri	+4.48±4.01(-7)	0	0	1.1(-7)	2.0(-6)	3.4(-7)
Z Her	-2.84±1.59(-8)	71	25	2.3(-12)	3.0(-11)	1.1(-7)
AW Her	-0.70±1.27(-6)					
MM Her	-1.40±2.03(-7)	0	0	0	5.9(-9)	1.6(-7)
PW Her	-4.92±6.70(-7)					
GK Hya	-2.28±1.27(-7)					
RT Lac	+2.89±0.32(-7)	0	0	5.9(-7)	7.3(-7)	4.4(-7)
AR Lac	-2.58±0.22(-7)	165	93	7.0(-10)	9.9(-10)	2.1(-7)
VV Mon	-2.23±0.54(-6)	156	91	4.8(-9)	1.3(-8)	5.2(-7)
AR Mon	-1.98±1.49(-6)	44	22	1.0(-10)	5.2(-9)	5.9(-6)
LX Per	+1.13±1.05(-6)	0	0	1.1(-7)	3.0(-6)	2.0(-7)
SZ Psc	-5.25±0.33(-6)	252	173	6.1(-8)	7.8(-8)	1.1(-6)
UV Psc	+1.46±0.42(-7)					
TY Pyx	+1.71±3.00(-7)	0	0	0	3.2(-7)	9.0(-8)
RW UMa	-2.65±0.75(-7)	245	79	4.4(-10)	1.4(-9)	3.5(-7)
XY UMa	+8.82±3.09(-8)	0	0	1.3(-7)	2.6(-7)	2.7(-8)
RS UMi	+1.46±2.96(-6)					
ER Vul	-2.47±0.92(-7)	723	295	3.6(-9)	1.8(-8)	2.7(-8)

tal semi major axis. R_A is given by

$$R_A/R_c = (B/\Omega)^{1/n}(U/\dot{M}_c)^{1/2n} , \qquad (2)$$

where B is the surface magnetic field, Ω is rotational (= orbital) frequency, n = 2 for a monopole field, and n = 3 for a dipole field. The first term in Equation (1) can produce a rocket effect, period changes in either direction depending on the value of θ(min). The last two give the long-term period change, which can be positive or negative depending on the size of R_A.

For a given value of B, dlnP/dt in Equation (1) has an absolute minimum at a certain value of $\dot{M}_c$. Therefore we derive minimum values of B required to produce the observed long-term period decreases and give them in the Table. Stars with blanks had no available absolute dimensions.

Also in the Table are $\dot{M}_c$ values derived with Equation (1) from the long-term period decreases by assuming n = 2, B = 1000 Gauss, U equal to the escape velocity from the cooler star, and from the long-term period increases by assuming $R_A = R_c$. They are from 10(-6) to 10(-11) $M_\odot$/yr, with individual values uncertain by typically an order of magnitude. The 10(-9) $M_\odot$/yr which we expect from evolutionary considerations is in good accord. Actually, since mass loss occurs over a fraction α of the cooler star's surface, R_A in Equation (1) should be replaced by $R_A \alpha \exp(1/2n)$, a change which would increase the values of $\dot{M}_c$ in the Table by a factor of ~ 2. Also $\dot{M}_c$ would be larger if n = 3 but would be smaller if U is greater than the escape velocity or if B > 1000 Gauss. The cooler star's thermal mass loss rate $\dot{M}_{th}$ is given in the last column of the Table. In most cases, if we consider the errors $\dot{M}_c$ is less than 1% of $\dot{M}_{th}$. Only in SZ Psc and the four short-period systems RT And, SV Cam, CG Cyg, and XY UMa is it comparable. The long-term increase for RT Lac might be produced by Algol-type mass transfer from the other (less massive) star, which fills its Roche lobe.

For SS Cam, SV Cam, RS CVn, CG Cyg, RT Lac, and AR Lac, which apparently show period changes in both directions, we consider nine possible explanations. Three which proved quantitatively unreasonable were supposing (1) that B varies between ~ 0 and ~ 1000 Gauss, perhaps during the course of a magnetic spot cycle, (2) that the limb of the star is deformed by its spot group, and (3) that Algol-type mass transfer is occurring. All six systems cannot be explained by any single mechanism. In some we cannot rule out or confirm apsidal motion (SS Cam) or a third body (SV Cam). Deciding between the rocket effect and light curve asymmetry is difficult because of the ironic and unfortunate coincidence that both lead to effects in the O-C curve which are identical in phasing and the maximum expected size is often comparable for both and comparable to the observed effect. Nevertheless we can conclude that not one of the six shows convincing evidence that the rocket effect is making the dominant contribution to the alternating excursions seen in their O-C curves.

The full version of this study, including tabulation of the times of minimum and plots of the O-C curves, will be published in Acta Astronomica.

DISCUSSION FOLLOWING HALL, KREINER AND SHORE

Naftilan: In your model for the period changes, how does the mass ratio affect the rate of change; and in particular would we expect RT Lac, which has a mass ratio considerably different from the other systems listed to have a greater period change rate?

Hall: The equation describing period change for conservative mass transfer would be very sensitive to mass ratio , specifically, near unit mass ratio (typical of inert RS CVns) would produce no period change. An equation for mass loss, however, is not sensitive to mass ratio. The size of the corotating Alfven radius is the controlling factor.

Stencel: In the interest of expanding the data base on RS CVn's, I am pleased to announce that on the basis of moderate and high dispersion spectra obtained at CTIO, the stars HD 39937, 101379, 155555 and 174459 are probable southern RS CVn systems. HD 155555 is also an X-ray source. Southern observers should be advised that these and fifty other southern candidates deserving observations will soon be available in a published list (Weiler and Stencel, 1979 Astron. J., in press).

van 't Veet: For the same system the sign of the period change may change on a short time scale. Do you explain this by a rapid change of the Alfven radius?

Hall: A variable Alfven radius was one mechanism we explored to explain the supposed period changes of alternating algebraic sign. It proved not to be a satisfactory explanation basically because, for very small values of R_A, the coefficient connecting dlnP/dt with dlnMc/dt always remains very small and the $\dot{M}$ required must be very large . Other mechanisms must be responsible, one of which is light curve distortion.

Walter: You stated that your computations of mass loss rate were consistent with that required to explain the X-ray emission. Our estimate of $\dot{M}$ was for a simple model assuming supersonic outflow with the X-ray emission arising in the flow. This is probably not the case. The X-rays most likely arise from solar-like magnetic flux tubes, which confine the hot gas. We do not require any mass loss. The mass would be lost from regions with open field lines, i.e., coronal holes. It may be possible to constrain $\dot{M}$ from measurements of the soft X-ray absorption, but this has yet to be done.

Dupree: Is there any spectroscopic evidence for mass loss in RS CVn systems?

Hall: The observations of asymmetries in Hα and velocity variations in MgII h and k indicate mass flows, albeit at lower than the escape velocity. The lines, however, are not formed that far out and one has no progressions to use for $v_{rad}(r)$.

COMMENT BY POPPER (Chairman of the Session) PRECEDING THE NEXT PAPER BY SHORE AND HALL

Popper: The next talk is a general topic concerning RS CVn systems. I wish to take advantage of this opportunity to make a general comment about them. A few months ago I received a call from a scientist at the Jet Propulsion Laboratory working in X-ray astronomy. For 30 minutes, we discussed whether σ CrB is a RS CVn system or not.

Why not a simple "yes" or "no"? The star is a double-lined spectroscopic binary with H and K emission in both components, late F or early G, period between one and two days. Furthermore it is an X-ray and radio source. What more does it take for a system to be clearly an RS CVn? σ CrB has a pretty good parallax showing that the two components are main-sequence stars, close to the zero-age main sequence. Most of us tend to consider that bona fide RS CVn are evolved, with at least one component a G or early K subgiant. Systems now being referred to as "RS CVn" include a wide variety in which a number of interrelated phenomena occur in various combinations. A class of systems should probably have common evolutionary status. Perhaps we should be referring to RS CVn phenomena rather than RS CVn systems.

A STARSPOT MODEL FOR THE RS CVn STARS

Steven N. Shore
Astronomy Department, Columbia University and
Astronomy Department, Case Western Reserve University

and

Douglas S. Hall
Dyer Observatory, Vanderbilt University

We employ a simplified dynamo model to describe the long term photometric spectroscopic behavior of the RS CVn stars. The essential feature of the model is that the stars are in nearly synchronous rotation, with the differential rotation slower and the rotational velocity higher than for a single evolved 1 to 2 solar mass star. The spot groups are formed by eruption of enhanced toroidal fields, which have areas at the photosphere of several tenths of the surface area; the sizes of these regions are shear limited. Estimates of the lifetimes of active regions and of flare energetics are presented. The RS CVn phenomenon is then related to both the evolutionary status and the orbital parameters of the binary system.

In order to model the dark wave behaviour of the RS CVn stars, we shall assume a simplified kinematic dynamo, of the type described by Babcock (1961). Although this is only a crude approximation of the complex physical processes which lead to regeneration of magnetic fields and appearance of spots on the solar surface, we feel that the observational evidence for starspot activity is not sufficiently complete to warrant a more comprehensive treatment at the present time. We hope that results of "preserving the appearances" by the Babcock dynamo model will serve as guides to more complete developments.

With the rather short orbital periods, near unit mass ratios, and lack of large scale mass transfer between components, we expect synchronous rotation of the component stars in the RS CVn systems. Tidal locking might be sufficient to insure that the less evolved component (F or G, V or IV) is nearly rigidly rotating, but the more evolved component (G or K, IV or III) has a deep convective envelope and it seems some residual differential rotation is preserved despite the effects of tidal coupling. We assume this is described by a rotation law of the form

M. J. Plavec, D. M. Popper and R. K. Ulrich (eds.), Close Binary Stars: Observations and Interpretation, 389–396.

$$\Delta\Omega = -\alpha \sin^2 \beta, \tag{1}$$

where β is stellar latitude, Ω is the rotational frequency, and α is the differential rotation rate.

A toroidal field is produced by wrapping of frozen-in poloidal field lines due to the differential rotation of the stellar photosphere. This poloidal to toroidal conversion is give by $B_\lambda/B_\beta = \cos\beta\ (d\lambda/d\beta)$, where B_β and B_λ are the poloidal and toroidal field components, respectively, β is stellar latitude, and λ is stellar longitude.

The build-up time Δt_* at a given latitude is then given by

$$\alpha\, \Delta t_* \sin 2\beta \cos\beta = (B_\lambda/B_\beta)_c \tag{2}$$

and $(B_\lambda/B_\beta)_c \simeq 3$ (Babcock 1961).

The latitude at which starspots first appear is that for which Δt_* is a minimum. For a generalized rotation law $\Delta\Omega = -\alpha \sin^n \beta$, where $n \geq 1$, this latitude is given by

$$\beta_*^o = \sin^{-1} \left(\frac{n-1}{n+1}\right)^{1/2} . \tag{3}$$

For the case n = 2, which we are assuming here for spotted stars in RS CVn systems, this yields $\beta_*^o = \pm 35\overset{\circ}{.}3$ for an equatorially symmetric dipolar initial field. (see e.g. Eaton and Hall 1979). This is the highest latitude at which spots will occur. The suggestion (Oskanyan et al. 1977) that spots appear at very high latitudes in some By Dra stars may be an indication that for those stars $n > 2$.

The value of α for an RS CVn star can be found from the observed migration curve. As the spot cycle progresses, the activity drifts towards the equator and the wave rotation frequency Ω increases. At the beginning of a spot cycle, the spots are at $\beta_*^o \approx 35^\circ$ and the wave's rotational frequency is minimal; at the end of a cycle the spots are very near the equator, and the wave's frequency is at its maximum. Since the wave's frequency Ω is given by $\Omega = 2\pi/P_{orb} + 2\pi/P_{migr}$, a direct application of the rotation law gives

$$\alpha = -(\Omega^o - \Omega^1)/(\sin^2 \beta_*^o - \sin^2 \beta_*^1) \approx 19/P_{mig} \text{ rad yr}^{-1}.$$

For the case of the prototype RS CVn, where $P^o_{migr} = 12$ yr and $P^1_{migr} =$ 8 yr (Hall 1972), $\alpha = 0\overset{\circ}{.}13$ day^{-1}, which is smaller than the solar case, as we anticipated above.

Once α is known, the build-up time Δt_*^o for the toroidal field at the latitude of first appearance, which corresponds to a dead time in the cycle of starspot activity, can be computed from eq. (2).

A toroidal field trapped at the base of the stellar convection zone will become unstable to buoyant eruption should it come into pressure equilibrium with the surrounding medium. The maximum field which the dynamo may produce by simple field wrapping is $B_\phi^2 \approx 16\pi\Omega\nu_c \ell$ where ν_c is the convective velocity, Ω is the rotational frequency and ℓ is the length scale. This may be compared with the equilibrium field $B_\phi^2 \approx 8\pi P_{G,BCZ}$ where $P_{G,BCZ}$ is the pressure at the base of the convective zone. For the typical parameter of an RS CVn star, we expect that the field will be strong. We can now crudely estimate the appearance of the starspot region which these flux tubes will generate at the photosphere as follows. Assuming that the tube has a characteristic cross section size of a pressure scale height, H_p, at the base of the convection zone we take the depth of the spot region to be also H_p at the photosphere. If the tube conserves flux and adiabatically expands during its ascent, its area should increase in order to maintain local pressure equilibrium giving for the area of the photospheric active region,

$$A_{RS} \approx \left(\frac{P_{G,1}}{P_{G,\odot}}\right)^{-1/2} \left(\frac{P_{G,BCZ}}{P_{G,BCZ\odot}}\right)^{1/2} \left(\frac{H_{p,BCZ}}{H_{p,\odot}}\right)^{2} A_\odot \approx \left(\frac{P_{G,1}}{P_{G,\odot}}\right)^{-1/2}$$

$$\left(\frac{T_{BC}}{T_\odot}\right)^{2} \left(\frac{g}{g_\odot}\right)^{-2} A_\odot$$

scaled to the solar values. This is typically $\approx 10^4 A_\odot$, of the order of magnitude required for the standard models of RS CVn systems (e.g. Hall 1976). In the solar case, the sizes of active regions are limited by internal processes, such as the decay of a given large spot or flaring. Shearing will be, however, the mechanism of destruction in the RS CVn systems, due to the enormous size of the active region. If we compare the area, we find that

$$A_{shear} \approx \left(\frac{\Omega}{\alpha}\right)^2 \lesssim A_{RS}$$

so that the stability of a region is on the timescale of the differential rotation. If there is one major region on the surface, then the depth of the dark wave should decrease with time, as the activity site is sheared and diffuses more uniformly over the photosphere, so that we expect that the magnitude of the dark wave should vary as A^{-1}, with the decay time being $\tau_S \propto \frac{2\pi}{\alpha \sin\beta}$.

We have used these values to fit an extremely simple starspot cycle to the observations of Haslag (1977 MSc Thesis, Vanderbilt) for RT Lac, shown in figure 1. The agreement for the rise time and decay time is reasonable, but it is clear that more than one active region is responsible for the dark wave.

The ratio of rise to decay time for a typical RS CVn spot cycle should be smaller than the solar case, $\Delta t_*^o/\tau_S \lesssim 0.3$, while for the sun, it is of order 1. Since the fraction of inactive stars is $\lesssim \Delta t_*^o/\tau_S$, we expect 1 or 2 non x-ray stars in the current sample. Z Her may be one star like this.

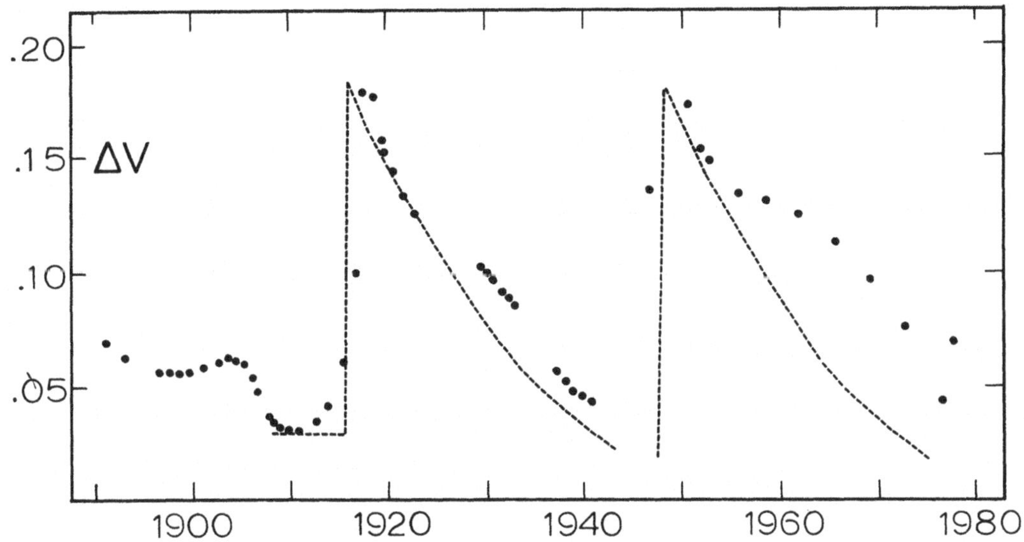

Figure 1. The wave amplitude in RT Lac (Haslag 1977). The dashed line is the theoretical light curve, in magnitudes, for a single active region $A^o_{RS} = 0.21$ being sheared out by differential rotation. Dilution by the secondary is assumed from light curve synthesis.

The individual active regions may consist of numerous aglomerated pores, as discussed in Meyer, et al. (1974) with characteristic sizes H_p. To attain the requisite areas, some 10^4 to 10^5 separate pores are needed. This implies stable dark waves and continuous flaring, in addition to strong and steady x-ray emission. Perhaps coincidentally, this is the order of magnitude of coronal loops required by Walter et al. (this volume) to explain the observed x-ray emission.

The longevity of an individual spot within an active region is limited by turbulent dissipation processes, and consequently the lifetime is approximately $\tau \sim \ell^2 \eta_T^{-1}$ where η is the turbulent diffusion coefficient and ℓ is the characteristic length within the spot. Taking $\ell^2 \sim A$, and scaling the convective velocities, we obtain a timescale of several years for an individual spot, which should be the shortest timescale for the variation of the shape of the dark wave in these systems. Observations of UX Ari (Hall, et al. 1977), λ And (Bopp 1978; Guinan 1979, private communication) and RS CVn (Luddington 1978, private communication) agree with this expectation. The overall amplitude of the wave is variable on the shearing timescale, while the shape of the wave, which is generally a saw-tooth, changes on the turbulent timescale.

Scharlemann (1979) has suggested a model in which the spots appear like floating, highly opaque but geometrically thin disks which are mobile in the presence of photosperic circulation currents. These floaters are assumed to cool the ambient photosphere setting up thermal convection currents which drag the spots together. If the wavelength of the disturbance tends to infinite length (vanishing wavenumber), then agglomeration occurs. The velocity of these floaters as estimated from the rise time for the spot cycle (typically of order ~ 3 to 5 years) is $v_{floater} \approx 0.1$ km s^{-1}. This is the same order of magnitude expected for thermally driven currents for a spot group with temperature deficit $\Delta T, v_{Th} \sim 0.1\ (\Delta T)^{1/3}$ km s^{-1}, evaluated using the atmospheric parameters in Kurucz (1979) for a (5500,3.0) star. The group will not be stable since meridional circulation driven by rotation should disperse its members. In addition, there are problems with the shape of the wave, which should be a sawtooth with a sharp leading edge, due to these currents, while the observed waves tend to be mirror images of this. It is an attractive model, with the analog of ice cubes in a very large martini glass being an inducement for experimentation.

The magnetic energy available for the generation of relativistic particles, assuming complete annihilation of the fields and direct conversion, is approximately the total magnetic energy contained within the active region, which is

$$\varepsilon_{RS}/\varepsilon_{\odot} = (B_{RS}/B_{\odot})^2\ (A_{RS}/A_{\odot})\ (g_{RS}/g_{\odot})^{-1} \approx 10^4.$$

Here we have taken the characteristic depth of the active region to be the pressure scale height, which is about ten times that in the sun. The total number of particles accelerated in reconnection by complete, efficient conversion scales as:

$$N_{rel} \sim B_s^2\ A\ \ell_a^{-2}\ \sigma^{-1},$$

where σ is the conductivity and ℓ_a is a scale length for field annihilation. This is assumed to be a function of the area of the active region only.

We have assumed a complex structure for the active region, not a single spot. While this is a difficult point to test observationally in these systems, the largest spot regions observed on the sun are usually highly complex, being composed of many smaller activity centers which vary in size and number on timescales shorter than the lifetime of the region as a whole. We use this analog to provide some estimate of the annihilation length ℓ_a. If we assume that the spot region is densely populated and that the scale height is larger than in the solar case, it is possible that ℓ_a is smaller for RS CVn stars than the sun. Close packing of spots creates flux tubes in the active regions so that assuming $\ell_a \sim A^{-1/2}$, we obtain $N_{rel} \sim A^2$.

The total energy emitted in flares, which is related to N_{rel}, may be as high as 10^7 times the solar case. Hence flaring activity in the RS CVn stars should be much more pronounced than in the sun. The use of the flux tube picture appears to be supported by the high coherence and circular polarization of the observed radio events.

Alternatively, the solar data is consistent with the view that the number of spots increases with the total area of the spot group. If N_{rel} is proportional to the number of annihilations, and the RS CVn flares are composed, in fact, of many subflares (as appears to be the case e.g. HR 1009) then $N_{rel} \sim A^2$ even if $\ell_a \sim \ell_{a,\odot}$.

As an additional astrophysically interesting point concerning the RS CVn and related stars, we note that if these stars contribute substantially to the background diffuse x-ray emission literature, they may also contribute to the background of low energy cosmic rays. The implication of this suggestion for the synthesis of the light elements, is that since the RS CVn stage in a close binary system happens only in low mass, close binaries, the contribution to the cosmic ray background is not prompt . That is, there is a delay of $\gtrsim 10^9$ years for this stage to occur after star formation, so that the rise in the low energy cosmic ray flux will not be immediate upon the formation of these stars, but will only have occurred "recently" in the evolution of the galaxy. The low energy flux might show a steep rise after $\sim 10^9$ years, remaining roughly constant thereafter, so that the abundances of Li, Be and B might have been frozen in relatively recently, and might differ substantially from the primordial values.

We wish to thank Drs. B. Bopp, L. Lucy, E. Spiegel, W. Ludington and most especially E.T. Scharlemann for numerous informative discussions on many aspects of this work.

REFERENCES

Babcock, H. W. (1961), Astrophys. J., 133, 572.
Bopp, B. (1978), IAU Comm. 42 Working Group Meeting, Socorro.
Eaton, J. and Hall, D. (1979), Astrophys. J. 227, 907
Hall, D.S. (1976), in Multiply Periodic Phenomena in Variable Stars. IAU Colloq. 29, Pt. 1 (ed. W.S. Fitch) Reidel, Dordrecht.
Hall, D.S., Montle, R. and Atkins, H. (1975), Acta Astron. 25, 125.
Hall, D.S. (1972), Publ. Astr. Soc. Pac., 84, 323.
Kurucz, R. (1979), Astrophys. J. Suppl, 40, 1.
Meyer, F., Schmidt, H., Weiss, N. and Wilson, P. (1974), MNRAS. 169, 35.
Oskanyan, V., Evans, D., Lacy, C. and McMillan, R. (1977), Astrophys. J. 214, 430.

DISCUSSION FOLLOWING SHORE AND HALL

Webbink: The evolutionary state of the RS CVn stars seems now well-defined as corresponding to binaries with nearly equal mass components in which the primary is now creeping up to the onset of rapid mass transfer. In the context of a dynamo model for the many peculiarities of these systems, it is not immediately obvious why their descendant binaries, namely those of Algol type, should not also show the same phenomena. Given that the Algol systems are much more dominated by the unevolved component, what evidence can be brought to bear concerning the existence or extent of the RS CVn phenomenon among them?

Shore: Yes, the progeny of the RS CVn stars will and probably do show evidence for surface activity. Gibson has made the point that the Algol radio events bear a striking resemblence to RS CVn flares. These stars are X-ray sources (at least β Per itself). However, you would not be able to observe the surface activity optically except perhaps H and K emission, which may be due to the stream. In the high energy phenomena, of course, we remove this obscuration and see the activity aright. A good analog is probably RT Lac.

Feldman: Rather than produce weak, infrequent radio flaring as you suggest it might, SZ Psc is perhaps the most radioactive RS CVn. (It is certainly in the same class as HR 1099, UX Ari, and RT Lac in this respect.) Moreover, Bopp (Univ. of Toledo) has found that it is similarly very active in Hα; he knows of no other RS CVn system that is more active.

Shore: It is true that pressure arguments give weak magnetic fields for this star. There is another possibility here. The fields which can be built up by equilibrium of Lorentz and Coriolis forces are quite large and almost certainly exceed the local pressure in the nearly semidetached systems. Therefore bouyancy will bring flux tubes quickly to the surface even as the dynamo operates at the base of the convection zone. What causes the problem I suggested is the straightforward interpretation of the pressure condition.

Fraquelli: I agree with Feldman's comment. I have several plates of SZ Psc obtained at DDO, which show highly variable H-alpha profiles. During Jakate's observations in October 1978, I obtained several spectra. At the same phase, one orbital cycle apart, there are drastic changes in the Hα profile. Not only is the emission variable, but whatever is going on in the system can happen on short time scales as well.

Shore: This system also shows one of the largest period changes of a secular nature, and my reply to Feldman should be added here as well. It would be very interesting, considering your comments, to check systems like SS Cam to see whether it too is as haywire a binary as SZ Psc.

Walter: As Walter *et al.* have pointed out in a paper submitted to Ap. J. Lett., the coronal surface fluxes in single stars with active chromospheres and in the weaker RS CVn systems are nearly identical. However, the most active RS CVn's show an order of magnitude more coronal activity. This shows that there may be something fundamentally different with these more active systems, although the basic mechanisms appear solar in nature.

Shore: I completely agree, and it is that these stars are in binaries, where a helper exists to keep the stellar photosphere well behaved. The energy sources for producing a hot corona are enhanced by the number of stable loops existing in the active regions of these stars compared with field differential rotators. However, I also agree that physical laws are covariant from solar physics to binary star papers.

Budding: Could you point out any special significance to be attached to the correlation of activity of the cool component to its passage across the Herzsprung Gap?

Shore: It is at this stage that the stellar envelope structure builds up sufficiently active convection at great enough depth to form flux ropes of the type required. In addition, field stars must be different from these binaries, which are nearly synchronous at this stage. The stage is of importance, but so is the fact that these stars are easily found.

THE CORONAE OF RS CVn SYSTEMS

Frederick M. Walter, Philip A. Charles, and Stuart Bowyer
Space Sciences Laboratory
University of California, Berkeley 94720

1. INTRODUCTION

RS CVn systems [1] are close binaries consisting of two late-type stars, one of which has evolved to the base of the giant branch. In most cases, the systems appear to be the product of normal stellar evolutionary processes [2,3]. These systems have attracted much recent attention because they exhibit greatly exaggerated stellar activity in the form of strong chromospheric emission [1]. There is evidence that a large fraction of the surface is covered by dark spots [4]. They also exhibit radio emission [5,6], and quiescent and flaring soft X-ray emission [7-12]. If these stars are undergoing normal evolution, why then are they so peculiar?

It is possible that their abnormalities are no more than a gross exaggeration of the activity associated with late-type convective stars, such as the Sun. In these systems, the active star has been forced into synchronous rotation at the binary period, which is much shorter than that of most late-type stars. Presumably the speeding up of the rotation has some effect upon the dynamo processes that ultimately generate the stellar activity. We have studied the coronae of the brightest RS CVn systems using soft X-ray observations, and we report here on the relation between the coronal activity in RS CVn's and that in more "normal" stars.

II. EXPERIMENT AND OBSERVATIONS

These observations were undertaken using the Low Energy Detectors (LED) of the HEAO-1 A-2 Cosmic X-ray Experiment[1] [13]. The LED's are sensitive to 0.15 to 2.8 keV X-rays. The data were searched for evidence of X-ray emission from fifty-nine RS CVn's and related objects listed by Hall [1], Eggen [14], and the Circulars of the Working Group on RS CVn systems of IAU Commission 42. The data analysis and the results are fully described by Walter *et al*. [10].

M. J. Plavec, D. M. Popper and R. K. Ulrich (eds.), Close Binary Stars: Observations and Interpretation, 397–402.

The fifteen positively detected sources are listed in Table 1 along with a probable detection of RW UMa [10]. The X-ray luminosities span an order of magnitude. The upper limits derived for the X-ray luminosities of the other sources are inconclusive since all but one are greater than 10^{31} erg s^{-1} indicating that these systems could well be bright X-ray sources but are too distant for detection with this experiment. Spectra of UX Ari [7], Capella [15], and broadband spectra of the weaker sources [10] show that all have temperatures of $\sim 10^7$K, but that some show no evidence for the line emission expected from a solar composition plasma. UX Ari appears underabundant in Fe by a factor of 30 relative to solar [7].

Table 1. Luminosities (.15-2.8 keV) of RS CVn Systems.

Source	L_x(x 10^{30} erg s^{-1})	Source	L_x(x 10^{30} erg s^{-1})
λ And	1.8 ± 1	HR 1099	12 ± 1.5
UX Ari	21 ± 1.5	HR 4665	8.4 ± 2.0
α Aur	4.0 ± .2	HR 5110	3 ± .9
RS CVn	62 ± 30	HD 5303	14 ± 4
σ CrB	3.9 ± .5	HD 155555	3.2 ± .7
σ Gem	21 ± 3	HD 224085	40 ± 12
AR Lac	15 ± 3	SAO 015338	70 ± 20 (flare)
HK Lac	100 ± 40	RW UMa	435 ± 31 (flare)

Three types of variability have been observed. Flaring, similar to radio flaring, on timescales of hours to days, appears commonplace [7,10,12,16]. Capella has been observed to vary by up to a factor of 2 on scales of hours [9,10], but no such variability is observed in UX Ari, HR 1099, or σ CrB. However, the emission from UX Ari was observed to decrease by at least a factor of three at the phase of maximum light; we interpret this to be an occultation by the body of the star of a corona localized above the spotted hemisphere. This observation allows us to put an upper limit on the scale height of the emission of ~ 0.1 of the stellar radius, as well as limiting the longitudinal extent of the emission to $\sim 120°$.

III. CORONAL MODELS

We have attempted to model the emission in terms of a loop model of the solar corona [10], wherein magnetic flux tubes emanating from the stellar surface in the active regions contain the coronal gas. Starting from the empirical relation between gas temperature T, pressure p, and loop length L [17], we derive expressions for L, f, the product of the number of loops N and the average cross sectional area of a loop relative to the stellar surface area, and the number of loops N times a geometrical factor α^2 ($\alpha^2 L^3$ = loop volume) in terms of three observable quantities, T, p, and the emission measure EM.

$$L = 3 \times 10^{11} (T_7)^3 \ p^{-1} \tag{1}$$

$$f = 340 \ (EM_{54})(T_7)^{-1} \ (R_{\odot}/R_*)^2 \ p^{-1} \tag{2}$$

$$N\alpha^2 = 50 \ (EM_{54}) \ p(T_7)^{-7} \tag{3}$$

where R_* is the stellar radius, p is in dyn cm^{-2}, T_7 is in 10^7 K, and EM_{54} is in 10^{54} cm^{-3}. Values for these parameters are given in Table 2 for UX Ari, Capella, the active Sun and a hypothetical active Sun, entirely covered with plages [18].

Table 2.

	T	EM	p	f	L	$N\alpha^2$
UX Ari	10^7	4×10^{54}	10	16	3×10^{10}	1700
Capella	10^7	1.2×10^{53}	1.5	.12	2×10^{11}	9
"Plage" Sun	10^7	4×10^{52}	10	1.3	3×10^{10}	20
Active Sun	3×10^6	10^{50}	1	.11	10^{10}	20

IV. DISCUSSION

From Table 2, we find that UX Ari (and the other active RS CVn's) have f of order unity and $N \sim 10^4$. This indicates that the corona covers most of the active hemisphere, unlike in the Sun where the corona is localized above small fraction of the surface, and that there are a large number of starspots. This is in good agreement with the optical results that require a large number of spots covering a large portion of one hemisphere. Capella is less active, with many fewer loops. This is a possible explanation for the absence of variability in UX Ari; changes in the loop structure of Capella have a more noticeable effect upon the total X-ray luminosity. We note that were the number of loops on the active Sun to be increased by 10^3, then f would also increase and the parameters would be comparable to those of UX Ari. Indeed, were the surface of the Sun entirely covered with active regions, and were it a subgiant, its X-ray luminosity would be $\sim 2 \times 10^{30}$ erg s^{-1} [18], comparable to that observed in the less active RS CVn's

We conclude that the coronae of RS CVn's and that of the Sun may well differ only in scale. If so, the mechanisms at work in RS CVn's are similar to that in the Sun, except that the rate of magnetic flux tube generation has greatly increased, perhaps due to forced rapid rotation. Results of a study of the coronae of stars with active chromospheres [19] have shown that these stars are some 100 times more luminous than the Sun in soft X-rays. When their more rapid rotation and greater surface area are taken into account, RS CVn's may not be particularly unusual, but just an extreme case of normal stellar activity.

ACKNOWLEDGMENTS

It is a pleasure to acknowledge useful discussions with other observers of RS CVn systems and with J. Linsky for stimulating discussions on coronal physics.

[1]The HEAO-1 A-2 experiment is a collaborative effort led by E. Boldt (GSFC) and G. Garmire (CIT) with collaborators at UCB,CIT,JPL, and GSFC.

REFERENCES

1. Hall, D.S., in "Multiple Periodic Variable Stars", Fitch, W.S., ed., (Reidel, Dordrecht), p. 287.
2. Morgan, J., and Eggleton, P.: 1979, M.N.R.A.S., 87, p. 661.
3. Popper, D., and Ulrich, R.: 1977, Ap. J. (Lett.), 212, p. L131.
4. Eaton, J.A., and Hall, D.S.: 1979, Ap. J., 227, p. 907.
5. Owen, F.N., and Gibson, D.M.: 1978, A.J., 83, p. 1488.
6. Feldman, P.A., Taylor, A.R., Gregory, P.C., Seaquist, E.R., Balonek, T.J., and Cohen, N.L.: 1978, A.J., 83, p. 1471
7. Walter, F., Charles, P., and Bowyer, S.: 1978, Ap. J. (Lett.), 225, p. L119.
8. Walter, F., Charles, P., and Bowyer, S.: 1978, A.J., 83, p. 1539.
9. ________________ 1978, B.A.A.S., 10, p.632.
10. Walter, F., Cash, W., Charles, P., and Bowyer, S.: 1979, submitted to Ap. J.
11. Charles, P., Walter, F., and Cash, W.: 1979, from "(COSPAR) X-ray Astronomy", Baity & Peterson, eds., (Pergamon, New York), p. 129.
12. Charles, P., Walter, F., and Bowyer, S.: 1979, submitted to Nature.
13. Rothschild, R. et al.: 1979, Sp. Sci. Instr., in press.
14. Eggen, O.: 1978, I.B.V.S., 1426.
15. Cash, W., Bowyer, S., Charles, P., Lampton, M., Garmire, G., and Riegler, G.: 1978, Ap. J. (Lett.), 223, p. L21.
16. White, N.E., Sanford, P.W., and Weiler, E.J.: 1978, Nature, 274, p. 561.
17. Rosner, R., Tucker, W.H., and Vaiana, G.S.: 1978, Ap. J., 220, p.643.
18. Vaiana, G.S., and Rosner, R.: 1978, ARAA, 16, p. 393.
19. Walter, F., Linsky, J., Bowyer, S., and Garmire, G.: 1979, submitted to Ap. J. (Lett.).

DISCUSSION FOLLOWING WALTER, CHARLES AND BOWYER

Eaton: I have two points about your interpretation of the X-ray observations of UX Ari that I wish you would clarify. First, you have interpreted the sudden appearance of X-ray emission on 15 February 1978 as the emergence of an active region from behind the limb. This fixed the height of the emitting loop structures at $\sim 0.1\ R_*$. How well does this interpretation agree with the optical light curve of UX Ari, i.e. is the light variation, presumably caused by spots, consistent with having no loops on the hemisphere seen on 14 Feburary 1978? Second, you presented observations obtained only ½ year earlier extending over nearly a complete rotational period of the binary which did not show significant variation in X-ray flux. Again, if you assume the spots giving rise to the optical light variation correspond to the feet of the coronal loops, is the lack of variation consistent with the optical light variation? What limits do you place on the extent of the corona in longitude at these two epochs?

Walter: First question: To the limits of our data we can't say there are no loops, but the number must be down by at least a factor of four. Due to the number of loops and the expected stability of the loops, it is extremely difficult to make enough disappear (and reappear) to explain the observation unless we postulate that they were indeed occulted by the body of the star. Since it happened during maximum light, when such an effect might be expected, it is very suggestive.

Second question: The earlier observation was centered near minimum light, and we scanned for five days, thus missing the phase of maximum light. Since the loops are optically thin, and since the corona is limb brightened, a corona covering 180° of longitude would be observable for nearly the entire period. Unfortunately we can't put good limits on the extent of the corona during the earlier observation because the aspect correction to the collimator response is large when the observed corona should be disappearing. The extent derived of $\sim 120^{\circ}$ of longitude covered by the bulk of the corona during the later observation is not inconsistent with earlier observations.

van 't Veer: The geometrical extension of the corona is limited by the Roche configuration of the system. Does this effect have to be taken into account for RS CVn stars?

Walter: I don't think so. For the most part the stars do not fill their Roche lobes and the derived coronal scale heights put the loops well inside the lobe. In the systems that may fill their Roche lobes, like RT Lac, this probably won't change things much because the magnetic flux tubes should be sufficiently rigid to keep the corona intact.

Feldman: Are there any bona fide RS CVn systems that definitely do not show X-ray emission, especially after observation by HEAO-2?

Walter: NO. Steve Shore has predicted that about 10% of them shouldn't be there, so it is getting interesting. The HEAO 1 A-2 upper limits are mainly greater than 10^{31}ergs^{-1} due to the large distances to these sources. Our first HEAO-2 IPC observations have shown nine new sources out of nine observed. This makes 24 definite detections, with one more possible. It has been suggested that Z Her is likely to be less active, but we haven't observed this yet.

Meyer: Since the thermal scale height of the 10^7K corona is more like a stellar radius in your case, the smaller extent of the corona must be due to the magnetic confinement in flux tubes which typically reach only out to 1/10 of the radius. This probably indicates that the characteristic separation of the two footprints of such flux tubes at the stellar surface is also only of this order of magnitude, which should then be the distance over which magnetic polarity at the surface changes.

Walter: In the case of UX Ari, we have calculated a loop length of 3 x 10^{10}cm, which is about one seventh the radius of the subgiant. Since the loop must be anchored to the surface, the footprints must indeed be separated by about .1 of the stellar radius and the height of the corona must also be of this order, as we have measured in UX Ari.

RADIO FLARE ACTIVITY OF RS CVn STARS

Paul A. Feldman
Herzberg Institute of Astrophysics
National Research Council of Canada, Ottawa

ABSTRACT

For several years, a systematic program to observe RS CVn and similar binaries has been undertaken with the 46-m telescope of the Algonquin Radio Observatory. A number of large radio outbursts at 2.8 cm wavelength has been found from HR 1099, AR Lac, SZ Psc, UX Ari, and HR 5110. In several cases, simultaneous, or nearly simultaneous, observations of these stars were made by cooperating observers at (other) radio, optical, UV, and X-ray wavelengths. It is now clearly established that the mechanism responsible for the cm-wavelength radio emission is nonthermal gyrosynchrotron radiation in a volume whose characteristic dimension is comparable with the binary star separation. More generally, a semi-quantitative model for the radio flare activity of these stars seems to be possible using greatly scaled-up analogues of solar magnetic activity.

DISCUSSION FOLLOWING FELDMAN

Bolton: Your comment regarding the problems with coordinated campaigns strikes a nerve. The trouble with radio and X-ray astronomers (or the people who schedule their observing facilities) is that they seem to consider a negative observation a waste of time. However, if optical observers are to interpret their data during flares, they must know what the system looks like in the quiescent state. Without simultaneous radio and/or X-ray observations to identify the quiescent states, they must be inferred in some way from the optical data. This procedure greatly increases the risk that our final conclusions will be determined more by the model assumed in the beginning rather than by the observations.

Feldman: Bolton is responding too strongly to my black (or gray) sense of humor regarding some of the recent coordinated campaigns. It is obvious that quiescent-state observations are necessary before any attempt can be made to interpret flaring systems.

M. J. Plavec, D. M. Popper and R. K. Ulrich (eds.), Close Binary Stars: Observations and Interpretation, 403.

OUT-OF-ECLIPSE LIGHT VARIATION OF RS CVn STARS

S. Catalano+, A. Frisina+ and M. Rodonò++
+Istituto di Astronomia, Università di Catania
++Osservatoria Astrofisico, 95125 Catania, Italia

ABSTRACT

The photoelectric observations of the close binary RS CVn carried out at Catania Observatory from 1963 to 1973 are presented. A new period of 9.48 years for the retrograde migration of the wave-like distortion on the light-curve has been determined. The wave amplitude is found to undergo cyclic changes with a period of about 4.74 years.

Possible correlation between the wave amplitude and the inclination of the rotational axis and the spectral type of the spotted component for several RS CVn and BY Dra binaries is investigated.

1. INTRODUCTION

One of the outstanding features of the binary system RS CVn is the retrograde migration on the light-curve of a wave-like distortion which was discovered in the very first years of observations carried out at Catania Observatory (Catalano and Rodonò 1967). Owing to this peculiar out-of-eclipse variability, RS CVn binaries need to be observed carefully for decades in order to ascertain and interpret correctly their physical characteristics.

In the present paper we present the results of Catania observations covering a complete cycle of variability of RS CVn and we analyze the main light-curve features, also for other RS CVn and BY Dra stars.

2. THE LIGHT CURVE OF RS CVn AND ITS VARIATIONS

Eleven out-of-eclipse V-band light-curves of RS CVn obtained at Catania Observatory from 1963 to 1973 are shown in Figure 1. The 1963 light-curve shape is repeated between 1972 and 1973. The retrograde migration of the wave-like distortion is better represented by the decreasing phase of the wave-light minimum as shown in Figure 2. A

M. J. Plavec, D. M. Popper and R. K. Ulrich (eds.), Close Binary Stars: Observations and Interpretation, 405–412.

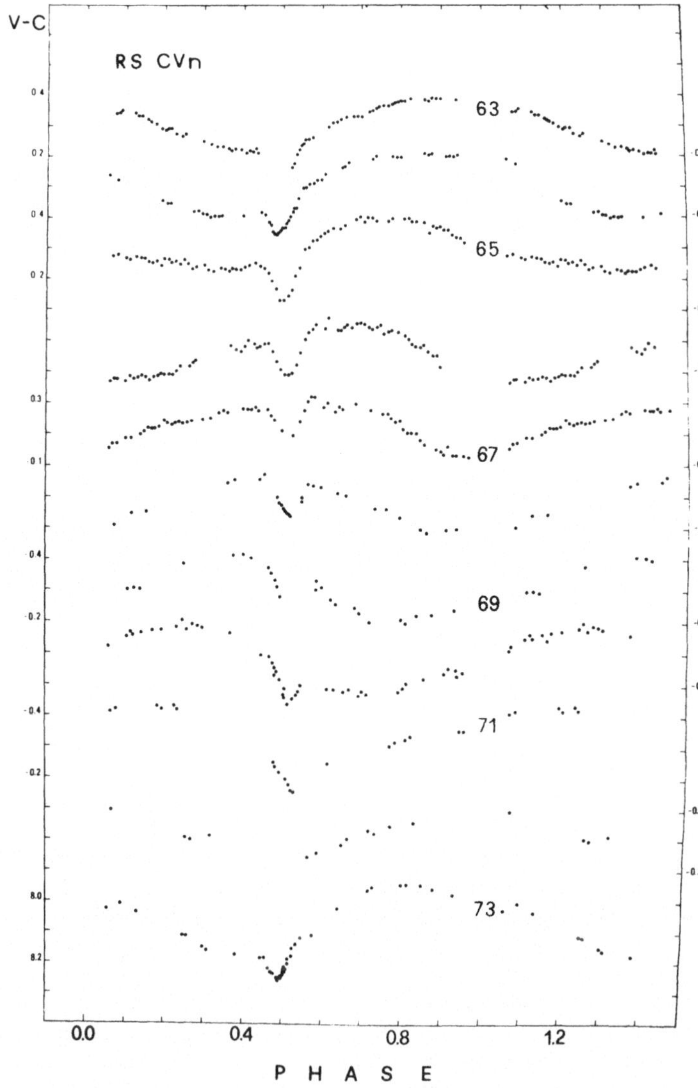

Figure 1. Out-of-eclipse light-curve of RS CVn from 1963 to 1973.

least square fit to the data gives an almost linear migration of the distortion wave with a period P_W = 9.48 years or 722 orbital periods. This indicates that the period of variability of the secondary component is $P_{orb} \times (1-1/722)$.

As before (Catalano and Rodonò 1969), each out-of-eclipse

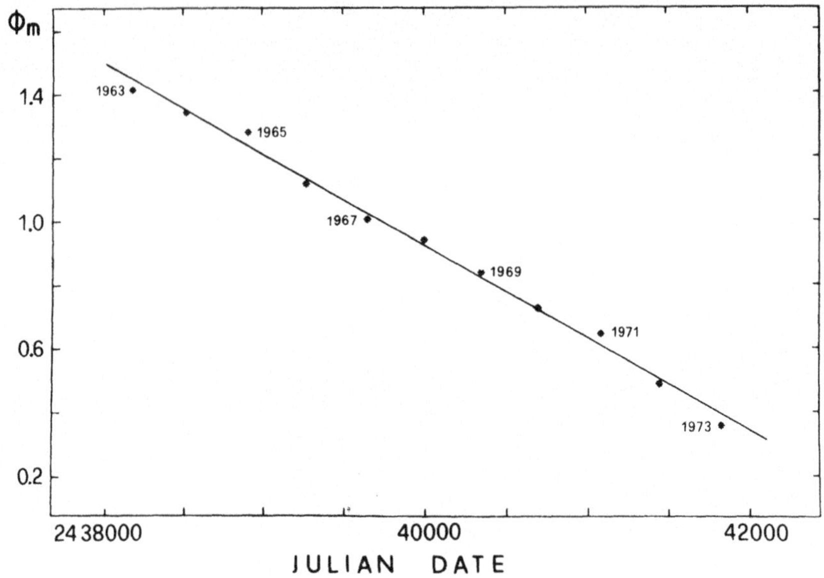

Figure 2. Phase of RS CVn wave distortion minimum versus Julian date. The continuous line represents a linear migration of the wave with a period P_W of 9.48 years or 722 orbital period.

light curve has been represented by a cosine curve in a least square fit to the data in order to determine the phase of the wave minimum (ϕ_{min}), its amplitude (Δ_{min}), and the mean light level (Δm_o). However, since in some light curves the distortion wave is not truly sinusoidal, the adopted least square fit gives a wave amplitude which is underestimated relative to the amplitude obtained as the difference between the maximum and the minimum of the out-of-eclipse light. The resulting data for each year and colour are listed in Table 1. The amplitude of the wave distortion fluctuates almost periodically in about 4.74 years around the mean value of $0^m.088$ with a half amplitude $0^m.013$

A previous analysis of the wave amplitude variation, including old visual and photographic observations (Hall 1972) has led to a considerably larger amplitude variation, from about 0.05 to 0.20, and a period of 23.5 years, much longer than the present one. Additional Catania observations and the 1975 Ludington light-curve shown in Zeilik et al. (1979) are consistent with our present result of a shorter period and a smaller amplitude variation.

3. VARIABILITY OF THE SECONDARY STAR

As already recognized (Catalano and Rodonò 1969) the out-of-eclipse light variation is due to the secondary star. Since the primary minimum

TABLE 1

Year	Δm_o			Δm_w			ϕ_{min}
	V	B	U	V	B	U	
1963	-0.304	-	-	0.089	-	-	0.413
1964	-0.309	-	-	0.101	-	-	0.339
1965	-0.310	-	-	0.081	-	-	0.278
1966	-0.257	-	-	0.085	-	-	0.118
1967	-0.236	-0.165	-	0.072	0.061	-	0.994
1968	-0.268	-0.171	-0.083	0.084	0.062	0.053	0.943
1969	-0.310	-0.199	-0.096	0.097	0.097	0.062	0.829
1970	-0.278	-0.171	-0.080	0.103	0.082	0.064	0.724
1971	-0.343	-0.222	-0.110	0.074	0.063	0.049	0.645
1972	-0.324	-0.210	-0.096	0.075	0.062	0.040	0.480
1973	-0.371	-0.234	-0.116	0.104	0.083	0.057	0.370

is total, by observing in different seasons, we get magnitude determinations of the secondary star at different phases of its variability. The available V, B-V and U-B at the totality in the standard system from 1967 to 1973 are presented in Figure 3 versus Julian date. The V magnitude covers the period of the intrinsic light variation of the secondary star from phase 1.00 to 0.36. The observed 0.6 magnitude increase of the secondary star is consistent with the observed average out-of-eclipse wave amplitude of the combined light of the entire system.

No large colour variations are observed. The B-V values show a little scatter around the value 0.93. The U-B values give some indication that the secondary star is redder by 0.07 magnitude at minimum relative to the maximum light. This reddening at minimum is not strong enough to be detectable in the combined light of the system. In fact it appears to be even bluer at the minimum of the distortion wave (Catalano and Rodonò 1969) due to the relatively low contribution of the redder star to the combined light of the system at this phase.

We have attempted to find solutions of the best covered light-curves following the Russell-Merrill method (1952) after removing the out-of-eclipse distortion wave. The radius of the primary star turns out to be constant while that of the secondary star shows a variation of about 20%, phase locked with the position of the distortion wave on the light-curve. However, to detect possible residual effects of the distortion-wave on the photometric eclipse, we have treated separately the descending and the ascending branches of the primary minimum, and an average of both. The three solutions show the largest disagreement when the minimum of the distortion wave falls at one of the quadratures, i.e. when the minimum suffers the maximum asymmetry. This would indicate that a residual effect of the distortion wave is still present on the rectified curves. Hence, the variation of the secondary star radius is likely to be a spurious photometric effect.

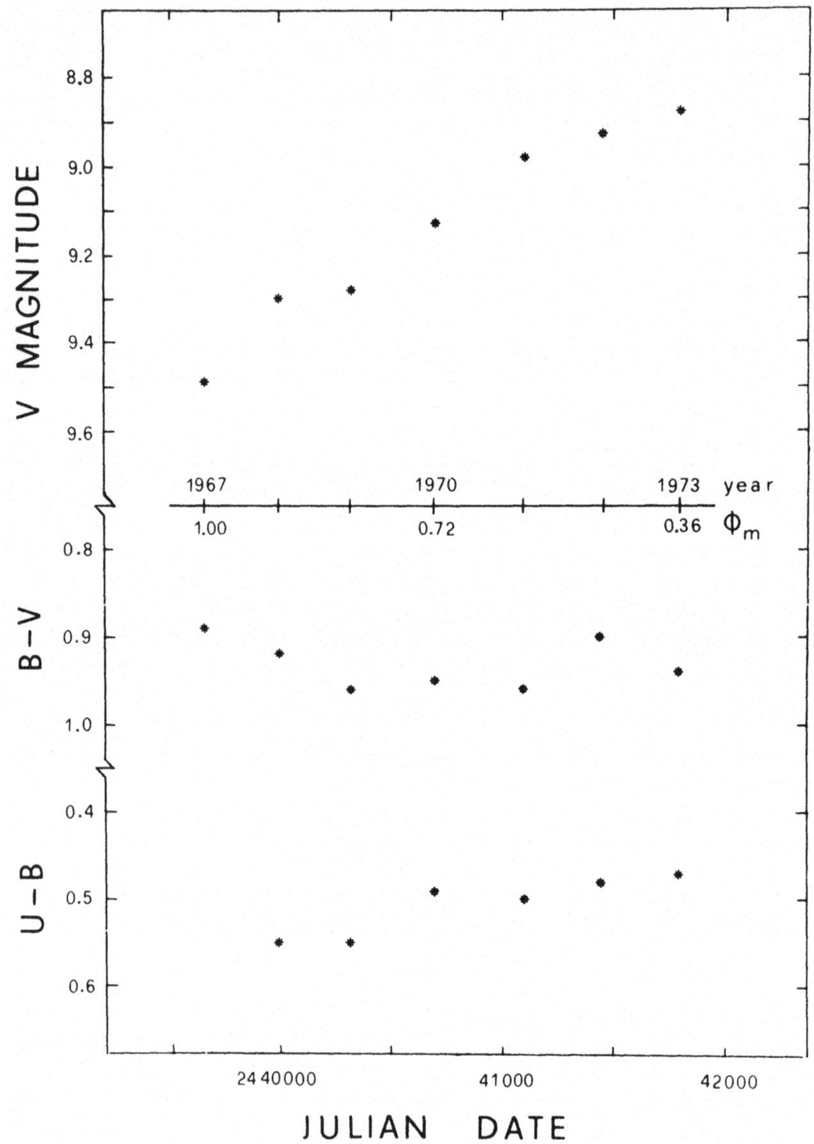

Figure 3. UBV magnitudes and colour indices of the secondary component determined at totality. Orbital phases ϕ_{min} of the wave distortion minimum and reference years are indicated in the middle scale.

Light-curve solutions of another RS CVn star (AR Lac) carried out by Theokas (1977) using Kopal's method also lead to variations of the fractional radii whose reality the author considered doubtful. Our Fourier analysis of the eclipses of RS CVn following Kopal's method have not led to meaningful solutions.

4. WAVE AMPLITUDES IN RS CVn AND BY DRA STARS

Slow periodic light variations of the stars are interpreted as uneven distribution of surface brightness, or starspots, in a rotating star which is not seen pole-on. This interpretation, reviving the early suggestion by Kron (1950), has been successfully applied to reproduce the observed out-of-eclipse light variations of RS CVn (Hall 1972) and BY Dra stars (see Kunkel 1975). However it is difficult to obtain unique solutions for the latitude, relative temperature, extent of the spotted area, and inclination of the rotational axis from the observed light curves. We have explored the possibility of the wave amplitude being statistically correlated with the angle between the rotation and the line of sight. Since this angle is not generally known, the orbital inclination readily gives the required angle by assuming the star rotation axis to be perpendicular to the orbital plane. The observed average wave amplitudes have been corrected by the light contribution of the unspotted companion for the purpose of extracting the actual light variation of the spotted star. No correlation between the average wave amplitude $(\Delta m_W)_c$, corrected by the light contribution of the unperturbed star, and the orbital inclination is apparent. All $(\Delta m_W)_c$ amplitudes are randomly distributed around the mean value of 0.18 (.09 s.d.) but for RS CVn which has the exceptionally large average wave amplitude of 0.82 magnitude. The lack of correlation between $(\Delta m_W)_c$ and $\underline{i}$ results from a random distribution in latitude and/or a wide range of the admissible fractional extent of the spotted region for the stars in our sample. This result has to be compared with that almost invariably found for both RS CVn and BY Dra stars, namely that the spotted regions are concentrated around 30-40 degrees latitude (Eaton and Hall 1979, Chugainon 1973, Torres and Ferraz Mello 1973, Bopp and Evans 1973, Vogt 1975).

Bearing in mind that numerous parameters affect the actually observed periodic brightness variation, we could expect different upper limits of the wave amplitude $(\Delta m_W)_c$ for different spectral types. In Figure 4 the average wave amplitude $(\Delta m_W)_c$ of RS CVn and BY Dra stars is plotted versus the spectral type of the spotted star. Although not well defined, a maximum is apparent at spectral type KO.

Chromospheric activity inferred from the absolute flux of the Ca II K emission line shows also a relative maximum at spectral type KO (Blanco et al. 1974).

These two concurrent results might indicate, as in the solar analogue, a close connection of the photospheric and chromospheric activity in late type stars.

According to Mullan (1974) starspots are convection cells penetrating throughout the entire depth of the convection zone and their size is expected to grow as the convection zone depth does, i.e. the later the spectral type is. However, as the convection zone thickens,

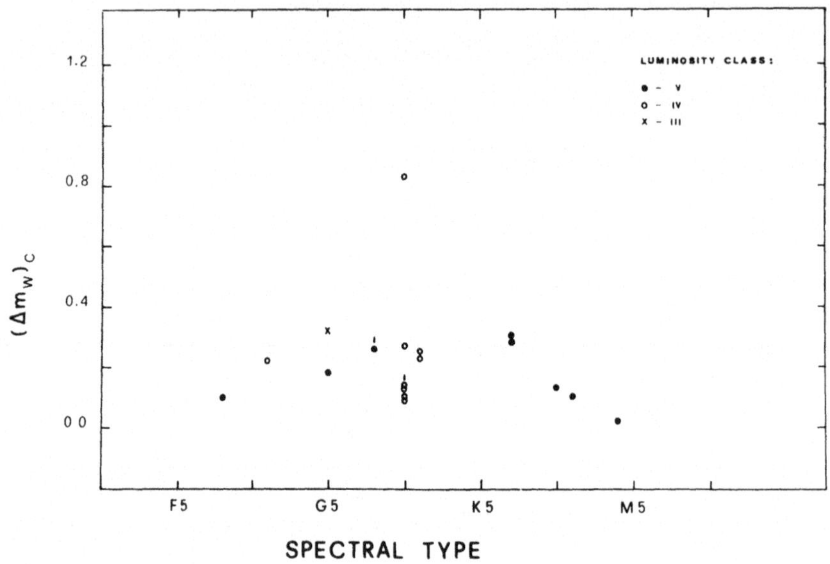

Figure 4. Average wave amplitudes, corrected for the light contribution of the unperturbed component, versus the spectral type of the variable component.

a low order multipole of the magnetic fields generated by dynamo action becomes more favoured. In stars of the latest spectral type, dipole fields would be more prominent. If it is aligned with the rotation axis, the largest field strength, and also the spots, should be confined to the rotation poles so that no periodic rotational modulation of the star brightness should result. Therefore, periodic brightness variations due to spots would be expected for stars with spectral types from F5-G0 (Mullan 1972), where the transition from shallow to deep convection zone occurs, to earlier than M5.5, i.e. before the star becomes completely convective (Mullan 1976). This picture is consistent with our result shown in Figure 4: the maximum periodic variations occur around spectral type K0.

Even if the various hypotheses postulated by Mullan (1976) are not entirely met, no rotational modulation of the star's brightness should result in late type stars because the spotted areas become so extended that they cover almost entirely the stellar photosphere.

REFERENCES

Blanco, C., Catalano, S., Marilli, E., Rodonò, M. 1974, Astron. Astrophys. 33, 257

Bopp, B.W., Evans, D.S. 1973, Monthly Notices Roy. Astron. Soc. 164, 343.

Catalano, S., Rodonò, M. 1967, Mem. Soc. Astron. Ital. 38, 345.
Catalano, S., Rodonò, M. 1969, in Non-Periodic Phenomena in Variable Stars, Ed. L. Detre, Academic Press, Budapest, pag. 345.
Chugainov, P.F. 1973, Izv. Krymskoj Astrofiz. Obs. 48, 3
Eaton, J.A., Hall, D.S. 1979, Astrophys. J. 227, 907.
Kron, G.E. 1950, Astron. J. 55, 69.
Kunkel, W.E. 1975, IAU Symposium 67, 15.
Mullan, D.J. 1972, Astrophys. Letters 12, 13.
Mullan, D.J. 1974, Astrophys. J. 192, 149.
Mullan, D.J. 1976, Irish Astron. J. 12, 161
Russell, H.N., Merrill, J.E. 1952, Princeton Contr. 26
Theokas, A.C. 1977, Astrophys. Space Sci. 52, 213.
Torres, C.A.P., Ferraz Mello, S. 1973, Astron. Astrophys. 27, 331.
Vogt, S.S. 1975, Astrophys. J. 199, 418.
Zeilik, M., Hall, D.S., Feldman, P.A., Walter, F. 1979, Sky Telesc. 57, 132.

DISCUSSION FOLLOWING CATALANO, FRISINA AND RODONO

Leung: Have you looked at the "wave" phenomena in the system of SV Cam?

Rodono: No, we haven't.

Oswalt: Have RS CVn or any similar systems on your observing program exhibited a constant light level during any phase interval between wave maximum and wave minimum?

Naftilan: On your graph of the wave amplitudes vs. spectral type, you had one point which was much larger than all the rest. Which system is this?

Rodono: It is RS CVn. The wave amplitude we actually observe on its light curve is about 0.20 magn. However you must take into account the light contribution of the primary star in order to get the intrinsic light variation of the secondary one, i.e. the quantity plotted in the figure vs. spectral type.

SPECTROPHOTOMETRY OF THE Hα REGION IN HR 5110.

Dorothy A. Fraquelli
David Dunlap Observatory
University of Toronto

Spectrophotometry of the bright RS CVn binary HR 5110 is presented. The data consist of 42 red spectrograms obtained with the DDO 74" telescope. All spectrograms were scanned using the PDS microdensitometer system at the DDO and rectified intensity tracings were produced.

Visual inspection of the plates reveals the occasional presence of Hα emission. The measures of the Hα absorption and emission equivalent width scatter between 4.9 and 2.4 Å. (For a standard Hα absorption line profile, the EW is 6.8Å). The EW is not correlated with orbital phase. Subtraction of a standard Hα absorption line profile (T_e = 7500 K, log g = 4.0, Kurucz, 1979) reveals an asymmetric emission profile. The emission profile exhibits a blue asymmetry over 270° of the orbit and a red asymmetry over 90°. This suggests that there is a general outflow of material from the system, which is blocked during the phase interval when the red asymmetry occurs.

Assuming that there is one source of emission in the system, constraints may be placed on its location. I find that the emission source lies no more than 1.03 $R_{sec.}$ from the center of mass of the secondary star. It is not located on the line of centers of the system. An obvious interpretation is that this emission source is an active region associated with starspots. A stream interpretation seems to be ruled out in a simple one source model.

Reference

Kurucz, R.: 1979, *Astrophys. J. Suppl.* 40, pp. 1-340.

M. J. Plavec, D. M. Popper and R. K. Ulrich (eds.), Close Binary Stars: Observations and Interpretation, 413–414.

DISCUSSION FOLLOWING FRAQUELLI

Feldman: Perhaps the "line source" of the Hα emission you describe can be accounted for as a rising Hα prominence, similar to what is found in the early stages of giant (Type IV) solar flares (reference: A.C. Riddle, Solar phys. 13, 448, 1970).

Fraquelli: I happen to have here the equivalent width data plotted against phase. You will notice that the radio flare data show no difference from the data obtained when I have no radio information. Either HR 5110 is active all the time, in which case you should have seen it frequently, or the radio flare has no effect on the Hα. Also, I have not shown any flare emission profiles. However there is no difference between flare and non-flare profiles. At the most, there might be a few percent difference, but I would have to wait until after the subtraction is redone before saying anything further.

Gregory: You showed both blue and red emission features with the red feature appearing as the blue feature disappeared. You argue that the blue feature is blocked but could you explain what gives rise to the red feature?

Fraquelli: The red feature is due to the material flowing away from the observer. You don't see this at phases 0.0 → .72 because the material is optically thick. If the secondary star was not blocking the emission source, you wouldn't see a red feature in the interval 0.72 → 0.0, but a blue one.

Zuiderwijk: I always feel a little uncomfortable about this type of extremely model-dependent analysis. Therefore, I would like to know if you are sure that you used the right radial velocity of the star to shift the theoretical spectra The shifting components you showed may easily be an artifact of a slightly wrong velocity correction.

Fraquelli: Yes, I am sure because the first time I did it, I did it wrong. The period is accurate in the seventh decimal place. In the ∿12 years since this period determination, the error in the period is not sufficient to produce the asymmetry shifts that I see. I appreciate your concern, as slight errors in the adopted velocity can significantly affect the central regions of the derived emission line profile.

THE STAR SPOT MODEL AND THE OBSERVATIONS OF SZ PISCIUM

Shyam M. Jakate
Astrophysical Institute
Free University of Brussel

Details of photometric observations during the totalities of the primary and secondary minima of SZ Psc are analysed in the light of the "star spot model" for RS CVn type binaries. Presence of scatter during the totality of primary minimum and its absence during the totality of secondary minimum of SZ Psc provide evidence towards one of the assumptions of the model,i.e. the spot activity confines to one of the hemispheres of the cooler star.

1. INTRODUCTION

The eclipsing binary system SZ Piscium is a member of RS CVn stars. Extensive photometric observations and radial velocities of SZ Psc are published by Jakate et al(1976, Paper I). SZ Psc was observed photomet--rically by the author during October and November 1978 season from the David Dunlap Observatory using its 61cm telescope. The present paper discusses the photometric observations of minima of SZ Psc in the light of the proposed "star spot model" for the RS CVn stars(Hall 1972).

2. DISCUSSION

Following are two basic assumptions, among others, of the "star spot model" proposed by Hall(1972) to explain various characteristics of RS CVn:

i) The surface brightness of the cooler star is non-uniform due the presence of star spots on it. ii) Furthermore, this star spot activity confines to only one of the hemispheres of the cooler star.

If the first assumption is right then one would expect some scatter during the totality of the secondary minimum, reflecting the non-uniform surface brightness as the hotter star scans the surface of the cooler star. Figure 1 shows the details of the secondary minimum of SZ Psc obtained in the year 1978. Outside as well as during the totality the

M. J. Plavec, D. M. Popper and R. K. Ulrich (eds.), Close Binary Stars: Observations and Interpretation, 415–418.

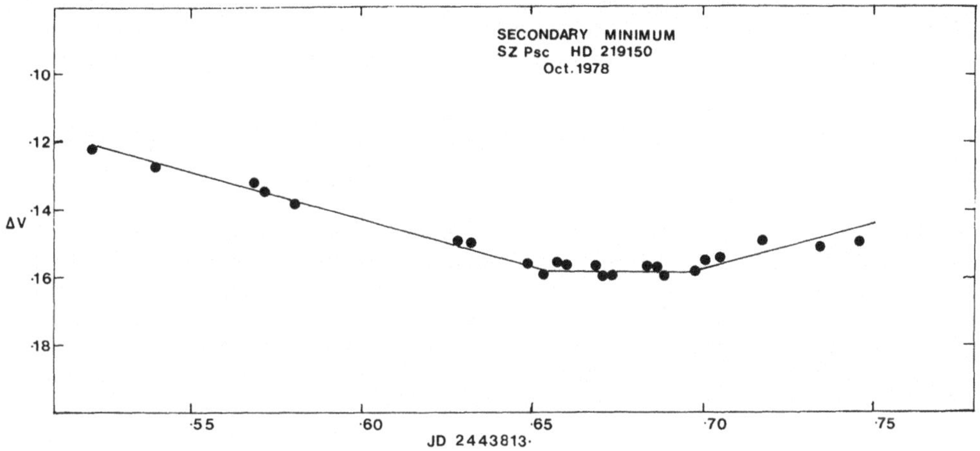

Figure 1: 1978 (JD 2443813.) secondary minimum of SZ Psc.

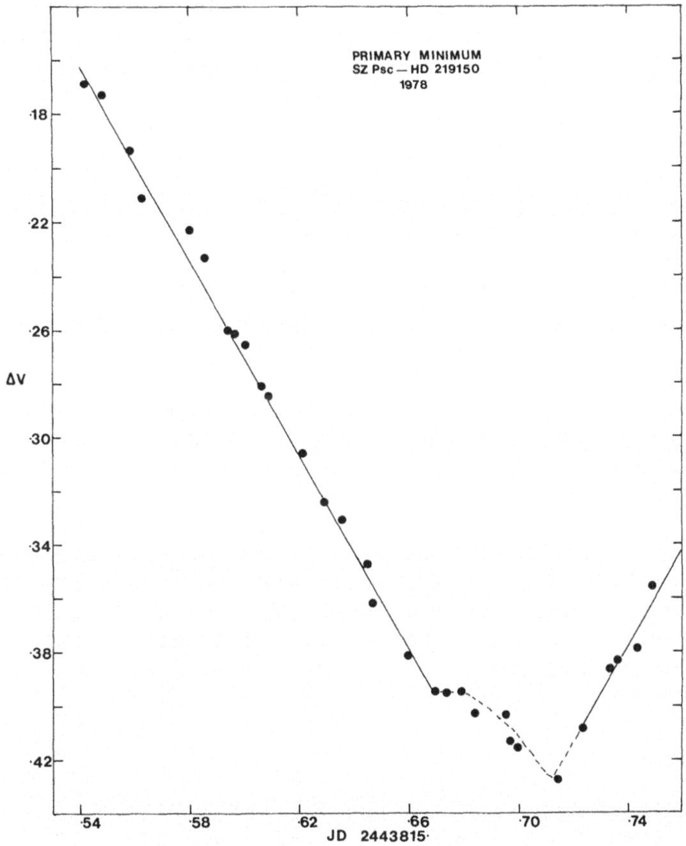

Figure 2: 1978 (JD 2443815.) primary minimum of SZ Psc.

scatter is not more than expected from the observational uncertainty (.003 mag.). The reason for the absence of the scatter could be that the star spot activity is not confined to this hemisphere of the cooler star facing the hotter star.

Figure 2 shows details of the primary minimum obtained two days later (JD 2443815). This time the scatter during the totality is significantly large (0.01 mag.). The scatter outside this totality is within the observational uncertainty. During the totality of the primary minimum of SZ Psc, we are looking at the opposite hemisphere of the cooler star. This could, perhaps, be the first direct evidence of the presence of non-uniform surface brightness preferentially on one of the hemisph--eres (in this case it is the opposite hemisphere) of the cooler star.

Durations of the totalities of the primary and the secondary minima of 1978 are equal, as expected. The ratios of the depths are in agreement with the spectral types and colours of these two stars.

Furthermore, the scatter seen during the totality of primary minimum has a definite pattern which is as suggested by the dotted line in Figure 2. This pattern seems to suggest that the crowding of the spots have taken place on the left side of the opposite hemisphere in 1978.

The primary minimum of SZ Psc observed in 1957 (Paper I), also shows scatter during the totality. The scatter has an equal amplitude but exactly opposite shape compared to what has been found in the case of the 1978 primary minimum. This suggests, that the situation of spots in 1957 on the opposite hemisphere of the cooler star was exactly opposite of 1978. Therefore, if this is a periodic phenomenon then this duration of 21 years(1978-1957) represents roughly, either a half cycle (P=42 years) or one and half cycle(P=14 years) or more number of cycles (P<5 years).

The scatter during the totality of the primary minimum of SZ Psc obtained in 1974 (Paper I) is small and represents a situation, some what intermediate compared to the extreme cases of 1957 and 1978, supporting 14 years period.

The picture would have been a lot more clear if we had complete light curves of SZ Psc for all these years. Nevertheless, the data presented here provides direct evidence for the major assumption of the "star spot model", i.e. the spot activity confines to one of the hemi--spheres of the cooler star.

REFERENCES

Hall, D.S.: 1972, Pub.Astron.Soc.Pacific, 84,p.323
Jakate,S.M., Bakos,G.A., Fernie,J.D., and Heard,J.F.: 1976, Astron.J., 81,p.250 (Paper I).

DISCUSSION FOLLOWING JAKATE

Geyer: Do you have an investigation of the behavior of the color index?

Jakate: For the 1978 observing program the aim was limited, just to obtain times of minimum. However, in our previous paper we did publish color curves.

Rajamohan: Isn't it articifial to assume starspots migrate periodically only on one hemisphere of the star?

Jakate: I did not say that star spot migration takes place only on one of the hemispheres. The assumption of the "star spot model" is that at a given time, the spot activity is confined to only one of the hemispheres.

Chambliss: Did you observe any variation in the depth of the primary minimum in SZ Piscium, and what is the duration of the total phase of the primary, i.e. how much time is there between second and third contacts?

Jakate: I cannot make definite comments on the depths of primary minima for the following reasons: 1) I do not have observations outside eclipses for 1978; 2) the comparison star used in 1978 is different. My observations suggest that the comparison star used during the 1957 and 1974 observations is probably a slow variable (IVBS 1578). The duration between second and third contacts is about 5½ hours.

van't Veer: What is the mass ratio of SZ Psc?

Jakate: About 1 as for most RS CVn stars.

Popper: The masses of the components of SZ Psc are 1.3 solar masses for the smaller, F star, and 1.6 solar masses for the cooler subgiant. They differ in mass more than in most typical systems of the group.

REMARKS ON THE COMPLICATED CLOSE BINARY SYSTEMS RT LACERTAE, CG CYGNI, AND RW CORONAE BOREALIS

E.F. Milone and S.A. Naftilan
Rothney Astrophysical Observatory,
The University of Calgary.

The three systems differ in period, scale and degree of contact: RW Com is a contact system, RT Lac is semi-detached and CG Cyg is a detached system. Yet they share the properties of a variable asymmetric light curve in the UBV, Ca II H and K emission, apparent infrared excesses, and period variability. If it may be argued that this ensemble of symptoms has similar origins in the different systems, then our intensive studies of these three binaries may have implications for the nature of all systems showing the RS CVn properties. We feel that the evidence points to the presence of circumstellar matter in the three binaries. We pose the questions: does this material create the photometric pecularities, is it incidental to them or does it arise from some other light-curve perturbing source?

Salient properties of the systems are reproduced in Table I and the evidence for mass loss or exchange in Table II. The systems are discussed individually below.

RT Lac has been included in the list of RS CVn systems or rather a list of close binaries with H & K emission by Popper (1976). A preliminary solution has been published by one of us (Milone, 1977) and a spot model for it published by Eaton and Hall (1979). RT Lac has a regressing sinusoidal distortion in its light curve. The amplitude of this distortion is greater at shorter wavelengths; for example, in the 1965 light curve the difference between maxima was 0.12, 0.17, and 0.24 mag. in V, B and U respectively. This and other circumstances indicate that the more massive star, seen at primary minimum, and having the later spectra type, *viz*, K1(IV), is more responsible for the distortion than is the other component. The most interesting property of the system is the apparent paradox raised by the blue primary minimum and the red secondary minimum; this datum alone suggests the presence of an envelope around the more massive (what we call the primary) component. Other arguments for circumstellar material in this system are as follows: 2) Our preliminary solution indicated that the secondary star is larger than the star it eclipses at secondary minimum but smaller than that which eclipses it at primary minimum; 3) The

M. J. Plavec, D. M. Popper and R. K. Ulrich (eds.), Close Binary Stars: Observations and Interpretation, 419–422.

secondary star is within 10% of filling its Roche lobe (Milone, 1976), providing the enabling condition for mass transfer; 4)Hall and Haslag (1977) have presented evidence of period changes and also 5)noted a change in light curve shape from an Algol-type light curve to a β-Lyrae type in the second decade of this century and again towards the end of the third; and 6)in 1972 and 1974 the system had both phase-independent and phase-dependent infrared excesses at JHK and L (Milone, 1976). A reasonable explanation for the phase-independent component is an anomalous interstellar reddening, brought about by material lost from the system. A simple spot model appears inadequate to explain these phenomena, although the presence of spots cannot be excluded.

TABLE I - Properties of the Systems

System	P	Sp	M_g/M_s	L_g/L_s	r_g, r_s
RT Lac	$5^d.07$	G9IV,K1IV	0.40	0.59,0.41	0.28,0.26
CG Cyg	0.63	G9V,K3V	1.0	0.46,0.54	0.28,0.22
RW Com	0.24	$\lesssim$G5V		[0.65,0.35]:	

The photometric observations of the dwarf system CG Cyg are discussed in detail by Milone et.al., (1979a); therefore the properties are only briefly summarized here. CG Cyg has increased in brightness by about 14% since 1965. Between 1965 and 1967 it suffered an abrupt change in period amounting to -2.5 (±0.8) dex (-6) day. Since then, the O-C curve has been following a parabola implying that $\dot{P} \sim 3.0$ (±0.6) dex (-10) day/day. In addition to variable H and K emission, CG Cyg has a migrating light curve distortion which is somewhat blue, and is <u>advancing</u> at an accelerating rate. This acceleration may be tied to the period increase and we are investigating this possibility. A study of the details of the light curve at every epoch, including a fine light curve of Jassur (1978) suggests that substantial variation may occur over periods of weeks or less. In particular, the regions in the maxima near external contact seem especially susceptible. Recent synthetic light curve runs, using Wood's code, have succeeded in fitting the minima from nearly every epoch but show varying degrees of departure at the maxima. In 1972 and 1974 and probably in 1977 as well, the apparent infrared excess measured in this system has been modest; $\lesssim 0.15$ magn; with new observations made at Mt. Lemmon in April 1979, we will be able to examine the long-term variability in the infrared although the phase behavior is not well known at any epoch. There is also evidence for a strong ultraviolet excess in the system: the primary minimum in U is too deep to fit any reasonable model consistent with B and V and the ultraviolet flux at maximum light predicted for a combination of stars of the same spectral types and with the same intrinsic V in combined light is less than what is observed. The H and K emission appears to be equally strong from both components, but since the eclipsing star at PM is contributing about 35% of the system's light in the maxima, it is relatively speaking, the stronger emitter at H & K. Present work indicates that the observed light curves could not be reproduced with a spot model alone. Finally, our spectroscopic study of this system (Naftilan and Milone 1979) has revealed 1) that the masses are equal; and 2) the presence of a weak but persistent emission component present at both quadratures near the systemic velocity, identifying material to be present at or near L_1 or around the system. These properties are consistent with the presence of circumstellar matter in the system although the period changes are not attributable to mass exchange in a simple way. Again they do not exclude spots from being present.

TABLE II - Mass Exchange/Loss Indicators

Indicator/System	RT Lac	CG Cyg	RW Com
Period change	Discrete, then smaller cyclic variations.	Discrete, then parabolic;	Parabolic
$\dot{P}$ or ΔP	+8.3 dex (-5) day.	-2.5 (±0.8) dex (-6) day; +3.0 (±0.6) dex (-10) day/day.	-0.40 dex (-10) day/day.
Roche lobe (y_g)	0.29	0.37	$\lesssim r_g$ (contact system).
Infrared Excess (2.2μ)	∿0.7 magn.	$\lesssim$0.15 magn.	∿0.6 magn.
Ca II H & K emission	From both comps., but peaks at P.M.	From both stars and a component at γ-velocity.	From both components.

All known light curves of the W UMa system RW Com were examined (Milone et.al., 1979b), and revealed a long-standing asymmetry - a brighter second maximum - with short term fluctuations. There is no evidence of a migration of this distortion feature, and it is not satisfactorily represented as a simple sine wave. The asymmetry at maximum, or the "O'Connell Effect", amounts to dV ∿ 0.07 and appears slightly blue; in the infrared, it is less than half this amount, and has the same sign. The system has a relatively large infrared excess at both maxima (Milone et.al., 1978). Significant variations in the Fourier representations of the light curve are found from epoch to epoch and the O-C data for the minima also differ significantly from epoch to epoch. The period has clearly varied and may be continuously decreasing; the weighted mean, $\dot{P}$ =-0.40 (±0.03) dex (-10) day/day. The H & K emission was discovered by Struve (1950, who noted that it appeared weak and complex at each maximum but strong at each minimum. He also noted that the Ca K_1 absorption feature was stronger from the approaching star at each quadrature. We have not yet been able to obtain sufficiently high-resolution, short-exposure plates to confirm this effect. Two spectrograms have revealed a spectral type of $\lesssim$ G5 for the combined light; the G5 assignment, however, rests on the weakness of the Balmer lines, so that Struve's G2 classification for each component may prevail, the discrepancy being attributable to in-filling by emission in Hγ and Hδ. Using Rucinski's (1973) A_4/A_2 criterion, the fill-out parameter is about 0.9, so that considerable space is available between the inner and outer Lagrangian surfaces for material to circulate about both stars. It is difficult to understand how spots can account quantitatively for the infrared or even qualitatively for the reported radial velocity effects in this system. The infrared excess determination based on 1972 and 1974 material has been confirmed with a preliminary examination of 1977 Mt. Lemmon data. The apparent excess depends on the adopted spectral types but it is out of the question that the spectra are later than G5 at maxima, though the minima are assumed to be later and redder. An assignment of G8 to the cooler component is appropriate for the ratio of visual surface brightnesses, $J_g/J_s = 1.17$, obtained in one of the limiting solutions: total secondary occultation.

All three of these systems have properties which strongly implicate the presence of circumstellar matter. All three show period changes and infrared excesses (which are computed using the latest possible spectral types - appropriate to the gravity-darkened elongated ends of the stars seen at the minima - and allowing for a reasonable amount of interstellar reddening). Because of the size of some of the effects it appears that circumstellar material is contributing in an important way to the photometric pecularities.

REFERENCES

Eaton, J. and Hall, D.S.: 1979. Astrophys. J. 227, p. 907.

Hall, D.S. and Haslag, K.P.: 1976, in IAU Colloquium No. 29, Multiple Periodic Variable Stars, Part II, ed. W.F. Fitch (Budapest: Academic Press), p. 331.

Jassur, D.M.Z.: 1978. IAU Com. 27, Inf. Bul. Var. Stars, No. 1458.

Milone, E.F.: 1976. Astrophys. J. Suppl. 31, p. 93.

Milone, E.F.: 1977. Astron. J. 82, p. 998.

Milone, E.F., Castle, K.G., Robb, R.M., Swadron, D., Burke, E.W., Michlovic, J.E., and Zissell, R.E.: 1979a. Astron. J. 84, p. 417.

Milone, E.F., Chia, T.T., Castle, K.G., Robb, R.M., and Merrill, J.E.: 1979b. Astron. J. submitted for publ.

Naftilan, S.A. and Milone, E.F.: 1979. Astron. J. 84, p. 1218.

Popper, D. 1976: IAU Com. 27, Inf. Bul. Var. Stars, Nos. 1083, 1090.

Rucinski, S.M.: 1973. Acta Astr. 23, p. 79.

Struve, O.: 1950. Astrophys. J. 111, p. 658.

OBSERVATIONAL EVIDENCE FOR LARGE-SCALE SPOT ACTIVITY ON THE PRIMARY COMPONENT OF XY URSAE MAJORIS

Edward H. Geyer
Sternwarte der Universität Bonn -
Observatorium Hoher List, D-5568 Daun, FRG

ABSTRACT

Extensive pe- and spectroscopic observations of the short period eclipsing binary XY UMa within the last four years confirm the star spot and stellar activity model, which was presented in 1975.

1. INTRODUCTION AND OBSERVATIONS

Four years agoe at the Cambridge IAU Symposium No.73 I presented a star spot model for the interpretation of the large light curve changes of the $0\overset{d}{.}48$-day eclipsing binary XY UMa. At that time and then in 1977 I predicted that this system should be a transient soft X-ray source, and a variable radio source as well (Geyer,1976,1977a). In the last few years I carried out intensive photometric and spectroscopic observations of this EB-binary, which not only confirm the spot model but also yield a lot of new insight into the stellar activity phenomenology.

The pe-photometry in U,B,V was carried out by means of a double beam photometer at the 106 cm - F/14.5 Cassegrain telescope of the Hoher List Observatory. This photometer allows the nearly continuous monitoring of the variable star with a typical time resolution of about 10 to 20 seconds per colour. Therefore many fine structures in the light curves become detectable, and also the time instants of the minima can be derived with an accuracy of $3 \cdot 10^{-4}$ days.

2. THE SYSTEM PARAMETERS

Out of the 11 light curves at my disposal since 1955, that of February / March 1976 is the most symmetric one. I used it to derive in a formal way the approximate system constants from the B- light curve: Primary minimum is a "transit";

$x = 0.8$ (adopted); $k = 0.54$; $r_g = 0.41$; $r_s = 0.22$; $i = 83\overset{\circ}{.}5$;

$z = \varepsilon^2 \sin^2 i = 0.138$; $J_s/J_g = 0.253$; $L_s = 0.07$; $L_g = 0.93$.

M. J. Plavec, D. M. Popper and R. K. Ulrich (eds.), Close Binary Stars: Observations and Interpretation, 423–427.

Therefore due to its low brightness, the secondary component is invisible on blue spectrograms with dispersions of 86Å/mm and 136Å/mm. From these we derived a mass function of about 0.12. If we adopt a mass of about one solar mass in accordance to its spectral type (G2V to G5V), it follows that the mass of the secondary component corresponds to a K5 star. This is also conform with the above given surface brightness ratio. Though the primary component seems to be oversized by about 20%, the system itself is still a detached one. From these facts we can conclude that the secondary component cannot be responsible for the observed light curve variations at all.

We used the calculated light curve from these system constants to obtain the residual curves from the observed light curves. These residual curves, which are freed now from the orientation and eclipsing light variations, represent the rotational variations due to asymmetric brightness distributions on the sphere of the primary component. These can now be analysed according the methods outlined already by Russell (1906) with the ambiguities that a great variety of distributions of brightness on the sphere may give rise to the same light curve. The maximum observed amplitude of the residual curve amounts to about $0^{m}_{\cdot}2$.

3. DISCUSSION

The observational facts for the evidence of strong spot activity on the primary component of XY UMa can be summarized as follows (figure 1):

i) Since the colour curve is in most cases strongly asymmetric during the primary minimum phase, and is oscillating around the time instant of the minimum, we have the strongest arguments for the existance of subluminous areas with reduced temperature on the stellar sphere.

ii) There exists a variable Nordmann - Tikhov effect which amounts up to 0.2% of the orbital period. It is the consequence of the described colour curve asymmetries. Thus the long debated and disbuted Nordmann - Tikhov phenomenon (Szafraniec, 1962) finds a simple explanation by star spot activity.

iii) The observed light curve flickering and the brightness spikes, which prevail sometimes for a couple of weeks, are indicative for the existance for star spot groups rather than extended subluminous areas. Furtheron it shows that there are certainly present chromospheric and flare activities.

iv) The symmetrical light curves always occur between the appearance of asymmetric light curves. Therefore no "migrating wave" like in the typical RS CVen variables is present.

v) This change over from symmetry to asymmetry repeats within 3.5 to 4 years ("spot cycle").

vi) The average system brightness of XY UMa varies on a time scale of 25 years. This is indicative for the modulation of the average spot frequency and/or the spot area on this time

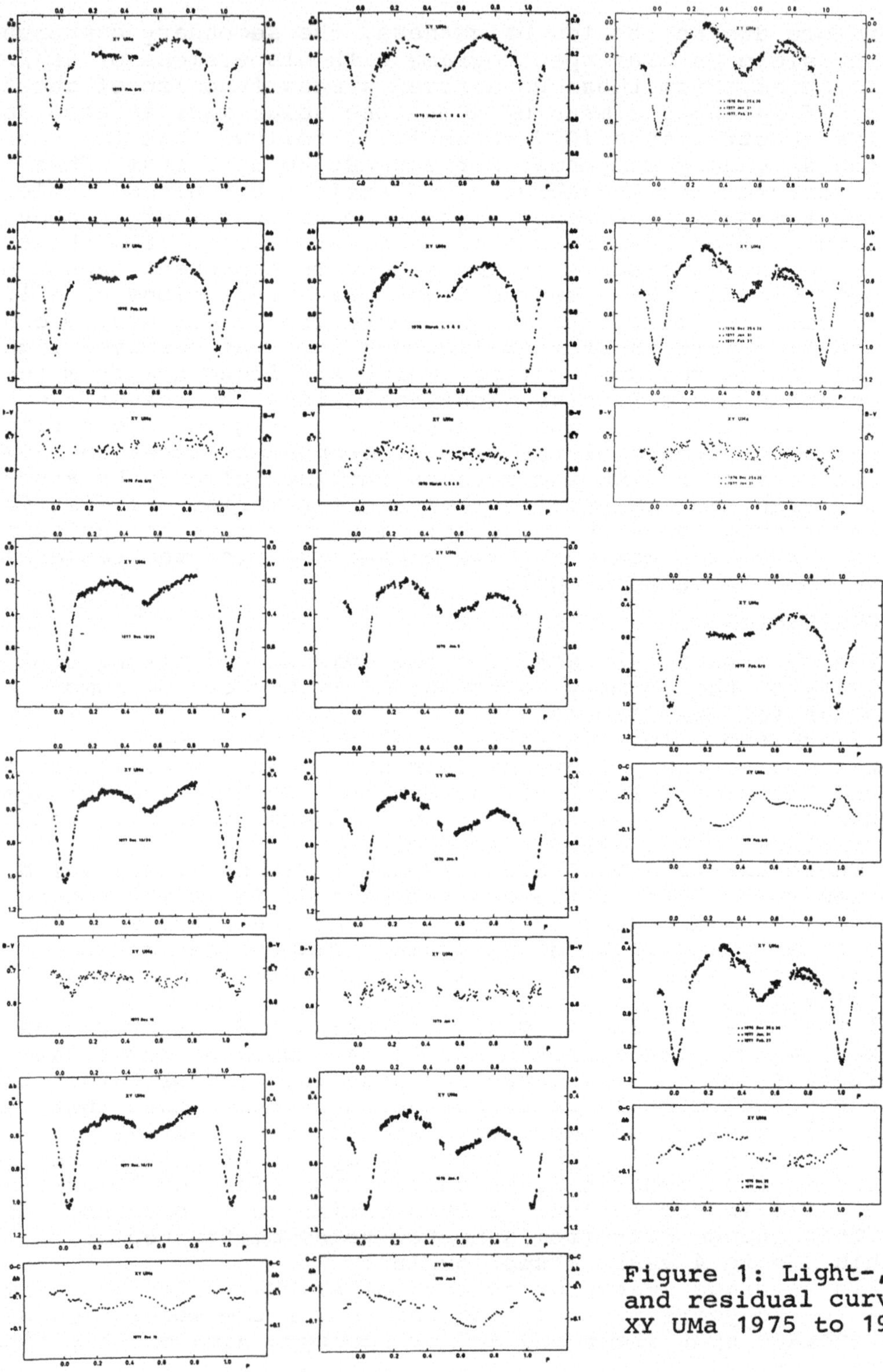

Figure 1: Light-,colour- and residual curves for XY UMa 1975 to 1979.

scale. It may be mere chance that the ratio of the spot cycle length to this modulation time scale is the same as in the Sun.
vii) The orbital period of 0.478995 day was constant during 25 years within $5 \cdot 10^{-7}$ day. In the (O - C)-diagram oscillations up to 2% occur (Geyer, 1977b).
viii) No circumstellar matter or matter within the system could be detected in the optical spectral range (Geyer and Metz, 1977).

4. CONCLUDING REMARKS

As earlier predicted, the strong stellar activity of the primary component of XY UMa should be especially prominent in the EUV spectral range and centimeter radio wavelength bands.

Finally some ideas about the evolution of such close binaries with strong magnetic star spot activities can be given: Since in analogy to the Sun the spot activity is certainly connected with prominence activity (Van't Veer, 1976) loss of rotational angular momentum according to the electromagnetic coupling mechanism proposed by Schatzman (1959, 1962) for the spotted star must occur. On the other hand in a close binary system the tidal effects force its components to rotate nearly synchronous to the orbital period at least in their outer layers. Thus there exists a feedback coupling between the rotation of the components and their orbital revolution, which is not only the motor for the differential rotation of the spotted star to induce a stellar dynamo, but also for the evolution of the binary system itself. It can take place in a twofold way:
a) If the total angular momentum of the binary system is not changed by the loss of rotational angular momentum of the spotted component, then the orbital angular momentum increases with a relevant increase of the orbital period. The system becomes more detached (Geyer, 1977a).
b) On the other hand, if the total angular momentum of the binary system is decreasing by the loss of rotational angular momentum, the orbital period is shortening, and the system may finally come into contact (Van't Veer, 1979). Such a system may be the WUMa type star TZ Boo about which extensive observations have been collected by Hoffmann (1978,1979).

REFERENCES

Geyer, E.H., 1976: Proc. IAU Sympos. No. 73, 313.
Geyer, E.H., 1977a: Proc. IAU Colloqu. No. 42, 292.
Geyer, E.H., 1977b: Astrophys.Space Sci. 48, 137.
Geyer, E.H. and Metz, K., 1977: Astrophys. Space Sci. 52,351.
Hoffmann, M., 1978: Astron. & Astrophys. Suppl. 33, 63.
Hoffmann, M., 1979: Private communication.
Russell, H.N., 1906: Astrophys. Jour. 24, 1.
Van't Veer, F., 1976: Proc. IAU Symp. No. 73, 343.
Van't Veer, F., 1979: Private communication.

DISCUSSION FOLLOWING MILONE-NAFTILAN AND GEYER

Eaton: I have questions about your solution of the light curves of RT Lac requiring a circumstellar envelope. Doug Hall and I have published a light curve solution for the same data which does not require such an envelope. As you know, we both analyzed the same data. Your solution was based on a modified light curve in which the distortion wave was subtracted out before a solution was attempted. Hall and I carried out essentially this same procedure, obtaining your results at V. However the eclipse depths ($k \approx 1$) and shapes made it impossible to use this sort of model to explain the anomalous color variation of the system. Instead, we have obtained relatively consistent solutions to the V light curves for both 1965 and 1968-69 using a spot model. These require that the more massive component have an anomalous energy distribution, as indeed seems to me to be the case for your solution. How confident are you of your rectification for the photometric wave? How does your model account for the anomalous color variations?

Milone-Naftilan: The solution is only for the V light curve. The envelope would have to be gray enough to permit the observed color indices at both minima. The rectification seemed to work at both 1965 and 1968 epochs but one can never rule out that the results are somewhat compromised by the rectification procedure. A bluer star underlying the envelope is, it seems to me, a reasonable explanation. The goodness of fit of either model to the maxima is not all that great, but of course we have not yet been able to come to grips with the apparent migration of the wave, as you have been able to do.

Sadik: Have you tried to plot a magnitude-color diagram (i.e. V vs. B-V) for out-of-eclipses points, that is to see if the system is getting redder or bluer when it's brighter? And secondly, is the V vs. B-V variation in the same sense from cycle to cycle or from one light curve to another?

Pakull: How do you solve your highly asymmetric light curves?

Geyer: The system parameters were obtained in the usual formal way from the most symmetric light curve I had observed. These parameters served to calculate the "theoretical light curve", which was used to obtain the residual light curves from the observed ones. In this way the light variation by orientation and eclipse is more or less removed, and the residual curve gives the intrinsic light variations due to stellar spots.

van 't Veer: I agree with you for the importance of prominence activity involving angular momentum loss and change of orbital parameters. I do not believe the detectability of the Tikhman-Nordhoff effect in your observations. The dispersion of the interstellar medium is insufficient to explain phase shifts of the light curve for different colors.

VIII. CATACLYSMIC VARIABLES AND POLARS

THE NATURE OF THE DIFFERENCE BETWEEN NOVAE, DWARF NOVAE, AND NOVALIKES

I.G. Mitrofanov
A.F. Ioffe Physical Technical Institute
Academy of Science of the USSR
Leningrad 194021, USSR

I. INTRODUCTION

Novae (N) and Dwarf Novae (DN) are considered to belong to the class of the Cataclysmic Variables (CVs). The Novalike stars and recently discovered Polars (the AM Herculis-type stars) seem to have the same nature as CVs and probably are members of this class. It is usually assumed that CVs are close binary systems, each system contains a degenerate dwarf and a nondegenerate second component, their period being usually less than 16^h (e.g. see: Robinson 1976). A second star is believed to overfill its Roche lobe and its matter flows onto the degenerated component. Then around the degenerated dwarf there arises an accretion disk. This disk and the hot spot where the interaction between the disk and outflowing stream occurs are the main sources of the blue and ultraviolet continuum and strong H and He emission lines.

The principal difference between CVs is connected with their outburst activity. The Novae and Recurrent Novae outbursts have an amplitude of $7\text{-}14^m$. The mean energy of a single outburst is $\lesssim 10^{45}$ ergs and the recurrence time between two successive outbursts is larger than 10 years. The outbursts are thought to arise from unstable thermonuclear burning of hydrogen accreted and accumulated on the surface of degenerated dwarf (Starfield et al., 1974). The outbursts of the DN have much smaller amplitudes $2\text{-}6^m$, a total energy output on the order of $10^{38}\text{-}10^{39}$ ergs and a recurrence time between roughly 10 days and 3 years. The DN are of U Gem and Z Cam type, but we assume that they are of similar nature and do not consider why they are different. Recently Szkody (1974, 1976 and 1977) has shown that during the outbursts the main sources of the DN radiation are the accretion disk and degenerated dwarf but between outbursts the main sources are the hot spot on the disk and the nondegenerated star.

Now there seems to be no generally recognized explanation of the DN phenomenon. Two groups of theories are well known. In the theories of the first group (Bath et al., 1974; Gorbatsky 1975) an instability of the envelope of the nondegenerated star was assumed to be the cause of the outbursts. This instability leads to a sharp increase in the luminosity of this star and/or an increase in the rate of outflow. This

M. J. Plavec, D. M. Popper and R. K. Ulrich (eds.), Close Binary Stars: Observations and Interpretation,431–437.

increased outflow is thought to lead to the increase in the accretion rate and, therefore, to expansion and brightening of the disk. However, remaining unsolved questions connected with these theories are: why is the hot spot on the periphery of disk brighter than the whole disk in the state of low luminosity? Why is the disk brighter than the spot in the state of high luminosity? And what causes the empirical relation between the total energy of the outburst E, the low state luminisoty L_{min}, and the recurrence time to the next outburst $\Delta t \sim E/L_{min}$? In the theories of the second group, by analogy with Novae, the outbursts are supposed to be connected with unstable thermonuclear burning of hydrogen accreted onto the surface of a degenerate dwarf (Starrfield et al., 1972). But in this case other important questions arise: what causes the difference between Novae and Dwarf Novae? And why is the intermediate class absent?

II. THE LOCAL THERMONUCLEAR BURNING OF THE HYDROGEN ONTO THE SURFACE OF A MAGNETIZED DEGENERATE DWARF

Let us assume that the outbursts of DN result from the unstable thermonuclear burning of hydrogen onto the surface of degenerate dwarf. Let us also assume that the dwarf has a strong magnetic field 10^6-10^8gs. The conditions on the dwarf's surface are not homogeneous because the accretion heats a small part of the surface to a high temperatures and a strong magnetic field leads to heterogeneities in the heat conductivity. In the presence of a magnetic field the conditions of the accumulation and nuclear burning of the accreted hydrogen are different in different points of the surface (Mitrofanov, 1978).

The area of the local region (LR) exposed to the accretion may be estimated by assuming that the plasma from the second component or disk flows along the lines of the dipole magnetic field $S \simeq 10^{14} \div 10^{15}\mathrm{cm}^2$. The hydrogen may spread over the whole surface of the dwarf due to mutual diffusion of the plasma and the magnetic field or due to instabilities. The diffusion time is $t_{dif} \simeq 4\pi\beta l^2/C^2 \sim 10^8(T/10^6)^{1.5}(\ell/10^6\mathrm{cm})\mathrm{s}$, where T is the temperature of the LR, ℓ is its characteristic size and β is the plasma conductivity. Local accumulation and ignition takes place if the inner edge of the hydrogen layer approaches the surface density n_{cr} for unstable burning in the LR faster than the hydrogen plasma spreads over the whole surface, i.e. if the time of accumulation $t_{acc} = S \cdot n_{cr}/\dot{M} = 10^7(S/10^{14}\mathrm{cm}^2)(n_{cr}/10^9\mathrm{g\ cm}^{-2})(10^{16}\mathrm{g\ s}^{-1}/\dot{M})\mathrm{s}$ is smaller than t_{dif}.

1) The LR may have a high temperature $\lesssim 10^7$. The critical density n_{cr} decreases with the increase of temperature. Therefore, the layer of hydrogen which is stable above the main cold part of the surface may appear unstable in the LR.

2) The LRs are situated near the magnetic poles of a degenerate dwarf. One can suppose that in such regions the heat conductivity is strongly decreased (e.g. due to suppression of the convection) and the conditions necessary for the unstable burning to occur appear much softer.

The mass of the burning hydrogen may be estimated as

$M = S \cdot n_{cr} \simeq 10^{23} - 10^{24}$g for $n_{cr} \simeq 10^9$g cm^{-2}. Assuming that W, the mean energy per gram released in the outburst dwarf novae is similar to W for novae, one can estimate W to be 10^{16} ergs g^{-1} which leads to a DN outburst energy E of $10^{39} - 10^{40}$ ergs. This magnitude agrees well with the observations.

If the difference between the LR and other parts of the dwarf's surface is not very large and the magnitudes of n_{cr} for these two regions are approximately the same, the hydrogen layer may become unstable on the whole surface. This may be the case of the dwarf's magnetic field or the accretion rate are sufficiently small $B \lesssim 10^6$gs or $\dot{M} \lesssim 10^{14}$g $\cdot$ s^{-1}. These stars should have their outbursts as Novae with a recurrence time $\gtrsim$10 yrs. In this way it becomes clear why there are few CVs with an outburst energy between 10^{40} ergs and 10^{43} ergs and with a recurrence time between 3 and 10 years.

The accumulation of the hydrogen mass occurs because accretion is more rapid than the destruction of hydrogen due to stable nuclear burning at the bottom of the layer. The burning becomes unstable if the rate of energy generation is larger than the rate of the cooling. It was shown (Paszynski and Zytkow, 1978) that the burning of a spherical shell is stable if $\dot{M} \gtrsim 10^{19}$ g s^{-1} because hydrogen burning occurs at a temperature where the rate of nuclear reactions is great enough to ensure equilibrium between the burning and the accretion. This result seems to be qualitatively appropriate also to local case and one can estimate the critical accretion rate to be $\dot{M}_{cr} \sim 10^{19} \cdot (S/S)$g $\cdot$ s^{-1} $\sim 10^{16}$ g $\cdot$ s^{-1}. This magnitude decreases as the dw of the magnetic field increases. Thus, the hydrogen burning is stable and outbursts are absent if the magnetic field and the accretion rate are sufficiently large. This regime is likely to take place in the case of AM Her-type stars ($B \gtrsim 2 \cdot 10^8$gs and $M \gtrsim 10^{16}$g $\cdot$ s^{-1}). These stars all have the properties of CVs with the exception of the most significant one - the outbursts themselves. They are the true Novalikes in contrast to the false ones which seem to be Novae with unrecorded outbursts.

III. THE POSSIBLE MECHANISM OF THE DWARF NOVAE OUTBURSTS

It is obvious that the local burning of hydrogen should differ significantly from the spherical case. It seems very difficult to investigate this difference quantitatively because of the influence of the magnetic field, the heterogeneity of the temperature and the possibility of flows into the layer. In the remainder of this paper a rough, qualitative approach will be used.

The optical depth of the layer before the onset of unstable burning, $\tau_o = \sigma_e n/m_p$, is very large (σ_e is the effective absorption cross-section). During the unstable burning, the layer usually begins to expand (Starrfield et al., 1972, 1974). In the spherical case its optical depth does not change. In the local case the matter from the LR expands almost isotropically and for $r \gg S^{1/2}$(r is the radius of the expanding hemisphere) its optical depth $\tau \simeq \tau_o \cdot (3S/2\pi r^2)$ becomes much smaller. From the equation of heat transport one can estimate the radiative cooling time $t_{cool} \simeq Cr^2/\kappa_r$, where $C = N_e k_B$ is the specific heat of the plasma and κ_r is the coefficient of radiative heat conductivity. This estimation has another form $t_{cool} \simeq \tau^2 k_B/\sigma_e \sigma T \simeq 4 \cdot 10^{-16} \tau^2 (10^6/T)$s ($\sigma$ is the Stephan-Boltzman's constant). When the cooling time appears comparable with the time of expansion r/V_{exp}, the internal pressure and the rate of the thermonuclear burning decrease, and the expansion stops. From the condition $t_{cool} \sim t_{exp}$ one can estimate r_{max} and t_{exp}. If $V_{exp} \simeq 10^7$cm/s they appear to be $r_{max} \sim 10^8$cm and $t_{exp} \sim 10$ s.

During the phase of the plasma expansion the frozen magnetic field becomes deformed. When the internal pressure decreases the field begins to restore its geometry. Therefore, expanded plasma has to fall down just onto the LR. Suitable conditions for the unstable thermonuclear burning are restored and the cycle is repeated.

This qualitative mechanism seems to explain the regular variation of the soft X-rays from SS Cyg at outburst with a mean period of 8.9 s and an amplitude of ~50% (Cordova et al. 1978). This mechanism also may explain the regular pulsation of the DNs optical luminosity during outburst (Robinson, 1973 a,b,c,). An additional fact in favor of this explanation is the relationship observed for Z Cam between the mean period of the pulsations and the energy of the outburst. It seems difficult to imagine another pulsation mechanism explaining this fact.

During the outburst the released energy is transported from the LR to the accretion disk. One possible transport mechanism is the generation of shock waves above the region of pulsing plasma. These waves may propagate through the disk and heat it. So, during the outburst the disk becomes the main source of the optical light of DN while the burning region is the main source of X-rays.

IV. CONCLUDING REMARKS

The distinction between DN and N may be a result of the local burning of hydrogen on the surface of a degenerate dwarf (Mitrofanov, 1978). The main cause of this localization seems to be the strong magnetic field $10^6 - 10^8$gs on the dwarf's surface. This supposition seems to agree with the observations of X-rays from some DN such as SS Cyg (Rappaport et al., 1974; Heise et al., 1978 and Cordova et al., 1978), U Gem (Mason et al., 1978, Swank et al., 1978) and EX Hya (Watson et al., 1978). If the field strength or the accretion rates are sufficiently small (e.g. $B \lesssim 10^6$gs or $\dot{M} \lesssim 10^{14}$g $\cdot$ s^{-1}) the localization has to disappear and the usual Novae outburst should be observed. Conversely, a very strong magnetic field together with a high accretion rate ($B \gtrsim 2 \cdot 10^8$gs and $\dot{M} \gtrsim 10^{16}$g $\cdot$ s^{-1}) is likely to lead to the steady burning with no outbursts. It seems to be the case of the true Nova-likes specifically the stars of AM Her-type.

The mechanism of local outburst with pulsation may explain the strong regular variations of SS Cyg X-rays and weak similar variations of the optical light from Z Cam, CY Cnc, AH Her, CN Ori, KT Per and SS Cyg. The essential decrease in X-ray flux at the late stage of the SS Cyg outburst (Ricketts et al., 1979) is possible due to the flow of matter from the burning region to the accretion disk instead of the normal accretion flow.

The above mentioned empirical relation, $\Delta t \sim E/L_{min}$, may be explained from the theoretical relation, $t_{acc} = M/\dot{M}$, by substituting instead of M and $\dot{M}$ the observable magnitudes E and L_{min} (E is a total energy of a burst and L_{min} is the low state luminosity produced by the flux from the hot spot on the disk). In this case we obtain $t_{acc} \sim E/L_{min}$. Using this relation two independent sequences of outbursts were discovered for some DNs. For example, Vogt (1974) and Bateson (1974) have found two types of outbursts for **VW Hyi (Table 1)**. It is natural to relate them with two different magnetic poles with slightly different conditions. In this case the estimations of the accretion rate $\dot{M}$ for the both sequences of the outbursts should be equal and indeed, it is easy to verify that the difference is less than 5%. This is a good confirmation of the proposed explanation of the nature of the Dwarf Novae.

ACKNOWLEDGEMENTS

The author thanks Prof. A.Z. Dolginov, Dr. A.I. Tsygan and Dr. D.G. Yakovlev for useful discussions.

REFERENCES

Bateson, F.M. 1974. IBSM No. 24, 26.
Bath G.T., Evans, W.D., Papaloizon, J., Pringle, J.E. 1974, Mon. Not. R. astr. Soc. 169, 447.
Cordova F., Garmire, G., Tnohy, I. 1978 IAU Arc. No. 3235.
Gorbatskij, V.G. 1975. Pis'ma Astr. Zh. 1, 23.
Heise B.M., den Boggende, A.J.F., Schrijver, J., and Grindlay, J.E., 1978. Astron. Astrophys. 63, L1.
Mason, K.O., Lampton, M., Charles, P., and Bowger, S. 1978. Astrophys. J. 226, L 129.
Mitrofanov, I.G. 1978. Pis'ma Astron. Zh. 4, 219 (Sov. Astron. Let. 4 (3) 119).
Paczynski, B. and Zytkow, A.N. 1978. Astrophys. J. 222, 604.
Rappaport, S., Cash, W., Doxsey, R., McClintock, J., and Moore, G. 1974. Astrophys. J. 187, L 5.
Ricketts, M.J., King, A.R. and Raine, D.J. 1979. Mon. Not. R. astr. Soc. (to appear).
Robinson, E.L., 1973 a. Astrophys. J. 180, 121.
Robinson, E.L., 1973 b ibid., 181, 531.
Robinson, E.L., 1973 c ibid., 183, 193.
Robinson, E.L., 1976. Ann. Rev. Astron. Astrophys. 14, 119.
Starrfield, S., Truran, J.W. and Sparks, W.M. 1972. Astrophys. J. 176, 169.
Starrfield, S., Sparks, W.M. and Truran, J.W. 1974. Astrophys. J. Suppl. 28, 247.

Swank, J.H., Boldt, E.A., Holt, S.S., Rothschild, R.E. and Serlemitsos, P.J. 1978, Astrophys. J. 226, L 133.
Szkody, P. 1974. Astrophys. J. 192, L 75.
Szkody, P. 1976. ibid., 207, 824.
Szkody, P. 1977. ibid., 217, 140.
Vogt, N. 1974 Astron. Astrophys. 36, 369.
Watson, M.G., Sherrington, M.R. and Jamson, R.E. 1978. Mon. Not. R. astr. Soc. 184, 79.

TABLE 1. THE OUTBURSTS OF VW HYA

	First type of outbursts	Second type of outbursts
Mean duration of outbursts, δt	4^d	11^d
Mean recurrence time, Δt	$28\overset{d}{.}7$	179^d
Mean amplitude Δm	$3\overset{m}{.}9$	$4\overset{m}{.}6$
Mean accretion rate, $\dot{M}$ [1]	$5.00\ L_{min}/W$	$5.05\ L_{min}/W$

[1] formulae for the estimation of the accretion rate

$$M = \frac{L_{min}}{W} \cdot \left(\frac{L_i}{L_{min}} \cdot \frac{\delta t_i}{\Delta t_i}\right), \quad i = 1, 2.$$

DISCUSSION FOLLOWING MITROFANOV

van Paradijs: Your model ascribes the difference between N and NL variables to a difference in the mean accretion rate of at least a factor of 100. You would then expect a systematic difference of about that factor in the steady luminosity of these systems (in between eruptions). How does this compare to the observations?

Mitrofanov: The difference between N and NL results not only from the difference in the accretion rate but also from the difference of the magnetic fields. Therefore there may be no difference in total luminosity. Even if the difference exists the energy is emitted in different spectral bands (e.g. ultraviolet and X-ray).

Nariai: What are the mass accreted and the rate of accretion?

Mitrofanov: The mass which is associated with the DN outburst is about 10^{22}- 10^{23}g; the rate of accretion seems to be about $10^{14}$to 10^{16} g/s.

Nariai: I wonder if the binary can maintain the rigid structure from the onset of accretion to flash. Otherwise the magnetic field will be winding the white dwarf and the localized accretion will not be possible.

Mitrofanov: The dipole approximation for the dwarf's magnetic field seems to be sufficiently good. This field is frozen into the core of the white dwarf. The diffusion time for accreted plasma is much smaller than the free-fall time. So, one can say that the accreted plasma is canalized by the magnetic field.

CURRENT WORK ON CATACLYSMIC VARIABLES AT OXFORD AND CAMBRIDGE

G.T. Bath*, C.C. Brunt+, P.P. Eggleton+, A.C. Fabian+,
R. Harkness*, S.K. Mayo+, J. Papaloizou+, J.E. Pringle+,
M. Rayne+, G.S. Savonije+, J.A.J. Whelan+, C.A. Whyte+,
and D.T. Wickramasinghe+
*Department of Astrophysics, South Parks Road, Oxford
+Institute of Astronomy, Madingly Road, Cambridge

This talk summarizes recent work that has been done on cataclysmic variables at Oxford and Cambridge. Full details should be sought elsewhere, either from the authors concerned or from the references given below.

On the observational side, WZ Sge has been observed spectroscopically before and after the December 1978 outburst, in August 1978 and July 1979, using the Image Photon Counting System on the Anglo-Australian telescope. The data are being used (C.C.B. and J.A.J.W.) to study changes in emission lines and in the broad absorption line profiles.

Ultraviolet spectra were obtained with the I.U.E. satellite during the 1978 December outburst (Fabian et al. 1979). The data yield information on the evolution of the spectrum during the decline and on the short timescale spectral variability.

Three dwarf novae (VW Hyi, EX Hya and BV Cen) have been observed spectrophotometrically in the wavelength range 1250-7500 Å, using spectra obtained simultaneously with the I.U.E. satellite and the Anglo-Australian Telescope. The spectra of EX Hya, seen in quiescence, and of VW Hyi, seen on the decline from outburst can be fit to theoretical steady accretion disc spectra (Bath, Pringle and Whelan, 1979).

Time-resolution spectroscopy of Z Cha round the orbit in quiescence has been carried out. The data demonstrate that the centre of the disc rotates faster than the outside. The final analysis will yield information on the orbital parameters (Rayne and Whelan, 1979; Rayne, 1980).

On the theoretical side, a detailed investigation has been started on the spectrum to be expected from an optically thick outflowing wind (R.H. and G.T.B.).

An investigation of the conditions under which the boundary layer between an accretion disc and a white dwarf can emit X-rays has been

M. J. Plavec, D. M. Popper and R. K. Ulrich (eds.), Close Binary Stars: Observations and Interpretation, 439–442.

carried out (Pringle and Savonije, 1979). For low accretion rates hard X-ray emission is possible. At high accretion rates the hard X-ray emission is suppressed and soft X-ray emission can be observed.

A study of the time-dependent behaviour of accretion discs during the outbursts of dwarf novae is being conducted (G.T.B. and J.E.P.). Adequate fits to the observed light curves can be obtained either by varying the mass-transfer rate or by varying the viscosity. The constraints on the types of variation which give rise to plausible light curves are being investigated.

Analytical and numerical computations of the oscillations of rapidly rotating white dwarfs are presented (Papaloizou and Pringle, 1978 and 1979a). The existence of the r-modes is confirmed, as is the confinement of the g-modes towards the equatorial regions. The results support the possibility that the coherent and the quasi-periodic variations seen in cataclysmic variables are caused by sets of low frequency oscillations excited in a rapidly rotating white dwarf envelope.

Theoretical continua ($3000 < \lambda(A^o) < 20,000$), UBV colours and Balmer absorption line profiles have been computed for the optically thick parts of steady-state accretion discs around white dwarfs. The results are compared with UBV colours taken from the literature and with new spectroscopic line profile data of several dwarf novae in outburst and a UX UMa type star, CoD -42^o14462. The overall agreement between theory and observation is good for UBV colours and reasonable for line profiles (Mayo 1979; Mayo *et al*. 1979).

A model is proposed to explain the super hump period seen in some cataclysmic variables during outburst (Papaloizou and Pringle 1979b). It is suggested that the orbits of these stars are slightly eccentric and that the superoutburst period, P_r, is the one on which the mass-transfer is modulated due to periodic variation of the stellar separation. Thus P_r is related to the orbital period P and the apsidal precession period U by $P_r - P = P^2/U$.

A simple model of binary star evolution has been produced and tested, including the effects of stellar mass loss and gain, of nuclear evolution and of loss of mass and angular momentum from the system (Whyte and Eggleton 1979). The program has been applied initially to the evolution of cataclysmic binaries. It is found that the 2 - 3 hour period gap and the concentration of systems at low binary period are due to evolution in systems with low mass white dwarfs.

REFERENCES

Bath, G.T., Pringle, J.E. and Whelan, J.A.J., 1979. Mon. Not. R. astr. Soc., in press.

Fabian, A.C., Pringle, J.E., Stickland, D.J. and Whelan, J.A.J., 1979, in preparation.

Mayo, S.K., 1979. Ph.D. thesis. University of Cambridge.
Mayo, S.K., Wickramasinghe, D.T. and Whelan, J.A.J., 1979. Mon. Not. R. astr. Soc., submitted.
Pringle, J.E. and Papaloizou, J., 1978. Mon. Not. R. astr. Soc., 182, 423.
Pringle, J.E. and Papaloizou, J., 1979a. Mon. Not. R. astr. Soc., in press.
Pringle, J.E. and Papaloizou, J., 1979b. Mon. Not. R. astr. Soc., in press.
Pringle, J.E. and Savonije, G.S., 1979. Mon. Not. R. astr. Soc., 187. 777.
Rayne, M., 1980. Ph.D. thesis. University of Cambridge.
Rayne, M. and Whelan, J.A.J., 1979, in preparation.
Whyte, C.A. and Eggleton, P.P., 1979. Mon. Not. R. astr. Soc., in press.

DISCUSSION FOLLOWING THE PAPER BY BATH *et al*. (GIVEN BY PRINGLE)

Mitrofanov: I would like to ask four questions about the model of dwarf novae:

(1) What is the cause of the sharp change in viscosity?

(2) How can you explain the existence of two types of outbursts?

(3) How can you explain the X-rays, in particular the anticorrelation with the optical curve during outburst, and the 8.9 s periodicity?

(4) How can you explain the sharp changes of the quasi-period of the optical luminosity?

Pringle: The nature of the viscosity itself is unknown, let alone any reason why it should or should not vary.

(2) I have seen no plausible explanation for this, but I would caution that the two types of dwarf novae may not be as distinct as has been thought.

(3) and (4) were more fully discussed in IAU Colloquium No. 53 which took place last week. I refer the reader to my contribution and to the references given at its end.

Mazeh: Did you observe the double peak emission in your UV spectra of WZ Sge? -- (After Pringle has answered positively):

Can you tell us about the separation between the components, in particular as compared with the separation between the outburst?

Friedjung: I can reply to the question concerning the emission peak separations of WZ Sge, as we have examined high-resolution IUE observations in Paris. On the 4th December 1978, the peaks of He II 1640 had a somewhat smaller separation than the pre-outburst Balmer separation, i.e. 800 - 1000 km/s. These results were presented in Rochester. Some observers who presented visual results there had found a separation of only 400 km/s for $H\alpha$, which they interpreted as due to an outer disk. Our discussions in Rochester showed that the interpretation may not be so clear-cut.

Ritter: With regard to the S-wave component in the spectra of WZ Sge, what is its amplitude and its phase shift with respect to the eclipse phase?

Pringle: As I explained, Clarke Brunt is still in the process of analyzing this data. All I can say is that the values given originally by Krzeminski and Kraft were pretty good.

CIRCUMSTELLAR MATERIAL IN CATACLYSMIC BINARIES

J. Smak
N. Copernicus Astronomical Center
Polish Academy of Sciences
Warsaw, Poland

ABSTRACT

This semi-review will concentrate primarily on geometric and photometric parameters of rotating gaseous disks in cataclysmic binaries. Dimensions, or more precisely - the outer radius of the disk can be estimated from three types of data: (a) From the analysis of eclipses of the hot spot, by assuming that its "radius-vector" is identical with the outer radius of the disk. (b) From the analysis of the eclipses of the disk itself. (c) From the radial velocity data giving the rotational velocity of the disk and the orbital velocity(ies) of the component(s). The resulting radius, corresponding to the peaks of the double emission lines, is not identical but presumably quite close to the outer radius of the disk. From a more detailed analysis of the disk's eclipse one can also obtain an information concerning the surface brightness distribution across the disk.

1. VARIABLE DIMENSIONS OF DISKS

Nearly all systems with sufficiently extensive coverage of their photometric behaviour show variable dimensions of their disks. The best studied cases are: U Gem (Smak 1971) in which the variations are related to the dwarf nova outburst cycle; Z Cha (Fabian, Pringle, Whelan, and Bailey 1978, Smak 1979a) in which two series of data (Warner 1974 and Bailey 1979) give different geometric parameters but with no obvious relation - so far - to its outburst or super-outburst cycles; RW Tri (Walker 1963, Winkler 1977, Smak 1979c) in which large and irregular variations occur on a time scale of days and longer. Three other systems with variable shapes of their eclipse curves, certainly due to variable parameters of their disks, are: UX UMa, DQ Her, and EM Cyg. In addition there is a very special case of HT Cas for which recent photometry by Patterson (1979) indicates that its disk is a transient feature: the hot spot in this system, while present between the outbursts, became undetectable following an outburst.

M. J. Plavec, D. M. Popper and R. K. Ulrich (eds.), Close Binary Stars: Observations and Interpretation, 443–446.

2. MINIMUM AND MAXIMUM DIMENSIONS OF DISKS

In all studied cases the outer radii of disks (or minimum radii in cases of variable dimensions) are larger than the minimum theoretical radii defined by the angular momentum carried by the stream coming from the secondary component. This implies that an angular momentum transfer mechanism(s) must operate in those disks either on a stationary basis or at least occasionally to redistribute the angular momentum. Only in the case of HT Cas the radius of its transient disk is nearly equal to the angular momentum radius and it may mean that in this case the angular momentum exchange mechanism is rather ineffective.

Turning to the largest observed radii we find the situation somewhat confusing. If we consider only five systems for which the evidence comes from the eclipse analysis, i.e. U Gem, Z Cha, RW Tri, HT Cas, and WZ Sge (Fabian, Lin, Papaloizou, Pringle, and Whelan 1978, Smak 1979b), we find that the outer radii are smaller than the largest non-intersecting orbits calculated within the three-body problem (Paczyński 1977, Smak 1976). Only in U Gem at outburst the disk is somewhat larger but even then it does not exceed the stability limit (Piotrowski and Ziołkowski 1970).

A different picture is obtained for three other systems: Z Cam (Robinson 1973a, 1973b, 1976), EM Cyg (Robinson 1974, 1976), and SS Cyg (Walker and Chincarini 1968, Robinson 1976) for which evidence comes from the radial velocity data. All three are double-line spectroscopic binaries with mass ratios near 1 and in all cases the rotational velocities of their disks imply radii comparable to or even larger than the Roche lobe. In all three cases the lines are not clearly double and therefore the resulting radii should be taken with caution.

3. SURFACE BRIGHTNESS DISTRIBUTION

In the case of two dwarf novae: U Gem (Smak 1971) and Z Cha (Smak 1979a) there is photometric evidence that the outer parts of their disks are brighter than the inner ones. On the other hand, in the case of one nova - DQ Her (Cherepashchuk 1979), and one nova-like - RW Tri (Ivanon 1969), the results of eclipse analysis indicate an increase of the surface brightness towards the center of the disk. This difference may imply that while the accretion does indeed operate in novae and nova-like systems, it does not take place between the outbursts of dwarf novae. Supporting this is a recent result by Kiplinger (1979) who finds that the spectral energy distribution of SS Cyg between the outbursts cannot be described by a steady-state accretion model.

4. RING/TORUS-LIKE DISK IN U GEM

Madej and Paczyński (1977) and Paczyński (1978) proposed a ring- or torus-like model for the disk in U Gem, in which no viscosity and accretion takes place between the outbursts. The only major effect of the exchange of the angular momentum is due to the fact that the incoming stream carries less angular momentum that that of the torus. Outbursts are due to an instability leading to a sudden accretion of the part of material stored in the torus onto the white dwarf (cf. Osaki 1974, Smak 1976).

With the parameters available for U Gem it is possible to test how self-consistent such a torus-like model is. If we identify the radius of the torus with the outer radius of the disk, we know that it shrinks between the outbursts from 0.38 to 0.34 in dimensionless units (Smak 1971). If this variation is exclusively due to the material supplied by the stream, we can calculate how much of the lower angular momentum material in the torus is needed to account for the observed decrease of the radius. We obtain that the total amount of mass added to the torus between the outbursts is about 1/7 of the amount present immediately after the outburst.

The amount of energy radiated by the torus-like disk between the outbursts, as being due to the decreasing mechanical energy, can be expressed in terms of the total amount of mass stored in the torus and its initial and final radii. Likewise, the amount of energy radiated by the hot spot between the outbursts is a function of the impact parameters and the total amount of material being supplied to the disk. The ratio of these two is equal to the ratio of luminosities of the disk and of the spot. With the 1/7 value obtained above we get that the disk/spot luminosity ratio should be about 3/2 in remarkable agreement with photometric data (Smak 1971, Table 3). True, it is not clear how much this agreement would be spoiled by applying appropriate bolometric corrections.

It has been pointed out earlier (Osaki 1974, Smak 1976) that the amount of energy released during a sudden accretion of the material supplied by the stream and stored between the outburst in the disk, is sufficient to explain the outbursts of U Gem. We can also consider the angular momentum balance. We assume that the 1/7 of the amount of mass available in the torus will be accreted. Its angular momentum must be added to the remaining 6/7. As a result we find that the radius of the torus should increase up to about 0.45. Again, this is consistent with the observational evidence (Smak 1971): during the outburst we indeed observe an expansion of the disk and 3-5 days after maximum we still have $r = 0.41$. It can only be added that the initial fast contraction of the disk, immediately after the outburst, may be partly due to tidal effects which should be particularly effective in the case of very large disks.

REFERENCES

Bailey, J. 1979 M.N.R.A.S., in press.
Cherepashchuk, A.M. 1979 Private information.
Fabian, A.C., Lin, D.N.C., Papaloizou, J., Pringle, J.E., and Whelan, J.A.J. 1978 M.N.R.A.S., 184, 835.
Fabian, A.C., Pringle, J.E., Whelan, J.A.J., and Bailey, J.A. 1978 I.A.U. Coll. No. 46, Hamilton.
Ivanov, L.N. 1969 Trudy Astr. Obs. Leningrad, 26, 33.
Kiplinger, A.L. 1979 Preprint.
Madej, J., and Paczyński, B. 1977 I.A.U. Coll. No. 42, Bamberg, p. 313.
Osaki, Y. 1974 Publ. Astr. Soc. Japan, 26, 429.
Paczyński, B. 1977 Ap. J., 216, 822.
1978 Nonstationary Evolution of Close Binaries, ed. A. Zytkow (Warsaw: PWN), p. 89.
Patterson, J. 1979 Private information.
Piotrowski, S.L., and Ziołkowski, K. 1970 Astrophys. Space Sci., 8, 66.
Robinson, E.L. 1973a Ap. J., 180, 121.
1973b Ap. J., 186, 347.
1974 Ap. J., 193, 191.
1976 Ann. Rev. Astron. Astrophys., 14, 119.
Smak, J. 1971 Acta Astr., 21, 15.
1976 Acta Astr., 26, 277.
1979a Acta Astr., 29, 309.
1979b Acta Astr., 29, 325.
1979c Acta Astr., 29, No. 4.
Walker, M.F. 1963 Ap. J., 137, 485.
Walker, M.F. and Chincarini, G. 1968 Ap. J., 154, 157.
Warner, B. 1974 M.N.R.A.S., 168, 235.
Winkler, L. 1977 A.J., 82, 1008.

IUE OBSERVATIONS OF WZ SAGITTAE IN OUTBURST

Albert V. Holm, Chi-Chao Wu, Warren M. Sparks,*
Francis H. Schiffer, III, and Albert Boggess*
Computer Sciences Corporation, Silver Spring, Maryland

*Goddard Space Flight Center, Greenbelt, Maryland

The International Ultraviolet Explorer was used to obtain spectra at both 7 Å and 0.2 Å resolution of the recurrent nova WZ Sagittae during its 1978 outburst. The first spectra were obtained on 1978 Dec 1.8 UT, approximately 0.7 days after discovery. The decay from outburst was followed until 1979 Jan 1, after which it was too near the Sun to be observed. A post-outburst spectrum was obtained on 1979 July 11. In this paper, preliminary results from the analysis of the low resolution spectra are discussed. These observations will be used to support the similarity between WZ Sge and dwarf novae and to test accretion disk models.

The images were processed in the standard manner (Boggess et al. 1978). There may be some photometric error in the 1978 Dec spectra in the wavelength range from 1150 Å to 2000 Å. An incorrect Intensity Transfer Function (ITF) was used to process IUE short wavelength spectra between 1978 April 22 and 1979 July 7. The nature of the ITF problem is such that spectra which were underexposed or which had high radiation-induced backgrounds might have photometric errors of 20% or more. Well exposed spectra of sources with temperatures of 15000 °K or more will have lower errors. WZ Sge was sufficiently bright in the UV that the exposure times were 200 seconds or less. As a result the background is low and most of the spectra are well exposed so that we expect the ITF error to be minor for these calibrated spectra. In any event, the largest errors would appear in the 1500 to 2000 Å region.

Throughout the outburst period there were few changes in the identity of the strong lines which appeared in the spectrum. N V 1240 Å, Si III 1300 Å, C II 1335 Å, Si IV 1400 Å, C IV 1550 Å, He II 1640 Å, and Al III 1855 Å persisted throughout the outburst. An illustration showing the spectra of WZ Sge at several times in the outburst period is given in the review paper by Sparks et al. (1979). The N V, C IV, and He II lines appeared as doubled emission lines in the earliest spectra. The C IV and N V emission lines strenghtened relative to the continuum as the visual magnitude faded from 8.0 to 10.5 and then weakened as the visual magnitude fell further. A strong absorption line of Si II 1265 Å

M. J. Plavec, D. M. Popper and R. K. Ulrich (eds.), Close Binary Stars: Observations and Interpretation, 447–451.

appeared after the first spectrum and persisted throughout the rest. By the time the post-outburst spectrum was obtained a broad Lyman α absorption line had appeared. At no stage did the spectrum of WZ Sge resemble that of Nova Cygni 1978 which evolved from an F or G type spectrum through a stage with strong chromospheric emission lines and eventually to a stage with high excitation forbidden lines and an undetectable continuum.

The continuum radiation of WZ Sge declined simultaneously at both visible and ultraviolet wavelengths. This is illustrated in Fig. 5 of Sparks et al. Which shows magnitudes derived from the IUE's Fine Error Sensor and from a 50 Å band about 1450 Å which contains no strong lines. The UV flux even reproduces a 1.5 mag dip in the visual light curve centered on 1979 Jan 1. There clearly is no phase lag between the visual and ultraviolet spectra such as seen for FH Serpentis (Code 1972) and for Nova Cygni 1978 (Sparks et al.). This behavior is similar to the apparent behavior of the dwarf nova SS Cygni (Holm and Gallagher 1974; Heap et al. 1978).

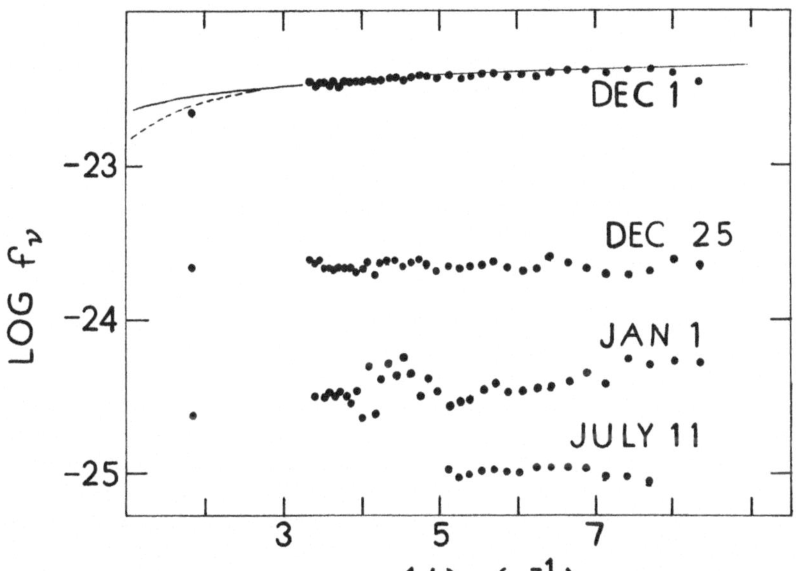

Fig. 1. The spectral energy distribution of WZ Sagittae

The ultraviolet energy distribution of WZ Sge on several occasions during and after outburst are shown in Figure 1. The data were averaged in 50 Å bins. There is little change in the overall energy distribution even though the luminosity fell by a factor of 400 between 1978 Dec 1 and 1979 July 11. This supports the conclusion that the visual and UV fluxes declined simultaneously.

The absence of an interstellar absorption feature at 4.6 μm^{-1} in Figure 1 implies that the extinction is low. Analysis using the method of Holm and Cassinelli (1977) shows that E(B-V) $\leq$ 0.025. The upper limit on the color excess can be used to derive an upper limit on the distance. Unfortunately, UBV colors have been measured for few stars in the direction of WZ Sge. The Photoelectric Catalogue (Blanco et al. 1968) gives A type stars, HD 191083 and BD+16 4120B, at a distance of about 300 pc (if main sequence) and having color excesses of 0.1 and 0.05 respectively. The implication from these stars is that WZ Sge is nearer than 300 pc. This estimate is in agreement with the estimate of 65 pc by Krzeminski and Kraft (1964) from a spectroscopic analysis which assumes the source of the visual light is the white dwarf and with the estimate of 83 pc by van Maanen (1926) from a trigonometric parallax. Although there are weakness in the groundbased data supporting the distance limit, the evidence suggests that the luminosity of WZ Sge at maximum light is, to an order of magnitude, similar to that for dwarf novae and much below that of classical novae.

Since the interstellar extinction is negligible, the observed energy distribution can be compared directly with models. We have used both stellar atmosphere models and model accretion disks. The Kurucz et al. (1974) line blanketed models give good agreement between the Dec 1 observations and a 20000 ^{o}K atmosphere and between the Dec 25 observation and a 16000 ^{o}K atmosphere. If the distance is 100 pc, the photosphere radius is 0.46 $R_{\odot}$ and 0.17 $R_{\odot}$ respectively on the two dates. It is probably fortuitous but interesting that the radius on Dec 1 is identical to the semi-major axis of the orbit derived by Robinson et al. (1978). We also considered accretion disk models because Robinson et al. (1978) find that at minimum an accretion disk accounts for most of the visual light. All the disk models of Herter et al. (1978) increase too steeply into the ultraviolet. A possible explanation of this disagreement between observation and theory, suggested by Herter et al. themselves, is that the disk is truncated at too small a radius. A simpler steady state disk model (Lynden-Bell 1969) predicts a power law spectrum, $\nu^{1/3}$, over a limited range of frequency. This power law spectrum is illustrated in Figure 1 by a solid line. There is surprisingly good agreement with the Dec 1 observed spectrum. The agreement is not as good with later observations in December, but there is a region of agreement with July 11 observation. Note that the dashed line in Figure 1 represents a modification of the power law spectrum which results from truncating the disk at a radius which does contribute a significant amount of radiation in the visual region.

The equations of Lynden-Bell (1969) can be solved formally to show that

$$R_{disk}/R_{*} \gg 1.3\ (\dot{M}_{max}/\dot{M}_{min})^{0.25}\ \nu_{max}/\nu_{min}$$

where R_{disk} is the outer radius of the disk at maximum, R_{*} is the radius of the white dwarf, $\dot{M}_{max}$ and $\dot{M}_{min}$ are mass transfer rate at maximum and minimum respectively, ν_{max} is the lowest frequency at maximum light at which the power law spectrum is a good fit, and ν_{min} is the highest frequency at minimum light at which the power law is good. From the 1978 Dec 1 and 1979 July 11 observations, this equation gives $R_{disk}/R_* \gg 70$. If this analysis is correct, the result implies that the mass of the white dwarf must be larger than estimated by Robinson et al. (1978) so that the stellar radius drops below 0.01 solar radii and the semi-major axis is increased above 0.5 solar radii. It should be noted that the data on which this conclusion is based may include photometric errors as discussed above. We shall re-evaluate this result when spectra can be corrected for the error.

REFERENCES

Blanco, V. M., Demers, S., Douglass, G. G., and Fitzgerald, M. P.:1968, Publ. U. S. Naval Observatory, 2nd series, Vol. XXI.

Boggess et al.:1978, Nature, 275, 377.

Code, A. D.: 1972, in Scientific Results of OAO-2, ed. A. D. Code, NASA SP-310, p. 535.

Heap et al.: 1978, Nature, 275, 385.

Herter, T., Lacasse, M. G., Wesemael, F., and Winget, D. E.: 1979, Ap. J. Suppl., 39, 513.

Holm, A. V., and Gallagher, J. S.: 1974, Ap. J., 192, 425.

Holm, A. V., and Cassinelli, J. P.: 1977, Ap. J., 211, 432.

Krzeminski, W., and Kraft, R. P.: 1964, Ap. J. 140, 921.

Kurucz, R. L., Peytremann, E., and Avrett, E. H.: 1974, Blanketed Model Atmospheres for Early Type Stars (Washington: Smithsonian Distribution Press).

Linsky et al.: 1978, Nature, 275, 389.

Lynden-Bell, D.: 1969, Nature, 223, 690.

Robinson, E. L., Nather, R. E., and Patterson, J.: 1978, Ap. J., 219,168.

Sparks, W. M., Wu, C.-C., Holm, A. V., and Schiffer, F. H., III: 1979, preprint.

van Maanen, A.: 1926, Pub. A. S. P., 38, 325.

DISCUSSION FOLLOWING HOLM, WU, SPARKS, SCHIFFER AND BOGGESS

Piirola: We observed the linear polarization of Nova Cygni 1978 in the UBV bands. The polarization increased from ∿1.5% to ∿1.9% between October 1 and 11. Also position angle was changed somewhat. This correlates with the infrared observations which show that a dust-formation phase was started between October 7 and 15.

Pringle: I would like to caution taking the formal value obtained for the outer radius of the disc. I have a feeling that this value should be taken rather as an upper limit since there are a number of effects that can help to deter the turn down of the theoretical disc spectrum at long wavelengths--for example if the outer regions of the disc become optically thin or if there is any reprocessing of the inner disc radiation by the counterparts of the disc or by the companion star.

Whelan: The disc atmospheres of Herter et al. are cut off at hot outer temperatures ($T \sim 20000K$) so they have a small value of R_{out}/R_{in}. That is why those models do not fit your data where you found $R_{out}/R_{in} \sim 100$. The black body Lynden-Bell model is good and the atmosphere models of Mayo et al. have larger values of R_{out}/R_{in} and go to cooler ($T \sim 5000K$) outer radii. Observationally we know that there are cool regions of the disc because we see low ionization potential neutral or once ionized species.

THE CYCLOTRON LINES IN THE OPTICAL SPECTRA OF THE STARS OF AM HERCULIS GROUP: OBSERVATIONS AND INTERPRETATION

I.G. Mitrofanov
A.F. Ioffe Physical Technical Institute
Academy of Science of the USSR
Leningrad 194021, USSR

1. Introduction

The stars of the AM Herculis group (AM Her, VV Pup, AN UMa and 2A 0311-227) are close binaries containing a mass losing, nondegenerate star and an accreting degenerate dwarf. Their main properties are: the large linear and circular polarization of the optical light, high and low luminosity states, the variable emission line, spectra of H, He and other elements and the identification of these objects with X-ray sources. It is generally accepted that the strong magnetic field of the degenerate dwarf is responsible for these peculiar properties and for the distinction between these objects and the cataclysmic variables (Mitrofanov 1978, 1979a). The polarized optical continuum may be emitted by the accreted magnetized plasma (e.g. Chanmugam and Wagner, 1979), by the magnetized photosphere of the degenerate dwarf (Mitrofanov et al. 1977), or by both sources (Mitrofanov, 1979b). To explain the observed X-rays, Lamb and Masters (1979) showed that a magnetic field about 10^8 gs is necessary. Unfortunately, the basic prediction of their model - the strong ultraviolet continuum in the spectrum of AM Herculis - appears to be absent (Raymond et al. 1979). Chanmugam and Wagner (1979) proposed a rather different estimate for B of $2 \cdot 10^8/m^*$gs ($m^* = 5 \div 25$). For further investigations of the AM Herculis-type stars it seems useful to find a direct observational method for measuring the dwarfs' magnetic fields.

When X-ray pulsars were associated with accreting magnetized neutron stars, Gnedin and Sunyaev (1973) showed that the hot accreted plasma is a possible source of the cyclotron radiation. A spectral feature which may be interpreted as an electron cyclotron line was discovered by Trümper et al. (1977) in the hard portion (~58 kev) of X-ray spectra of Her X-1. The electron cyclotron frequency, ν_{cyc}, corresponding to the magnetic field $B \gtrsim 10^8$gs falls in the region of visual light $\nu_{cyc} = 2.81 \cdot 10^{14} \times (B/10^8\text{gs})$Hz. Consequently, the optical spectra of AM Herculis type stars may possibly contain electron cyclotron features.

In Section II the theoretical results concerning the generation of

M. J. Plavec, D. M. Popper and R. K. Ulrich (eds.), Close Binary Stars: Observations and Interpretation, 453–465.

the cyclotron radiation are discussed. The cyclotron interpretation for the strong emission feature, λ3470, discovered by Liebert et al. (1978) in the spectrum of VVPuppis is proposed in Section III. The preliminary results of the search for cyclotron lines in the spectra of AM Herculis and MV Lyrae are considered in Sections IV and V.

II. The Cyclotron Emission from the Heated Magnetized Plasma

The electromagnetic radiation is known to propagate through the magnetized plasma as ordinary and extraordinary waves. If the wave vector $\vec{k}$ is parallel to the direction of the external magnetic field $\vec{B}$, the electric vector $\vec{E}$ of extraordinary wave rotates similarly with the Larmor precession of electrons, but the electric vector of the ordinary wave rotates in the opposite direction. Therefore, the absorption of the extraordinary wave has a resonance for $\nu = \nu_{cyc}$ and is much stronger than the absorption of the ordinary wave. In the simple case ($\vec{k} \,||\, \vec{B}$) the ordinary wave absorption has no resonance and is comparable with the absorption in a plasma without a magnetic field. In all papers known to the author which have dealt with the intensity, spectrum and polarization of the radiation emitted by the hot magnetized plasma (e.g. Fabian et al., 1976; Masters et al., 1977; Chanmugam and Wagner, 1979; Lamb and Masters, 1979) only the extraordinary wave has been taken into account. Details of the ordinary wave absorption are given by Mitrofanov et al., (1979).

In the paper by Mitrofanov et al. (1979) the case of low density and nonrelativistic plasma was considered with $\omega_p^2/\omega_{cyc}^2 \ll \beta$ and $kT_e/m_e c^2 \ll 1$. Here ω_p is the plasma frequency, β is V/c and V is the thermal velocity of electrons with the temperature T_e. The temperature T_e was taken to be sufficiently high to ensure the validity of the classical approach $kT_e/\hbar\omega_{cyc} \gg 1$. The effects of both the free-free and the cyclotron absorption were included. In the presence of the strong magnetic field the effective collision frequency determining the free-free absorption depends on the mutual orientations of the external magnetic field B and the E-vector of the electromagnetic wave

$$\nu_{||,\perp} = \nu_o \frac{\Lambda_{||,\perp}}{\Lambda_o} \cdot \frac{kT_e}{\hbar\omega}\left(1 - e^{-\frac{\hbar\omega}{kT_e}}\right), \tag{1}$$

where

$$\nu_o = \frac{4}{3}\left(\frac{2\pi}{m_e}\right)^{1/2} \cdot \frac{N_e^4}{(kT_e)^{3/2}} \cdot \Lambda_o$$

is the effective collision frequency for the plasma without magnetic field, N is the electron density and $\Lambda_{||,\perp}$ are the quantities which for $B \to 0$ become the usual Coulomb logarithm Λ_o. They may be calculated using the formulae of Pavlov and Panov (1976). The differences between $\nu_{\parallel}$, $\nu_{\perp}$ and ν_o are significant only for $\omega \simeq \omega_{cyc}$. But even for these frequencies the approximation

$$\nu_{\parallel} \sim \nu_{\perp} \sim \nu_o \simeq 4 \cdot 10^7 \Lambda_o \left(\frac{N}{10^{16}\ \mathrm{cm}^{-3}}\right) \cdot \left(\frac{10^6}{T_e}\right)^{3/2} \quad \mathrm{Hz} \tag{2}$$

is sufficiently good if the cyclotron absorption is essential.

The absorption coefficients for the each wave were calculated using the dispersion equation method. For the most interesting cases analytical expressions appear to be available. If the mean effective frequency of the electron-proton collisions ν_o is much larger than the Doppler broadening $\beta\omega_{cyc}$ of the cyclotron resonance, the known formulae obtained for the absorption coefficients (Pavlov and Shibanov, 1977) are obtained:

$$k_{1,2} \simeq \frac{\omega_p^2 \nu_{\perp}}{4\omega(\omega+\omega_{cyc})^2}\left[\frac{1}{2}(1 + \cos^2\alpha) \pm P_V\cos\alpha \mp \frac{1}{2}P_Q\sin^2\alpha\right] +$$

$$+ \frac{\omega_p^2 \nu_{\perp}}{4\omega(\omega-\omega_{cyc})^2}\left[\frac{1}{2}(1 + \cos^2\alpha) \mp P_V\cos\alpha \mp \frac{1}{2}P_Q\sin^2\alpha\right] +$$

$$+ \frac{\omega_p^2 \nu_{\parallel}}{4\omega^3}\sin^2\alpha(1 \pm P_Q), \tag{3}$$

where

$$P_Q = \frac{\omega_{cyc}\sin^2\alpha}{(4\omega^2\cos^2\alpha + \omega_{cyc}^2\sin^4\alpha)^{1/2}}, \qquad P_V = \frac{2\omega\cos\alpha}{(4\omega^2\cos^2\alpha + \omega_{cyc}^2\sin^4\alpha)^{1/2}}$$

In this expression the upper sign corresponds to the ordinary ($\iota = 2$) wave and α is the angle between $\vec{k}$ and $\vec{B}$. The formulae (3) are valid also in the case when $\nu_o \ll \beta\omega_{cyc}$, but $\Delta\omega = |\omega - \omega_{cyc}|$ is so large that the cyclotron absorption appears negligible or if the angle $\alpha \to 90^o$ and the linear Doppler broadening becomes so small that $\omega_{cyc}\beta\cos\alpha \ll \nu_o$.

If $\beta\omega_{cyc} \gg \nu_o$, $\omega \sim \omega_{cyc}$ and $\alpha \lesssim \frac{\pi}{2} - \arccos\frac{\nu}{\sqrt{2}\beta\omega_{cyc}}$ the cyclotron absorption is essential and the coefficients $k_{1,2}$ equal

$$k_1 \simeq \frac{\sqrt{\pi}}{\sqrt{8}} \frac{\omega_p^2}{\omega_{cyc}^2 \beta} \cdot \frac{1 + \cos^2\alpha}{2|\cos\alpha|} \cdot \mathrm{Re}W(Z); \quad (4)$$

$$k_2 \simeq \frac{1}{2\sqrt{2\pi}} \frac{\omega_p^2}{\omega_{cyc}^2}\beta \cdot \frac{\sin^4\alpha(2 + \cos^2\alpha)^2}{|\cos\alpha|(1 + \cos^2\alpha)^3} \cdot \frac{\mathrm{Re}W(Z)}{|W(Z)|^2}, \quad (5)$$

where

$W(Z) = e^{-z^2} \cdot (1 + \frac{2i}{\sqrt{\pi}}\int_0^z dx\, e^{x^2})$ is the Kramp function and $Z = \frac{\omega - \omega_{cyc}}{\sqrt{2}\omega\beta\cos\alpha}$. The expression (4) for k_1 is well known. It has the strong resonance $\sim \omega_p^2/\omega_{cyc}\beta$ with the width $\sim \beta\omega_{cyc}$. The expression (5) for k_2 as far as we know had not been obtained elsewhere before for the low density plasma. This coefficient has a much smaller resonance $\sim \beta\omega_p^2/\omega_{cyc}^2$ and broader width. It is clear that this resonance is significant if $\beta \gg \nu_o/\omega_{cyc}$. Its magnitude is comparable with the resonance in the absorption coefficients at the second harmonic $\omega = 2\omega_{cyc}$. So, the resonant lines resulting from the ordinary wave and from the second harmonic should be emitted in similar conditions.

To illustrate our results we calculated the spectrum and polarization of the radiation from magnetized homogeneous slabs. The optical depth τ for the extraordinary wave with $\omega = \omega_{cyc}$ and $\vec{k} \parallel \vec{n}$ was used as a parameter ($\vec{n}$ is a normal to the slab's surface). Some examples are presented at the Figure 1. For $\tau \lesssim 1$ the spectrum of outcoming radiation is similar to the resonant profile of k_1. If $\tau \gg 1$ but $\lesssim \beta^{-2}$ the broad cyclotron continuum due to extraordinary wave and the narrow line due to ordinary one are emitted. If $\tau \gg \beta^{-2}$ and $\omega_{cyc}/\nu_o\beta$ the spectral distribution becomes Planckian. The polarization of radiation disappears when the slab becomes optically thick for ordinary waves because in this case the opposite polarization of the two waves compensate each other.

So, one comes to the conclusion that depending on the optical depth the hot magnetized plasma may radiate a cyclotron line, a broad cyclotron continuum with a narrow line or only the continuum.

III. The Possible Identification of the Emission Feature λ3470 in the VV Puppis Spectra

Although the VV Pup spectrum usually has strong H and He II emission lines, Angel et al. (1977) have observed that sometimes this star has an unusual state of steady low luminosity with no strong H and He II emission. According to a more detailed description of these observations (Liebert et al., 1978), in March and April of 1977 VV Pup was in two different states. In March the magnitude at minimum phase was small, $m_{min} \simeq 17.5$, brightening was large, $\Delta m \simeq 2.5$, and the H and

He II emission lines were strong. The average Hβ flux was about 1% of the continuum 3500-6000Å flux. In April the minimum light was approximately the same, but brightening was small, $\Delta m \simeq 0.5$, and the spectra had no strong H and He II emission (Hβ flux did not exceed 0.02% of the 3500-6000 Å continuum). Liebert et al. (1978) found a strong unidentified emission line near 3470 Å which was statistically significant in spectra obtained both by a single-channel prism scanner of the Steward 2.3 m telescope and by a Kitt Peak 4 m intensified image-dissector scanner. This feature corresponded to an energy flux of about 5-10% of the 3500-6000 Å continuum, i.e. it was very strong. When VV Pup was in the active state (March 1977), we can conclude from Liebert's data (1978) that the emission feature λ3470 was absent.

It seems possible to interpret the strong emission feature λ3470 as a cyclotron line (Mitrofanov, 1979). The magnitude of a magnetic field B corresponding to the wavelength of this line is equal to $3.1 \cdot 10^8$gs. The accuracy of this determination depends on the spectral width of the feature, and the error may be estimated to be $\Delta B = B \cdot \Delta\lambda/\lambda = 10^7$gs.

In April 1977, the continuum source in the 3500-6000 A band was accepted to be the degenerated dwarf's surface with a temperature of about 9000 K (Liebert et al. 1979). The distance of VV Pup is estimated to be about 100 pc. Thus in the state of steady low luminosity $L_{cont} \simeq 10^{30}$ erg/s and $L_{cyc} \simeq 10^{29}$erg/s. It seems natural to assume that the cyclotron continuum was absent and nearly the total gravitational energy released was radiated in the cyclotron line. To provide for the observed L_{cyc}, a weak accretion at a rate of $\sim 10^{12}$g/s $(M_{dwarf}/M_\odot) \cdot (5\ 10^8 \mathrm{cm}/R_{dwarf})$ is necessary. The radius of the Alfven surface appears to be much larger than the distance between the VV Pup components. This means that the dwarf's magnetic field at every point determines the dynamics of the transferring plasma. Assuming a dipole field configuration, one can estimate the area S of the dwarfs surface where the accretion occurs. If the radius of the outflowing region on the nondegenerate component is of the order of its radius $2\ \ 10^{10}$ cm and the distance between the dwarf and the outflowing region is about $2.5 \cdot 10^{10}$cm, then $S \simeq 5 \cdot 10^{15}\mathrm{cm}^2$. The hypothesized cyclotron line should arise in the plasma slab heated by the stand-off shock above this area. The electron temperature may be estimated from the condition that the linewidth of λ3470 is not smaller than Doppler broadening $\Delta\lambda/\lambda \simeq 4 \cdot 10^{-2}$. Hence $V \simeq 6 \cdot 10^8$cm/s and $T_e \lesssim 10^6$.

Taking into account the absence of cyclotron continuum one should assume that the heated plasma is optically thick relative to the cyclotron absorption of the extraordinary wave. This condition is satisfied $\tau_1 = 2\omega_{cyc}k_1 d/c \lesssim 1$ if $\langle Nd \rangle \lesssim 10^{15}\mathrm{cm}^2$, where $\langle Nd \rangle$ is the mean product of the electron concentration N and the thickness d of the slab. An increase in the accretion rate leads to an increase in the temperature and density of the heated region. The optical depth τ_1 becomes >> 1 and as a result the cyclotron continuum is emitted. The periodic brightening of VV Pup in the active state is thought to be associated with this continuum. Similarly, increased accretion results in the strong H and He II emission. Thus one may predict

Table 1

Spectral lines	Mean FWHM	Maximal FWHM	Mean equivalent width	Maximal equivalent width	$\frac{\overline{W}}{\overline{W}(4686)}$	$\frac{W_{max}}{W_{max}(4686)}$	$\frac{W^{(1}}{W(4686)}$	$\frac{W^{(2}}{W(4686)}$
He II 4686	13.2	16.5	22.9	35				
He II 4199			0.3	0.6	0.013	0.017	0.007	0.006
He II 4541.6			1.0				0.017	0.010
? 4550	20.3	39.4	4.6	11	0.20	0.30		
H_β	17.1	20.4						

1) in accordance with the Tables of Atomic Spectral Lines

2) in accordance with the theoretical estimations; the local thermal equilibrium and kT >> Hν are assumed (ν is the frequency of spectral lines)

an anticorrelation between the periodic brightening and the strong H and He II emission on the one hand and the presence of the cyclotron line on the other.

For equilibrium between the pressure of the infalling plasma $\dot{M}V_{ff}/2S$ and the thermal pressure inside the hot region $2NkT_e$ it is necessary to have $N \simeq 10^{14}cm^{-3}$ for $T_e \simeq 10^6$. Therefore, d should be about 10 cm. The plasma heated by the standoff shock has to provide the observed luminosity of the cyclotron line

$$L_{cyc} \simeq \frac{B^2}{4\pi}\delta_T\beta^2 cS\langle Nd\rangle \simeq 10^{14}\langle Nd\rangle\left(\frac{B}{3\cdot 10^8 gs}\right)^2 \cdot \left(\frac{T_e}{10^6}\right) \cdot$$

$$\cdot\left(\frac{S}{5\cdot 10^{15}cm^2}\right)\frac{ergs}{s} \qquad (6)$$

where δ_T is the Thomson cross-section of the electron. The value of L_{cyc} estimated from (6) agrees with the observed luminosity $\sim 10^{29}$ergs/s if $\langle Nd\rangle$ is just the same, i.e. $\sim 10^{15}cm^{-2}$. Some part of the energy emitted in the cyclotron line should be absorbed by the surface of the degenerate dwarf giving rise to a hot spot with the luminosity $L_{sp} \lesssim 10^{29}$ergs/s.

IV. The Possible Cyclotron Line in the Optical Spectra of AM Herculis

The observations of AM Her-type stars were carried out by the A.F. Ioffe Physical Technical Institute, the Special Astrophysical Observatory and the Crimea Astrophysical Observatory. The spectroscopy was performed at the 6-m telescope. The apparatus and the first results were described by Burenkov and Voikhanskaya (1979) and we will discuss only several (about 60) spectra (Figure 2) obtained in 7-8 of July, 1978. Besides the usual emission lines of H (from Hβ to H_{14}), He, C, N, etc., some of the spectra also have strong and broad emission features at 4550 A accompanied by absorption at 4610A. The equivalent width W and the spectral profile of λ4550 depends on the phase of the orbital period. The phases were calculated in accordance with elements used by Priedhorsky et al. (1978). The magnitude of W is very small << 1 Å and the wavelength of the features is rather indeterminate at the phases 0.85-1.0. At the other phases W appears to be much larger ~ (5 ÷ 10) Å. We shall discuss below the interpretation of the feature λ 4550 only at phases for which its equivalent width is sufficiently large.

A large number of atoms and ions (N I, Fe II, Ne I etc.) have weak spectral lines near 4550 Å. Nevertheless they may be rejected because the stronger lines of the same spectral series of these atoms and ions are absent in the observed spectra. The nearest possible atomic line seems to be He II 4541.6. For the identification of λ4550 with this line one should assume that it has the regular red Doppler shift ~8 Å which corresponds $V_{\parallel} \simeq 500$ km/s. This velocity is comparable with the amplitude of the observed velocity curve of He II 4686 (Greenstein et al., 1977), but this curve is quite symmetrical with approximately equal positive and negative shifts. So, it seems difficult to identify λ4550 with He II 4542. Other arguments against this identification are given below.

The full width at half-maxima (FWHM) of λ4550 is approximately twice the FWHM of the strongest line λ4686 of He II. The ratios of the line strengths W(4550)/W(He II 4686) and W(4550)/W(HeII 4199) are approximately an order of magnitude larger than the estimated ratios W(He II 4542)/W(He II 4686) and W(He II 4542)/W(He II 4199). The line strength of λ4550 appears to be much larger than the possible strength of the line He II 4542. The equivalent widths of the all H and He lines depend on the orbital phase and their variations are well correlate with each other. The variation of λ4550 does not correlate with any of them. Finally, one should take into account that the spectral feature λ4550 lies approximately at the spectral boundary ~4500 Å between the polarized red and nonpolarized blue light (Kruzewski 1977).

So, the spectral feature λ4550 is essentially distinguished from the other atomic lines and it may be interpreted as an electron cyclotron line (Voikhanskaya and Mitrofanov, 1979). It seems to be emitted by the heated slab of the magnetized plasma above the surface of the degenerate dwarf. The absorption feature λ4610 is likely to be due to cyclotron absorption in the higher and colder regions of the accretion column with a smaller magnetic field. The wavelength and the linewidth correspond to a magnetic field $B = (2.3 \pm 0.1) \cdot 10^8$ gs in the emitting region, i.e. near the pole of the magnetic degenerate dwarf.

In contrast to VV Puppis the cyclotron line and the polarized cyclotron continuum seem to be present in the optical spectra of AM Herculis. This is possible (see Section II) if the optical depth τ_1 of the plasma slab for the extraordinary wave is >> 1 and for the ordinary wave $\tau_2 \sim \tau_1 \beta^2 \lesssim 1$. The electron temperature T_e and the parameter β may be estimated from the condition that the Doppler broadening is smaller than the mean observed linewidth of λ4550-4610. They appear to be $\beta \lesssim 10^{-2}$ and $T_e \lesssim 3 \cdot 10^5$. If the observed cyclotron line was emitted almost perpendicularly to the magnetic field the linewidth is proportional to $\beta^2 \omega_{cyc}$. In this case β and T_e may be as much as $\sim 3 \cdot 10^{-2}$ and $\sim 10^6$, respectively. The parameter <Nd> of the emitting region is estimated about $10^{18} \div 10^{19}$ cm^{-2}.

V. The Possible Broad Cyclotron Feature in the Spectrum of MV Lyrae

The novalike star MV Lyrae is a possible member of the AM Herculis group (Voikhanskaya et al., 1978). The variable linear and circular polarization of its light achieved ~1.5% and ~4%, respectively. In the spectrum of MV Lyr Greenstein (1954) reported broad hydrogen and faint λ4471 He I emission lines. Current observations indicate that these lines have very variable strengths and profiles (Voikhanskaya 1979).

Among all spectra of MV Lyrae obtained in 1977-78 two large groups were picked out. The spectra of the first group had strong emission

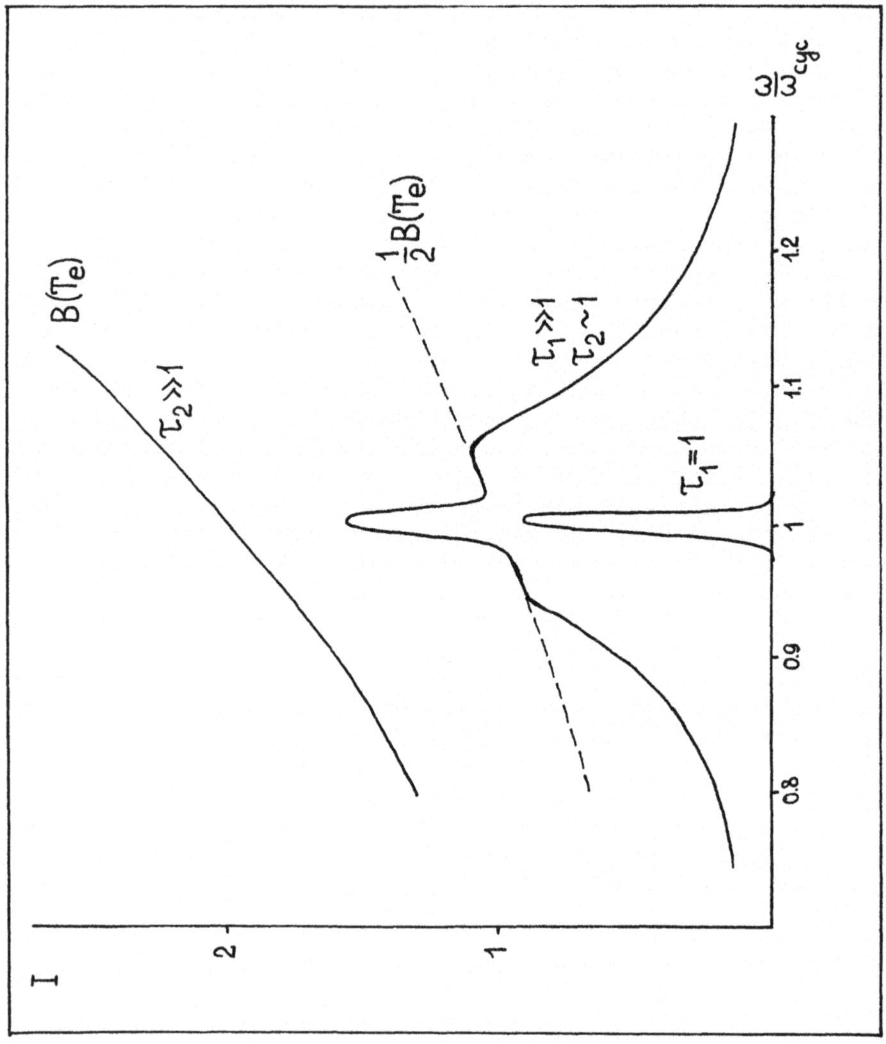

Fig. 1. The spectrum of radiation from a magnetized homogeneous slab with $\vec{n} \,||\, \vec{B}$, $kT_e/\hbar\omega_{cyc} = 5$, $\beta = 0.01$, $\nu_o/\omega_{cyc} = 10^{-4}$ and $\alpha = 60^o$; $B(T_e)$ is the Planck function.

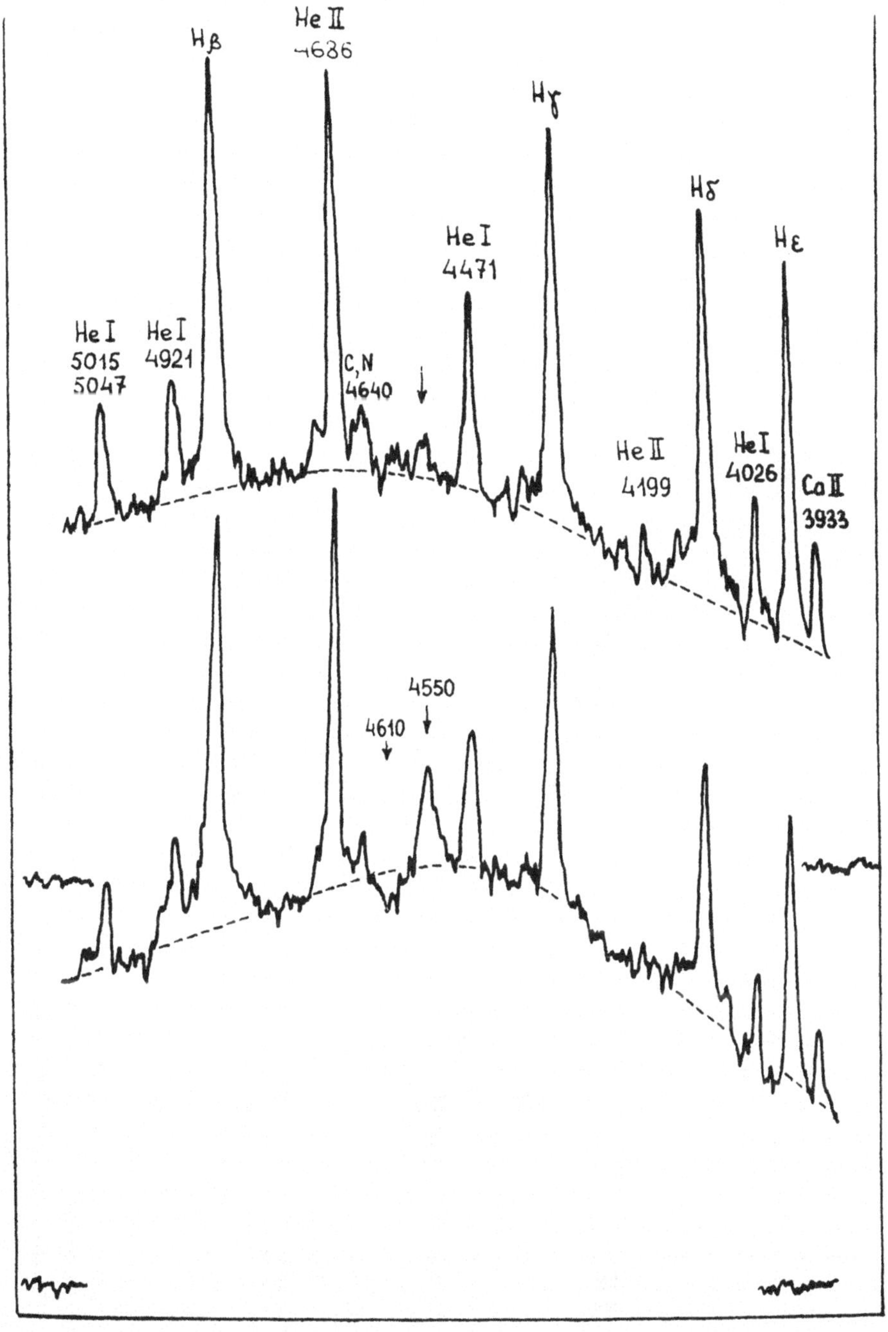

Fig. 2. Two spectra of AM Herculis obtained on June 7/8, 1978. Dashed are the continua. Noise levels are indicated at left and right.

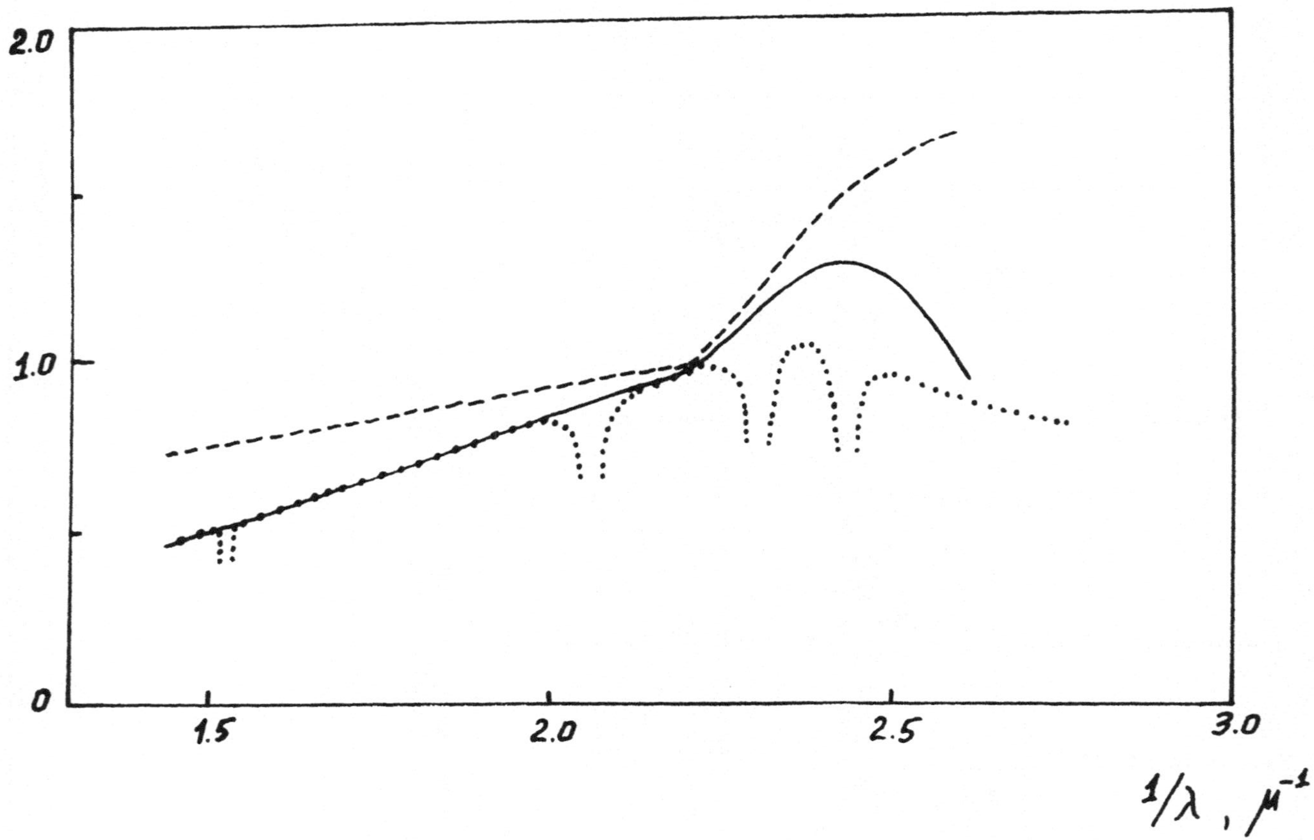

Fig. 3. The generalized spectra of MV Lyrae at the states with strong hydrogen lines (dashed) and without them. The dotted line represents the model spectrum of a DA dwarf.

lines and the second one contained spectra without any noticeable emission lines. In Figure 3 the solid and dashed lines correspond to the generalized spectra from the second and first groups, respectively. In the state with strong lines the spectrum of MV Lyrae has a significant excess in the ultraviolet region. In the state without atomic emission lines the spectrum of MV Lyrae fits with the model spectrum of DA-dwarf ($T_{eff} \simeq 2 \cdot 10^4$) (Voikhanskaya, 1979) but with two distinctions: the absence of the absorption hydrogen lines and the presence of the broad emission feature $\lambda 4200$ which has an equivalent width ~140 Å and a FWHM of about 450 Å.

It seems natural to suppose that this feature is also the electron cyclotron line (Voikhanskaya and Mitrofanov, 1979). Using the line-width one can estimate the electron temperature to be $\sim 10^6$. The corresponding magnetic field is about $(2.6 \pm 0.2) \cdot 10^8$gs.

MV Lyrae is known to be a spectral binary with a period $\sim 2^h$ (Walker, 1968). It undergoes irregular light variations up to 0.4 mag in the ultraviolet band. As the wavelength increases the amplitude decreases. The characteristic time scale of the light variations is $2\text{-}5^{min}$. Walker (1954) supposed the light increases to be due to the brightening of the some parts of the dwarf's surface, i.e. due to the origin of a number of hot spots. The peculiarity of MV Lyrae may result from the nonstationary regime of the accretion into the strong magnetic field of the degenerate dwarf. The variable and multicomponent profiles of the emission lines indicate that the plasma is accreted as separated drops. Every falling drop heats a local region onto the dwarf's surface and a bright spot appears. This may lead to the irregular variations of light and polarization and to the broadening of the cyclotron line.

Acknowledgements

Many results discussed in this paper were obtained by the author together with Dr. N.F. Voikhanskaya (in Sections IV and V) and Drs. G.G. Pavlov and Yu. A. Shibanov (in Section II). I thank them for useful and pleasant collaboration.

REFERENCES

Angel, R., Liebert, J., Stockman, H., and Bond, H.: 1977, IAU Circ. 3049.
Burenkov, A.N. and Voikhanskaya: 1979, submitted to Astron. Zhur. (USSR).
Chanmugam, G. and Wagner, R.L.: 1979, submitted to Astrophys. J.
Fabian, A.C., Pringle, J.E., and Rees, M.J.: 1976, Mon. Not. 175, 43.
Gnedin, Yu.N. and Sunyaev, R.A.: 1973, Astron. Astrophys. 25, 232; 36, 379.
Greenstein, J.L.: 1954, Publ. Astron. Soc. Pacif. 66, 79.
Greenstein, J.L., Sargent, W.L.W., Boroson, T.A., and Boksenberg, A.: 1977, Astrophys. J. 218, L 121.

Kruszewski, A.: 1978, in "Nonstationary Evolution of Close Binaries", ed. A.N. Zytkow, p. 55.
Lamb, D.Q. and Masters, A.R.: 1979, submitted to Astrophys. J. Lett.
Liebert, J., Stockman, H.S., Angel, R., Woolf, N.J., Hege, K., and Margon, B.: 1978, Astrophys. J. 225, 201.
Masters, A.R., Pringle, J.E., Fabian, A.C., and Rees, M.J.: 1977, Mon. Not. R.A.S. 178, 501.
Mitrofanov, I.G.: 1978, Pisma Astron. Zhur. 4, 219 (Sov. Astron. Letters 4 (3) 119.
Mitrofanov, I.G., Pavlov, G.G., and Gnedin, Yu.N.: 1977, Pisma Astron. Zhur. 3, 341 (Sov. Astron. Lett. 3 (4), 182.)
Mitrofanov, I.G., Pavlov, G.G., and Shibanov, Yu.A.: 1979, submitted to Astron. Zhur.
Mitrofanov, I.G.: 1979a, see previous article in these Proceedings. 1979b, submitted to Nature.
Pavlov, G.G. and Panov, A.N.: 1976, JETF 71, 572.
Pavlov, G.G. and Shibanov, Yu.A.: 1978, Astron. Zhur. 55, 373.
Priedhorski,W.C., Krzeminski, W., and Tapia, S.: 1978, Ap.J. 225, 542.
Raymond, J.C., Black, J.H., Davis, R.J., Dupree, A.K., Gursky, H., Hartmann, L., and Matilsky, T.A.: 1979, Ap.J. 230, L95.
Trümper, J., Pietsch, W., Reppin, C., Sacco, B., Kendziorra, E., and Staubert, R.: 1976, Proc. VIII. Texas Symp. on Relativistic Astrophysics, Boston.
Voikhanskaya, N.F., Gnedin Yu.N., Efimov, Yu.S., Mitrofanov, I.G., and Shakhovskoi, N.M.: 1978, Pisma Astr. Zhur. 4, 272 (Sov. Astron. Lett. 4(3) 148.
Voikhanskaya, N.F., Mitrofanov, I.G.: 1979, submitted to Pisma Astr. Zh.
Voikhanskaya, N.F.: 1979, submitted to Astron. Zhur.
Walker, M.F.: 1954, Publ. Astron. Soc. Pacif. 66, 71.
Walker, M.F.: 1966, Adv. in Electronics and Electron Phys. 22B, 761.

ULTRAVIOLET OBSERVATIONS OF AM HERCULIS

J.C. Raymond, G. Branduardi, A.K. Dupree, G. Fabbiano and L. Hartmann
Harvard-Smithsonian Center for Astrophysics, Cambridge, MA

Spectra of AM Herculis taken with IUE show a continuum which is partially eclipsed in phase with the X-ray eclipse and can be separated into two components; a black-body component (kT_{BB} = 25 - 30 eV) which accounts for the ¼ keV X-rays and the eclipsed UV continuum; and a component roughly described by $F_\nu \propto \nu^{-1}$ which is not eclipsed.

Am Herculis is a binary X-ray source believed to contain a strongly magnetic white dwarf accreting matter from an M dwarf (cf. Stockman et al. 1977; 1978). The ultraviolet continuum of AM Her consists of a component having a Rayleigh-Jeans spectrum, which is eclipsed in phase with the X-rays, and a flatter component which is not eclipsed (Raymond et al. 1979). We obtained thirteen low dispersion, short wavelength spectra with the IUE satellite on March 17, 1979 providing uniform coverage over nearly two cycles. The IUE instrument is described by Boggess et al. (1978).

By assuming that the "black body" component is completely eclipsed and that the spectral shape of the flat component is constant ($F_\lambda \propto \lambda^{-1.16}$) we can separate the contributions of the two components of the UV continuum. Figure 1 shows the black body and flat contributions to the flux at 1400 Å. The units are 10^{-13} ergs cm^{-2} s^{-1} Å^{-1}. Phases are based on the primary minimum in V and the period of Szkody and Brownlee (1977).

The similarity of the black body light curve to the soft X-ray light curve (Tuohy et al. 1977) strengthens the hypothesis that a single emission region accounts for both. If the emission is indeed black body in nature, $T_{BB} \sim 30$ eV. A power-law fit to the difference between the continua at maximum and minimum of the black body light curve yields $P_\lambda \propto \lambda^{-3.75}$. The uncertainty in spectral index is difficult to estimate, but ± 0.5 is probably realistic.

Models currently available (Lamb and Masters 1979; King and Lasota 1979) do not match the observed Rayleigh-Jeans component in the U.V.

M. J. Plavec, D. M. Popper and R. K. Ulrich (eds.), Close Binary Stars: Observations and Interpretation, 467–469.

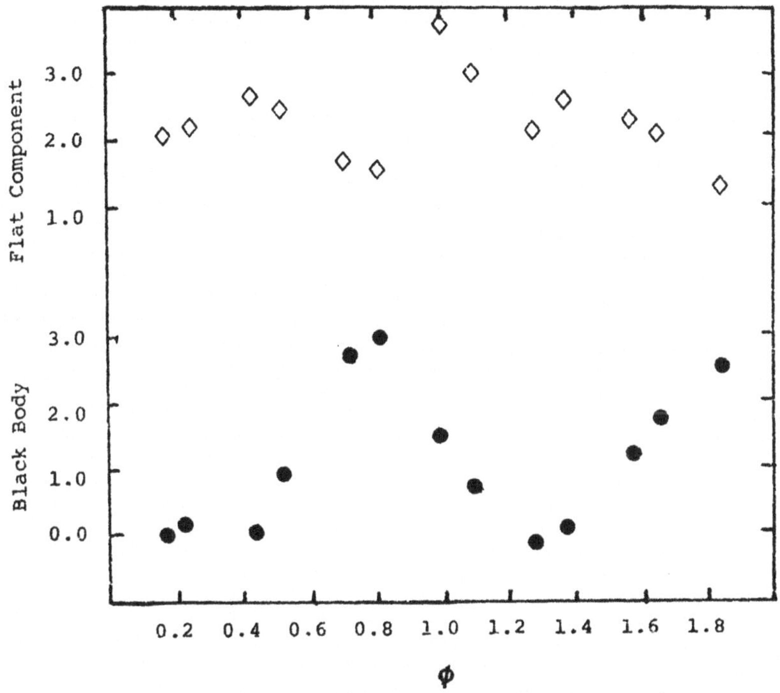

We gratefully acknowledge the assistance of the IUE Observatory staff in the acquisition and reduction of these data. This work has been supported by the National Aeronautics and Space Administration under Grant NSG5370 to Harvard University.

Boggess, A., Carr, F.A., Evans, D.C., Fischel, D., Freeman, H.R., Fuechsel, C.F., Klinglesmith, D.A., Krueger, V.L., Longanecker, G.W., Moore, J.V., Pyle, E.J., Rebar, F., Sizemore, K.O., Sparks, W., Underhill, A.B., Vitagliano, H.D., and West, D.K.: 1978, Nature, 275,372.

King, A.R., and Lasota, J.P.: 1979, preprint.

Lamb, D.Q., and Masters, R.A.: 1979, preprint.

Raymond, J.C., Black, J.H., Davis, R.J., Dupree, A.K., Gursky, H., Hartmann, L., and Matilsky, T.A.: 1979, Ap. J. (Letters) 230, L95.

Stockman, H.S., Schmidt, G.D., Angel, J.R., Liebert, J., Tapia, S., and Beaver, E.A.: 1977, Ap. J. 217, 815.

Szkody, P., and Brownlee, D.E.: 1977, Ap. J. (Letters) 212, L113.

Tuohy, I.R., Lamb, F.K., Garmire, G.P., and Mason, K.O.: 1977, Ap. J. (Letters), 226, L17.

DISCUSSION FOLLOWING RAYMOND, BRANDUARI, DEPREE, FABBIANO AND HARTMANN

Pringle: These stars seem to have a split personality since there was a certain amount of discussion on them at the IAU Colloquium on White Dwarfs held last week in Rochester. At that meeting Dr. Don Lamb seemed happy that his work could accommodate the ultraviolet observations you now report by taking a less massive white dwarf. I regret that he is not here to tell us more about the model.

Tapia: At the risk of sounding rather conformist I would like to say that the theoretical efforts to interpret AM Herculis objects appear to me more than satisfactory. We should not forget that our geometrical models are naive and we can't avoid it. One example is the sketch of the accretion column that I presented in my talk. Perhaps we should be as demanding of the geometry as we are of the physics.

UPPER LIMITS FOR THE LUMINOSITY OF A dM COMPANION IN AM HERCULIS OBJECTS

S. Tapia and G. V. Coyne
Steward Observatory and Lunar and Planetary Laboratory
Tucson, Arizona

SUMMARY*

The quest for optical identification of X-ray sources has revealed a new class of variables currently formed by AM Her, AN UMa, VV Pup, and the most recent member, 2A 0311-227 (for which a variable star designation will be available soon). These objects show many of the variations common to close binary systems where mass transfer is under way. However, they are distinguished by the high degree (10 to 35 %) of linear and circular polarization observed in the optical and near infrared spectral regions. Apparently the polarization is due to cyclotron emission of electrons located in a region where the magnetic field strength is about 10^8 gauss.

Synchronized periodic variations of the polarization and the X-ray flux suggest that the magnetic field originates in an accreting degenerate dwarf star. In such systems formation of an accretion disk is prevented by the strong magnetic field which channels the inflow of plasma towards the magnetic polar caps, forming accretion columns. The X-ray luminosity and probably the bulk of the optical luminosity are produced in the accretion columns.

Sporadic states of low luminosity, up to 5 mag fainter than the bright state, are also observed in AM Herculis objects. According to the model of magnetically confined accretion, the low states indicate drastic reductions of the mass transfer rate with comparable reductions off efficiency from the emission mechanisms taking place at the polar caps and in the accretion columns. The optical and infrared polarization should be especially affected by the low states because, aside from the reduction in efficiency of cyclotron emission, the polarization is expected to be diluted by the relatively larger contribution of unpolarized flux from stellar components. Nevertheless, observations of AM Her, AN UMa, and VV Pup obtained during states of low luminosity reveal that the degree of circular polarization remains practically unchanged

M. J. Plavec, D. M. Popper and R. K. Ulrich (eds.), Close Binary Stars: Observations and Interpretation, 471–472.

and the values measured at 0.70 and 0.85 μm place strict upper limits on the luminosity of a dM type companion.

*This summary was submitted before the meeting; no update has been received. (Editorial remark).

DISCUSSION FOLLOWING THE PAPER BY TAPIA AND COYNE

Mayo: The secondary might show up at longer wavelengths, i.e. in the IR bands in the light curves. In particular, those systems that are observed to eclipse in the optical, render it possible to observe a secondary eclipse at IR wavelengths.

Tapia: I am not sure that we observe star eclipses in the optical light curves of these systems. I agree that your suggestion should be pursued by infrared observers.

Friedjung: Can you give a minimum absolute magnitude for the red component?

Tapia: If we use the current information on the distance of these objects, that is approximately 100 pc, the derivation of absolute magnitudes is trivial. However, I do not think that we step into safe grounds if we assume that they are bona fide M dwarfs.

Whelan: One way of seeing if the disc changes is to measure its electron density, possibly by use of the Balmer emission lines. Since you have access to excellent detectors, will you attempt this?

Tapia: The data is in the drawer!

OBSERVATIONAL STRUCTURE OF DISK-LIKE ENVELOPE IN THE SYSTEM DQ HER

A.M. Cherepashchuk and E.S. Dmitrienko
Sternberg Astronomical Institute
Moscow University
Moscow, U.S.S.R

ABSTRACT

The analysis of the eclipses in the system DQ Her based on the photometric observations of different authors during 1954, 1956 and 1975 allows us to obtain the photometric elements of the system and the structure of disk-like envelope surrounding the white dwarf. The orbit inclination of the system is $i = 80^\circ \pm 3^\circ$, the major axis of the secondary component is 0.40 ± 0.03, the orbit radius is $(10 \pm 2)10^{10}$cm, the masses of the components are: $m_{wd} = (0.65 \pm 0.35)M_\odot$, $m_2 = (0.50 \pm 0.15)M_\odot$. A decrease in the optical luminosity of $\sim$20% and a total decrease in size of the disk-like envelope of $\sim$10% is observed from 1954 to 1975. The brightness of the disk-like envelope increases towards its center. The radiation temperature increases toward the center according to the laws $T \sim r^{-0.20}$ and $T \sim r^{-0.35}$ for 1954 and 1975, respectively. The average value of the radiation temperature of the disk-like envelope is about 10^4K. The bolometric luminosity of the envelope is (1 to 2)$L_\odot$. The bolometric luminosity of the central white dwarf is anomalously high: $\sim$(20 to 200 $L_\odot$) and the effective temperature of its surface is $\sim$(1.0 to 1.7)10^5K. All these results indicate that the disk-like envelope in the system DQ Her is an accreting disk.

M. J. Plavec, D. M. Popper and R. K. Ulrich (eds.), Close Binary Stars: Observations and Interpretation, 473.

IX. CONTACT BINARIES

THEORIES OF CONTACT BINARY STARS

Frank H. Shu
Department of Astronomy, University of California
Berkeley, California 94720

ABSTRACT

We review and contrast two current theories for the structure of contact binary stars: discontinuity theory and thermal relaxation oscillation theory. We find that the two theories are complementary with the crucial theoretical issue to be resolved being the secular stability of the temperature inversion layer. Critical observational tests remain to be performed.

1. INTRODUCTION

Today I wish to review a fairly controversial subject, namely, current theories of contact binaries. There are two main contending schools of thought, which Lucy and Wilson (1979) have conveniently labelled the DSC and TRO theories. DSC, or discontinuity theory, was advanced principally by Shu, Lubow, and Anderson (1976, 1979), but some elements of it had already appeared in the earlier work of Biermann and Thomas (1972) and Vilhu (1973). TRO, or thermal relaxation oscillation theory, was advanced by Lucy (1976) and was also worked upon by Flannery (1976) and Robertson and Eggleton (1977).

Two conceptually distinct and yet ultimately related issues underlie the discussions. I shall refer to them as:

(a) Kuiper's paradox, which is the lowest order issue, concerns the structure of contact binaries. How can the structure of two stars in physical contact be made compatible with the requirements of the Roche geometry?

(b) Lucy's paradox, which is the next order issue, concerns the luminosity redistribution in contact binaries. How do the luminosities emerging from the interiors get transformed to become the surface brightness distribution characteristic of contact binaries?

M. J. Plavec, D. M. Popper and R. K. Ulrich (eds.), Close Binary Stars: Observations and Interpretation, 477–484.

The ordering parameter is the actual energy flux F which is required to carry the heat out of the interior divided by the hypothetical enthalpy flux $\rho h a_s$ which would result if matter were to flow at the speed of sound a_s:

$$\delta = F/\rho h a_s \ . \quad (1)$$

The value of the parameter δ at the base of the common envelope (just above the inner critical surface) must be much less than the ratio of the depth of the common envelope, H, to the orbital separation, D, if the system is to achieve the good thermal contact required by the light curves of W UMa stars. Indeed, the light curve theory of Anderson and Shu (1977) derived from this assumption is virtually identical, except for minor details, to the theory of Lucy (1968b) which is well known to give a good fit to the observations. Models of ZAMS contact binaries with filled fractions $f \sim 0.5$ show δ to be $\sim 10^{-8}$ in models of low total mass and to be $\sim 10^{-5}$ in models of high total mass (Lubow and Shu 1977, 1979). There is some observational evidence that the filled fraction f is systematically less than 0.5 in the former contact binaries (Lucy 1973, Rucinski 1973), and this fact may be related to the issue of the maintenance of the temperature inversion layer (§ 5).

2. KUIPER'S PARADOX

Kuiper (1941) argued that the sizes of (what we now know to be) isolated main sequence stars are generally incompatible with the mechanical requirements of the Roche geometry for contact binaries unless the two stars are identical. A paradox arises because W UMa stars are main sequence stars which have mass ratios typically about 0.5 (Lucy 1968a).

Shu, Lubow, and Anderson (1976, 1979) have extended Kuiper's paradox to nonspherical stars of arbitrary amounts of core evolution. Given the masses of the two stars, M_I and M_{II}, and the orbital separation, D, plus the zero-order equations of stellar structure, we find that, except for special eigenvalue solutions of little physical interest, there are too many boundary conditions to satisfy if all thermodynamic variables are continuous across the inner critical surface. This extension brings us to the two proposed escapes from Kuiper's paradox: DSC theory and TRO theory.

3. THE CENTRAL FEATURES OF DSC THEORY

To lowest order in δ, horizontal and vertical hydrostatic equilibrium in the corotating frame requires the stellar fluid to be barotropic, i.e., the pressure P and the density ρ to be functions only of the effective potential Φ. In DSC theory, P is continuous across the inner critical surface but ρ (and the temperature T) need not be. In particular, Shu, Lubow, and Anderson (1976) argued that Kuiper's paradox can be resolved by allowing the specific entropy s (and the temperature T) below the Roche lobe of one of the stars be less (for mechanical stability) than the corresponding value at the base of the common envelope (contact discontinuity). To satisfy energy conservation, the thermodynamic variables are continuous across the Roche lobe of the other star, but their derivatives are not (weak discontinuity).

DSC theory applies to contact binaries of all masses and stages of stellar evolution as long as synchronism is maintained and $\delta << 1$ at the base of the common envelope. The results of Lubow and Shu (1977, 1979) for ZAMS contact binaries show:

(a) ZAMS contact binaries of low total mass have common convective envelopes, with the contact discontinuity residing on the secondary (star II). Thus, the specific entropy of the convection zone beneath star II's Roche lobe is less than the specific entropy of the common convective envelope.

(b) ZAMS contact binaries of intermediate total mass have common radiative envelopes, with the contact discontinuity residing on the secondary. Thus, the specific entropy of the convection zone beneath star II's Roche lobe is less than the specific entropy at the base of the common envelope.

(c) ZAMS contact binaries of high total mass have common radiative envelopes, with the contact discontinuity residing on the primary. Thus, the specific entropy of the <u>induced</u> convection zone beneath star I's Roche lobe is less than the specific entropy at the base of the common envelope.

Our studies showed a contact discontinuity to be <u>sufficient</u> to

(a) resolve Kuiper's (1941) paradox,

(b) explain Eggen's (1967) period-color relation for W UMa stars,

(c) account for the existence of contact binaries with common radiative envelopes (Rucinski 1973, Leung 1980).

Shu, Lubow, and Anderson (1979) also claimed that a contact discontinuity is generally also <u>necessary</u> for all viable theories of contact binaries (with synchronous spins), with or without evolved cores, and with or without thermal equilibrium.

4. THE CENTRAL FEATURES OF TRO THEORY

To see that a contact discontinuity would also naturally arise even in TRO theory, let us review Lucy's (1976) model. I shall confine my remarks to Lucy's version of TRO theory because only Lucy's version uses commonly accepted zeroth-order equations of stellar structure (cf. Flannery 1976 and Robertson and Eggleton 1977).

Lucy's theory applies only to contact binaries of low enough total masses to have common convective envelopes. To motivate his model in the context of what we have discussed so far, let us imagine what would happen if we initially force the specific entropy in the outer convection zones of the equilibrium model of low total mass in Lubow and Shu (1977, Fig. 2) to have a single value above and below the inner critical surface. The imposition of a higher value of s beneath the Roche lobe of star II would lead to an expansion of those layers. To compensate so that star II can still fit inside its Roche lobe, we would need to decrease the volume occupied by the deep radiative interior. Lucy (1976) performs this task elegantly by artificially lowering the nuclear energy generation rate in the secondary's core for time $t < 0$. For $t > 0$, the energy generation rate is allowed to have its natural value. The overcompressed secondary responds by expanding and transferring mass to the primary. With a uniform value imposed for s in the outer layers, the mass ratio is then determined as an eigenvalue of each subsequent instant of time t. Mass transfer from the less massive star to the more massive with the conservation of total angular momentum increases the orbital separation, which leads eventually to the loss of contact (see Fig. 3 of Lucy 1976).

When contact is lost, the secondary detaches from its Roche lobe and shrinks toward its naturally smaller ZAMS state for isolated stars. In the meantime, the primary tries to expand toward its naturally larger ZAMS state for isolated stars. Because the primary is then constrained to fill its Roche lobe, this expansional tendency leads to mass transfer from the lobe-filling primary to the detached secondary. This decreases the orbital separation and eventually causes the two stars to come once again into contact. Lucy (1976) ends his formal calculations just before contact is reestablished with the comment that hydrodynamic events would accompany the formation of the contact configuration. Thus Lucy's calculations end with the hope that an oscillatory cycle would ensue.

Let us pursue in our imagination what the subsequent events might be. At the instant of contact, the primary has a higher outer value of the specific entropy (proportional to -log K in Table 1 of Lucy 1976) than the secondary. The outer layers of the primary, being therefore buoyant, would flow to cover the secondary and to form a common convective envelope. The entropy of this outer convective envelope would continue to maintain a relatively high value because it is determined by photospheric boundary conditions. The entropy of the convection zone beneath the Roche lobe of the secondary would maintain a lower value (i.e., the secondary would maintain a contact discontinuity) for at least the thermal timescale required to heat up this layer. The crucial difference in opinion of DSC theorists and TRO theorists centers precisely on this issue. DSC proponents speculate that after initial transients have died away, the contact discontinuity so established could continue to be maintained by fluid flow for nuclear timescales. TRO proponents speculate that the contact discontinuity so established cannot be maintained and would heat up the interior of the secondary -- presumbably to establish an oscillatory cycle. Let us merely note here that even if TRO theory is correct, the original model must be modified so that the oscillations take place about a state with a contact discontinuity. In other words, the zero-order models of DSC theory constitute the equilibrium states about which TRO models oscillate.

5. LUCY'S PARADOX

From the above discussion, we see that the crucial issue boils down to whether the equilibrium models of DSC theory are secularly stable. We believe that this issue is intimately related to the next order issue: Lucy's paradox. A generalization of Lucy's (1968a) arguments which lead to this paradox is given below (see also Fig. 2 of Shu, Lubow, and Anderson 1979). To lowest order in δ, the stellar fluid is barotropic with all thermodynamic variables functions only of the effective potential Φ. For such a barotorpic state, the luminosities <u>perpendicular</u> to equipotentials can be shown to satisfy:

$$L_{rad} \propto -dT/d\Phi, \; L_{conv} \propto (-ds/d\Phi)^{3/2}, \; L_{fluid} = 0, \tag{2}$$

with known proportionality constants. Applied to an equipotential surface B which lies just above the inner critical surface, equation (2) yields a definite ratio of the luminosities L_I' and L_{II}' which enter the common envelope on the sides of stars I and II. In general, the luminosities L_I and L_{II} which cross the equipotential surfaces W and C just below the Roche lobes of stars I and II will not equal L_I' and L_{II}'. (For example, in thermal equilibrium L_I and L_{II} equal the nuclear energy generation rates in the cores, which are unrelated to the distribution laws [2] required by the barotropic state above the inner critical surface.) If the surfaces W, C, and B can be chosen to lie arbitrarily closely (as in non-DSC theories), then the conversions $L_I \rightarrow L_I'$ and $L_{II} \rightarrow L_{II}'$ would require <u>infinite</u> horizontal fluid fluxes (a finite luminosity redistribution through a vanishingly small cross-sectional area). It is precisely the realization that the energy carrying capacity of fluid flow is finite (although very large) that led Shu, Lubow, and Anderson (1976, 1979) to propose the following resolution of Lucy's paradox:

The uneven heating of the base of the common envelope by the emergent interior luminosities leads to a spatially thin baroclinic layer which straddles the inner critical surface. To lowest order in δ, the thin layers which sandwich the inner critical surface are modelled by a contact discontinuity and by a weak discontinuity. To next order, the fractional thickness of these layers can crudely be estimated to be of order $\delta^{2/5}$, with fluid flow hypothesized to maintain the "discontinuity" structures in a quasi-stationary state.

The topic of the long-term maintenance of the temperature inversion layer excites all the present controversy. The detailed arguments are too complex to discuss here; let me merely state here the crucial issues and refer the reader to the original discussion in the literature. How does the long-term maintenance of the temperature inversion layer work in theory (cf. Hazlehurst and Refsdal 1978)? The basic requirements involve a "heat engine" and a "refrigerator" (Shu, Lubow, and Anderson 1979). Can it work in principle (cf. Papaloizou and Pringle 1979)? Yes, the principle involves a locally time-dependent form of "negative eddy conductivity" (Shu, Lubow, and Anderson 1979, 1980). Why should it work in detail (cf. Smith, Robertson, and Smith 1979)? Because the zero-order theory allows three independent parameters (taken for convenience to be M_I, M_{II}, and D), the thickness of the common envelope can be adjusted to yield the critical value of δ at the base of the common envelope required for the "heat engine" to drive the "refrigerator" by exactly the right amount to maintain the transition layer in a steady state for given total mass $M_I + M_{II}$ and total angular momentum J (Shu, Lubow, and Anderson 1980). The likelihood of solutions satisfying this requirement is large because the possible range of δ corresponding to the range of filled fractions from f = 0 to f = 1 is large. This last conjecture is important for answering the objection of Lucy and Wilson (1979) that DSC models do not provide unique evolutionary sequences. Would the proposed solution be secularly stable? Shu, Lubow, and Anderson (1980) have given a physical argument that the solution would be secularly stable, but more detailed calculations are needed.

6. SUMMARY

Although DSC and TRO theory seem superficially to be quite distinct, they are actually complementary. Without proving secular instability of the temperature inversion layer, TRO theory remains incomplete because the existence of a <u>complete</u> cycle has not yet been demonstrated with standard assumptions. On the other hand, without proving secular stability, the long-term maintenance of the contact discontinuity of DSC theory remains an unproven <u>assumption</u>. Lucy and Wilson (1979) argue that observations favor the TRO theory although the evidence at present is not yet decisive. More specific comparisons will become possible after the DSC theory is made complete by a (perhaps semi-empirical) evaluation of the critical value of δ at the base of the common envelope for a given total mass and angular momentum. We believe that critical observational tests will come in two different directions:

(a) Specific predictions by TRO and DSC theories concerning filled fractions, ratios of systems in the two phases of relaxation oscillation, and the properties of contact binaries with common radiative envélopes.

(b) Evolutionary sequences: Webbink (1977) has claimed that TRO theory requires contact binaries to evolve into single stars. We have hopes that the complete DSC theory would ultimately allow the evolution of W UMa stars into cataclysmic variables (Kraft 1965).

REFERENCES

Anderson, L., and Shu, F.H. 1977, *Ap. J.*, 214, 798.
Biermann, P., and Thomas, H.-C. 1972, *Astr. and Ap.*, 16, 60.
Eggen, O. J. 1967, *Mem. R. A. S.*, 70, 111.
Flannery, B. P. 1976, *Ap. J.*, 205, 217.
Hazlehurst, J., and Refsdal, S. 1978, *Astr. and Ap.*, 62, L9.
Kraft, R. P. 1965, *Ap. J.*, 142, 1588.
Kuiper, G. P. 1941, *Ap. J.*, 93, 133.
Leung, K. C. 1980, this volume.
Lubow, S. H., and Shu, F. H. 1977, *Ap. J.*, 216, 517.
Lubow, S. H., and Shu, F. H. 1979, *Ap. J.*, 229, 657.
Lucy, L. B. 1968a, *Ap. J.*, 151, 1123.
Lucy, L. B. 1968b, *Ap. J.*, 153, 877.
Lucy, L. B. 1973, *Ap. Sp. Sci.*, 22, 381.
Lucy, L. B. 1976, *Ap. J.*, 205, 208.
Lucy, L. B., and Wilson, R. E. 1979, *Ap. J.*, 231, 502.
Papaloizou, J., and Pringle, J. E. 1979, *M. N. R. A. S.*, in press.
Robertson, J. A., and Eggleton, P. P. 1977, *M. N. R. A. S.*, 179, 359.
Rucinski, S. M. 1973, *Acta Astr.*, 23, 79.
Shu, F. H., Lubow, S. H., and Anderson, L. 1976, *Ap. J.*, 209, 536.
Shu, F. H., Lubow, S. H., and Anderson, L. 1979, *Ap. J.*, 229, 223.
Shu, F. H., Lubow, S. H., and Anderson, L. 1980, *Ap. J.*, submitted.
Smith, D. H., Robertson, J. A., and Smith, R. C. 1979, *M. N. R. A. S.*, in press.
Vilhu, O. 1973, *Astr. and Ap.*, 26, 267
Webbink, R. F. 1976, *Ap. J.*, 209, 829.

DISCUSSION FOLLOWING SHU

Sugimoto: In the region of the temperature inversion near the contact discontinuity, there should be a great jump in entropies. Is convection transporting the correct amount of energy across it which is coming from the deep interior of the star?

Shu: The "correction" in the temperature inversion layer arises not from any internal instability but from being driven from above (by "dredging").

Sugimoto: When the contact is realized as a result of rapid mass transfer, the primary star below the common envelope is greatly out of thermal equilibrium and much mass outflow, say $10^{-4}M_{\odot}yr^{-1}$, is required in order to keep the radius of the star smaller than the critical Roche lobe, as was seen in the case of SV Cen. Are the theories discussed by you applicable even in such cases of mass transfer, or does such a case require a third theory?

Shu: I don't know. It depends on how badly out of thermal equilibrium is the deep interior, I would guess.

van 't Veer: Your computations are based on the hypothesis of conservation of angular momentum. However it becomes more and more evident from the reports we heard this morning and yesterday that for solar-type binaries great angular momentum losses must occur. So I believe that your and Lucy's models are not valid for the majority of contact binaries.

Shu: Our computations are for equilibrium configurations; we did not consider the effects of evolution.

Wilson: What does DSC presently say about the marginal contact of the W-type systems?

Shu: If there really is a "magic value" for delta, say 10^{-5} or so, low-mass contact binaries would have f considerably less than 0.5 while high-mass contact binaries would have f typically substantially more in contact. This is, however, a very speculative answer which requires more careful theoretical and observational considerations.

Lucy: I believe that Frank Shu has given us a fair-minded view of the current theoretical situation. In particular, I agree that the existence of the TRO cycles depends on an unproved assumption that the equal entropy condition will be re-established when the components come back into contact.

With regard to observational tests of the competing theories, Bob Wilson and I have a recent paper (Ap. J., 1979) discussing all the tests we could think of.

ROTATION BROADENING AND THE SHAPES OF W URSAE MAJORIS STARS

Lawrence Anderson,
University of Toledo, Toledo, Ohio U.S.A.

Malcolm Raff, and Frank H. Shu
University of California, Berkeley, California U.S.A.

Abstract. We extract rotation broadening functions from the spectra of W Ursae Majoris Stars. Using a fast Fourier transform we deconvolve photographic spectra, covering some 500 A including dozens of strong lines, with equivalent spectra from non-rotating stars of similar spectral type. The resulting rotation functions contain information about global features such as the shape of the stellar surface (e.g. mass ratio and degree of contact), gravity brightening and limb darkening. We present preliminary data on the stars VW Cep and ER Vul. The rotation function of the former reveals the presence of the third component found visually by Heintz (1975), while that of the latter shows it to be detached and have mass ratio 0.9.

1. INTRODUCTION

It is now well recognized that W Ursae Majoris stars are pairs of main sequence or near main sequence stars in physical contact. The defining characteristics of the W UMa group are that the light curves exhibit curved maxima and nearly equal minima. The former characteristic indicates extreme aspherical distortion of the components while the latter indicates nearly equal effective temperatures for each component despite their having quite different masses. Nearly equal temperatures strongly suggests that a common envelope completely redistributes the energies generated in the separate cores (Osaki 1965; Lucy 1968a,b; Shu, Lubow, and Anderson 1976). Such a common envelope must have a photosphere which lies on an equipotential between the inner and outer critical surfaces of the Roche model for binary stars (Kuiper 1941). Unfortunately, the light curves alone do not provide the degree of contact (what we henceforth call the filled fraction, of the volume between the two critical surfaces filled by the common envelope) unambiguously. Specifically, an atlas of theoretical light curves (Anderson and Shu 1979) shows that there exist loci with nearly identical light curves running from a given mass ratio, low filled fraction, and radiative gravity brightening to higher mass ratio, higher filled fraction, and convective gravity brightening with reduced limb darkening (c.f. $q = 0.5$, $f = 0.0$, radiative →

M. J. Plavec, D. M. Popper and R. K. Ulrich (eds.), Close Binary Stars: Observations and Interpretation, 485–489.

q = 0.6, f = 0.5, convective, zero limb darkening). In view of this ambiguity, we have chosen to investigate the shapes of W UMa stars through the rotational broadening of spectral features. (c.f. Shajn and Struve 1929, Rucinski 1971).

2. METHOD

We compare observed rotation broadening functions with the grid of theoretical functions by Anderson and Shu (1979). The calculations of the theoretical functions are thoroughly discussed in the reference. Both the calculations and the reduction of the observations share an important assumption which we refer to as the uniform profile model. We assume that, relative to the continuum, the spectral profile remains uniform over the projected view of the star except for the rotational Doppler shift (c.f. Gray 1976).

Each spectrum was recorded on baked 127-04 behind a Varo Mk II single stage image tube on the 20" spectrograph at the coudé focus of the Shane telescope at Lick Observatory. The dispersion was about 16 A/mm in the second order. The spectra were widened to about 0.7 mm. We then a) digitized each record to 4096 points spaced equally in $x = \ln(\lambda/\lambda_0)$ over the range $\lambda\lambda$5000-5500 Å with the PDS microdensitometer at the U.C. Berkeley Astronomy Department, b) converted to intensities relative to the continuum $[(I_c(x) - I(x)) / I_c(x)]$, and c) determined complex fourier transforms of each spectral record.

Under the uniform profile assumption the observed spectrum of a W UMa star is a convolution of its rotation function with the spectrum of a non-rotating star of the same spectral type (Simkin 1974, Gray 1976). The Fourier transform of the rotation function is then the complex quotient of the W UMa transform divided by the appropriate standard transform, wavenumber by wavenumber. We chose for the denominator the non-rotating star which gave the cleanest "continuum" in the resulting functions. Slight variations in spectral type (±1/2 class) of the standard star do not alter the overall shape of the resulting rotation function, but do alter the "noise". Before transforming back into spectral wavelengths, we filtered the transform to provide a spectral resolution ~0.0002.

3. RESULTS

We present four observations of two stars classed EW in the Russian General Catalogue of Variable Stars (Kukarkin, et al. 1969): VW Cephei and ER Vulpeculae. The observational data are given in Table I. The rotation broadening functions appear in Figure 1. Zero phase is defined as the transit minimum. All plates were reduced with an average of the spectra of two non-rotating stars, the sun (a plate exposed to the daylight) and HR 483.

TABLE I
Exposure Data (all plates 23 August 1978 UT)

Plate	UT	Exposure	Phase
VW Cep 027	06:21	6 min	0.27
VW Cep 037	09:52	6 min	0.79
ER Vul 026	05:58	4 min	0.73
ER Vul 033	08:19	4 min	0.87

The best fit theoretical broadening functions from Anderson and Shu (1979) also appear in Figure 1. For VW Cephei we found a mass ratio $q = 0.4 \pm 0.05$ and a filled fraction f not much larger than 0.0 (marginal contact). We also clearly detected spectroscopically the third component discussed by Hershey (1975). Its contribution to the rotation function, about 1/16 the area, agrees very well with Hershey's determination of the magnitude difference $\Delta m = 2.9$. Since this component is presumably not rotating, its width in the rotation function is a measure of our resolution. For ER Vulpeculae, we found that the components are actually detached with mass ratio $q = 0.9 \pm 0.05$. The EW classification results from the nearly unit mass ratio, rather than contact. Being detached, ER Vul lies outside the range of the grid of theoretical functions; each star rotates more slowly than it would if it were in contact with the same separation.

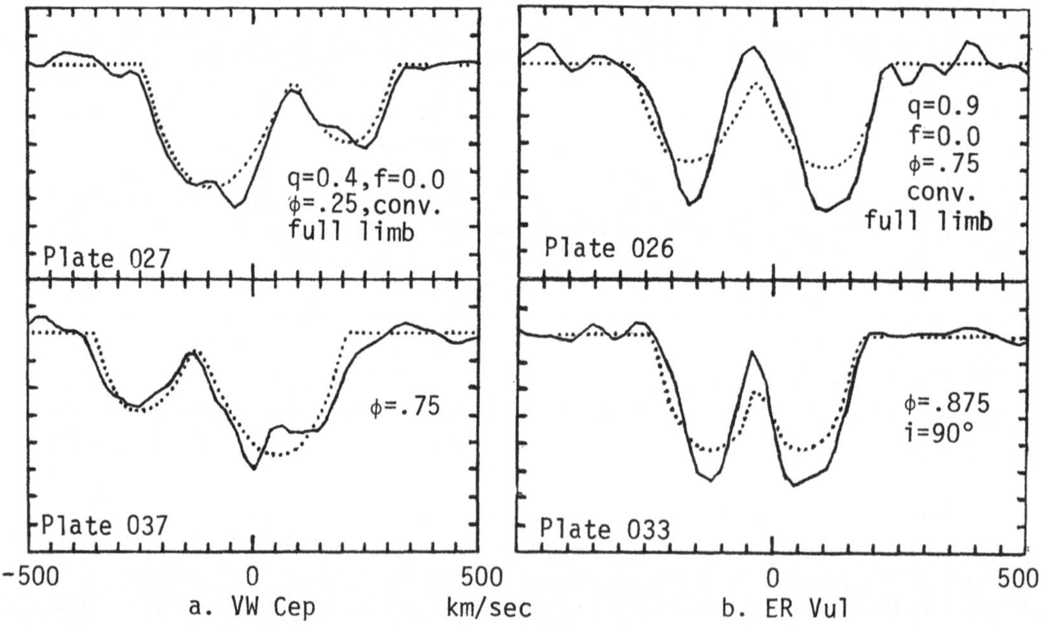

Fig. 1. Rotation broadening functions at phases given in Table I.
Solid lines: observed functions.
Dotted lines: best fit models from Anderson and Shu (1979).

4. DISCUSSION

The mass ratios agree very well with previous determinations. Binnendijk (1967) gives q = 0.41 for VW Cephei (remarkably close to our value given that he was probably unaware of the third component). Northcott and Bakos (1956) give q = 0.93 for ER Vul.

We stress that the theory and observations are kept entirely separate until the final comparison is made; the observations are not reduced in some parametric form with a physical model in mind. The only assumption common to both calculation and reduction is the uniform profile assumption. To be consistent with our introductory comments about the light curves, we should mention that there are also loci of equivalent rotation broadening functions in Anderson and Shu's (1979) atlas. These loci run from low filled fraction and convective gravity brightening to higher filled fraction and radiative gravity brightening at constant mass ratio. If we assume the gravity brightening is fixed (as it ought to be) these loci present no problem. However, there is some question as to the value of the convective brightening exponent ($T_e^4 \propto g^{0.0}$ according to Anderson and Shu 1977, $\propto g^{0.32}$ according to Lucy 1967). If one lets this exponent be a free parameter, one must rely on the intersection of the light curve and rotation function loci.

We thank S. Vogel for his advice on Fourier transforms and W. Hoffman and S. Vogel for their assistance with the PDS microdensitometer. We also thank S. Rucinski, S. Worden and J. Whelan for providing us with plates prior to our own observations.

REFERENCES

Anderson, L. and Shu, F.H.: 1977, Astrophys. J., 214, pp. 798-808.
Anderson, L. and Shu, F.H.: 1979, Astrophys. J. Suppl., 40, pp. 667-697.
Binnendijk, L.: 1967, Publ. Dominion Astrophys. Obs., Vol. XIII, No. 3.
Gray, D.F.: 1976, "The Observation and Analysis of Stellar Photospheres," John Wiley and Sons, New York, pp. 394-400.
Heintz, W.D.: 1975, IAU Inf. Circ. Visual Double Stars No. 64.
Hershey, J.L.: 1975, Astron. J., 80, pp. 662-667.
Kuiper, G.P.: 1941, Astrophys. J., 93, pp. 133-177.
Kukarkin, B.V., Kholopov, P.N., Efremov, Yu. N., Kukarkina, N.P. Kurochkin, N.E., Medvedeva, G.I., Perova, N.B., Fedorovich, V.P., and Frolov, M.S.: 1969, "General Catalogue of Variable Stars", Moscow.
Lucy, L.B.: 1967, Zs. f. Ap., 65, pp. 89-92.
Lucy, L.B.: 1968a, Astrophys. J., 151, pp. 1123-1135.
Lucy, L.B.: 1968b, Astrophys. J., 153, pp. 877-884.
Northcott, R.J., and Bakos, G.A.: 1956, Astron. J., 61, p. 188.
Osaki, Y.: 1965, Publ. Astron. Soc. Japan, 17, pp. 97-105.
Rucinski, S.: 1971, Bull. Am. Astron. Soc., 3, p. 237.
Shajn, G., and Struve, O.: 1929, Monthly Notices R.A.S., 89, pp. 222-239.
Shu, F.H., Lubow, S.H., and Anderson, L.: 1976, Astrophys. J., 209, pp. 536-546.
Simkin, S.M.: 1974, Astron. Astrophys., 31, pp. 129-136.

COMMENTS FOLLOWING ANDERSON, RAFF AND SHU

Collins: The use of the "standard line" approximation over such a distorted stellar system is an extremely dangerous one and probably will change the "ambiguity locus" of your line fits. Variations in the gravity, associated statistical equilibrium and local value of μ (i.e. $\cos \theta$) over the surface of the system can and do greatly affect line strengths and profiles. Various lines are affected differently and hence averages over many lines are subject to systematic error.

Anderson: Our uniform profile assumption is not really defensible, except on practical grounds. However, center-to-limb variations in line strengths are likely to be the most serious departures (consider your absorption line which reduces to virtually zero strength at the limb). Such variations mimic limb darkening in the rotation function. Comparing entries with and without limb darkening in the theoretical grid reveals little significant difference. Such variations would be buried in our observational "noise".

Wilson: Twigg and Robert (in press) have determined and collected values of the gravity darkening, limb darkening, and bolometric albedo parameters for at least several dozen contact binaries of both late and early type. Plotting these vs. surface temperature, they find much consistency in the results, thus indicating that they are finding the correct values. Therefore it seems that one can reliably determine these quantities from light curves alone if they are solved by a full differential correction procedure. Perhaps that would also happen with the velocity profiles if they were analyzed by differential corrections. In any case, it is clear that it is not necessary to continue light curve and velocity profile information to get unique answers.

Anderson (response and question to Wilson): I have a two-part answer, in part a comment in favor of light curve results, and in part a question about light curve fitting procedure. A) Although VW Cephei is an individual case, it is interesting that the rotation profile agrees with the general result of low filled fractions found for late-type W UMa stars through the light curves. B) Have you or others examined regions in parameter space other than the converged solution for solutions which also fit the data within the statistical error of the data?

Wilson: Twigg and Robert tell me that they have investigated "local minimum" problems rather thoroughly for most of their binaries, and find that their results are unique.

Rucinski: The problem of the limb darkening in the line being different than that in the continuum should be partially removed if you simultaneously use many lines formed by different mechanisms; how many lines contribute actually to your observational profiles?

Anderson: About one dozen strong lines and many more weaker ones.

W UMa STARS AND ANGULAR MOMENTUM LOSS

Osmi Vilhu and Timo Rahunen
Observatory and Astrophysics Laboratory
University of Helsinki, Finland

The structure and evolution of W UMa stars is still unsolved although considerable progress has been achieved in recent years. Most theoretical studies are based on the common convective envelope model by Lucy (1968a,b), which almost satisfactorily explains the nearly equal minima of the light curves. All A-type (Wilson, 1978) and some W-type systems (Whelan et al., 1979) may contain an evolved primary. In this case stable models exist (Hazlehurst, 1970; Moss and Whelan, 1970). Computations performed for the subsequent evolution (Moss, 1971; Hazlehurst and Meyer-Hofmeister, 1973; Rahunen and Vilhu, 1977) show nuclear time scale evolution towards more extreme mass ratio, supplemented by possible thermal time scale oscillations.

Many (or most) W-type systems are, however, thought to be unevolved. They have been found in young galactic clusters (Van't Veer, 1975), and usually the system TX Cnc, which is 4 magnitudes below the turn-off point in Praesepe's main sequence, is cited as an unevolved system. However, it is not easy to construct zero-age contact systems in thermal equilibrium without departing significantly from normal abundances or reaction rates (Lucy, 1968a; Whelan, 1972a,b), or without allowing entropy differences between the components (Biermann and Thomas, 1972 and 1973; Vilhu, 1973). Models with entropy differences have been shown (Hazlehurst, 1974) to be secularly unstable, but after the suggestion of the contact discontinuity model (Shu, Lubow and Anderson, 1976; Lubow and Shu, 1977; Shu, 1979) discussion on this interesting possibility continues. This could be noticed also at this symposium. The evolution of the contact discontinuity and the unequal entropy model is probably the same, and this may give the possibility to the scenario W UMa → CV (Vilhu, 1974).

Departing significantly from thermal equilibrium and allowing the mass transfer to be the dominating mechanism, zero-age models with equal entropies can be constructed (Lucy, 1976; Flannery, 1976; Rahunen and Vilhu, 1977; Robertson and Eggleton, 1977). Such models lead in thermal time scale ($\sim 10^7$ y) to break of contact when the inner critical surface is reached. Following Rucinski's idea (1973, 1974), Lucy (1976) and Robertson and Eggleton (1977) suggested that

M. J. Plavec, D. M. Popper and R. K. Ulrich (eds.), Close Binary Stars: Observations and Interpretation, 491–494.

after the break of contact the system evolves in reversed direction back into contact and that this continues in a cyclic manner. So far it is not clear how long this process can continue. Without any net mass transfer from the secondary to the primary, the system should rapidly evolve towards unequal entropies (or contact discontinuity). Of course, beside the mass transfer there may be some other mechanism supporting the oscillations.

The break of contact can be avoided if the angular momentum is allowed to decrease. This possibility was discussed already by Robertson and Eggleton. With angular momentum loss the separation between the components decreases, and both stars can more easily fill their Roche lobes. On rather general arguments Struve (1950), Webbink (1976) and Van't Veer (1979) present scenarios where a W UMa star evolves into a single main sequence star. Webbink and Van't Veer find that 70-80 % of the angular momentum should be lost during the process, and Struve thinks that the rest of angular momentum leads to the formation of a planetary system (the solar system has about 10 % of the angular momentum of a typical W UMa star).

According to Van't Veer, solar type magnetic flare activity is the main reason for this angular momentum loss. Great amounts of angular momentum can be lost without significantly changing the total mass of the system. We can speculate that the flare activity increases when the contact becomes shallower and shallower. This would accelerate the angular momentum loss which in turn increases the degree of contact. Then the flare activity diminishes and the whole process will start again. In this way it is possible that a mechanism exists which keeps the system in contact with a suitable loss of angular momentum. The rapid period changes observed can be identified with such an activity, and the extremely strong coronas found by IUE (Dupree, 1979) may also tell about this.

The aim of the present work is, preserving the contact conditions ($\Delta C = \Delta S = 0$) to find out whether it really is possible to obtain more extreme mass ratios, what is the angular momentum loss needed and what is the time scale (The system B from Rahunen and Vilhu, 1977 and total mass = 1.8 $M_{\odot}$ are used). Preliminary results show that rather extreme mass ratios (at present as far as 0.3) can indeed be reached in a time scale of 10^8 y. The rate of angular momentum loss needed to keep the system in contact was about 10^{43} g cm^2 s^{-1} per year, leading to 20 % loss of the total angular momentum during the process. This loss rate lies well within the general tendency of decreasing angular momentum loss with increasing age, as deduced by Van't Veer (1979) with the help of T Tauri-stars, Pleiades and Hyades dwarfs, and the sun.

We can only speculate what happens when a rather extreme mass ratio has been reached. Because the secondary becomes too radiative (the internal luminosity of the secondary tends to zero with decreasing mass) the contact conditions can no longer be satisfied. This, probably, can be avoided by assuming that a small part of the luminosity transfer takes place below the convective envelope. For extreme mass ratios the concept of a common convective envelope also seems artificial. It is expected that after some critical limit the system catastrophically coalesces into one single star.

Finally, it is interesting to note (following Struve and Van't Veer) that, if a considerable part of all W UMa stars follows the above scenario, then the W UMa state should be a normal state in the life of any solar type dwarf (the ratios of space densities and of life times of G dwarfs and W UMa stars would both be equal to 100). Anyhow, the angular momentum loss should be considered as a serious possibility in addition to the current problems connected with the cyclic and contact discontinuity models.

REFERENCES

Biermann,P., and Thomas,H.-C.: 1972, Astron. Astrophys., 16, p. 60.
Biermann,P., and Thomas,H.-C.: 1973, Astron. Astrophys., 23, p. 55.
Dupree,A.K.: 1979 (this symposium).
Flannery,B.P.: 1976, Astrophys.J., 205, p. 217.
Hazlehurst,J.: 1970, Mon. Not. R. astr. Soc., 149, p. 129.
Hazlehurst,J.: 1974, Astron. Astrophys., 36, p. 39.
Hazlehurst,J., Meyer-Hofmeister,E.: 1973, Astron. Astrophys., 24, p. 379.
Lubow,S.H., and Shu, F.H.: 1977, Astrophys.J., 216, p. 517.
Lucy,L.B.: 1968a, Astrophys.J., 151, p. 1123.
Lucy,L.B.: 1968b, Astrophys.J., 153, p. 877.
Lucy,L.B.: 1976, Astrophys.J., 205, p. 208.
Moss,D.L.: 1971, Mon. Not. R. astr. Soc., 153, p. 41.
Moss,D.L., and Whelan,J.A.J.: 1970, Mon. Not. R. astr. Soc., 149, p. 147.
Rahunen,T., and Vilhu,O.: 1977, Astron. Astrophys., 56, p. 99.
Robertson,J.A., and Eggleton,P.P.: 1977, Mon. Not. R. astr. Soc., 179, p. 359.
Rucinski,S.M.: 1973, Acta Astr., 23, p. 79.
Rucinski,S.M.: 1974, Acta Astr., 24, p. 119.
Shu,F.H.: 1979 (this symposium).
Shu,F.H., Lubow,S.H., Anderson,L.: 1976, Astrophys.J., 209, p. 536.
Struve,O.: 1950, Stellar Evolution, Princeton University Press, p. 234.
Van't Veer,F.: 1975, Astron. Astrophys., 44, p. 437.
Van't Veer,F.: 1979 (a preprint).
Webbink,R.F.: 1976, Astrophys.J., 209, p. 829.
Whelan,J.A.J.: 1972a, Mon. Not. R. astr. Soc., 156, p. 115.
Whelan,J.A.J.: 1972b, Mon. Not. R. astr. Soc., 160, p. 63.
Whelan,J.A.J., Worden,S.P., Rucinski,S.M., Romanishin,W.: 1979, Mon. Not. R. astr. Soc., 186, p. 729.
Vilhu,O.: 1973, Astron. Astrophys., 26, p. 267.
Vilhu,O.: 1974, in R.J. Tayler (ed.), Late Stages of Stellar Evolution, IAU Symp. 66, p. 168.
Wilson,R.E., 1978, Astrophys.J., 224, p. 885.

COMMENT FOLLOWING VILHU AND RAHUNEN

Sugimoto: How much mass has been transferred from the secondary to the primary before the breakdown of the conservative mass transfer? If it has been an appreciable fraction of the stellar mass, what is the effect of your approximation of the luminosity jump ΔL which was used instead of computing the hydrodynamic flow of specific entropy?

ENERGY TRANSFER BY CIRCULATION IN W URSAE MAJORIS SYSTEMS

David H. Smith, Robert Connon Smith and J. Alistair Robertson
Astronomy Centre, Physics Building, University of Sussex,
Falmer, Brighton BN1 9QH, U.K.

1. INTRODUCTION

After Lucy (1968) introduced the contact-binary model with a common convective envelope, it was envisaged by Hazlehurst & Meyer-Hofmeister (1973) that a sideways flow of convective elements would carry energy from the more luminous star, the primary, to the less luminous star, the secondary, as a result of horizontal pressure variations. Webbink (1977) extended this picture by noting that the interaction between vertical entropy gradients and large-scale smooth circulation currents in the common envelope would provide the necessary redistribution of flux. That is, energy is absorbed by the flow during its vertical motion in the primary and is released during its vertical motion in the secondary. Webbink (1977) mentioned two mechanisms by which a large-scale circulation could be generated: (1) the non-spherically symmetric force field due to rotation and tides which will drive an analogue of classical Eddington-Sweet circulation and (2) differential heating of the base of the common envelope. Although these mechanisms are conceptually different, they are not in practice easy to disentangle, and will certainly both be operating in contact binaries.

2. EDDINGTON-SWEET CIRCULATION

Any non-spherically symmetric perturbation to a star will give rise to departures from radiative equilibrium and result in circulation currents. If the combined influence of rotation and tides is to cause a flow rising in one component and sinking in the other, then energy transfer will result. For the Eddington-Sweet solution to be valid the effects of viscosity, inertia and Coriolis forces must be negligible, and Webbink (1977) has argued that to a first approximation they are. Because of this and the schematic flow patterns sketched by Webbink (1977) showing a flow rising in one component and falling in the other, some confusion exists in the literature on the role of the Eddington-Sweet currents. To clarify this we have numerically calculated the velocity component of the Eddington-Sweet circulation normal to equipotential surfaces for mass ratios $q = 1.0$, 0.4, and 0.1 (Smith & Smith, in preparation). From the topology of the flow we conclude that there

M. J. Plavec, D. M. Popper and R. K. Ulrich (eds.), Close Binary Stars: Observations and Interpretation, 495–499.

are no tidally and rotationally induced currents linking the two stars. This result confirms that the classical Eddington-Sweet theory is inadequate for describing the energy transfer.

3. DIFFERENTIAL HEATING OF THE COMMON ENVELOPE

Shu *et al.* (1979) were the first to use Webbink's idea of differential heating as the driving force for circulation. But Hazlehurst & Refsdal (1978), Papaloizou & Pringle (1979) and Smith, Robertson & Smith (1979) have all shown that the discontinuity model of Shu *et al.* (1979) is unacceptable and we will not consider it here. Nonetheless we believe that non-uniform heating of the base of the common envelope is the best candidate for the driving mechanism. We have solved the hydrodynamic equations in a two-dimensional box of length a_1 and depth a_3 in a uniform gravitational field $\underline{g}$, containing a perfect gas with viscosity ν and thermal diffusivity κ. The full, compressible, hydrodynamic equations are written as follows for ease of numerical solution:

$$\partial\rho/\partial t + \nabla.(\rho\underline{u}) = 0, \quad (1)$$

$$\partial(\rho\underline{u})/\partial t + \nabla.(\rho\underline{uu}) + \nabla p + \rho\underline{g} - \nu\nabla^2(\rho\underline{u}) = 0, \quad (2)$$

$$\partial T/\partial t + \nabla.(T\underline{u}) + (\gamma-2)T\nabla.\underline{u} - \kappa\nabla^2 T = 0, \quad (3)$$

and $$p = R_*\rho T \quad (4)$$

where the symbols have their usual meanings. The boundary conditions used are:

$$T = T_o \text{ on top, } T = T_B + T_1 \cos(\pi x/a_1) \text{ on base} \quad (5)$$

$$w = \partial u/\partial z = 0 \text{ on top and base.} \quad (6)$$

The variables ρ, T, u and w are assumed to be periodic in x. It was found easier to evolve time-dependent equations to a steady state than to solve time-independent equations implicitly. The steady-state solutions obtained are characterized by 7 dimensionless parameters:

γ, $P \equiv \nu/\kappa$, $R \equiv g a_3{}^3/\nu\kappa$, $C \equiv g a_3/R_*(T_B+T_1-T_o)$,

$A \equiv a_1/a_3$, $\tau_B \equiv T_B/(T_B+T_1-T_o)$, and $\tau_1 \equiv T_1/(T_B+T_1-T_o)$.

The range of values explored to date is:

$\gamma = 5/3$, $1/256 < P < 8$, $46.875 < R < 48000$,

$1/2 < A < 3$, $C = 3.75$ and 5, $\tau_B = 2$ and $0.05 < \tau_1 < 0.45$.

A typical solution for $R < 10^4$ consists of a single cell with flow rising at the hot end, as would be expected. The stagnation point is

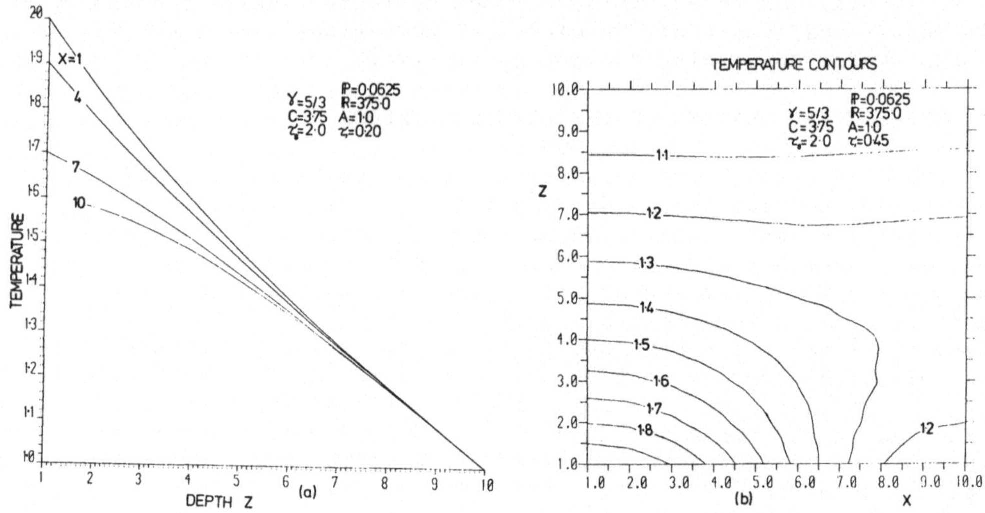

FIGURE 1.
(a) Temperature as a function of the vertical coordinate, z, for 4 different values of the horizontal coordinate, x; $\tau_1 = 0.20$.
(b) Temperature contours for the case $\tau_1 = 0.45$. Note the temperature inversion for $x > 7$ and the uniform temperature for $z > 7$.

located approximately one pressure scale-height, H_p, above the base and is nearer to the cool end for $A > 1$. The whole cell spans about $3H_p$ when $C = 3.75$. For $R > 10^4$ there is a second thin cell near the surface.

In general, decreasing P increases $|\underline{u}|$, and not surprisingly the maximum velocity is related to the temperature perturbation τ_1: $|\underline{u}|_{max} \approx \alpha\tau_1 c_s$ where $\alpha \approx 1$ for $A = 1$ and c_s is the sound speed. The temperature distribution shows several interesting effects. The horizontal temperature variations disappear after about one scale height (see fig.1). It is possible to form stable temperature inversions when the temperature perturbation is large (see fig.1(b)), i.e. $\tau_1 > 0.25$ for $A = 1$. But as the aspect ratio increases a larger value of τ_1 is needed to cause this. The size of the inversion increases with the ratio of the thermal to dynamical timescales, and with R, but is almost independent of P.

4. CONCLUSIONS

1) Differential heating will cause a circulation linking the two stars, which will operate throughout the common envelope, smoothing out horizontal temperature variations in about one pressure scale-height.

2) For realistic temperature perturbations, the horizontal velocity will be a considerable fraction of the sound speed. But the vertical

velocity will be much less due to the thin-layer geometry of the common envelope, i.e. high A.

3) Temperature inversions are unlikely to be formed in the common envelope, due to the high aspect ratio, A $\sim$ 100. If an inversion is formed then the system will not be in thermal equilibrium, since there will be a net flow of energy into the secondary.

Although these results are only preliminary, we believe that they will form a useful basis for further study of energy transfer in contact binaries.

REFERENCES

Hazlehurst, J. & Meyer-Hofmeister, E.: 1973, Astron.Astrophys.24, pp. 379-392.
Hazlehurst, J. & Refsdal, S.: 1978, Astron.Astrophys. 62, pp. L9-L11.
Lucy, L.B.: 1968, Astrophys.J. 151, pp.1123-1136.
Papaloizou, J. & Pringle, J.E.: 1979, Monthly Notices Roy.Astron.Soc. in press.
Shu, F.H., Lubow, S.H. & Anderson, L.: 1979, Astrophys.J. 229, pp. 223-241.
Smith, D.H., Robertson, J.A. & Smith, R.C.: 1979, Monthly Notices Roy.Astron.Soc. in press.
Webbink, R.F.: 1977, Astrophys.J. 215, pp.852-863.

COMMENTS FOLLOWING SMITH, SMITH AND ROBERTSON

Nariai: I have presented the model of circulation a few years ago (Nariai, K.: Publ. Astron. Soc. Japan, 28. 587, 1976). Is your model essentially the same as mine?

D. Smith: Yes, many of the details are similar.

Shu: I would like to suggest that your calculation is more appropriate to Los Angeles than to the interiors of contact binaries. Our parameter δ is, in order of magnitude, given by your conbination $(C/PR)^{\frac{1}{2}}$, which is typically 0.1 or larger in your calculations. On the other hand, the true value of δ is four or more orders of magnitude smaller at the base of the common envelope of most contact binaries. Nevertheless, it is interesting that even with a relatively large value of δ, you get a temperature inversion.

D. Smith: The temperature inversions are a consequence of our lower boundary conditions and may not have any physical significance.

Shu: In reply to Webbink's comment, let me state that our definition of δ is $F/\rho h a_s$. Webbink's definition is different and associates the thermal timescale of a convection zone with the travel time of a convective blob. Webbink's definition is irrelevant to our discussion because it ignores the blobs have only a slight entropy excess or deficit with respect to their surroundings.

Anderson: In order to determine whether a temperature inversion over the cooler boundary can be maintained by circulation, you should extend a thermally non-conducting wall (representing the potental ridge between the two stars) up from the lower boundary. Then both flux and temperature should be specified at the two new lower boundaries.

LOW COLOR TEMPERATURE VARIATIONS OVER THE COMMON ENVELOPE OF W URSAE MAJORIS AND THE CAUSE OF ITS W-TYPE LIGHT CURVE

J. A. Eaton*, C. - C. Wu**, and S. M. Rucinski***
* Copernicus Astronomical Center, Polish Academy of Sciences
** Computer Sciences Corporation, Silver Spring, Maryland
***Warsaw University Observatory

We discuss ultraviolet photometry of the eclipsing binary W Ursae Majoris obtained with the Groningen instruments on the Astronomical Netherlands Satellite (ANS). Fifty measurements of this star's brightness were obtained for each of the bandpasses at 2200, 2500, and 3300 Å. This new data is significantly more precise than Rucinski's OAO-2 photometry, enabling us to draw definitive conclusions about the distribution of brightness on the surface of W UMa. The ultraviolet light curves are generally similar in shape to those for optical passbands; the amplitudes are greater in the UV, while the depths of the primary and secondary eclipses are nearly equal at both 2200 and 2500 Å. We have analyzed the (2200 - 3300) color curve to determine how much the color temperature varies over the common envelope, concluding that any such temperature variations must be small. In terms of the traditional surface-brightness parameters for gravity darkening, temperature excess of the smaller component, and bolometric albedo, we find $\beta = 0.03 \pm 0.01$, $X = \Delta T/T = 0.000 \pm 0.009$, and $A_{bol} = 0.4 \pm 0.4$. Further, we conclude that the star is limb darkened considerably more at 2200 than at 3300 Å.

The new results strongly restrict the possible ways of giving W UMa a W-type light curve, i.e., one for which the eclipse of the smaller component is the deeper eclipse at optical wavelengths. They rule out the light curve model based on a general temperature excess which has been employed by a number of workers as well as the model featuring large gravity darkening espoused chiefly by R. E. Wilson. Also, we find no evidence for the high albedo or very low limb darkening needed by the light curve model of Anderson and Shu or for the temperature's being abnormally high in the region between the components as suggested by Berthier. Presently, the leading explanation of the ultraviolet and optical light variations seems to be a spot model in which at least 20% of the disc of the larger component would have to be covered with dark spots to give the W-type light curve observed at 'V'.

A full account of this work is being submitted to the Astrophysical Journal.

M. J. Plavec, D. M. Popper and R. K. Ulrich (eds.), Close Binary Stars: Observations and Interpretation, 501–503.

DISCUSSION FOLLOWING EATON, WU AND RUCINSKI

Webbink: After seeing Andrea Dupree's UV spectra of some W UMa stars, I wonder how confident we can be that we are actually seeing continuum variations in your observations.

Rucinski: The rich emission spectrum becomes important at shorter wavelengths than used here but it will be necessary to carefully check the spectra for the presence of (variable) lines in the ANS bandpasses.

Shu: I would like to point out that in Lawrence Anderson and my light-curve paper, we suggested that spots on the primary were the most viable explanation for the W-type phenomenon. We even gave a speculative explanation of this expectation on the basis of the contact discontinuity theory.

Guinan: The light curve of W UMa appears to be unstable with changes occurring on time scales of days to years. Wouldn't this somewhat diminish the certainty of the determination of the gravity brightening exponent from an analysis of a single-epochal light curve?

Rucinski: It seems that the instabilities are caused by relatively dark spots with almost no influence on colors like the (2200-3300) color used in this analysis.

Geyer: The W UMa system TZ Boo is an extreme case for extreme light curve changes. My colleague Dr. Hoffmann has obtained a fine collection of light curves in UBV at the 42" Cassegrain reflector of Hoher List Observatory. They show not only the interchange of the depth of primary and secondary variations within a cycle of about 3.5 years, but also the color index is changing (Astron. Astrophys. Suppl., 33, 63).

Anderson: If those color observations were in the visible region, they may well show color variations at the same time your ultraviolet colors show none, since the spot appears black independent of λ for sufficiently short λ.

Rucinski then asked what wavelength bands Geyer observed, and he answered "UBV".

Budding: I would question whether your results are as clearly formed as they would appear to be. First, I think there may be some uncertainty about the relationship between color and effective temperature for these atmospheres. Second, in order to determine the gravities the actual geometric parameters have first to be known--presumably through the normal photometric solution, which I think would then be convoluted with what was assumed about gravity darkening.

Eaton: We used a flux-temperature scale based on the temperature-spectral type-bolometric correction scale of Johnson and on Wu's color-spectral class calibration for the ANS photometric system. As you suggest, this temperature scale could lead to slight systematic errors, but not to the extent of letting us fit the light wave with large gravity darkening or general temperature difference between components.

Rucinski: The geometry used was that obtained from the optical solution by Whelan, Worden, and Mochnacki (which is practically identical to solutions by other authors). The (2200-3300) color curve does not depend much on the geometry but is very sensitive to the distribution of temperature over the contact configuration.

THE PHYSICAL STATUS OF VW CEPHEI

Albert P. Linnell
Dep't. of Astronomy and Astrophysics, Michigan State Univ.,
E. Lansing, Michigan, U.S.A.

Abstract. The decreasing period of VW Cep, interpreted on conservative case mass transfer, implies mass flow from primary to secondary. The TRO theory then predicts semi-detached status. The system b-v color is the same at the two minima, to within observational error. This is in marginal disagreement with a hot secondary. It also implies close thermal contact, in disagreement with the postulated semi-detached status.

Unanimous agreement is lacking among research workers concerning the evolutionary status of W UMa stars and the interpretation of their light curves. The evolutionary status protagonists include the supporters of thermal relaxation oscillations (TRO) in one group, led by Lucy (1976), Flannery (1976), and Robertson and Eggleton (1977). The other group supports a thermal equilibrium model, made possible by an entropy discontinuity at the inner contact surface. This group is led by Shu, Lubow, and Anderson (1976).

In the study of W UMa light curves, the usual interpretation of the W-subclass (Binnendijk, 1970) systems has required a secondary (less massive) component whose photosphere is hotter than that of the primary. This accounts for the observed association of primary minimum with the eclipse of the less massive component. Rucinski (1974) used the hot secondary model in his study of W UMa light curves, as have several other investigators. The W-subclass systems were a problem for the Lucy (1973) light curve study, which otherwise was successful in representing the A-subclass systems. The Lucy common convective envelope model asserts a gravity brightening coefficient of 0.08 for convective envelopes, and incorporates a Rucinski (1969) bolometric albedo factor of 0.5.

The most extensively developed light curve synthesis program is that of Wilson and Devinney (1971). Wilson and Devinney (1973) were able to fit the W-type light curve of RX Com successfully, but required a value of gravity brightening much larger than the Lucy coefficient.

M. J. Plavec, D. M. Popper and R. K. Ulrich (eds.), Close Binary Stars: Observations and Interpretation, 505–509.

Wilson and Biermann (1976) have made a detailed light curve synthesis study of the W-subclass system TX Cnc. They found that if the temperature at the pole of the primary is taken as a free parameter, allowing a temperature discontinuity between components, in disagreement with the Lucy common convective envelope model, a fair fit could be obtained with a hot secondary and a bolometric albedo that agrees with the Rucinski value 0.5. If no temperature discontinuity between components is allowed, the solution produced a hot primary, with a bolometric albedo of 1.0. However, the gravity brightening coefficient in both solutions is much larger than the Lucy value, reaching eleven times larger in the second one. Significant residuals remain near secondary minimum in both solutions.

Anderson and Shu (1977) have developed a new theory of W UMa light curves which differs from the model used by Lucy, Wilson, and others in several important respects. The Anderson-Shu (hereafter called AS) theory obtains a gravity brightening coefficient of 0.0. This theory also asserts the existence of a "reflection effect" in which the local temperature rises as the amount of unobscured sky decreases, an effect not present in the Lucy or Wilson models. A notable feature of the AS theory is its ability to produce W-subclass light curves without requiring a hot secondary.

Thus, the theoretical interpretation of W-subclass light curves remains obscure. It would be most desirable if observational means could be found, reasonably model-independent, which could directly test whether the secondary has a higher temperature than the primary or not. We return to this point later.

VW Cep is a W-subclass system which has been studied extensively. Reliable times of minima are available spanning 55 years. Photometric results of a two-year international campaign have been summarized by Kwee (1966). More recently it has been shown to be a triple system (Hershey, 1975), with a third component moving in a relative orbit of substantial eccentricity, but far enough away to be a negligible tidal or thermal influence on the eclipsing system. Light of the third component does contaminate the binary light curve, and motion of the binary about the common center of mass with the third component modulates the observed times of minima. The system has gained special interest recently. Dupree (1978) has found high excitation far ultraviolet lines in IUE spectra, and Cruddace (1978) reports that a low flux level soft X-ray source has a position error box which includes VW Cep. Observations of the system began with the MSU automated photometer in the fall of 1978. A discussion of the entire light curve will appear elsewhere. This paper considers times of minima and system color.

The times included in the period study by Hershey (1975), IBVS times published since Hershey's paper, and minima obtained at MSU have been combined in a new period study. The period has been decreasing monotonically for the 55-year interval during which reliable observations have been reported. Available light curve solutions for VW Cep (Lucy, 1973; Rucinski, 1974) agree on shallow contact, a common property of W-subclass systems. It is unlikely the system is losing mass at L_2. The period variation then implies conservative case mass transfer from primary to secondary.

The most detailed study of TRO effects is in the paper by Robertson and Eggleton (1977). During close thermal contact mass is transferred from secondary to primary at a rate $\sim 10^{-8} M_\odot$/yr. During the semi-detached phase, mass flows from primary to secondary at a rate $\sim 10^{-7} M_\odot$/yr. The rate of period variation for VW Cep and the component masses obtained by Hershey lead to a mass transfer from primary to secondary of $2 \times 10^{-7} M_\odot$/yr. The implied semi-detached status is consistent with the Lucy (1973) solution for the surface potential, compared with the potential of inner critical surface. It is of interest that hydrodynamic mass flow could deliver enough energy to the secondary to account for its entire luminosity. This is an alternative to the Whelan (1972) mechanism of energy transfer in the superadiabatic region as a means to produce a hot secondary. It then is an attractive speculation to attribute the observed high energy phenomena to the impacting stream.

Earlier mention was made of the need for model-independent means to measure the component temperatures. The B-V color of a thermal stellar source has a close and well-calibrated correlation with color temperature. The color at elongation phases has a different interpretation in different theories, and so is model-dependent. At conjunction the "neck" region is hidden if the orbital inclination is high. At maximum phase of transit eclipse the visible annulus of the eclipsed star experiences relatively small irradiation from the eclipsing star, so different assumptions concerning the bolometric albedo should have a small effect. The silhouette of each star is much closer to a circle than at elongation. The difference between limb darkening in V and B is much smaller than their individual center-limb variation. Thus, B-V color data at conjunctions should be interpretable on the basis of uniform temperature circles, in a relatively model-independent manner.

Table 1 compares the b-v data at the two conjunctions. Four sets of data are from Kwee mean light curves. The last two lines are from MSU observations, uncorrected and corrected for contamination by the third component. To within observational accuracy ($\pm$ 0.01 mag.), the difference in color between the two minima is zero. This implies the secondary is not hotter than the primary, and that the components are in very close thermal contact. On an approximation modelling the stars as uniform temperature circles, but attributing the difference in depths of primary and secondary minima to a hot secondary, the color difference should be about 0.01 mag.

TABLE 1

(b-v) (comp. star) - (b-v) (VW Cep)

Source	Phase 0.0	Phase 0.5	Diff.
Kwee I	0.207	0.201	+0.006
Kwee II	0.201	0.194	0.007
Kwee III	0.196	0.191	+0.005
Kwee IV	0.203	0.210	-0.007
Linnell (U)	0.192	0.192	0.000
Linnell (C)	0.160	0.162	+0.002

This result is in disagreement with the semi-detached status imputed from the period variation. It seems unlikely that the close temperature equality would result from thermalization of hydrodynamic flow. Robertson and Eggleton (1977) calculate an appreciable difference in component surface temperatures during the semi-detached stage of TRO oscillations. This difference would be easily detected.

These results suggest that the TRO model for close thermal contact may need further elaboration, possibly along the lines of Webbink's (1977) model which includes substantial circulation currents.

Further details concerning these results will appear elsewhere.

REFERENCES

Anderson, L. and Shu, F.H.: 1977, *Astrophys. J.* 214, pp. 798-808.
Binnendijk, L.: 1970, *Vistas in Astronomy* 12, pp. 217-256.
Cruddace, R.: 1978 (private communication).
Dupree, A.K.: 1978 (private communication).
Flannery, B.P.: 1976, *Astrophys. J.* 205, pp. 217-225.
Hershey, J.L.: 1975, *Astron. J.* 80, pp. 662-667.
Kwee, K.K.: 1966, *Bull. Astron. Soc. Neth.* 18, pp. 448-458.
Lucy, L.B.: 1973, *Astrophys. Space Sci.* 22, pp. 381-392.
Lucy, L.B.: 1976, *Astrophys. J.* 205, pp. 208-216.
Robertson, J.A. and Eggleton, P.P.: 1977, *Monthly Notices Royal Astron. Soc.* 179, pp. 359-375.
Rucinski, S.M.: 1969, *Acta Astron.* 19, pp. 245-255.
Rucinski, S.M.: 1974, *Acta Astron.* 24, pp. 119-151.
Shu, F.H., Lubow, S.H. and Anderson, L.: 1976, *Astrophys. J.* 209, pp. 536-546.
Webbink, R.F.: 1977, *Astrophys. J.* 215, pp. 851-863.
Whelan, J.A.J.: 1972, *Monthly Notices Royal Astron. Soc.* 156, pp. 115-128.
Wilson, R.E. and Devinney, E.J.: 1971, *Astrophys. J.* 166, pp. 605-619.
Wilson, R.E. and Devinney, E.J.: 1973, *Astrophys. J.* 182, pp. 539-547.
Wilson, R.E. and Biermann, P.: 1976, *Astron. Astrophys.* 48, pp. 349-357.

COMMENTS FOLLOWING LINNELL

Leung: If there is enough time in my talk, I will show some slides on VW Cep. It is suggested that VW Cep may be a semidetached system from the photometric solution.

Webbink: In the model of energy exchange I published, some temperature excess of the secondary envelope was required, assuming it were convective, to close the circulation and establish good thermal contact. I don't think there is then necessarily any inconsistency in having equal surface temperatures and mass flow from primary to secondary (as indicated by the period derivative), both of which could be symptomatic of poor thermal contact in such a late-type W UMa system.

Linnell: I'm glad to know that!

MASS LOSS FROM CONTACT BINARY SYSTEMS AND THEIR DYNAMICAL EVOLUTION

Kyoji NARIAI
Tokyo Astronomical Observatory, University of Tokyo
Mitaka, Tokyo 181, Japan.

1. INTRODUCTION

When there is mass loss from a binary system, the lost mass carries energy and angular momentum out of the system. Therefore, the remaining system must adjust its orbital parameters to the changing values of the total kinematic energy E and the total angular momentum N as the total mass M decreases. The parameters concerned here are : the fractional mass μ , the semi-major axis a , and the eccentricity e .

Mass loss from binaries takes place preferentially through the second Lagrangian point which we will denote by L . Kuiper (1941) has shown that two asymptotic orbits exist for particles leaving L with zero relative velocity. Most particles that pass L outward converge to one of the asymptotic orbit that leads a particle far from the system. Therefore, the specific energy f (in units of GM/a) and the specific angular momentum K (in units of $(GMa)^{1/2}$) of escaping particles are almost the same as those of a particle leaving L in the direction of the line connecting two centers of mass ; f=0 and K=1.7. These values should be compared to the mean values of the system and to the values at L . For the case of two point masses of equal weight moving in circular orbits, the mean values are : $f=-\infty$, K=0.25 and at L : f=-0.29, K=1.44. The infinity of the mean energy is caused by the point structure of the components and will be removed if realistic models are introduced.

Some of the particles leaving L in the direction opposite to the motion of rotation converge to another asymptotic orbit that hits the Lagrangian lobe immediately after the departure from L . These particles have smaller azimuthal velocity than those converging to the orbit that permits escape from the system. Therefore, the centrifugal force cannot balance the gravity. In the case of mass loss from a contact binary, we have to think only of the former particles as far as the losses of energy and angular momentum from the system are concerned. This situation is not altered for non-zero initial relative velocity at L (Nariai 1975, 1977).

M. J. Plavec, D. M. Popper and R. K. Ulrich (eds.), Close Binary Stars: Observations and Interpretation, 511–515.

We may call the mechanism of acceleration of particles by two rotating bodies as "the gravitational stellar accelerator". The rotating non-axisymmetric gravitational field works as a sling. If we increase the initial relative velocity at L , the escaping particle spend less time in the region where the stellar accelerator functions effectively. However, it should be noted that the ratio of the specific angular momentum and the mean specific angular momentum is about 6 at L and this ratio becomes 7 at infinity. Therefore, as far as the transport of angular momentum is concerned, we may say that the most important part lies within the Lagrangian lobe. The behavior of the remaining system caused by the loss of angular momentum is not affected very much by the efficiency of the "stellar accelerator" outside the Lagrangian lobe as long as mass loss takes place through L . When mass loss is caused by stellar wind originating from corona surrounding the binary, it is supposed that the velocity of the wind at L , is larger than the velocity of the orbital motion. In such a case, a particle will escape from the system before the stellar accelerator works effectively, and K will remain near its initial value.

Nariai and Sugimoto (1976) have studied the effects of mass loss on the remaining system with some assumption on the fractional mass and without taking account of the angular moment of rotation. They have found that the eccentricity increases and the semi-major axis decreases very rapidly if there is mass loss from L . According to Zahn (1976), the characteristic times for circularization and synchronization caused by the tidal effects are quite small if the component stars have convective envelopes. They are 2×10^4 years and 10^6 years, respectively, for a system of $\mu=0.5$ and the orbital period of 1 day. Assuming that the characteristic time of mass loss through the stellar wind is smaller than the life of main sequence stars but larger than the tidal characteristic times, we will study the dynamical evolution of a contact binary in the following section.

2. DYNAMICAL EVOLUTION

We assume that the orbit of the binary is circular and the both components rotate at the synchronous rate. Then the total angular momentum N is written as

$$N = M(GMa)^{1/2} \left(m_1 m_2 + \sum m_i s_i^{\,2} \right) , \tag{1}$$

where m_i is the fractional mass of the i-th component and s_i is the ratio of the gyration radius d_i to the semi-major axis a . $\sum$ represents the sum $\sum_{i=1}^{2}$. The total kinematic energy E is the sum of the potential and kinetic energies of the orbital motion and the rotational energy of the component stars. It is written as

$$E = N(GM/2a) \left(-m_1 m_2 + \sum m_i s_i^{\,2}\right) . \tag{2}$$

The gravitational energy and the internal energy of the component stars, the production of energy by nuclear reactions and the loss of energy from the photosphere do not appear in equation (2). The variations δN and δE are written with the orbital parameters and with the total mass as

$$\delta N=M(GMa)^{1/2}\{ P\delta \ln M + Q\delta \ln a + (R-S)\delta\mu \} , \tag{3}$$

$$\delta E=-M(GM/2a)\{ T\delta \ln M + Q\delta \ln a + (R+S)\delta\mu \} , \tag{4}$$

where

$$P=1.5m_1m_2 + \sum (1.5+2\alpha_i)m_is_i^2) , \tag{5}$$

$$Q=0.5(m_im_2 - 3 \sum m_is_i^2) , \tag{6}$$

$$R=m_1 - m_2 , \tag{7}$$

$$S=(1+2\alpha_1)s_1^2 - (1+2\alpha_2)s_2^2 , \tag{8}$$

$$T=2m_1m_2 - \sum 2(1+\alpha_i)m_is_i^2 , \tag{9}$$

and

$$\alpha_i=d \ln d_i/ d \ln M . \tag{10}$$

When the process of mass loss is very slow and the structure of the components do not change, we have $r \propto M$, therefore $\alpha_i=1$. α_i is a quantity of the order of 1, and it appears always as a factor of s_i^2. The ratio of the gyration radius to the radius of the star is 0.27 for a polytrope of index 3. For a star with the polytropic index of approximately three, s_i^2 is given by $0.1(r_i/a)^2$.

We can write δN and δE as functions of the lost mass δM using two parameters K and F as

$$\delta N=K(GMa)^{1/2}\delta M , \tag{11}$$

$$\delta E=F(GM/2a)\ \delta M . \tag{12}$$

We will take K=1.4 as we are thinking of mass loss by stellar wind through L . If we are concerned with the quiet ejection, we have to take K=1.7. If we include the effect of magnetic field, more angular momentum can be carried away by the escaping mass and K will be larger than the corresponding values without magnetic field.

Three mechanisms contribute to F . The first contribution is given by the stellar accelerator and will be written as $v^2(a/GM)$ for ejection with zero initial relative velocity. v denotes the velocity at infinity. For ejection from the corona, it is written as $(v^2-2U_c)(a/GM)$ where U_c represents the specific internal energy of coronal gas. In both cases, this quantity is almost zero. The second contribution comes from dissipation through the tidal effects. This quantity is always positive. The third contribution comes from the structural change of the component stars. Let us think of what happens in one of the components which is compressed by accreting mass. As mass is accreted on it, the star is compressed and rotates more rapidly than before violating the synchronism. A part of the excess rotational energy is transferred to the orbital motion by the tidal coupling. Thus, a binary can convert a part of the difference in the potential energy into the kinematic energy when there is mass transfer between the components. F can be either positive or negative.

From equations (3), (4), (11), and (12), we can derive the derivatives of $\ln a$ and μ as functions of $\ln M$;

$$\frac{d \ln a}{d \ln M} = \frac{R(K+F-P+T)+S(K-F-P-T)}{Q(3R-S)} \tag{13}$$

$$\frac{d\mu}{d \ln M} = \frac{2K-F-2P-T}{3R-S} \tag{14}$$

3. DEGREE OF CONTACT

It is interesting to know how the degree of contact changes with mass loss. The degree of contact c is given by

$$c=(\phi-\phi_2)(\phi_1-\phi_2) \tag{15}$$

where ϕ represents the potential at the surface of the common envelope and ϕ_1 and ϕ_2 are the potentials at the inner and the second Lagrangian points, respectively. It should be noted that $(da/d\mu)_c$ is positive for the region of μ [0, 0.5]. If F is very large, we have negative gradient

$$d \ln a/d\mu=-(R-S)/\ Q \tag{16}$$

and the binary will evolve from the stage of thin contact (c=1) to the stage of contact at L(c=0) rapidly. If F is close to zero, $(da/d\mu)$ is positive. Therefore, we can expect that the degree of contact of a contact binary remain unchanged for a long time. Detailed calculations show that the degree of contact of contact binaries should be close to 1(thin contact). For details, see Nariai (1979).

REFERENCES

Kuiper, G. P. : 1941, Astrophys. J., 93, 133.
Nariai, K. : 1975, Astron. Astrophys., 43, 309.
Nariai, K. : 1977, Publ. Astron. Soc. Japan, 29, 263.
Nariai, K. : 1979, Publ. Astron. Soc. Japan, 31, 311.
Nariai, K. and Sugimoto, D. : 1976, Publ. Astron. Soc. Japan, 28, 593.
Zahn, J.-P. : 1976, Astron. Astrophys., 57, 383.

DISCUSSION FOLLOWING NARIAI

Vanbeveren: In the calculations of particle motions in binary systems one usually uses a Roche geometry. It may be that radiation forces and shadow effects change the Roche geometry (see e.g. two papers published in A & A, 54, 877 and Astrophys. and Space Sci. 57, 41). Could you comment on that in connection with your calculations?

Nariai: I do not think that radiation changes the situation for systems with low luminosity such as W UMa-type binaries.

Eaton (in reply to Vanbeveren's comment): Comment on modification of Roche lobes by radiation pressure: This will be important only for the hottest stars which should be gravity-darkened. With von Zeipel gravity darkening, the shape and size of the gravitational equipotentials will not be changed. Hence, radiation pressure should not affect the shape of the Roche lobe which determines the limiting size of a binary component. I made this comment in an article appearing in Acta Astronomica, 28.

Wilson: I'd like to point out that I also made this same comment at last year's eclipsing binary symposium in Victoria.

Zuiderwijk (comment on a remark by Vanbeveren): Apart from the fact that the concept of separating radiation pressure from the total pressure in the stellar interior and atmosphere is basically wrong, I would like to make another comment on Schuerman's (1972) model. The major result from his computations, concerning the "opening of the critical surface" near L_1, is only an artifact of his assumption of a spherically symmetric radiation field. This assumption also means that gravity darkening is neglected in the model. As gravity darkening is directly related to hydrostatic equilibrium, we can conclude that Schuerman's model is also not in hydrostatic equilibrium.

Re: Schuerman, D.W., Astrophys. Space Sci. 19, 351, 1972.

Kondo: There is no time now for me to discuss whose radiation pressure model is appropriate. It is a complex problem. For instance, the shadow effects as discussed by Vanbeveren must be taken into account. I simply wish to point out that radiation pressure and other effects must be properly treated for discussions of the critical equipotential surfaces.

SOLAR TYPE CONTACT BINARIES - THE INITIAL MASS RATIO DISTRIBUTION FUNCTION.

Frans Van 't Veer
Institut d'Astrophysique, Paris.

From a study of the mass ratio function of magnetically active contact binaries it is shown that the great majority of newly formed systems must possess nearly equal components.

We define the mass ratio (q) of a binary as the ratio $q = m_2 / m_1 < 1$. The distribution function of the q's is given by what we call the mass ratio function (MRF). The determination of the MRF for contact binaries (CB) is one of the major problems which need to be solved, when we wish to obtain a better knowledge about their formation and evolution. In a previous study (Van 't Veer, 1978) we determined from the best information available for 27 W UMa stars, taking into account observational selection effects, the MRF of solar type CB's. This function which is displayed in fig 1 shows a gradual increase of the observed number of CB's towards the lower mass ratios. We call this function the actually observed or actual mass ratio function (AMRF). It represents the observed result produced by now existing CB's formed at different epochs with possibly different q's, and different physical and dynamical evolution since their formation. Clearly, for our understanding of CB's it is much more interesting to know the distribution function of the q's at the moment of formation. We call this function the initial mass ratio function (IMRF); it represents the mass ratio distribution function of CB's at the very beginning of their existence. It is the aim of this paper to investigate the IMRF by using our results concerning the AMRF and the dynamical evolution of solar type CB's studied earlier in a theoretical paper (Van 't Veer, 1979).

In general this problem may be solved if we possess reliable information about what we call the mass ratio evolution function(MREF). This new function defines the relation between q and time (t) for an arbitrary CB. It may be formulated by the general expression $q = q(t,q_i)$ where q_i is the initial mass ratio the CB had at its time of formation $(t = 0)$. The actual mass ratio then becomes $q_a = q(t',q_i)$ where t' is the age of the CB, that means the time necessary for its evolution to q_a. Obviously the knowledge of both the IMRF and the MREF, for every q_i, combined

M. J. Plavec, D. M. Popper and R. K. Ulrich (eds.), Close Binary Stars: Observations and Interpretation, 517–520.

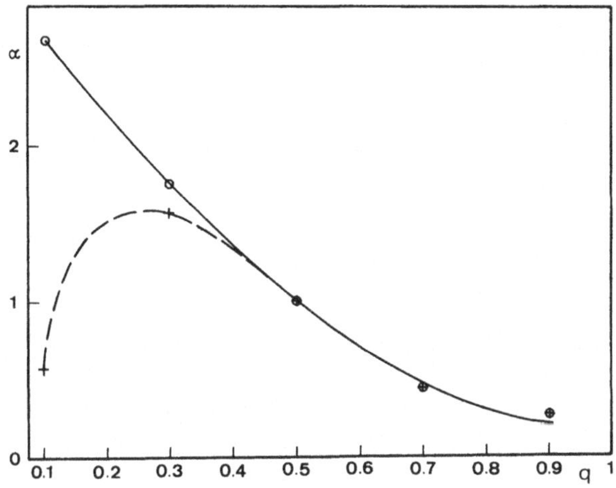

fig. 1 - The mass ratio function of solar type W UMa binaries (solid curve).

with an hypothesis on the rate of formation of CB's permits us to derive the AMRF. However we already know the AMRF and we are trying to investigate the IMRF. So our problem is precisely the inverse and cannot be solved by direct integration. We can, for the moment being, only try to see which IMRF may be considered as the best acceptable in order to give, in combination with a theoretically justified MREF, the observed AMRF. This may be done by an integration procedure we symbolically represent as AMRF = $\int$ (IMRF) (MREF) dt.

We do not enter into the mathematical details of this procedure, which will be published elsewhere, but we need to say some words about the MREF. In the litterature on this subject we find two ways of considering the dynamical evolution of a CB : 1. By using the hypothesis of conservation of angular momentum (AM). Most evolutionary models are based on this principle, 2. By considering the dynamical evolution as controlled by continuous loss of AM. This principle was defended mainly by Huang (1966), Okamoto and Sato (1970) and recently formulated quantitatively by Van 't Veer (1979).

Clearly the first way makes it possible to maintain a CB without changing the q during a very long time. In that case the IMRF may be about the same as the AMRF, that means a high formation rate of low - q CB's. The second way, which seems the more realistic for solar type CB's exhibiting magnetic activity, predicts a gradual but non linear decrease of the q for every synchronized CB, as long as the AM loss continues. The solid curve of fig 2 applies to the case of constant AM loss with time, it clearly shows the general trend of the decrease of q with time (on an arbitrary time scale) with the most rapid changes occuring for high q values (nearly equal components). The function of fig 2 is obtained by calculating the mass transfer $m_2 \rightarrow m_1$ necessary to equilibrate the AM loss of a synchronized CB involved by magnetic activity

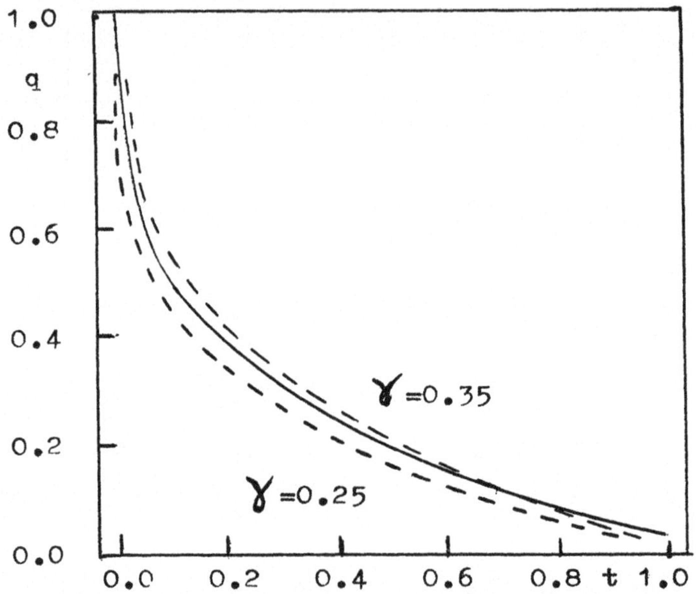

fig. 2 - The MREF of a synchronized CB with constant rate of AM loss.

(see Van 't Veer, 1979 for a detailed discussion of this matter). The general character of the resulting q variation may be given by an expression of the type $q(t) = 1 - t^{\gamma}$. A reasonably good fit is obtained for $0.25 < \gamma < 0.35$. With this theoretically derived MREF in hand we may now try to obtain information about the best IMRF that gives an AMRF compatible with the relation derived from the observational material. We only need to suppose that both the total number of CB's formed, as their frequency distribution are independent of time. So we reasonably consider the formation process of CB's as quantitatively unaltered since at least the birth of the oldest still existing CB.

We are aware of the fact that our development contains some weaknesses (constant loss of AM, discussed below, and lack of precision in our AMRF, which needs to be confirmed by new and more precise observation of q), but it seems nevertheless possible to draw some conclusions which may be briefly summarized as follows : the great majority of newly formed BC's must have high q's with at least 50% of them higher than $q = 0.8$. When we take into account the plausible tendency of a decreasing AM loss rate during the life of a CB, we are forced to admit a still higher production of the high-q CB's with respect to the low-q CB's. This effect may become so important that only the hypothesis of an exclusive production of CB's with equal components may explain the actually observed mass ratio distribution. For the time being however it is difficult to give any information on the change of AM loss rate and we summarize with the tentative conclusion that most, if not all, CB's will begin their life with nearly equal components.

References.

Huang, S.-S. : 1966, Ann. d'Astrophys. 29, 331
Okamoto, I. Sato, K. : 1970, Publ. Astr. Soc. Japan 22, 317.
Van 't Veer, F. : 1978, Astron. Astrophys. 70, 91
Van 't Veer, F. : 1979, Astron. Astrophys.

ON SUDDEN PERIOD CHANGES IN CONTACT BINARIES

Ove Havnes
Institute of Mathematical and Physical Sciences
The Auroral Observatory, University of Tromsø,
Norway

ABSTRACT

We propose that sudden period jumps in contact binaries can be caused by the action of a third companion in a highly eccentric orbit of small periastron distance. Time jumps as observed are easily reproduced for a wide range in mass of the third body but observations may restrict the possible mass range to less than a few times 0.1 $M_{\odot}$.

INTRODUCTION

Contact binaries of solar type may exhibit rather sudden changes in their periods (e.g. Kreiner 1977) which are usually interpreted in terms of mass transfer within the system or away from it (e.g. Van't Veer 1979).

We here argue that it is, at least, physically possible to produce similar period changes through the action of a third companion orbiting around the binary in a highly eccentric orbit of small periastron distance. The effect of this is different from that of a remote companion in an orbit of low eccentricity which may cause a light-time effect as in, for example, AK Herculis (Schmidt and Herczeg 1959). A third body in an orbit of intermediate eccentricity has been suggested by Mazeh and Shaham (1976) to produce a long term variability in binary stars. In our case, however, we examine what usually is termed a close passage during which one may transfer energy away from or into the binar depending on the conditions of the close passage.

M. J. Plavec, D. M. Popper and R. K. Ulrich (eds.), Close Binary Stars: Observations and Interpretation, 521–526.

NUMERICAL COMPUTATIONS AND RESULTS

Our numerical integration program is a modified version of a regularized N-body program by Peters (1968, 1969).

The masses of the binary members are in all cases M_1 = 1 $M_\odot$ and M_2 = 0.5 $M_\odot$ with radius $R_1 = R_2$ = 0.005 a.u. They are initially in circular orbits with mutual distances of $2R_1$, i.e. they are just in contact. The companion is given an orbital period of 6.5 years in all cases. We have computed a comparatively large number of cases for masses of the third body M_3 = 1 $M_\odot$ and 0.1 $M_\odot$ and fewer cases for M_3 = 0.01 $M_\odot$.

For a fixed value of M_3 the energy transfer (period change) increases with decreasing periastron distance q. The dependence on the phase angle of the binary also becomes more marked at smaller q. At a sufficiently small q the third body will in most cases be thrown into a hyperbolic orbit and disappear from the system. In Fig. 1 we show for M_3 = 1 and 0.1 $M_\odot$ the fractional change of the binary period $\Delta P/P$ as function of its phase angle for different values of q.

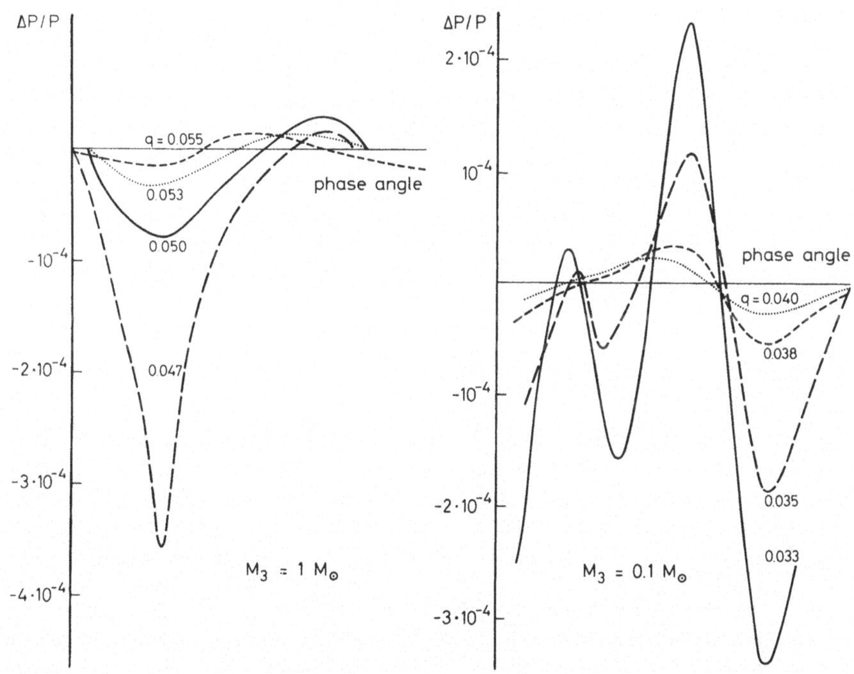

Fig. 2. The period changes as function of the binary phase angle for different periastron distances q.

As q decreases the probability for obtaining a negative $\Delta P/P$ far exceeds that of obtaining a positive $\Delta P/P$. Such a situation is evidently unstable and the orbital evolution of M_3 would be toward a progressively smaller period of the binary and a correspondingly longer period of M_3 with its eventual escape from the system. We note that as the periastron distance q increases the probability of positive and negative period jumps becomes practically equal when the relative period jumps decrease below 10^{-5} which is in the region which Kreiner (1977) finds for the observed period jumps in W UMa binaries. If the period jumps are caused by close passages of a third body it is essential that the observed jumps corresponds to periastron distances giving about equal probabilities for positive and negative ΔP to ensure the stability of M_3 over a prolonged period of time.

We have calculated a sufficient number of passages for $M_3 = 0.01\ M_\odot$ to find at what values of q the period jumps of the binary becomes of the order of those observed. This happens for $q \lesssim 0.035$ a.u. Extrapolating our results for $M_3 = 1$, 0.1 and 0.01 $M_\odot$ toward lower masses indicates that the required period jumps can be the result of close passages of a third body of mass as low as $M_3 \sim 10^{-4}\ M_\odot$ if the periastron distance is small enough.

Some restrictions on the mass of the third body may be set. $M_3 \sim 1\ M_\odot$ are unlikely mainly because such stars, of the same magnitude as the binary members, should be directly observable. There should also in most cases be an observable light-time effect. With $M_3 \sim 0.1\ M_\odot$ the light of M_3 is so weak as to escape detection and the maximum light-time effect (for a period of 6.5 years) is now $\sim$ 5 min. Due to orientation effects the apparent light-time may escape detection.

CONSEQUENCES OF PERIOD JUMPS

The action of the third body which may cause a change in the orbital period of the binary - and the third body - also influences on the binary orbit to make it slightly elliptical. Further, the intrinsic rotation of the binary members are not influenced noticeably by the close passage whence the binary is brought out of synchronization. Such changes in the binary orbit and rotation may have a profound influence on the characteristics of the binary. One effect will be that some sort of pumping action is set up in the contact region as this is stretched and compressed during the noncircular motion of the binary. When $|\Delta P/P| \sim 10^{-5}$ the difference between maximum and minimum distance between the binary members is $\Delta R \lesssim 1.5\ 10^3$ km resulting from a binary orbit eccentricity of e $\sim$ 0.001. This will most certainly generate surface waves which may have an observable influence on the light curve of the binary and may also possibly trigger mass transfers from one binary member to the other.

The lack of synchronization will cause the contact point to "slip" and move over the surface of the binaries. We have found that this motion of the atmosphere of the one binary relative to the other is too slow to give rise to any of the familiar drift instabilities (e.g. Hasegawa 1975). There may, however, be another effect of the non-synchronization which can result in a rather violent instability. Assuming that the stars have magnetic fields this is dragged along by the contact point as it moves over the surface. A deformation of the field results and a neutral sheet is formed along the line traced out by the contact region. An acceleration of surface plasma may eventually take place and may be visualized as a "snapping back" of field lines. This is essentially the same mechanism as suggested by Parker (1963) to accelerate matter in solar flares.

All these effects, which may well be induced by close passages of a third body, act to disturb the system and may possibly also lead to disturbances in the light-curves of the binary.

FORMATION AND STABILITY OF THE SYSTEMS

Three body systems as the ones we have envisaged with their high eccentricity may seem rather exceptional in view of our statistics on multiple systems. We can of course not exclude the possibility that the chances for the formation of such systems are so small as to be of no practical interest. However, we would like to suggest some possible ways of forming these systems. It is well known from numerical simulations of the evolution of stellar clusters that close passages between the members of the cluster have a profound effect on its evolution. This leads, among else, to the ejection of stars from the cluster. In small clusters with an initial number of stars of the order of 10 most cluster members are ejected and the eventual result is one double - or triple - system with some other stars in wide orbits (Van Albada 1968). Wielen (1967) has studied larger systems and finds the same general results but with the formation of several double - or triple - systems as the final state of the cluster. We here argue that if a close binary was member of such a cluster and experienced one or more close encounters with other cluster members it could possibly aquire one or more companions with the required highly eccentric orbits. This could happen both if it had its own "planetary" system which was strongly disturbed during the close passage of another star or possibly also through capture of "planets" from other cluster members during their close passages. If the presence of planetary systems are a common feature around stars it may be that highly eccentric orbits are not uncommon due to reasons given above.

DISCUSSION

The periastron distance q which is required to obtain time jumps in the interval $10^{-5} \gtrsim |\Delta P/P| \gtrsim 10^{-6}$ varies from grazing encounters up to several tens of stellar radiae depending on the third body mass. A large q results in a smaller period jump than observed. We cannot, however, exclude the possibility that a passage resulting in a small direct period jump creates disturbances which in turn trigger mass transfer events leading to larger changes in the binary periods. We are unable to specify beyond what periastron distances the binary will be unaffected by a close passage of a third body.

REFERENCES

Albada, T.S. Van: 1968, Bull. Astron. Inst. Neth. 19, 479.
Hasegawa, A.: 1975, Plasma Instabilities and Nonlinear Effects. Springer-Verlag.
Kreiner, J.M.: 1977, IAU coll. No. 42, Eds. R. Kippenhahn, W. Strohmeier, Bamberg, p. 393.
Mazeh, T., Shaham, J.: 1976, Astrophys. Journ. 213, L17.
Parker, E.N.: 1963, Astrophys. J. Suppl. 8, 177.
Peters, C.F.: 1968, Treatment of binary encounters in stellar dynamics. Naval Weapon Lab. Tech. Rep. TR-2163.
Peters, C.F.: 1969, Private communication.
Schmidt, H., Herczeg, T.: 1959, Zeitschr. f. Astrophys. 47, 106.
Van't Veer, F.: 1979, Preprint.
Wielen, R.: 1967, Veroff. Astron. Rechen-Instituts Heidelberg, No. 19.

COMMENTS FOLLOWING HAVNES

Guinan: Many of the brighter W UMa-type systems appear to have companions moving in rather wide orbits and with orbital periods of 30 years or greater. Examples are VW Cep, i Boo, W UMa, and AM Leo. These companions have orbits that appear to be too wide and have periods too long to produce the observed short term changes in periods. In addition would not the presence of an interacting third component produce more or less periodic changes in the orbital period of the W UMa system?

Havnes: It is true that the companions you mention cannot produce the period changes I have considered. This requires much larger eccentricities (and shorter periods). If the period changes were solely produced during the close passages of a third body, the interval between each period jump could change quite a bit if $M_3 \lesssim 0.01M_\odot$. This is because a transfer of energy to/from the binary from/to the third body has an effect on the orbit of M_3 which increases as the mass of M_3 decreases. However, if such large changes in the M_3 orbit took place, I think one could get problems with its orbital stability. For this reason somewhat larger M_3 may be required, in which case the intervals between the periastron passages would be practically constant. Unequal intervals between the period jumps could still result, as there is always the possibility that the effect of one passage can be too small to be detected. Another possibility is the presence of several "planets" in eccentric oribits; this may, however, encounter problems when it comes to the formation of such systems.

In my opinion the best thing, both from stability and formational considerations, would be if a third body generally produced smaller period changes than what one apparently observes and that these act as disturbances of the binary, which in turn experiences larger period jumps as it attempts to restore synchronization (through mass transfer?).

Shu: I would like to support the general philosophy that one should be careful not to interpret every period change as a mass transfer event. Such effects may arise for reasons that have nothing to do with binary evolution. You give one example; Doug Hall gave another yesterday on the basis of magnetic activity. Indeed, mass loss need not even be involved in the magnetic activity. Very minor changes of stellar structure associated with a magnetic cycle may give rise to period changes of the observed magnitude.

Rucinski: As a comment to Dr. Shu's remark I would like to point out that the intervals of period constancy as inferred from the O-C diagrams (5-15 years typically) might roughly correspond to the time scales of magnetic-field structure reorganization as observed in solar-type phenomena.

Geyer: In a recent paper by me (Astrophys. Space Sci., 48, 137, 1977) I gave a few other causes which give rise to erratic "period changes" which have nothing to do with mass loss or mass exchange in a binary system. For example if there is a non-symmetric brightness distribution which is slowly changing by itself, this gives rise to cumulative errors in the O-C diagram. Therefore whenever light curve variations are known, one must be skeptical about period changes.

CURRENT OBSERVATIONAL DEVELOPMENT IN EARLY-TYPE CONTACT BINARY SYSTEMS

Kam-Ching Leung
Behlen Observatory, Department of Physics and Astronomy, University of Nebraska, Lincoln, NE 68588, U.S.A.

This is a summary of current observational results of early-type contact binary systems. The classical approaches to derive a photometric solution of a very close binary are found to be incapable of distinguishing whether a system is detached, semi-detached, or in contact. A survey of contact systems based on the more modern approach of Wilson and Devinney (1971) and others are summarized in the following. Contact systems with spectral type O are: V382 Cyg (Devinney 1976 private communication, Bloomer et al. 1979), V729 Cyg (Leung and Schneider 1978c), UW CMa (Leung and Schneider 1978b), AO Cas (Schneider and Leung 1978); with spectral type B are: SV Cen (Wilson and Starr 1976, Rucinski 1976), V701 Sco (Wilson and Leung 1977, Anderson et al. 1979), BH Cen (Leung and Schneider 1977), RZ Pyx (Devinney 1976 private communication); with spectral type A are: AU Pup (Leung and Schneider 1978a), V1073 Cyg (Leung and Schneider 1978a), V535 Ara (Leung and Schneider 1978a), BV 845 (Leung and Darland 1978), V1010 Oph (Leung and Wilson 1977).

Figure 1 is a diagram showing the relation between the surface potential and mass ratio of the known early-type systems. In cases where the mass ratios were not certain, multiple solutions for different mass ratios were derived. It was found that the solutions remained in contact configuration and the degree of overcontact remained at about the same value. Thus a point in Fig. 1 will shift roughly parallel to the inner or outer contact boundary (Leung and Schneider 1978a and 1978b). It is interesting to note that only V701 Sco and BH Cen have mass ratios about unity.

The majority of the systems are located near the inner contact potential boundary (small degrees of overcontact), but a few of them have contact potentials approaching the outer contact potentials (large degrees of overcontact).

M. J. Plavec, D. M. Popper and R. K. Ulrich (eds.), Close Binary Stars: Observations and Interpretation, 527–531.

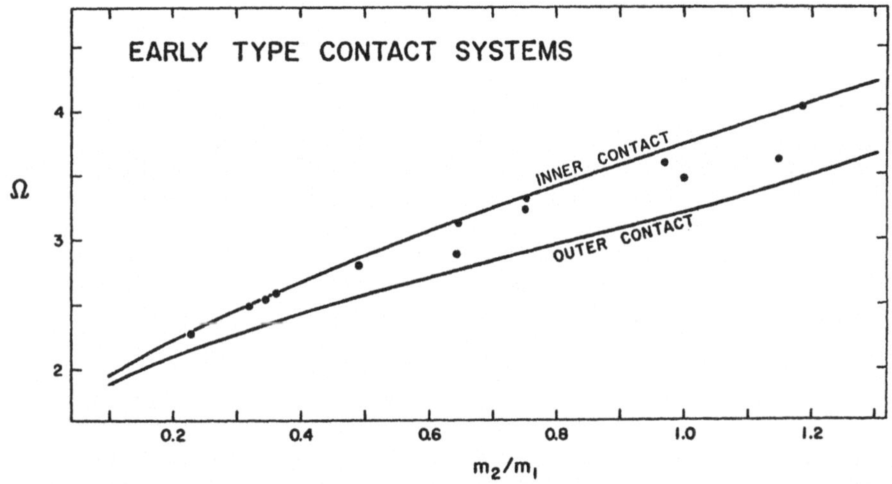

Figure 1. Surface potential - Mass ratio diagram for early-type contact systems.

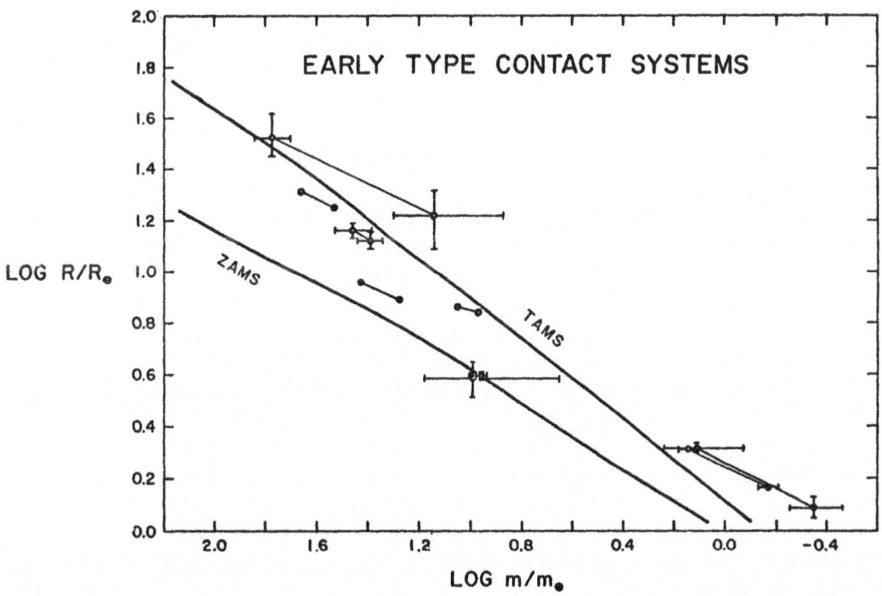

Figure 2. Mass - radius diagram for early-type contact systems.

More than half of the systems have sufficient spectroscopic information for determination of absolute dimensions of the components of the binaries. Their locations in the log $R/R_{\odot}$ and log $m/m_{\odot}$ diagram are shown in Fig. 2 together with the ZAMS and TAMS lines of Stothers (1972). The two systems located near the ZAMS are V701 Sco and BH Cen. (Since the mass ratio of each system is about unity, both components occupy the same location.) They are also members of very young galactic clusters. Their locations in the H-R diagram indicate that they are zero-age systems. It is believed that the systems which fall between the ZAMS and TAMS lines are evolved contact systems through case A mass exchange. The three systems lying above the TAMS line are most likely to be evolved contact systems through case B mass exchange.

It is interesting to note that both of the zero-age systems found so far have mass ratios near unity. Does this mean that only mass ratio of unity is allowed for radiative common envelope zero-age contact systems as Lucy suggested? Obviously, this important question deserves further observational investigation. Hopefully, more zero-age contact systems will be discovered for observational tests between the Lucy and Shu theories. The system TU Mus deserves immediate attention, since, based on its location in the period-spectral type diagram (Leung and Schneider 1978a), this is a potential candidate for a zero-age contact system this system has a mass ratio of 0.67!

It is also appropriate to draw our attention to the system VW Cep even though it is a late type system. This eclipsing binary has a large secondary variation in its light curve. A successful attempt had been made (Leung unpublished), employing a numerical technique, to decouple the observed light curve into a binary light curve and a periodic secondary variation light curve. The binary light curve was analyzed by the Wilson and Devinney method (1971). Two photometric solutions were found (see Fig. 3), a semidetached solution and a detached solution. The latter assumed the gravity darkening exponents and the bolometric albedos equal to unity, which may not be realistic for late type stars. The former solution (semidetached) was derived by letting these coefficients be adjustable parameters. The final adopted values were: $g_1 = 0.6$, $g_2 = 0.7$, $A_1 = 0.3$, and $A_2 = 0.7$. These values are reasonable for late type stars. Therefore, the semidetached configuration solution for VW Cep was adopted. (Note that no contact solution was found even though attempts were made.) Thus, this system may be an example of a system after contact has broken. Lucy's thermal relaxation oscillations theory may be favored in this case!

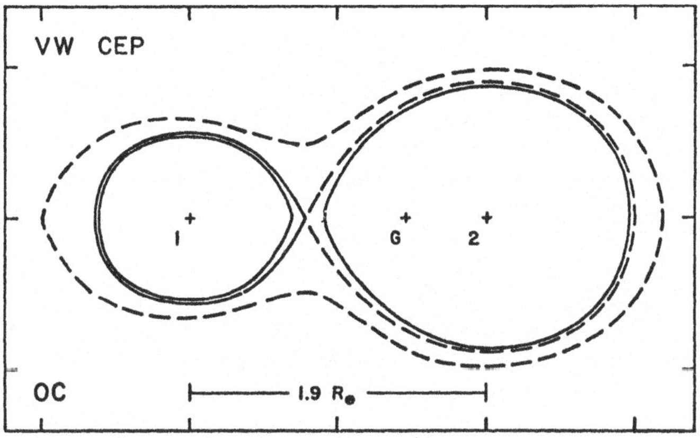

Figure 3. Semidetached and detached configurations of VW Cepei.

REFERENCES

Anderson, J., Nordstrom, B., and Wilson, R. E. : 1979, Astron. Astrophys. (in press)

Bloomer, R. H., Burke, E. W., King, C., and Millis, R. L. : 1979, Bull. Amer. Astron. Soc. 11, pg. 439.

Leung, K. C., and Darlan, J. J. : 1978, Publ. Astron. Soc. Pacific 90, pp. 97-100.

Leung, K. C., and Schneider, D. P. : 1977, Astrophys. J. 211, pp. 844-852.

Leung, K. C., and Schneider, D. P. : 1978a, Astrophys. J. 222, pp. 917-923.

Leung, K. C., and Schneider, D. P. : 1978b, Astrophys. J. 222, pp. 924-930.

Leung, K. C., and Schneider, D. P. : 1978c, Astrophys. J. 224, pp. 656-569.

Leung, K. C., and Wilson, R. E. : 1977, Astrophys. J. 211, pp. 853-859.

Rucinski, S. M. : 1976, Publ. Astron. Soc. Pacific 88, pp. 244-250.

Schneider, D. P., and Leung, K. C. : 1978, Astrophys. J. 223, pp. 202-206.

Stothers, R. : 1972, Astrophys. J. 175, pp. 431-452.

Wilson, R. E., and Devinney, E. J. : 1971, Astrophys. J. 166, pp. 605-619.

Wilson, R. E., and Leung, K. C. : 1977, Astron. Astrophys. 61, pp. 137-140.

Wilson, R. E., and Starr, T. C. : 1976, Monthly Notices Roy. Astron. Soc. 176, pp. 625-632.

COMMENTS FOLLOWING LEUNG

Linnell: Although I have not seen the details of this solution, I recall the solution of TX Cnc by Wilson and Bierman, using the same light synthesis program. In that case, there were two solutions, one allowing a thermal discontinuity between components, the other producing a hot primary. Especially in the latter case, the gravity brightening coefficient is much larger than even the Lucy theory allows. Is the same true of the VW Cep solution?

Leung: In the case of VW Cep, $g_1 = 0.60$, $g_2 = 0.65$. These values are expected for late type stars.

Anderson: The rotation broadening function for VW Cep, although indicating marginal contact, does fit our atlas entry corresponding to the correct spectroscopic mass ratio. This fit implies that the surface temperature is the same on both stars (in agreement with Dr. Linnell's results) which in turn implies good thermal contact.

X. ... AND THE REST (SYMBIOTICS, SUPERGIANTS, PLANETARIES, POPULATION II SYSTEMS)

THE SYMBIOTIC BINARY SYSTEM AG PEGASI

Charles D. Keyes and Mirek J. Plavec
Department of Astronomy, University of California
Los Angeles, CA 90024, U.S.A.

ABSTRACT

A complete coverage of the spectrum of AG Peg has been obtained between wavelengths 1200 - 7000 A by us, and can be supplemented by IR photometric observations by others. Our IUE observations yield a lower value of E(B-V), about 0.12. The two stellar components are easily recognized, but their characteristics are still rather uncertain.

The cool component may be a normal M1.7 III giant, but the temperature and luminosity of the hot star remain largely indeterminate. Firstly, there are no good models for a hot subdwarf, and secondly, it is difficult to determine the relative contribution of the star itself and the surrounding hydrogen cloud.

The emission lines observed in the UV have double or triple structure, indicating two or three distinct emitting regions.

INTRODUCTION

AG Pegasi is a prominent representative of the symbiotic stars, although it lacks one typical characteristic, namely persistent flare activity. It brightened only once, about 100 years ago, and has been on the decline all the time since. It is probably an extremely slow nova. Its smaller-scale light fluctuations, and in particular its spectrum peculiarities and variations have over the years attracted the attention of many investigators, notably of Merrill, Boyarchuk, Hutchings and Redman, Cowley and Stencel. References to their work can be found in the important paper by Hutchings, Cowley, and Redman (1975), which concluded the period during which no direct evidence existed for the presence of a hot component, although the binary nature of the object was no longer in doubt.

A new era started when observations in the satellite ultraviolet brought about direct evidence for the presence of a hot star. Gallagher et al. (1979) discussed broad-band photometry from the OAO-2 satellite. They concluded that the hot component is a very luminous UV source with

M. J. Plavec, D. M. Popper and R. K. Ulrich (eds.), Close Binary Stars: Observations and Interpretation, 535–542.

a luminosity probably lying between 1000 $L_\odot$ (if the cool component is a normal M3 III star) and 8100 $L_\odot$ (if the cool component fills its critical Roche lobe). They preferred the higher luminosity as it can better explain the nova-like eruption.

AG Pegasi has been studied at UCLA for several years as part of CDK's thesis on symbiotic stars. Our contribution to be discussed here consists of scans obtained with the 120-inch Shane telescope of Lick Observatory, and of MJP's IUE scans made with low and high dispersions in 1978.

We re-examined the interstellar reddening of AG Peg and found the color excess to be most likely $E(B-V) = 0.12 \pm 0.03$, from the 2200 A extinction bump. The higher value of 0.2 obtained by Gallagher et al. is probably due to emisssion lines contaminating the broad-band OAO-2 photometry. With the adopted color excess, we derived the overall energy distribution between 1200 and 7000 A shown in Fig. 1. The photometric accuracy of the Lick IDS scanner is well known. It is very gratifying to see that the three scans made with three different grating settings, and the two scans with the two IUE cameras match very satisfactorily: no arbitrary shifts were made.

THE COOL COMPONENT

For both components, the scans cover only a small fraction of their total luminosities. There appears to be less ambiguity in the case of the cool component. Infrared photometry of AG Peg was published by Szkody (1977) and Swings and Allen (1972). Their excellent mutual agreement suggests that there probably has been little variability in the IR. An independent estimate of the total flux from the cool component was based on our Lick spectrophotometric data from the interval 4260 - 6800 A. The scanner data from the interval 3200 - 4260 A cannot be used, since it contains an unknown admixture of the radiation of the hot component and/or b-f and f-f hydrogen continuum. The two estimates give for the total observed flux (corrected for interstellar extinction) a value $f_c = 5.5 \times 10^{-8}$ erg/cm^2/s, with a 15% estimated uncertainty.

The colors obtained from Szkody's magnitudes were used to determine the spectral type of the cool star, using Lee's (1970) calibration. Assuming that the luminosity class is III, we find an encouraging agreement of four different estimates, and adopt the mean, M1.7 III. Comparison of our scans with two standards, 104 Her (M1 III) and 26 Peg (M3 III) indeed places AG Peg in between them, and suggests that the cool component may be normal, contrary to some previous conclusions presumably affected by the transition region shortwards of about 4200 A. The effective temperature of M1.7 III in Lee's calibration is 3570 K. We will consistently express the distance d in kpc and the stellar parameters in solar units. Then the general formula

$$R = 5.89 \times 10^{12} \ T^{-2} \ d \ f^{1/2} \quad (1)$$

leads to the following equations for the cool component:

$$R_c = 108 \ d \ , \ L_c = 1720 \ d^2 \ , \ M_c = -3.3 - 5 \log d \ . \quad (2)$$

If the cool star is really M1.7 III, then $d = 0.5$ kpc and $R_c = 56 \ R_\odot$. Luminosity class II giants are 1.9 mag brighter, so the radius would be 134 $R_\odot$ and $d = 1.24$ kpc ($z = 0.6$ kpc below the galactic plane). Even in this case, the star would be substantially smaller than its critical Roche lobe, estimated to be 285 $R_\odot$ after Hutchings, Cowley, and Redman (1975).

THE HOT COMPONENT

The observed (and corrected for extinction) flux between 1200 and 3260 A is $f_h' = 1.26 \times 10^{-8}$ erg cm^{-2} s^{-1} . (Emission lines in the same region contribute an additional 3.45×10^{-9} erg cm^{-2} s^{-1}, i.e. 27% on top of the observed continuum). The observed continuum is most likely a blend of a true stellar continuum and continuous radiation of a hydrogen nebula.

Two extreme cases can be considered. In the first alternative, the radiation contaminating the flux from the cool star around 4000 A (in fact, with traces going up to 5200 A) is attributed mainly to the hot star directly. In this case, the whole observed "hot" continuum between 1200 A and 5000 A can be reasonably well fitted by a Kurucz model with $T_{eff} = 25{,}000$ or 30,000 K. Only a relatively weak b-f contribution shortward of the Balmer limit is needed in this case. For $T = 30{,}000$ K we then get from (1):

$$R_h = 1.03 \ d \ , \ L_h = 773 \ d^2 \ , \ M_h = -2.46 - 5 \log d \ . \quad (3)$$

Also, for the surface gravitational acceleration we get

$$\log g = 4.4 - 2 \log d \ , \quad (4)$$

so that the good fit found for $\log g = 4.5$ is not inconsistent with the rest of the data, if we assume that the mass of the hot star will not be much different from one solar mass. However, the fact that $T_{eff} = 25{,}000$ K and $\log g = 3.0$ gives an even better fit casts doubts on the procedure. It would indeed be surprising if a standard plane-parallel LTE and normal-composition atmosphere represented well the hot subdwarf. The most seriuos argument against this fit is, of course, that the star must radiate enough photons capable of ionizing He II, since the He II recombination spectrum is strong.

In the second extreme alternative, we may assume that the contaminating radiation around 4000 A is all due to a hydrogen cloud. This implies a much higher temperature for the central star, but enables us to avoid assigning high luminosity to it, because we may assume that the star's flux falls off very rapidly already in the far UV and becomes negligible longward of, say, 2000 A. Then the observed (corrected) flux from the star is only $8.8x10^{-9}$ erg cm^{-2} s^{-1}. Assume, quite tentatively, that the star's temperature is 10^5 K and that it is nearly represented by a black body. Then the total flux is $f_h = 1.6x10^{-7}$ erg cm^{-2} s^{-1} and we have, in analogy to (3),

$$R_h = 0.24\ d \quad , \quad L_h = 5000\ d^2 \quad , \quad M_h = -4.5 - 5 \log d \quad . \qquad (4)$$

In principle, an analysis of the strength of the He II lines, combined with properly matching the two components (stellar and nebular) of the continuum, should determine the star's parameters. Although we are trying hard, we are not sure that the solution will be well determined. There are no good models of hot subdwarfs that would reliably give the number of photons emitted shortward of 227 A (Bohlin, Harrington and Stecher, 1978), and the continuum fitting will be largely arbitrary since the stellar continuum will probably not be adequately modelled.

THE EMISSION LINES

Besides a well-developed recombination spectrum of He II, the UV spectrum of AG Pegasi shows a number of strong emission lines such as N V, C IV, Si IV, and lower ions of these elements. The high-dispersion IUE spectra show that the line profiles are quite complex: we can distinguish two or three components in most lines.

One component comes from a highly ionized region near the hot object and shows broad P Cygni profiles of a Wolf-Rayet character. The N V resonance lines at 1239 and 1242 A and the permitted N IV 1718 line show this structure. The N IV] intercombination line at 1486 A (Fig. 2) displays both the broad Wolf-Rayet structure and a sharp nebular component (FWHM = 0.3 A) which is apparently formed in a more distant, quiescent region.

All the other intercombination lines, C III] 1909 A, Si III] at 1892 A, and the N III] λ 1750 group, display only the sharp, nebular component. The resonance doublet of C IV at 1548 and 1550 A (Fig. 3) shows the nebular emission component and several narrow, blue-shifted absorptions superposed on a broad λ 1550 emission. The 1548 A component is quite weak, probably due to a broad P Cygni absorption component of the 1550 line.

The presence of C III] 1909 A, the absence of C III] 1906 A, and evaluation of critical densities for collisional de-excitation of the relevant energy levels establishes a density range between 10^7 and 10^9 cm^{-3} for the nebular emitting region. Maximum densities implied by the other intercombination lines are all less than 10^{11} cm^{-3} .

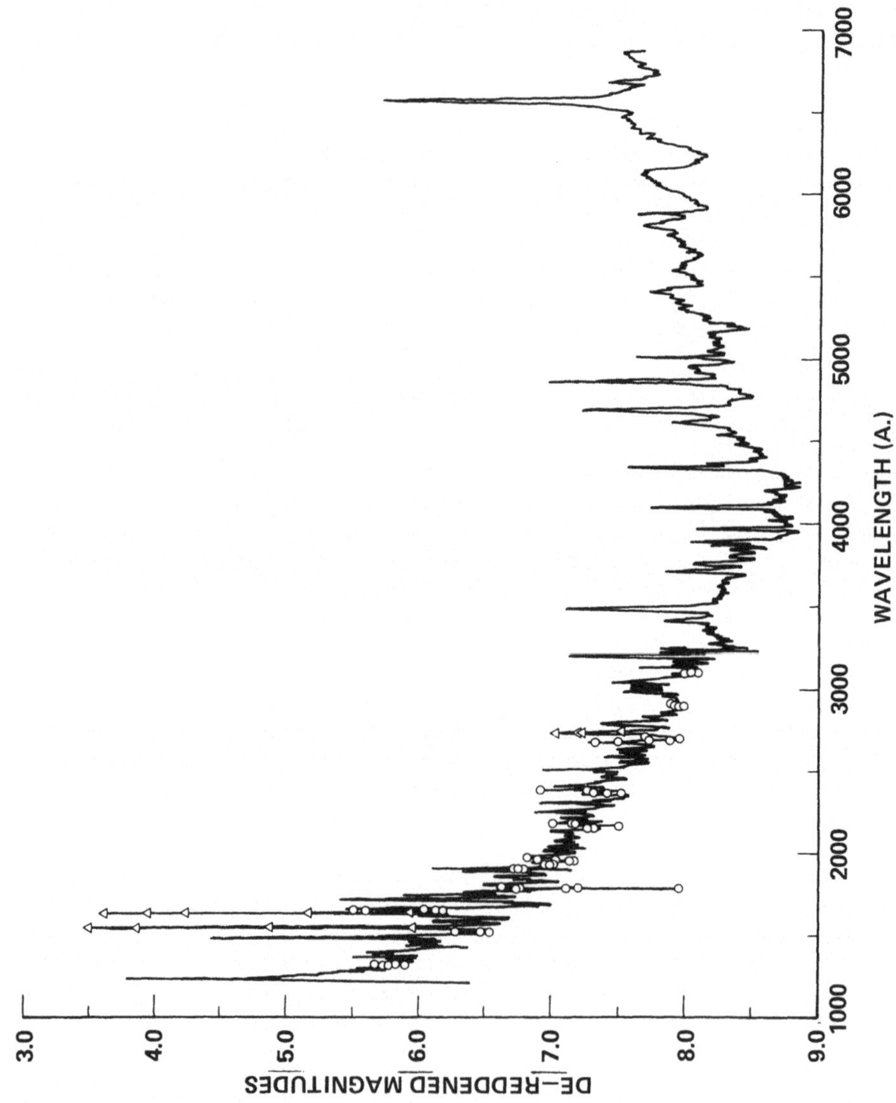

Fig. 1: Spectrum of AG Pegasi, 1200-7000 A.
Circles mark reseau contamination, triangles are saturated pixels.

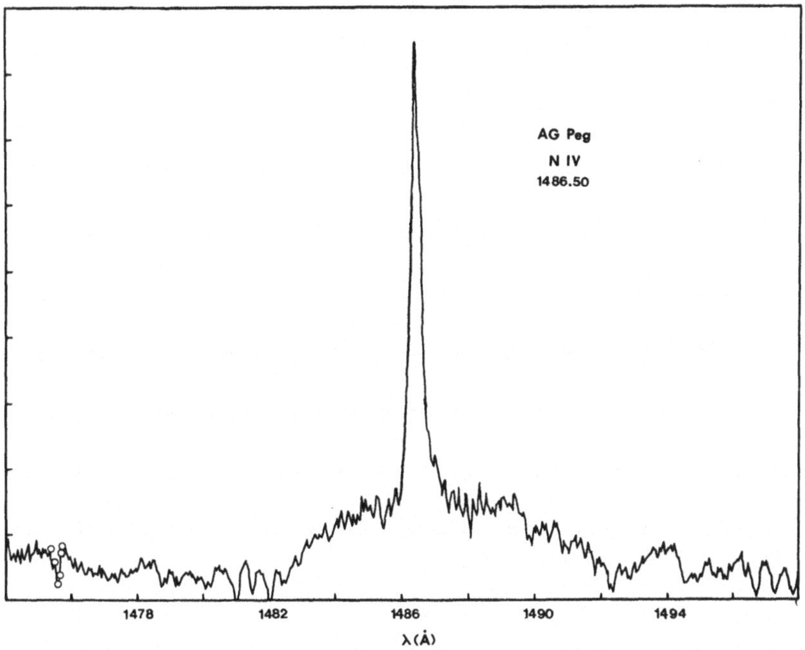

Fig. 2: The N IV] line in AG Peg.

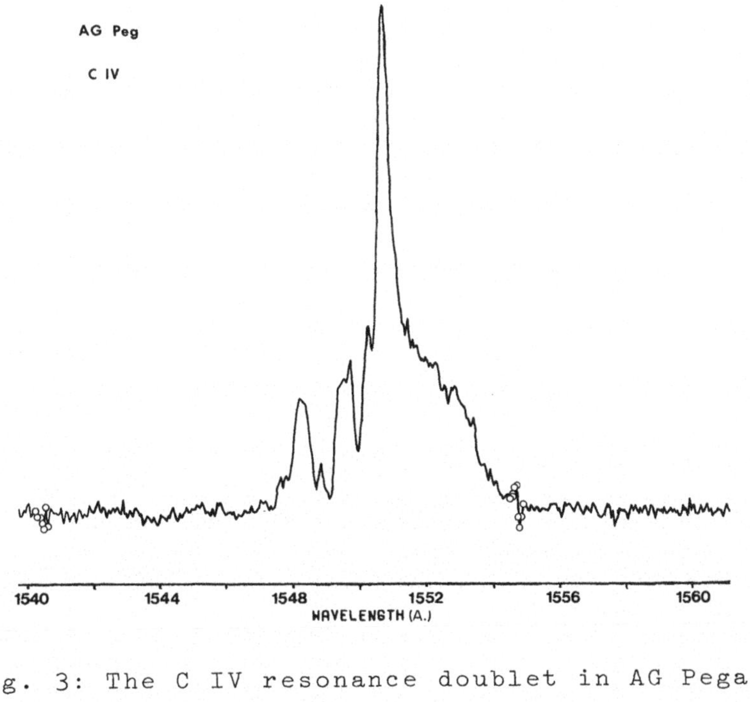

Fig. 3: The C IV resonance doublet in AG Pegasi.

A FEW COMMENTS -- RATHER THAN A CONCLUSION

The cool star appears to be much smaller than its Roche lobe. Indeed, Hutchings, Cowley, and Redman (1975) postulate a stream flowing from the hot star towards the cool component. Yet the M star must play (or have played) an important role in the system, and probably was the ultimate source of the material that triggered the nova-like outburst about a century ago. One can obtain an M giant in a long-period binary system by letting a K giant lose mass via Roche lobe overflow (Plavec, Ulrich, and Polidan, 1973). However, such a mass transfer is almost inevitably catastrophic as it comes from a deep convective envelope, and would probably give the system a quite different character. Mass loss via a strong stellar wind is more likely for the M star.

The M star is probably more massive than the hot component, which appears to be more advanced in evolution. This indicates that we may be witnessing the second mass transfer phase. In the first phase, which could have been a case B, the initially more massive star lost most of its mass and was transformed into a helium star located now not far from the white dwarf region.

Although the low log g we have found for the hot star is probably quite spurious, the object may be extended. It is in the related star AR Pavonis (Thackeray and Hutchings 1974). Since AG Peg behaves like an extremely slow nova, Bath's model of an effective photosphere formed in an optically thick wind (Bath, 1978) may well apply.

A number of binary systems do not meet all criteria for being called symbiotic, but bear distinct resemblance to them and must be considered as possibly related to the symbiotics. AR Pavonis is a bona-fide symbiotic star, but may also be classified as a long-period Algol. It is in turn similar to RX Cas, a member of the "W Serpentis group" described by Plavec at this Symposium. Those stars display strong emission lines of C IV, N V, Si IV etc., apparently formed by collisions in a stellar wind. This strongly suggests that the emission lines observed in AG Peg may in part be formed in a similar regime.

REFERENCES

Bath, G.T.: 1978, Mon. Not. R.A.S. 182, 35.

Bohlin, R.C., Harrington, J.P., and Stecher, T.P.: 1978, Ap. J. 219, 575.

Gallagher, J.S., Holm, A.V., Anderson, C.M., and Webbink, R.F.: 1979, Astrophys. J. 229, 994.

Hutchings, J.B., Cowley, A.P., and Redman, R.O.: 1975, Ap. J. 201, 404.

Lee, T.A.: 1970, Ap. J. 162, 217.

Plavec, M., Ulrich, R.K., and Polidan, R.S.: 1974, Pub. A.S.P. 85, 769.

Swings, J.-P. and Allen, D.A.: 1972, Pub. Astr. Soc. Pacif. 84, 523.

Szkody, P.: 1977, Astrophys. J. 211, 432
Thackeray, A.D. and Hutchings, J.B.: 1974, Mon. Not. R.A.S. 167, 319.

DISCUSSION FOLLOWING KEYES AND PLAVEC

Guinan: Has AG Pegasi been detected as an X-ray source with the Einstein telescope? The strength of some of the emission features would indicate that it would have a substantial X-ray luminosity.

Keyes: To my knowledge, AG Pegasi has not been detected by X-ray detectors prior to Einstein. I do not know if it has been observed with Einstein.

I U E OBSERVATIONS OF Z ANDROMEDAE

M. Friedjung
Institut d'Astrophysique, Paris, France.

Abstract -

The preliminary results of a joint study of this symbiotic star with 6 Italian astronomers are reported. The relative intensities of the intercombination emission lines indicate electron densities of 10^{10} and an upper limit to the dilution factor between 10^{-4} and 10^{-6}. The resonance emission lines give a maximum thickness in the line of sight. A hot (T between 19000° and 26000°) continuum is interpreted as coming from a hot subdwarf, but the high excitation emission lines appear to be formed near the cool giant component of what is probably a close binary.

I- Introduction.

The study here described is a joint collaboration with A. Altamore, G.B. Baratta, A. Cassatella, A. Giangrande, O. Ricciardi and R. Viotti. The calculations are not yet complete and the results are preliminary.

Z Andromedae is often considered to be the prototype of symbiotic stars. Such stars have periods of intense nova like activity folloved by quiescent phases. In the latter the visual spectrum shows both the presence of high excitation emission lines and of the spectrum of a cool giant. Cyclic variations of visual luminosity, emission line intensity and radial velocity are often observed.

Different kinds of physical process have been invoked to explain the properties of symbiotic stars. They are most commonly thought to be interacting binaries (Boyarchuk, 1969), but active chromospheres and coronae around single stars, and also in some cases processes associated with the origin of planetary nebulae have also been invoked. In any case the physics of different symbiotic stars is probably not the same. In my conclusions I shall not contradict the binary hypothesis (otherwise I might be kicked out of this binary meeting), but the situation seems to be more complex than indicated by simple forms of the hypothesis.

The IUE short wavelength region was observed at low resolution on May 5, 1978 and at high resolution on February 17, 1979, whime the long wavelength region was observed at low resolution on June 13, 1978

M. J. Plavec, D. M. Popper and R. K. Ulrich (eds.), Close Binary Stars: Observations and Interpretation, 543–546.

and February 17, 1979. The data were converted to absolute fluxes using the sensitivity curves of Bohlin et al (1978). The UV spectrum is seen to be dominated by a large number of narrow emission lines often of high ionization on a weak continuum, clearly visible in the short wavelength region.

II - Analysis of the emission line flux ratios.

The line intensity ratios for the $^2P^o$ - ^4P NIII] (intercombination) multiplet as a function of electron density N_e have been calculated by my colleagues using the level populations kindly provided by H. Nussbaumer. The result which is fairly insensitive to electron temperature T_e for values of the latter between 3×10^4 and 10^5 °K give a weighted mean over the line ratios of $N_e = 2.0 \pm 0.5 \times 10^{10}$.

Calculations for CIII were also performed by O. Ricciardi and A. Viotti. A comparison of the CIII] 1908.7 Å and NIII] 1749.7 Å line fluxes assuming cosmic abundances and no outside radiation gives $Ne = 2 - 3 \times 10^{10}$ for $T_e = 6 - 8 \times 10^4$ °K.

The line fluxes for these ions also give a condition concerning the dilution of radiation where the lines are formed. Preliminary calculations by my colleagues of the effect of radiation on the sublevel populations suggest a maximum dilution factor of between 10^{-4} and 10^{-6}.

In the case of the OIV] lines the already published line flux ratio calculations of Flower and Nussbaumer(1975) and of Vernazza and Mason (1978) were compared with observation. The former calculations give at $T_e = 1.5 \times 10^5$ °K a log N_e between 9.8 and 10.5 from two line ratios. The latter calculations made assuming not quite the same physical data only give definate limits to log N_e between 10.0 and 11.5 for one line ratio, for T_e between 1.8 and 2.0×10^5 °K. The closeness of the N_e values found for this ion to those found for the two others suggest line formation in similar regions.

The study of the high ionization resonance lines gives more information.At the high electron densities found, electron scattering becomes important at quite small geometrical thicknesses. If the line optical thickness of the line emitting region is τ_e and the electron scattering optical thickness τ_c, the effective continuum optical thickness encountered by a photon in its random walk is $\sqrt{\tau_\ell \tau_c}$. The line flux ratios in a resonance multiplet excited by collisions unaffected near $\tau_\ell = 1$ when no sublevel mixing occurs,are changed near $\sqrt{\tau_\ell \tau_c} = 1$ (for the physical conditions here calculated, collisional de- excitation is negligible). The measured line flux ratios are significantly smaller than in the optically thin case; this could however be due to blending with interstellar absorption as well as to the effect here considered. In any case the lines would disappear for $\sqrt{\tau_\ell \tau_c} > 1$, as photons would then be scatterred out of them because of the Doppler effect. Taking log $N_e = 10.4$ the whole half widths of lines supposed broadened by the Doppler effect equal to 60 km/s, and assuming cosmic abundances (justified by Boyarchuk, 1970), a line of sight maximum log thickness of 10.7 is obtained when the fluxes of the NV and SiIV resonance doublets are considered.

The emissivities of the regions giving the different lines can in principle also be used to calculate volumes, if certain assumptions are made concerning their physical conditions. Such calculations are not yet complete. However if the solar model calculations of Yang et al (1975) are used for the SiIV 1393.8 and the NV 1242.7 Å lines assumed formed where excitation and ionization are collisional at log T_e = 4.40 and 5.30 respectively, the volumes found are the same as those of spheres with log radii of 11.5 and 11.7 respectively.

III - The Continuum.

The faint continuum seen cannot be due to free-free and free-bound emission from where the high ionization line emission is produced. It is best interpreted as coming from an optically thick region. If E(B-V) is taken as zero, my colleagues found that the continuum is fitted by black body radiation at 19000° from a sphere of radius 2 x 10^{10} cm while if E(B-V) is taken as 0.5, one has a temperature of 26000° and radius of 4 x 10^{10} cm. The 2200 Å feature is hard to measure because of the presence of many emission lines, but leads to a conservative E(B-V) of 0.25.

IV - The model.

The hot continuum is most easily interpreted as coming from a subdwarf companion, as suggested by Boyarchuk (1967).

The high ionization line emission however probably comes from a region around a cool component. The reasons are :

a) The low maximum dilution factor indicates formation far from any hot star.

b) The narrow lines if formed in a wind expanding at at least the local escape velocity would have to come from at least 7 x 10^{12} cm from a 1$M_\odot$ star. This is hard to reconcile with the thickness and preliminary emitting volume calculations. If the lines were formed in a rotating disk, one could only avoid high ionization line formation in a low velocity region of the order of the size of a cool giant, if the inclination was small. Boyarchuk (1967) suggested an inclination of < 10° to explain the lack of observed orbital radial velocity variations of lines supposed due to the hot star; the probability of such a geometry is only 1/66 .

c) The thickness and preliminary volume calculations are hard to reconcile with formation in an extended nebula.

Formation near the atmosphere of a cool giant of the high ionization lines during the present quiescent phase is most compatible with the data. This conclusion may contradict the results of Linsky and Haisch (1979) concerning the lack of regions of high ionization for cool stars to the right of the Hertsprung-Russel diagram. However in the present state of knowledge it is not certain whether this is a difficulty.

References -

Bohlin, R.C., Holm, A., Carnochan, D., Snijders, M.A.J., 1978, IUE Project Report.

Boyarchuk, A.A., 1967, Astr. Zh. 44, 1016.

Boyarchuk, A.A., 1969, in "Non Periodic Phenomena in variable Stars" Academic Press, Budapest, p. 395

Boyarchuk, A.A., 1970, Iz Krim. Astrophys. Obs. 42, 264.

Flower, D.R., Nussbaumer, H., 1975, Astron. & Astrophys. 45, 145.

Linsky, J.L., Haisch, B.M., 1979, Astrophys. J., 229, L 27
Vernazza, J.E., Mason, E.H., 1978, Astrophys. J., 226, 720
Yang, L.K., Nicholls, R.W., Morgan, F.J., 1975, Solar Phys., 45, 351.

DISCUSSION FOLLOWING FRIEDJUNG

Hack: Is it possible that the difference in intensity between the two resonance lines of NV, of CIV, and of SiIV is due to partial blends of the two components of the doublet? How broad are these lines?

We have observed CI Cyg last June with IUE and the spectrum looks very similar to that of Z And you have shown.

Friedjung: We have observations at high resolutions, where the resonance lines are well resolved. They are narrow with a width (before correction for instrumental resolution) of the order of 0.3 A.

Stencel: Do not completely dismiss the Linsky-Haisch thermal locus (Astrophys. J. Letters, 1979). It will be discussed with new observations at IAU Colloquium 51: Turbulence in Stellar Atmospheres.

Also, how does the hot component cause the hot M star atmosphere? Will there be phase dependent emission line variations?

Friedjung: I would not like to speculate at the present stage on the modulation of the hot region lines.

A BINARY MODEL FOR THE SYMBIOTIC STAR V 1329 CYGNI

D. Chochol, J. Grygar, L. Hric
Astronomical Institute of the Slovak Academy of Sciences
Tatranská Lomnica and Astronomical Institute of the
Czechoslovak Academy of Sciences
Ondřejov, Czechoslovakia

This emission-line object was discovered by Kohoutek in 1969. Its light curve was subsequently reconstructed since 1891. Before 1964, the object fluctuated at around 15^m, with occasional deep minima as deep as 2.3^m. Its spectrum was late M. In 1964 the object brightened up to 14^m and then it reached 11.5^m in the middle of the year 1966. Since 1966 the object has been slowly decreasing in brightness. Stienon et al. (1974) suggested that the photometric behaviour prior to the outburst could be explained in terms of an eclipsing binary with a period of 960 days. This is now confirmed by the analysis of the post-outburst photometric data and independently by the investigation of the radial velocities of the emission-line component. We have derived the following spectroscopic elements: Period = 950.07^d; γ = 37 ± 6 km/s; K_1 = 62 ± 8 km/s; f(M) = 23 $M_\odot$; $a_1 \sin i$ = 1160 $R_\odot$. We propose a binary model for the symbiotic object consisting of a bright red giant M5 II with a mass of 25 $M_\odot$, and of a very hot compact star of 1 $M_\odot$. Line profiles of the nebular lines as well as Balmer lines suggest the existence of ejected "polar caps" and an equatorial ring around the hot component, much like in novae. It seems that the hot component accretes the material from the giant via strong stellar wind. Changes in the accretion rate can cause the observed optical fluctuations. The major outburst in 1964-1966 was probably caused by a shell flash which also ionized the expanding nebula surrounding the hot component.

M. J. Plavec, D. M. Popper and R. K. Ulrich (eds.), Close Binary Stars: Observations and Interpretation, 547.

THE UV SPECTRUM OF VV CEP.[x]

R. Faraggiana
Astronomical Observatory, Via Tiepolo 11, Trieste, Italy

Abstract

IUE spectra of VV Cep have been obtained in 1978 and 1979 during the chromospheric phase of the eclipse. The profiles of the Mg II resonance lines differ from one another and this peculiarity, common to other M type supergiants, is examined. The value of M_V derived from the extension of the Wilson and Bappu relation to Mg II lines is compared with other determinations. The variation of the absolute flux in the 1200-2000 Å range is presented.

Introduction

The last eclipse of the peculiar binary system VV Cep (P = 20.34 yrs) occurred in 1976-78. The long chromospheric eclipse phase has been covered spectroscopically both for ingress and for egress at the two main telescopes of the Haute Provence Observatory. The phenomena observed on these ground spectra down to 3300 Å are complex; some of the physical parameters of VV Cep determined up to now (e.g. the radii of the stars, the temperature of the secondary star) are still uncertain. No spectroscopic features that can be confidently related to the hot companion have been observed in the region 3300-7000 Å and in order to obtain more information on the chromospheric layers of the primary star and on the spectrum of the secondary VV Cep has been observed by IUE in 1978 and 1979. Two high resolution spectra covering the long wavelength region were obtained on July 31, 1978 and on March 13, 1979 respectively. The low resolution spectra obtained

x -- Based on observations by Dr. P.L. Selvelli obtained with the International Ultraviolet Explorer at the Villafranca Satellite Tracking Station of the European Space Agency

M. J. Plavec, D. M. Popper and R. K. Ulrich (eds.), Close Binary Stars: Observations and Interpretation, 549–553.

with the large aperture and covering the short wavelength range (1175-2000 Å) have been used mainly to determine the absolute flux of the star.

The high resolution spectra.

Due to the crowding of the lines only these two spectra could be used for line identifications. The spectrum in this region is complex, since both absorption and emission lines are present and a reliable continuum cannot be drawn. Most of the sharp absorption lines are chromospheric lines due to once-ionized metals and the general behaviour is very similar to that of the ground spectra in the region 3300-3600 Å at the same phase. The most interesting ions are Fe II, for which no UV lines can be observed from the ground and the two Mg II resonance lines.

The strongest Fe II lines are present as emission lines with a central absorption core, indicating that they, as the Mg II lines, are formed in the extended chromosphere.

The h and k lines of Mg II show the typical self-reversed profiles expected for chromospheric resonance lines. The two profiles are different from each other. The R/V ratio of the emission components in higher for λ 2795 than for λ 2802; the absorption core of λ 2795 is symmetrical and violet-displaced by about 0.25 Å with respect to the centre of the emission line, while the absorption core of λ 2802 is asymmetrical, having a red wing steeper than the violet wing, and at the bottom is centred on the emission component.

These peculiarities are not characteristics of only VV Cep; at least some of them are shared by other late-type supergiants. In α Ori (M2Iab) (Kondo et al., 1972), α Sco (M1Ib) (Bernat and Lambert, 1978) only λ 2795 shows asymmetric emission components, while λ 2802 is symmetric; the same is true for Boss 1985, another VV Cep system (Cassatella et al., 1979). On the other hand, in giant stars such as α Boo (K2III) Moos et al., 1974) and CH Cyg (M6III) (Hack, 1979) the two line profiles do not differ from each other.

The fluorescence mechanism in Fe I and Mn I lines through coincidences with Mg II resonance lines has been investigated by Bernat and Lambert (1976) and seems to offer the key to interpret the different profiles of Mg II lines in M supergiants. In fact the Fe I 4307 is weaker than expected from its f value compared to λλ 4271 and 4325 of the same multiplet 42 in the stars with different Mg II profiles investigated by Bernat and Lambert; from our ground spectra we can add that λ 4307 is weaker than expected also in VV Cep and Boss 1985 and it is of the expected intensity in CH Cyg where the two Mg II lines have the same profile.

The asymmetry of the absorption core of the h line present on both spectra of VV Cep at our disposal seems not to have been observed in other stars; we cannot say whether the fact that the profile is more complex than those of other supergiants observed by Copernicus is due to an unidentified peculiarity of VV Cep or to the higher resolution of IUE or on the other hand to an instrumental effect since the h line is at the edge of two echelle orders of IUE.

It is well known that a relation between the line widths and the absolute visual magnitude can be derived from Mg II lines in analogy to the Wilson and Bappu relation derived from the Ca II lines. The calibration of this relation for Mg II is still improvable; some discrepancy exists between the calibration by different authors mainly as regards the most luminous stars. Moreover we can surmise that in VV Cep these lines are variable with phase as is the Ca II K line; in fact no M value could be derived from K line by Wilson (1976) for VV Cep. Finally we have at our disposal only two UV spectra taken during the chromospheric phase of the eclipse, and owing to all the above-mentioned uncertainties not much weight can be assigned to the M_V derived from the Mg II emission widths. Neverthless, using the M_V versus log W relation by McClintock et al. (1975), a value of $M_V = -4.3$ is derived from the mean value of the lines on both spectra; using the most recent relation of Weiler and Oegerle (1979) a value of $M_V = -4.9$ is derived from the K line. A value of $M_V \approx -4$ is derived from the Ca II K line measured on a spectrum taken during the totality. We can compare these values with the most recent determination of the absolute magnitude of VV Cep, $M_V = -4.0$, obtained by van de Kamp (1977) from a long series of astrometric measures.

The low resolution spectra.

The flux distribution derived from the three low resolution spectra of the short wavelength region is not the same on the three spectra and the increase of the mean flux for $\lambda > 1450$ Å between the June and July 1978 spectra had been interpreted as due to the progress of the eclipse (Faraggiana and Selvelli, 1979). However the next spectrum, obtained on March 1979, shows unexpectedly that in the same region the mean flux decreased to the value of June 1978 except in the range 1450-1620 Å where it is slightly higher on the more recent spectrum. In the region below $\lambda = 1420$ Å the mean flux is about the same on the three spectra; the only remarkable difference is a steadily incresing absorption feature at $\lambda = 1300$ Å; OI 1302.174 or, more probably, the blend of Si III Mult. UV 4 is responsible for it.

In the region around 1800 Å the strong emission lines visible on the first spectrum are filled up by the increased continuous emission on the next spectrum and are dominant again on the last spectrum, where unfortunately crosses on the tracing indicate the poor quality of the data.

The mean flatness of the spectrum on the whole range does not have a counterpart in theoretical models or on the observed S2/68-spectra of any normal star. The VV Cep spectrum is probably highly distorted by the presence of variable emissions not directly related with the phases of the eclipse.

At this stage we are not able to choose between the various ways of explainig the observed flux distribution; for example it might be due to the contribution of the secondary star distorted by an envelope surrounding it as well as by the envelope surrounding the whole system and responsible for the forbidden emission lines on the ground spectra, or to high and not homogeneous layers of the chromosphere of the primary and to the large envelope.

References.

Bernat, A.P. and Lambert, D.L.: 1976 Astrophys. J. 204, 830

Bernat, A.P. and Lambert, D.L.: 1978 Montly Notices Roy. Astron. Soc. 183, 17p

Cassatella, A., Giangrande, A. and Viotti, R.: 1979 Private Communication

Faraggiana, R. and Selvelli, P.L.: 1979 Astron. Astrophys. 76, L 18

Hack, M.: 1979 Private Communication

Kondo, Y., Giuli, R.T., Modisette, J.L. and Rydgren, A.E.: 1972 Astrophys. J. 176, 153

McClintock, W., Henry, R.C., Moos, H.W. and Linsky, J.L.: 1975 Astrophys. J. 202, 733

van de Kamp, P.: 1977 Astron. J. 82, 750

Weiler, E.J. and Oegerle, W.R.: 1979 Astrophys. J. Suppl. 39, 537

Wilson, O.C.: 1976 Astrophys. J. 205, 823

COMMENTS FOLLOWING FARAGGIANA

Milone: The difference in asymmetry between the Mg II H and K lines is also seen in the NRL echelle (solar) atlas data. (We also worried about whether this was real. These features were reduced to ∿0.030Å but the spectrum was "holey" and had much scattered light.)-- This wasn't said but probably should have been!

Faraggiana: I am glad to know that different stars observed with different instruments still show the same "error"--It makes us confident that the "error" is in the stars.

Stencel: You found M_V = -4.2 which is somewhat underluminous for an M supergiant--what causes this?

New M_V correlations should be checked: Weiler & Oegerle (Astrophys. J. Suppl. 39, 1979) have re-evaluated M_V-log W_o(Mg II) with seventy Copernicus spectra and improved on all earlier estimates; Stencel (Astrophys. J., 215, 1977) has provided an additional M_V-log W_o relation for the H-K wing emission lines among cool evolved stars. Also, I have checked Weiler-Oegerle against Wilson-Bappu with IUE for 50 cool stars and find agreement to $\pm 0^m.7$ which is close to the $\pm 0^m.5$ error of each separate correlation.

Faraggiana: The value I quoted has been derived using Weiler and Oegerle relation following the suggestion you gave me two days ago. Even in this case the most luminous stars deviate from the mean relation. I can say that: 1) The value derived for VV Cep is a lower value of the luminosity owing, as I said, to the presence of the underlying continuum of the hot companion; 2) A large difference between h and k profiles is observed in M supergiants and is absent in M giants. VV Cep which shows a small difference between the two profiles may have an intermediate luminosity.

ULTRAVIOLET OBSERVATIONS OF 31 and 32 CYGNI

Robert E. Stencel and Yoji Kondo
NASA - Goddard Space Flight Center;
Andrew P. Bernat
Kitt Peak National Observatory, and
George McCluskey
Lehigh University

Ultraviolet observations of the Zeta Aurigae systems appear to have several advantages over comparable visual wavelength studies. A wide range of large optical depth resonance lines of abundant species permit the study of the supergiant atmosphere and circumstellar environment at virtually all phases. The International Ultraviolet Explorer satellite (Boggess, et al., 1978) is well suited to obtaining spectra between 1150 and 3200 A, although the competition for observing time is non-negligible.

Our initial observations of 31 and 32 Cygni were made in Sept. 1978, as part of the observing program of Kondo and McCluskey on interacting binary stars. 31 Cygni was observed at phase 0.62, and 32 Cygni at phase 0.17. Detailed reports on these observations are forthcoming (Stencel, et al., 1979; Bernat et al., 1979), but some of the highlights of the observations are presented here.

A. 32 CYGNI

Qualitatively, the UV spectrum of 32 Cyg is best described as a superposition of P Cygni features among strong lines on a bright B star continuum. Numerous lines of Si II, O I, C II, Al II and III and Fe II appear with P Cygni characteristics. Higher excitation lines, Si III 1298A and Si IV 1402A, are also present in the SWP2491 image. The LWR-2275 image is dominated by the Mg II 2800A resonance doublet and numerous Fe II lines in the 2200 - 2700 A region, all showing P Cygni profiles. Generally, the P Cygni-like features are centered symmetrically about the rest wavelength of the line.

Further, the detailed profiles of the stronger lines which exhibit P Cygni shapes indicate a complex of several absorption components separated by tens of km/s, with shifts ranging above -200 km/s of the rest velocity, and possibly as high as -400 km/s (O I lines). Satellite absorption components in the Ca II K line have been previously reported

M. J. Plavec, D. M. Popper and R. K. Ulrich (eds.), Close Binary Stars: Observations and Interpretation, 555–559.

(Wright and Hesse, 1969), but only near eclipse and at displacements under -150 km/s. The cores of the Mg II resonance lines exhibit these absorptions nicely (Figure 1). We have measured the wavelengths of well defined absorption components in the entire IUE spectrum and computed relative velocities with respect to the absorption-emission crossover wavelength, which was itself generally within 15 km/s of the lab wavelength for the transition. The means of the velocity groupings are 54, 147, 250 and 392 km/s, all shortward shifted with respect to the rest wavelength. The latter three are greater than the escape velocity along the line of sight to the B star. Neither velocities of this magnitude nor strong absorption satellites so far from eclipse have been seen in previous visual studies. We suspect the total excitation might be dominated by the distance from the B star along the line of sight, and indeed, the incidence of the largest velocities is among the lowest excitation lines, which are also the weakest features. The temperature and density thus may fall off with the distance squared in the circumstellar envelope, as suggested among other cool supergiants. The discrete absorption components could be a consequence of the acceleration mechanism for the mass outflow, tempered by the gravitational dynamics within the system. High excitation lines like Si IV show a more gradual transition from absorption to emission than is the case for the lower excitation lines like Mg II and Fe II which exhibit a large gradient of flux with wavelength at the absorption to emission crossover in their P Cygni profiles. This gradient is controlled by the density and velocity dispersion in the line forming region, which suggests the hotter, denser material is near the B star, while the reduced velocity dispersion but increased blueshift material is farther away, possibly in response to some form of radiative acceleration. The image is that of a hot turbulent region near the B star, and a cooler calmer but faster moving wind somewhat removed from the binary.

Wilson and Abt (1954) and Wright (1959) deduced on the basis of curve of growth studies that discrete condensations were possible in the atmosphere of the K supergiant. They discussed their findings with a solar prominence analogy. This result must be re-examined since over-ionization (NLTE) effects are likely in the outer atmosphere. Further, it now appears that K supergiants do not possess coronae, and the role of strong magnetic fields is not resolved. Although the details of atmospheric structure revealed near eclipse might not compare in detail with our out-of-eclipse spectra, the discrete absorption components may reveal some temporal variation connected with the structure and evolution of such "condensations". To investigate this we have pursued the repeated observation of 32 Cyg each few months. The first two spectra were obtained by Kondo and McCluskey, the third in collaboration with R. Wing. Pending further analysis, it appears as though a slight increase in the strength of the most shortward component (in Mg II) has occurred, despite no change in displacements. Geometry might suggest that as phase 0.5 is approached, we see the purely radial outflow component (-150 km/s) as the radial vector converges with the line of sight and the optically thick path at its velocity is maximized.

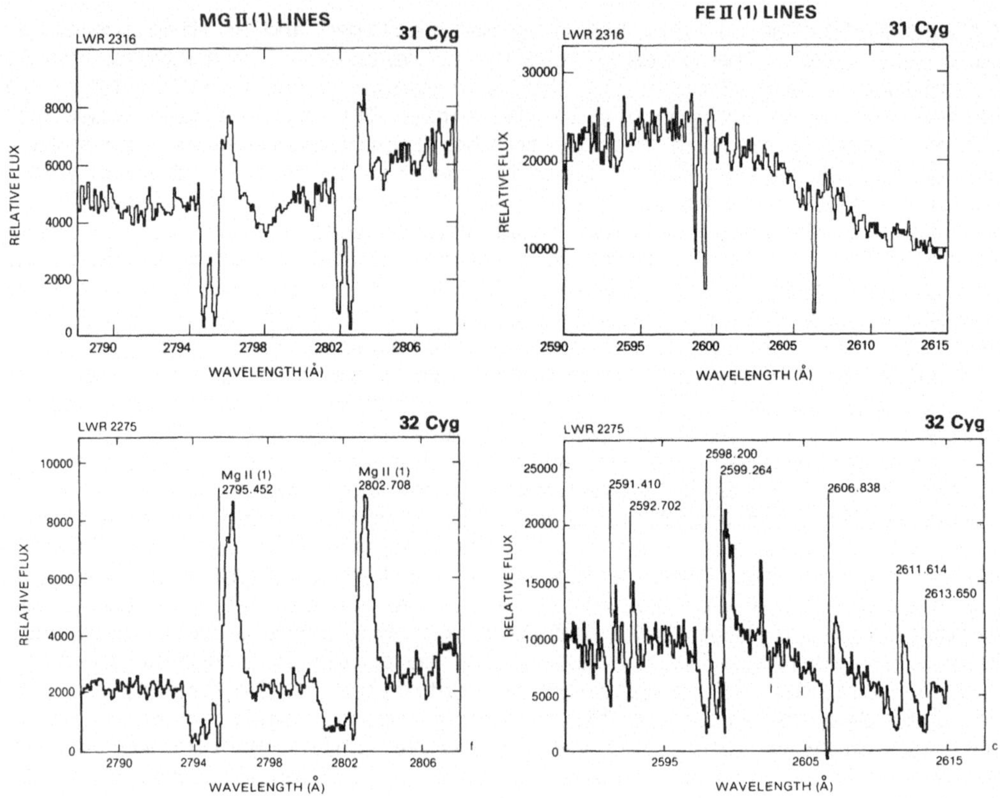

Figure 1. UV lines of Mg II and Fe II in 31 and 32 Cygni.

More quantitatively, ionization modeling for the B star environs was made by comparing observed and computed equivalent widths to deduce density, temperature and turbulence values. From the known geometric configuration of the B star in the system, its UV radiation field and separation in the plane of the sky, we used photoionization-recombination detailed balancing to compute ionization equilibrium and densities at each point along the line of sight for a specified density and inverse distance squared falloff. A Doppler and wind outflow velocity was also assumed to obtain the equivalent widths. Then by iterating on the specified density, one converges to the observed values. At the projected separation between the stars, the log particle density obtained was 7.2, neglecting dust and molecule formation. Assuming continuity and spherical symmetry, the estimated log of the mass loss rate is -6.4 (solar masses per year). This value is consistent with other estimates for this sort of system and the general mass loss rates for K supergiants (saito, 1970, 1973; Sanner, 1976).

In 32 Cyg, the mutual interaction between the B and K star may be greater than previously anticipated. The strong P Cygni profiles of the UV spectrum suggest that the B star resides within the upper atmosphere of the K supergiant and that its ionizing radiation penetrates substantially, possibly setting up moving shocks along the egress side of the chromosphere. The wide range of velocities observed in the circumstellar components presumably arise from the combined energetics of the K supergiant stellar wind, the B star radiation field and orbital motions. Further monitoring of the UV spectrum of this system is desirable.

B. 31 CYGNI

This system is similar to 32 Cyg except that its orbital period is about three times longer. Its Mg II and Fe II lines are shown in Fig. 1 also, and while they show a P Cygni shape, both the intensity, and number of absorption components is reduced. This occurs because of the larger orbital separation. Careful analysis shows that generally higher ionization stages are seen compared to 32 Cyg, including more lines of Si III and IV and stronger Al III. No Fe II lines other than zero volt are seen, suggesting an interstellar-like contribution. Since the phase (0.62) places the B star in front of the K supergiant and the separation is greater than in 32 Cyg, the lower density may favor the higher excitation. In any case, the Mg II lines still clearly show the presence of strong outflow near 31 Cyg.

We thank the IUE Observatory for their tremendous assistance, and R.E.S. also thanks the NAS-NRC for research support.

REFERENCES

Bernat, A., Stencel, R., Kondo, Y. and McCluskey, G., in preparation.
Boggess, A., et al. 1978 Nature 275, 2.
Saito, M. 1970 Publ. Astron. Soc. Japan 22, 455.
________ 1973 Astrophys. and Space Science 22, 133.
Sanner, F. 1976 Astrophys. J. Suppl. 32, 115.
Stencel, R., Kondo, Y., Bernat, A. and McCluskey, G. 1979 Astrophys. J. (In Press, October).
Wilson, O. and Abt, H. 1954 Astrophys. J. Suppl. 1, 1.
Wright, K. 1959 Publ. Dominion Astrophys. Obs. 11, 77.
Wright, K. and Hesse, K. 1969 Publ. Dominion Astrophys. Obs. 13, 301.

DISCUSSION FOLLOWING STENCEL, KONDO, BERNAT AND MC CLUSKEY

Guinan: 32 Cyg is in an eccentric orbit with $e \simeq 0.3$. Perhaps you may see changes in the spectra as a function of the varying separation between the stellar components in addition to those changes expected from the eclipses. Have you detected any spectral changes that could arise from this?

Stencel: Not yet. We did observe near periastron and after, and perhaps temperature changes might be seen. This is a good idea. Also your recent photometry (Publ. Astron. Soc. Pacific, 91, 1979) is valuable for new dimensions to incorporate in this analysis.

CLOSE BINARY SYSTEMS IN GLOBULAR CLUSTERS: A PRELIMINARY REPORT

R.F. Webbink
University of Illinois

ABSTRACT

A brief survey of known eclipsing binaries and cataclysmic variables in globular cluster fields is presented. None of the 47 known or suspected eclipsing variables is a promising candidate, although a very few remain possible members. V101 in M5 is a good candidate for membership, among 5 known or suspected dwarf novae. Three novae have been discovered in globular cluster fields, of which two are almost certainly members. Attention is also called to the eclipsing binary V80, a system which appears to contain an RRc variable, in the dwarf spheroidal galaxy in Ursa Minor.

1. INTRODUCTION

The subject of the binary frequency in globular clusters has in recent years acquired new importance, owing principally to the discovery of X-ray sources in several clusters (Giacconi, et al. 1974; Clark, Markert, and Li 1975; Grindlay 1978), and to the realization that the presence of significant numbers of binaries in cluster cores could, via superelastic encounters, pump enough energy into the stellar system to delay or prevent core collapse (Heggie 1975; Hills 1975).

2. DISCUSSION

The difficulty in directly detecting eclipsing binaries in globular clusters has recently been reviewed by Alexander and Budding (1979). Searches for such variables in globular clusters have a long history of frustration. Baade searched for W Ursae Majoris systems on deep plates of M3 without success (Payne-Gaposchkin 1979), as did Trimble (1976) in M55. Budding and Alexander (1978) were likewise unable to identify any promising candidates in a survey of main sequence stars in NGC 6397. A search of ω Centauri by Niss, Jørgensen, and Laustsen (1978) turned up a few possible candidates, but the light curves of these objects, apart

M. J. Plavec, D. M. Popper and R. K. Ulrich (eds.), Close Binary Stars: Observations and Interpretation, 561–565.

from one certain non-member (Liller 1978b), are far too fragmentary to establish with any certainty that they are in fact eclipsing binaries. The three most viable candidates listed in the Catalogue of Variable Stars in Globular Clusters (Hogg 1973) have now been eliminated as members on the basis of radial velocity studies: V65 in NGC 3201 (Geyer 1975), V78 in ω Cen (Liller and Lichten 1978; Geyer and Vogt 1978), and V3 in M71 (Liller 1978a). A search by the author of the General Catalogue of Variable Stars (GCVS: Kukarkin et al. 1969) and its supplements for eclipsing binaries and cataclysmic variables lying within the apparent tidal radii of globular clusters (Webbink 1979) revealed 26 eclipsing systems omitted from Hogg's catalogue. Of this number, however, only three (HI Ara in NGC 6397, V1555 Sgr in NGC 6528, and FX Sge in M71) satisfy the most elementary membership criteria (in terms of apparent brightness, form of the light curve, and apparent colors). All three of these binaries lie far from the cluster cores, however, and since all three clusters lie within 600 parsecs of the galactic plane, membership for even these variables is doubtful. A final discouraging note is added by Gunn and Griffin (1979), who, in a spectroscopic survey of 111 known members of M3, failed to find any evidence of duplicity, and set an upper limit of 6 percent to the binary frequency among giants lying above the horizontal branch.

A different impression emerges of the frequency or cataclysmic variables, however. Five apparent dwarf novae are known in globular cluster fields: V101 in M5, V9 in NGC 6712, V4 in M30, V2359 Sgr in M69, and V1431 Sgr in NGC 6522. This last variable is listed as a possible nova in the GCVS, but judging from its extreme faintness and rapid decline, it is more likely of U Gem type in the author's opinion. Of these dwarf novae, that in M5 stands as a very promising candidate: this cluster lies at high galactic latitude ($b = 47°$), and the variable conforms closely in range and magnitude with typical disk dwarf novae (cf. Warner 1976). Curiously, the other four candidates are all approximately 2½ magnitudes brighter at maximum (if members) than disk systems; unfortunately, the details of the outburst mechanism in dwarf novae are too poorly understood to use this apparent coincidence as a criterion for or against membership.

The most convincing cases for membership occur among the three novae which have been observed in globular cluster fields: T Sco in M80; an anonymous nova in M14; and V1148 Str in NGC 6553. Indeed, this number is itself very suggestive, as a straightforward convolution of the apparent distributions of known novae and globular clusters yields a probability of less than 10 percent that <u>any</u> nova should have been seen in the field of any of the clusters which have been searched for variables (Hogg 1973), much less than two should have been seen within the cluster <u>cores</u>. Of these three novae, there can be little doubt that T Sco, although positively sighted by only three observers (independently) nearly 120 years ago (Sawyer 1938), is indeed a member of M80: it appeared only 4".80 from the cluster center, and conforms so closely to the rate of decline-absolute magnitude relationship for classical novae that it is frequently used as a standard in its calibration (e.g., van den Bergh 1975). The nova in M14 (1938) was not recorded until well after maximum, but it too appears projected within the cluster core. Only in the case of V1148 Sgr (1948) is membership highly suspect: this nova appeared well out in the half of NGC 6553, only 6° from the galactic center. It is unique in having shown a K-type absorption spectrum at maximum (Mayall 1949).

In terms of the true nova rate in globular clusters, the very circumstances attending the discovery of these three lead one to suspect that, in comparison to the galactic disk, it is very high indeed. T Sco had the good fortune to lie in the same field as the long-period variables R and S Sco, two of only about four dozen variable stars known at the time of its discovery. It happened to be in outburst when Auwers observed M80 at the beginning of a three-year astrometric survey of bright northern nebulae and clusters. The nova in M14 was detected by Hogg and Wehlau (1964) on the first few plates in a series taken more than two decades earlier for the specific purpose of searching for variables in this cluster. Only V1148 Sgr, as it happens, was discovered during a routine patrol. In terms of the effect novae may have on the gas content of globular clusters (Scott and Durisen 1978), this inference that the observed rate in these clusters is very incomplete is clearly of profound importance.

3. A FOOTNOTE

The present discussion would be remiss without mention of one eclipsing system observed not in a globular cluster, but in an even more extreme Population I stellar system, the dwarf spheroidal satellite of our own galaxy in Ursa Minor. Unless it is a freak superposition of a foreground binary and a cluster variable, V80 in this galaxy is a $2^{d}.07$ binary with a primary which is itself an RRc type variable (Kholopov 1971). Only one other eclipsing binary is known of this type (RW Ari). This object deserves study in its own right, but also as the most promising candidate known at present for an eclipsing binary in a primordial stellar system.

This research was supported by National Science Foundation grant AST 78-12309.

REFERENCES

Alexander, M.E., and Budding, E. 1979, Astr. Astrophys., 73, 227.
Budding, E., and Alexander, M.E. 1978, paper presented at the NATO workshop on Globular Clusters, Cambridge, England.
Clark, G.W., Markert, T.H., and Li, F.K. 1975, Astrophys. J. (Letters), 199, L93.
Geyer, E.H. 1975, in "Variable Stars and Stellar Evolution", IAU

Symposium No. 67, ed. V.E. Sherwood and L. Plaut (Dordrecht: D. Reidel), p. 559.
Geyer, E.H., and Vogt, N. 1978, Astr. Astrophys., 67, 297.
Giacconi, R., Murray, S., Gursky, M., Kellog, E., Schreier, E., Matilsky, T., Koch, D., and Tananbaum, H. 1974, Astrophys. J. Suppl., 27, 37.
Grindlay, J.E. 1978, Astrophys. J. (Letters), 224, L107.
Gunn, J.E., and Griffin, R.F. 1979, Astr. J., 84, 752.
Heggie, D.C. 1975, Mon. Not. R. Astr. Soc., 173, 729.
Hills, J.G. 1975, Astr. J., 80, 809.
Hogg, H.S. 1973, Publ. David Dunlap Obs., 3, No. 6.
Hogg, H.S., and Wehlau, A. 1964, J. R. Astr. Soc. Canada, 58, 163.
Kholopov, P.N. 1971, Perem. Zvezdy, 18, 117.
Kukarkin, B.V., Kholopov, P.N., Efremov, Yu.N., Kukarkina, N.P., Kurochkin, N.E., Medvedeva, G.I., Perova, N.B., Fedorovich, V.P., abd Frolov, M.S. 1969, General Catalogue of Variable Stars, 3rd edition (Moscow: Astr. Sov., Akad. Nauk S.S.S.R.).
Liller, M.H. 1978a, paper presented at the NATO workshop on Globular Clusters, Cambridge, England.
Liller, M.H. 1978b, Inf. Bull. Var. Stars, No. 1527.
Liller, M.H., and Lichten, S.M. 1978, Bull. Am. Astr. Soc., 10, 411.
Mayall, M.W. 1949, Astr. J., 54, 191.
Niss, B., Jørgensen, H.E., and Laustsen, S. 1978, Astr. Astrophys. Suppl., 32, 387.
Payne-Gaposchkin, C. 1979, Stars and Clusters (Cambridge, Mass.: Harvard University Press), p. 230.
Sawyer, H.B. 1938, J. R. Astr. Soc. Canada, 32, 69.
Scott, E.H., and Durisen, R.H. 1978, Astrophys. J., 222, 612.
Trimble, V.L. 1976, Bull. Am. Astr. Soc., 8, 443.
Van den Bergh, S. 1975, in "Galaxies and the Universe", Stars and Stellar Systems IX, ed. A. Sandage, M. Sandage, and J. Kristian (Chicago: University of Chicago Press), 509.
Warner, B. 1976, in "Structure and Evolution of Close Binary Systems", IAU Symposium No. 73, ed. P. Eggleton, S. Mitton, and J. Whelan (Dordrecht: D. Reidel), p. 85.
Webbink, R.F. 1979, in preparation.

DISCUSSION FOLLOWING WEBBINK

de Loore: Just a small remark concerning the globular X-ray sources. These are most probably binaries, but they should not be considered as normal binaries. Probably they consist of a normal star and a captured neutron star, and were originally not binary systems. The fact that these "X-ray binaries" are found in clusters should not be used as an argument for the presence of binaries in clusters, and the fact that no binaries are found in globular clusters may not be used as an argument against the binarity of these X-ray sources.

Hack: The novae for which we have reliable evidence that they are binaries are about 50-60% of the total number of novae, however we assume that all novae are binary systems, just because we do not have any better explanation for the nova phenomenon. So I am glad to see that you pose the question: is it true that all novae are binary systems? and that your data on globular clusters suggest that perhaps the answer is: no.

Whelan: Could you learn anything about binaries in globular clusters from blue stragglers? These may or may not be binaries but today we heard a theory that they are triple stars.

Webbink: I would rather not speculate about these stars. I have the impression that the less the contamination from field stars in the color-magnitude diagrams of these cluster, the less apparent is the blue straggler extension of the main sequence. Certainly in those cases where blue stragglers are apparently present, these stars should be carefully examined for evidence of duplicity.

Rajamohan (in response to Dr. Whelan's remark): The blue stragglers have probably nothing to do with close binary systems for the following reasons: (1) Not all of them are binaries; (2) Pendl and Seggewiss in IAU Coll. No. 32 "The Physics of Ap Stars" showed that a fairly large number of these objects are Ap stars. It is possible that blue stragglers are more magnetic than other cluster members and hence linger on the main sequence a little longer than do the others.

Budding: Can W UMa systems live as long as 10^{10} years?

Webbink: This is a question I attempted to address in a paper this January in the Astrophysical Journal. Certainly we do see W UMa systems in very old open clusters, for example, AH Cnc in M67, which is roughly half the age of the globular clusters. One cannot exclude that some contact binaries may arise out of the evolution of initially detached systems, but apart from that possibility, my inclination is that, though primordial contact binaries may survive 10 billion years, the mass ratio will necessarily have become so extreme in that time (to accomodate growth of the primary), that their low amplitudes and short periods will make them very difficult to detect.

A SEARCH FOR SPECTROSCOPIC BINARIES AMONG PLANETARY NUCLEI

Roberto H. Méndez*
Instituto de Astronomia y Fisica del Espacio

The present paper is a progress report on a search for spectroscopic binaries among central stars of planetary nebulae. In the last few years it has become clear that such a program may result in a significant improvement of the available information about the central stars, and, whatever the outcome, it will certainly help to test current theoretical work on the origin of planetary nebulae.

The spectrograms used in this project are being taken mostly by the writer and Virpi S. Niemela at the Cerro Tololo Inter-American Observatory (CTIO) with the image-tube spectrographs of the 1-m and 4-m telescopes, on IIIa-J baked plates, at dispersions around 45 Å mm^{-1}. Several earlier spectrograms were obtained from 1970 to 1976 with the conventional spectrographs of the CTIO 90- and 150-cm telescopes, on IIa-O plates, at dispersions from 40 to 125 Å mm^{-1}. The radial velocity measurements are made with a Grant comparator-microphotometer.

I have chosen the nebula IC 418 as the velocity standard. A detailed analysis of the image-tube nebular velocities will be published elsewhere; the main result is that no significant deviations have been found. The errors in the stellar radial velocities depend mainly on the amount and quality of the available lines. For A-type stars we would be able to detect orbital motion with a total amplitude $\geq$25 km s^{-1}; for O subdwarfs the limit is of the order of 50 km s^{-1}.

Table I shows the present status of the program; some comments on individual stars follow.

---

*Visiting astronomer, Cerro Tololo Inter-American Observatory, operated by the Association of Universities for Research in Astronomy, Inc., under contract with the National Science Foundation.

*Member of the Carrera del Investigador Cientifico, CONICET, Argentina.

M. J. Plavec, D. M. Popper and R. K. Ulrich (eds.), Close Binary Stars: Observations and Interpretation, 567–570.

TABLE I

OBJECT	SPECTRAL TYPE	NUMBER OF PLATES	ORBITAL MOTION	PERIOD
NGC 246	O VI	18	not detected	
NGC 1360	sdO	79	confirmed	30 d?
NGC 1535	O	18	not detected	
IC 418	O7fp	31	not detected	
NGC 2346	A5	34	confirmed	16 d
He 2-36	A2 III	19	not detected	
NGC 3132	A2 V	49	not detected	
NGC 4361	O	5	probable	
A 36	sdO	7	very probable	
He 2-131	O8fp	9	not detected	
He 2-138	BCO Ib	7	not detected	

NGC 1360. The 8-day period reported 2 years ago (Méndez and Niemela 1977) is not correct. Figure 1 shows the distribution of radial velocities. The observed distribution is consistent with a very eccentric orbit, with periastron near the ascending node. Our wrong estimate of the period can be understood as a consequence of assuming a circular orbit. The new data bring the system velocity in agreement with the nebular velocity, which is approximately +60 km s^{-1}. At least another velocity maximum (we have two) is needed to determine the period.

IC 418. This is an interesting by-product of the program. The stellar emissions do not seem to vary, but the velocitiesof the strong C IV absorptions λλ5801, 5811 sometimes depart from the nebular velocity by as much as 70 km s^{-1}, always towards more negative values. The time scale of the variations is one day or less. Other stellar absorptions (He II, O III) would appear to show the same behaviour, but with a much smaller amplitude. Apparently there is a variable velocity field in the atmosphere of this Of central star.

NGC 2346. The A-type central star was reported last year to be a spectroscopic binary (Méndez et al. 1978). New spectrograms indicate a period of 16 days with a semiamplitude K=18 km s^{-1}. Unfortunately the system would seem to be a foreground object (Méndez 1978).

NGC 4361. On two plates the H and He II stellar absorptions are found to be shifted redwards relative to the superimposed nebular emissions by about 60 km s^{-1}.

A 36. The radial velocity of this star appears to be variable, with a period of perhaps several days (see Figure 2).

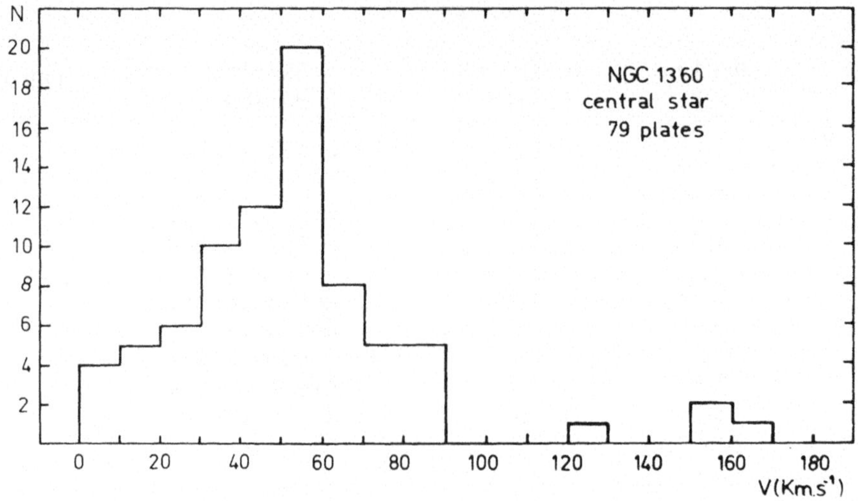

Fig. 1. NGC 1360: the distribution of radial velocity determinations, ½(Hδ+Hγ). N is the number of plates.

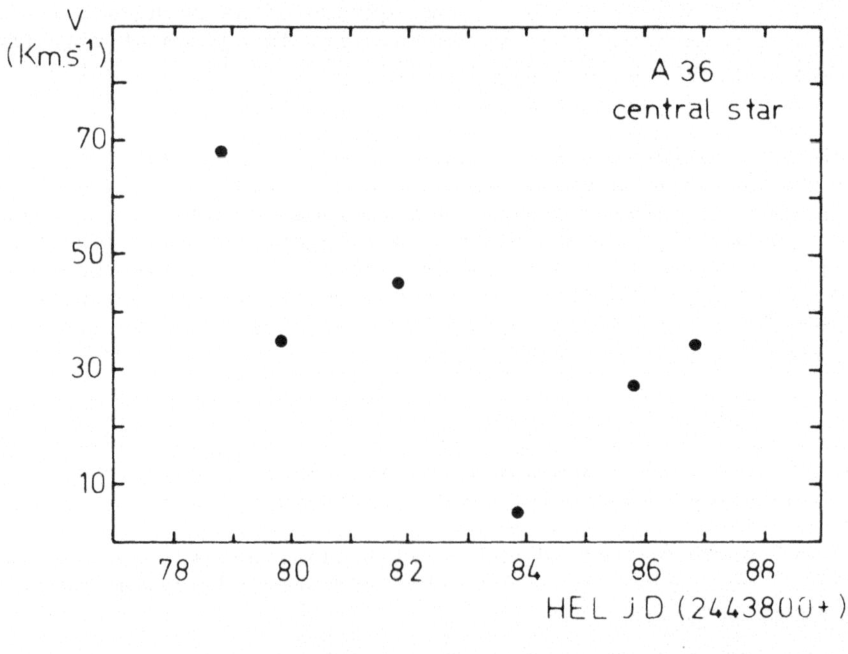

Fig. 2. A 36: the heliocentric radial velocities, ½(Hδ+Hγ), of the central star.

REFERENCES

Méndez, R.H. and Niemela, V.S.: 1977, Mon.Not.R.Astr.Soc. 178, pp. 409-414.
Méndez, R.H., Niemela, V.S. and Lee, P.: 1978, Mon.Not.R.Astr.Soc. 184, pp. 351-354.
Méndez, R.H.: 1978, Mon.Not.R.Astr.Soc. 185, pp. 647-660.

DISCUSSION FOLLOWING MÉNDEZ

Popper: What is the perception of those working in this field? Are binary star processes considered to be important contributors to the totality of planetary nebulae?

Shaviv (response to Popper's question): The PN stage originates in a very special situation when the mass losing star heats up and succeeds to ionize the nebulae before it fades away. Hence you refer to a special stage and circumstances.

We believe that all single RG stars of mass $\lesssim 6M_\odot$ become WD via the formation of PN. Actually, the statistics of WD formation and PN formation seem to confirm this assumption.

The PN with a CBS should therefore be a rare event and it happens only for a special combination of parameters.

Popper(response to Shaviv): It appears to be widely accepted that all M giants (int he appropriate mass range) produce planetaries. If a large fraction of planetaries is produced by binary processes, is there a possible contradiction?

Niemela (responses to Shaviv/Popper): The reason why many people have been saying that the central stars are single simply is that nobody had observed them for radial velocity variations, mostly because they are so faint. Only recently with the more effecient image-tube spectrograph has it been possible to begin a search for spectroscopic binaries among planetary nuclei.

Mayo: What can you say about the spectral types of these suspected (and confirmed) binaries in planetary nebulae nuclei? Do you see the spectra of both the companion star and the hot component? If the companion is an early type star, it could contribute significantly to the irradiation of the nebula. Finally what type of binary do these seem to be, or are they perhaps some new type of binary?

THE FORMATION OF PLANETARY NEBULAE WITH CLOSE BINARY NUCLEI

J. Saltzman
Dept. of Physics and Astronomy, Tel-Aviv University,
Ramat-Aviv, ISRAEL

M. Livio
Dept. of Physics and Astronomy, University of Florida,
Gainesville, FL 32611

G. Shaviv*
Joint Institute for Laboratory Astrophysics, University of
Colorado and National Bureau of Standards, Boulder, CO 80309

Recent observations of nuclei of planetary nebula have shown that some of the central stars are binaries and even close binaries with periods of hours to few days.

Theories of PN formation discuss only the formation of a nebula from a single star. Recently Livio et al. (1979) (LSS) have discussed the possible formation of a PN with a close binary central system. The basic idea is that the system started as a widely separated binary. The process of mass transfer from the red giant to the companion main sequence star, the ejection of gaseous nebula and the decrease in separation occur simultaneously. The sequence of events is as follows: the red giant starts to lose mass to the secondary when they are widely separated. The secondary expands and overflows the outer critical surface passing through the second Lagrange point L_2. An instability arises in which matter flows out of the system through L_2. A steady state is established in which matter flows through L_1, the secondary tries to expand and matter flows out via L_2. A crucial point is that the timescale for the whole process described above is the same as the timescale for the disappearance of the ejected nebula.

Here we report on some further calculations on the form of mass loss. LSS assume particle trajectory for which the angular momentum loss per unit mass $d\ \ell n\ J/d\ \ell n\ m = 1.7(1+q)^2/q$ where q is the mass ratio. This expression was derived by Kuiper (1941) and verified

*On leave of absence from Dept. of Physics and Astronomy, Tel-Aviv University, Tel-Aviv, ISRAEL.

M. J. Plavec, D. M. Popper and R. K. Ulrich (eds.), Close Binary Stars: Observations and Interpretation, 571–573.

later by Nariai (1975) and Lin (1977). The treatment was, however, criticized by Prendergast (1960). Prendergast claims that since the mean free path of the particles is very much smaller than the orbital separation the single particle approximation is not justified.

We have therefore reformulated the Lubow and Shu (1975) hydrodynamic equations and series expansion and adopted them to the case of outflow from L_2. The basic assumptions are: (1) the expansion parameter is $\varepsilon = v_{sound\ isothermal}/v_{orbital} << 1$, (2) the zero order equation is the single particle equation. (3) Assume the following nondimensional units: length=separation, velocity~orbital velocity so that $v_{sound} \cong \varepsilon$. Then streamlines originate from an ε neighborhood of L_2 with $|\vec{v}| = \varepsilon$. The ejection angle is given by the asymptotic solution of the linearized three-body problem. (4) The system of coordinates is distance from L_2 (along the streamline), the normal to the flow in the plane, and the normal to the plane of motion. (5) Hydrostatic equilibrium perpendicular to the orbital plane was assumed.

Our preliminary results are that for low <u>initial ε</u> the alignment and subsequent motion are practically independent of $\vec{v}_{initial}$. <u>No disk is formed</u>, the streamlines do not intersect for some range of parameters used in the model. There are few percent corrections to ν.

Finally we would like to draw attention to a few possible candidates. Basically we should expect a nonspherical PN. Livio (1979) has compiled a list of possible candidates; let me bring three examples of such objects:

<u>NGC 1360 (220-53°1)</u>

CPD-26 389 is a single line spectroscopic binary with a period of 8 days and mass function of 0.19 M_o. The nebula is non-spherical.

<u>UU Sagittae, Abel 63</u>

This is an eclipsing binary with $P = 0.465^d$, $M_1 = 0.9\ M_o$, $M_2 = 0.7\ M_o$. The nebula is very non-spherical.

<u>NGC 2346</u>

This is an A0 plus an O subdwarf with variable radial velocity. Non-spherical nebula.

Why do we see few PN with CBS? There exists a compromise between the reduction in separation and the ejected mass. When the ejected mass is large a very close contact binary is formed. The less massive star may dissipate completely and the hot central star may cool before the nebula becomes transparent. On the other hand, if the ejected mass is small the nebula becomes transparent very quickly and the stars remain widely separated.

RERERENCES

Kuiper, G.P.: 1941, Ap. J. 93, p. 133.
Lin, D.N.C.: 1977, M.N.R.A.S. 179, p. 265.
Livio, M.: 1979, preprint.
Livio, N., Saltzman, J. and Shaviv, G.: 1979, M.N.R.A.S. 188, p. 1.
Lubow, S.H. and Shu, F.H.: 1975, Ap. J. 198, p. 383.
Nariai, K.: 1975, Astron. Astrophys. 43, p. 309.
Prendergast, K.H.: 1960, Ap. J. 132, p. 162.

DISCUSSION FOLLOWING SALTZMAN, LIVIO AND SHAVIV

Shu: In our paper (Shu, Lubow, and Anderson, Ap. J., 229, 223, 1979) on mass and energy flows in contact binaries, we presented our 1975 unpublished results on the same problem. You may be interested in comparing your results with those tabulated in our paper. In particular, we found that mass could be ejected all the way to infinity only for a range of mass ratios between 0.064-0.78.

Tutukov: L_2 was determined only for solid body rotation systems But to have solid body rotation at least several thermal time scales must pass. The envelope of the primary (at least in the case B systems) expands in the same or shorter time scale The original big common envelope seems rather possible in this case. Did you consider this case?

Nariai: υ Sgr and HD 30353 that are hydrogen-deficient binaries eject mass through the Lagrangian point. When I was studying these stars twelve years ago, I thought that a similar mechanism could be applied to the formation of planetary nebulae. But I could not follow the problem further because of the difficulty in observing the planetary nuclei at that time. I am very happy today to hear that Dr. Shaviv is trying to apply the mechanism to the formation of planetary nebulae with the recent results of observations.

HD137569, A POPULATION II REMNANT OF MASS EXCHANGE EVOLUTION

C.T. Bolton and J.R. Thomson
David Dunlap Observatory, University of Toronto
P.O. Box 360, Richmond Hill, Ontario, CANADA L4C 4Y6

The temperature, surface gravity, and abundances derived for this halo B star by Danziger and Jura (1970, Ap.J., 161, 997) show that it is a field suprahorizontal branch star very similar to B29 in M13. The only difference is that HD137569 has a higher (solar) C abundance. We find that HD137569 is a spectroscopic binary with P=529.8(2) days, V_o=45.0(2) km/s, K=16.2(3) km/s, e=0.12(2), ω=233°(10), and T=JD2441922(14). The parenthesized quantities are the uncertainties (probable errors) of the last digits of the elements.

These elements combined with the atmospheric parameters argue that the unseen secondary is probably the more massive. The evolutionary history of the system is difficult to derive because of the great uncertainty in the stellar models. The derived asini (=168 $R_\odot$) indicates that if the orbital elements have not changed during the evolution the primary probably filled its Roche lobe when it was near the tip of the asymptotic giant branch. Since mass exchange usually leads to a net increase in the semi-major axis, it could have started earlier on the asymptotic giant branch or at the tip of the giant branch. The normal C abundance suggests that mass exchange did not begin before the helium flash.

COMMENT FOLLOWING BOLTON AND THOMSON

Kwee: Where in the HR diagram are your stars situated with respect to the Cepheid instability strip?

Bolton: Most of them are on the blue side of the instability strip.

M. J. Plavec, D. M. Popper and R. K. Ulrich (eds.), Close Binary Stars: Observations and Interpretation, 575.

CONCLUDING REMARKS

J. Smak
N. Copernicus Astronomical Center
Polish Academy of Sciences
Warsaw, Poland

During the last four days we have seen a number of interesting and important results concerning problems which will certainly attract our attention in the years to come. To give examples, I would mention: formation of binaries, structure of contact systems, evolution with a common envelope, planetary nebulae with CBS nuclei, starspots and magnetic fields. As for individual objects, SS 433 would be obviously No. 1 on our list of highlights.

What seemed the most characteristic feature of this Symposium, however, was the relatively large number of papers and significant amount of discussion devoted to rather "old-fashioned" subjects. To be more specific - to the first mass-exchange phase of CBS and to objects produced during that phase. Any statistics would indeed show that U Cep or Algol were mentioned by name more frequently than Cyg X-1 or even SS 433! I have used the term "old-fashioned" to emphasize that these objects have been with us for many decades and also to remind you that we seemed to understand their evolutionary significance over 10 years ago, when the first evolutionary models brought us the solution of the Algol-paradox and provided clues to the origin of WR stars. I do not imply that we are old fashioned. It seems to me that we are taking another look at these problems for many good reasons, our main motivation being that a much better understanding of the first mass-exchange/mass-loss phase is a necessary condition for a complete, meaningful and consistent treatment of further phases responsible for the existence of all other, more exotic objects.

Under these circumstances I feel excused to concentrate my remarks just on those few old-fashioned problems. And I will start with a truly classical subject, namely with those good, clean, and apparently no longer interesting, detached main-sequence binaries. It is worthwhile to remember that they are still one of the main sources of our information on fundamental parameters of main sequence stars. Some of them show the apsidal motion (not even mentioned at this Symposium!), the only case when direct observations tell us something

M. J. Plavec, D. M. Popper and R. K. Ulrich (eds.), Close Binary Stars: Observations and Interpretation, 577–580.

about the internal structure of the stars. Let me discuss in more detail one possible use of these systems. The Vogt-Russell theorem tells us that the mass and chemical composition uniquely determine all other parameters of the star; to include the evolutionary effects, we can add here another free parameter - the age. It was this basic principle that was used some 2-3 decades ago by Martin Schwarzschild and others to determine the helium content and the age of the Sun exclusively via the stellar model calculations. The same principle can be applied to the components of main sequence binaries. As for their masses, radii and luminosities, we know them much better than 20 years ago. We now know, or at least have the possibilities to determine their chemical composition; with the spectroscopic data now available not only from the visible region, but also from the ultraviolet, the abundance of helium or of any important heavier element needs not to be an unknown. In principle we are left with only one free, i.e. unknown parameter - the age, while four boundary conditions (two luminosities and two radii) are supplied by observations. Thus we should be able to test our models to see how good is our theory or - more specifically - how good is the physics which we put into our models. We may learn something of general importance for the theory of stellar structure. Even if we don't it is worth trying.

The phase of the mass-exchange following the main sequence was first dealt with in the 60s and it may now seem incredible how really successful we were at that time confining ourselves to the conservative, pure mass-exchange case. Part of the secret was, of course, that while most of the final parameters of the system may depend critically on the amount of mass and momentum lost from the system, those of the mass-loosing component depend primarily on its structure at the onset of the mass outflow. In the case of very massive binaries, we now know that the continuous mass loss via stellar wind is as important factor in their evolution as the most classical mode of mass outflow through L_1. As we have seen during the last few days, there is a fairly complete, although in many aspects only qualitative, picture starting with the main-sequence binaries, through the WR phase, up to the massive X-ray binaries. (It may be interesting to note that, although the X-ray binaries have been discussed quite adequately, the supernova explosion has not even been mentioned during this Symposium!) It has become evident that a much better understanding of the outflow mechanisms, flow patterns, and the efficiency of mass and momentum loss is badly needed to calculate improved evolutionary sequences. We still depend mostly on observational data on stellar wind from single stars. Similar data are needed for various types of binaries, while on the theoretical side further studies should be made of the modifications introduced by the duplicity effects (proximity of the Roche lobe, flow patterns near the mass-collecting component, etc.).

In the case of binaries of lower masses and luminosities we still believe that the mass outflow through L_1 occurring on a thermal time scale is the dominant mode. But compared to the original conservative case (total mass and angular momentum = const.) we seem to have moved to another extreme. Indeed, several papers have argued for nearly total mass loss from the system. Needless to say, we have known already for some time that there are Algols (e.g. AS Eri) that cannot be evolved back to the main sequence within the conservative case. In some cases there is observational evidence for the mass outflow from the system. Obviously this effect has to be incorporated into our picture. However, while doing so we should not rely too heavily on fragmentary evidence. To balance the arguments presented for the extremely non-conservative case, I wish now to make the following "conservative" remarks. We know that the low velocity outflow from L_1 leads to formation of a disk/ring and/or deposition of the material onto the surface of the mass-gaining component. While it is known - practically since Kuiper's time - that in the case of very large disks only a very small increase of velocity is needed to drive the material through the outer Lagrangian point, it is not obvious what could cause such an effect on a major scale; and we also know that the build-up of very large disks is seriously limited by the tidal effects. Of course, the situation may not be as simple during the rapid phases. To that end it is worthwhile to quote, however, one of the results presented during this Symposium, namely that when we do not allow the mass-gaining component to collect mass during that rapid phase, we end with a star which significantly deviates from the standard main sequence. Since this is not confirmed by observations we must conclude that even during the rapid phase we are closer to the conservative situation than to the other extreme. In fact we know systems (such as Beta Lyr) in which the effects of a major mass transfer during the rapid phase can still be seen. It is almost trivial to say that much can be learned from further studies of the mass-gaining primaries of Algols (in particular it is worthwhile to mention the problem of their chemical composition which - surprisingly - has been given very little attention during the last few days). The most important, however, is the problem of how effective the mass outflow can be in carrying away the angular momentum. For it is quite clear that our evidence for the non-conservative case refers primarily to the angular momentum loss rather than to the mass loss.

Finally, I wish to comment on what seems to be a renaissance of spectroscopic studies of many types of binaries, particularly those of the Algol type. This is mostly due to tremendous new technical possibilities, particularly in the ultraviolet, but also due to the stars themselves, which - like U Cep - do their best to attract more of our attention. And while the type of data and the resulting flow patterns are so similar to those published long ago by Struve and his collaborators, we should now rely more strongly on our present physical understanding of these phenomena, in particular on results obtained from hydrodynamical or even purely mechanical calculations. On the other hand we should remember that what often seems to be a single spectral line may in fact be a blend of several absorption and emission components. And even in the case of "clean" lines it is not always possible to identify uniquely the place of their origin. This is not to imply that there are no cases involving other effects, such as the magnetic fields or radiation pressure. We have to study them to make sure that our picture becomes more complete.

I now come to the most obvious and non-controversial part of my remarks. I believe it will be on behalf of all of us if I express our appreciation and gratitude to Mirek Plavec and members of the scientific organizing committee, to Don Fernie, Tom Bolton, and members of the local organizing committee, as well as to all astronomers, graduate students, secretaries, and all others from the David Dunlap Observatory and the Astronomy Department of the University of Toronto, for their energy, enthusiasm, hard work, and patience which made this Symposium a success and our stay here so pleasant. Let us thank them all.

SOME THOUGHTS ON INTERACTING BINARY SYSTEMS

Roger K. Ulrich
Department of Astronomy
University of California, Los Angeles

ABSTRACT

I present here some thoughts on the theory and observation of interacting binary systems. The complex physical processes possible in these systems make our present understanding inconclusive. New types of observation (X-ray, EUV, radio) present new challenges to the theoretician. I discuss here those problems which seem to me to hold the most promise for future progress and relate the papers in this conference to these problems.

The interacting binary systems as a class have characteristics which make them both easy and difficult to study and understand. They are easy to study because they are nearby and relatively bright. They rotate so we can view them from different angles to derive the geometry of the interaction. They vary on time scales from milliseconds to months, giving us a chance to study processes. They are difficult to study because they are fully three-dimensional; the gas dynamics involves both thermal convection and probably shear flow turbulence; the time variability means that theoretical models cannot postulate a strictly steady state.

Some of the major topics of this conference have been: the physics of the mass transfer process, especially in Algols; the theory of evolution of binaries, including mass and angular momentum loss from the system; the RS CVn systems; contact systems; and the application of observations with IUE to the study of a wide variety of interacting systems. Although the reports on polarimetry represent a much smaller number of pages in this volume, they deserve special attention because of the potential power of this technique.

The important new techniques discussed in this volume are IUE observations and polarimetry. IUE data are presented by Dupree et al. (1.7), Drechsel et al. (4.7), Kondo et al. (5.10), Plavec (5.13), Hack et al. (5.16), Polidan and Peters (5.18), Kippenhahn (6.7), Holm et al.

M. J. Plavec, D. M. Popper and R. K. Ulrich (eds.), Close Binary Stars: Observations and Interpretation, 581–583.

(8.4), Raymond et al. (8.6), Faraggianna (10.4), Stencel et al. (10.5), Keyes and Plavec (10.1) and Friedjung (10.2). These spectra typically have lines typified by transition region temperatures with species such as CIV and SiIV producing both absorption and emission. The absorption lines indicate expansion due to some form of gas streaming or stellar wind. The high optical depth in the absorption and the strength of the emission imply a significant population of these species which suggest the presence of a harder radiation source which perhaps cuts off in the soft X-ray spectral domain. Alternatively, the extended transition region could be powered by mechanical energy transport as is the sun. This seems less likely because of the implied volume which must be involved. The harder radiation could come from shocks in the system where highly supersonic gas streams intersect. The polarimetry techniques are well described by McLean (1.9) with applications presented by Piirola (5.12) and Simmons et al. (6. 5). I can only hope that this method is vigorously pursued since it seems capable of extracting information such as the inclination that we are accustomed to believe for non-eclipsing systems is inaccessible to a direct observational determination.

The theory of mass transfer and its application to Algol systems is discussed primarily in chapters 3 and 5. The presence of gas streams and their influence on the spectra is well documented. These stars are accepted to be the products of mass exchange with the bulk of the transferred matter staying in the system. I am concerned that the importance of the Benson expansion of the mass gaining star has not yet been evaluated for these systems. The paper by Webbink (3.5) and subsequent discussion clearly brings out this problem. Perhaps the difficulty lies with our assumption that the matter accreted on the mass gaining star is uniformly distributed over the stellar surface. If instead it arrives in a dense stream, the matter can cool radiatively to the stellar surface temperature before it expands and thus achieve a lower state of entropy. The Benson expansion might then not occur at all because of the heat sink this overcompressed matter represents.

The RS CVn systems have received considerable attention because of their soft X-ray and radio emission (papers 7.3 and 7.4). The starspot model along the lines discussed by Shore and Hall (7.2), Jakate (7.7) and Geyer (7.9) is attractive because it simultaneously provides a mechanism for the photometric irregularities and the non-thermal emission. This model may well be the correct description of the systems. However, I believe that several important theoretical questions must be answered. First, the large magnetic field implied by the starspot model almost certainly should involve detectable magnetic fields. This should be checked and it is conceivable that the starspot model could be severely restricted. Second, a clear mechanism favoring starspots in binary systems instead of single stars with similar rotation rates must be put forward. There are single stars with similar rotation rates but which lack the RS CVn characteristics. The suspicion in paper 7.2 is that this difference can be accounted for by a Babcock-type model. A quantitative demonstration that this idea works is presently lacking. I believe that we ought to recognize that forced non-radial oscillations are also a possible mechanism for both the photometric irregularities and the extra mechanical energy. In order for the oscillations to couple to the orbital motion, low-order g modes must be the ones involved. The square of the frequency of these modes is proportional to a mass weighted average of $(\nabla_{ad} - \nabla)$. Since most of the envelope is convective with this quantity negative, the modes must extend into the interior radiative zone and will have low frequencies. Slightly non-synchronous core rotation could be the source of energy for the oscillations. These ideas are admittedly rather speculative, but may be worth further investigation. Until it can be shown that a model of this type doesn't work, the starspot model cannot be proven by elimination.

The subject of contact binary theory has been discussed at length in the recent literature. The contact discontinuity theory by Shu and his collaborators has been the center of much of this discussion. The theory is quite startling in that it involves a temperature inversion which appears to be unphysical. The paper by Shu (9.1) does not come to grips with this particular issue because of the length of discussion which an adequate presentation would require and because the fact that the ideas are presented in earlier papers. The complexity of the hydrodynamic flow requires caution for an accurate treatment. The small scale of the temperature inversion means that the differences between isopotential, isobaric and isoentropic surfaces are crucial to the energy flow. I believe that the various disproofs of the Shu, Lubow and Anderson theory have not been adequately careful in defining appropriate closed surfaces. Shu in his review does point out other areas where the theory is inadequate. I hope that this interesting theory can be given a fair hearing and accepted or rejected for valid reasons.

Clearly, there is much which can be profitably studied in the field of interacting binaries. I have touched on only a few of the problems which I find interesting. Undoubtedly, the last chapter in the story of these systems is not yet even begun.

SUBJECT INDEX

absolute dimensions 55, 81-88, 203-204, 315
abundances 53, 172, 221-222
accretion
  observation 39, 253, 256, 258, 457
  theory 13, 139-143, 211-215, 329-332, 432-437
accretion disks
  in Algol and W Ser systems 229, 241, 243-244, 247, 249, 257-258, 268, 269-270, 287-292, 300-302
  in X-ray binaries, novae, and dwarf novae 45, 443-445, 449
  theory 156-159
Algol (semi-detached) systems
  evolution 211-216
  hydrogen lines 60, 287-292
  masses and radii 203-204
  mass flow 35, 207-210, 211-216, 217-218, 243-244, 269-270, 305-310, 579, 582
  models 152-154, 263-264
  origin 130, 132
  polarization 249-250
  spectroscopic observations 53, 233-236, 237-241
AM Her stars 318, 431, 453-464, 467-469, 471-472
Am stars 97
angular momentum 326, 445, 491-493
Ap stars 95-98
atmospheric eclipses 73-77
Be stars 60, 61, 69, 199, 235, 294-296, 316, 361-364, 367-373
black hole candidates 313-314
blue stragglers 145-148, 545
capture 146-147
cataclysmic variables (see dwarf novae, novae)
circulation by differential heating 495-499
coalescence 145-148
common envelope 145-148
contact binaries (not exclusively W UMa stars)
  evolution 127-132, 133-137, 139-143, 511-515
  models
    discontinuity theory 478-479, 481-484, 583
    thermal relaxation oscillation theory 479-484
  observations 60, 527-529
coronae 397-402
cyclotron lines 453-464
detached systems
  absolute dimensions 81-88, 577-578
  hydrogen-line photometry 60
  period changes 89-93
disks (see accretion disks)
distortion wave in light curves
  observations 392, 405-412, 419-421
  theory 389-394

dwarf novae
observations 318, 439-441, 443-445, 562
theory 155-157, 431-435, 439-441
evolution of binary systems
contact binaries 511-514
high mass systems 115-119
intermediate mass systems 109-114, 133-137, 211-215
pre-main sequence systems 149-154
W UMa systems 127-132
Wolf-Rayet binaries 170-173
flares
optical 229-232
radio 347
formation of binary systems 7-14
gas streams
in Algol systems, models 231, 305-310
in Algol systems, observations 219, 238-240, 294, 295
in Vela X-1, observations 359
globular clusters, binaries in 561-565
Hα emission
in Algol systems 225-227, 233-236, 287-292
in X Per 233, 236, 368
in HR5110 413-414
hot spots
on accretion disks 445, 459
on photospheres 308
Hubble-Sandage variables 155-159
interstellar extinction 280, 447, 536
limb darkening 76
line profiles 225-227, 359
magnetic fields 35, 308-310, 391, 393, 432-434, 437, 453-457
main-sequence binaries 81-88
mass flow (exchange and loss)
in Algols and W Ser systems 29, 207-210, 217-218, 244, 256, 259, 578-579, 582
in contact binaries 491-494, 511-515
in O-type systems 163-167, 195-197
in planetary nebulae 571-573
in radio binaries 35
in RS CVn binaries 383-387
theory 103-107, 110-111, 115-119, 123-126, 133-137, 156-157
mass ratios 163-167, 182, 492, 517-519
mixing mechanisms 214-215
M dwarfs 36, 471-472
multiple systems 55
neutron stars 316, 325, 329-334, 337-338, 364
novae 60, 444, 447, 562-565
orbits, photometric 87, 193, 413
orbits, spectroscopic 177-180, 355
O-type stars 42, 115-119, 163-167, 195-198
period changes 89-93, 383-386, 421, 506-507, 521-526

period determination 368-370
photometry
b, v 507-508
UBVRI 349-350, 361-363, 405-412, 415-418, 423, 425
Hα and Hβ 59-62, 299-302, 415-418
ultraviolet 501
planetary nebulae 571-573
planetary nuclei 567-570, 571-573
polarization
observations 249-250, 343-345, 450
theory and analysis 65-70, 71-72, 77, 343-345, 456
polars (see AM Her stars)
radial velocities 83, 177-180, 275-278, 319, 355, 367-373
radio emission 31-37, 229-232, 347, 395, 403
rings 233-234
rotation 27-29, 223, 263-264, 267-268, 485-489
RS CVn systems
hydrogen-line photometry 60, 63, 413
light curves 405-412
mass loss 383-387
radio emission 32, 36, 403
starspot model 389-396, 415-418, 582-583
ultraviolet emission 46
X-ray emission 318, 397-402
semi-detached systems (see Algol systems)
spectrum analysis
time resolved 233-236, 544-546
ultraviolet 39-50, 195-198, 271-284, 447-448, 550, 582
SS Cyg systems (see dwarf novae)
star spots 389-396, 410, 415-418, 423-426
statistics of close binaries 15-21, 23-26, 96-97, 103-107, 163-165
stellar wind 34, 45, 123-126, 163-167, 286
sun 399
supergiants 47-50, 60, 63, 316, 549-553, 555-559
supernova explosions 324-325
symbiotic systems 60, 535-542, 543-546, 547
third body 521-526
thermonuclear burning 330-332, 432-435
tidal braking 27-29
tidal distortion 36
Tikhov-Nordmann effect 424, 427
ultraviolet observations
Algols and W Ser systems 237-240, 251-259, 271-284, 582
B stars 293-297
cool binaries 46-47, 501-503
novae 447-448
O-type systems 195-198
supergiants with hot companions 47-50, 549-553, 555-558
symbiotic stars 538-540, 543-546
X-ray sources 39-46, 350-352, 467-468
V/R variations 289, 372-373

white dwarfs 139-143, 432-434, 439-440, 450
Wolf-Rayet stars 68, 116, 120-121, 169-175, 177-180, 181-184, 187-191
W Serpentis systems 252-259
W UMa systems
  mass ratios 517-519
  models 127-131, 477-484, 491-494, 495-499
  period changes 521-526
  photometry 421-422, 501-503, 505-509, 529-531
  rotation 485-489
  ultraviolet emission 46-47
X-ray binaries
  evolution and origin 323-326, 335-341
  models 375-379, 453-459
  optical photometry 349-352, 361-365
  optical spectroscopy 359, 456-464
  orbits 355-356
  survey 313-321
  ultraviolet spectroscopy 39-46, 349-352, 467-468
X-ray bursters 329-334
X-ray emission 39, 232, 392, 397-402
X-ray pulsars 41, 335-341, 361-365
X-ray sources 357, 569

INDEX OF STARS AND OTHER OBJECTS

Star Names

Algol (see β Per)
Capella (see α Aur)
Pleione (see 28 Tau)

Constellation Designations

ζ And 60
λ And 32, 35, 46, 47, 48, 60, 392, 398
Z And 260, 543-546
RT And 91, 93, 384, 385
TW And 203, 204, 209, 308
XZ And 209
KX And 251

RY Aqr 203, 204

R Aql 32
KP Aql 54
V889 Aql 288

RW Ara 209, 308
HI Ara 562
V535 Ara 527

RW Ari 563
UX Ari 32, 392, 395, 398, 399, 401, 402, 403

α Aur (Capella) 56, 398, 399
β Aur 90
ζ Aur 60
π Aur 32
WW Aur 90
AR Aur 90, 95
CQ Aur 384
EO Aur 117
IM Aur 152, 209

α Boo 550
44 Boo 46, 47, 48, 525
SS Boo 384
SU Boo 209
TZ Boo 426, 502

54 Cam 32
Y Cam 209
Z Cam 434, 444
SS Cam 384, 385, 395
SV Cam 384, 385, 412

S Cnc 53, 203, 204
RU Cnc 384
RZ Cnc 209, 384
TX Cnc 491, 506, 531
WY Cnc 384
AH Cnc 131
CY Cnc 434

RS CVn 32, 60, 63, 384, 385, 390, 392, 398, 405-410, 412

R CMa 57, 60, 61, 209
UW CMa 60, 116, 117, 195-198, 527
CW CMa 54
EZ CMa 68, 182, 183, 184

YZ CMi 32

δ Cap 69

η Car 159

γ Cas 60, 61, 69, 70
RX Cas 252, 269, 270, 541
RZ Cas 53, 60, 209
SX Cas 60, 252-261, 303
TV Cas 209
TW Cas 209
YZ Cas 54, 55
AO Cas 60, 116, 527
CC Cas 32, 33, 34, 117
HT Cas 443, 444

SV Cen 133-136, 527
BH Cen 527, 529
BV Cen 439
V747 Cen 527
ω Cen-V78 562
Cen X-3 117, 314, 315, 323, 324, 325, 335
Cen X-4 318

U Cep 53, 150, 203, 209, 234, 237-241, 243-247, 249-250, 252, 261, 264, 265, 305, 307, 208, 209, 577
RS Cep 261
VV Cep 39, 47, 48, 49, 50, 60, 549-553
VW Cep 46, 47, 48, 60, 486-489, 506-509, 525, 529, 531
XX Cep 209

CQ Cep 177, 179, 180, 182
CX Cep 182
EK Cep 53
GK Cep 17
GP Cep 116, 120, 182

UV Cet 32

Z Cha 439, 443, 444
RS Cha 93

Cir X-1 314, 347

RW Com 419-422
RX Com 505
UX Com 384

α CrB 53
σ CrB 32, 398
U CrB 53, 209, 252
T CrB 257
RT CrB 384
RW CrB 209

W Cru 252

28 Cyg 60
31 Cyg 60, 555, 557, 558
32 Cyg 60, 61, 555-558
P Cyg 60
SS Cyg 434, 444, 448
SW Cyg 209, 264, 266, 308
VW Cyg 209
WW Cyg 209
ZZ Cyg 209
CG Cyg 384, 385, 419-422
CH Cyg 60, 550
CI Cyg 546
EM Cyg 443, 444
MR Cyg 209
V367 Cyg 252, 253, 260
V380 Cyg 60
V382 Cyg 527
V442 Cyg 54
V444 Cyg 75, 77, 78, 116, 182
V448 Cyg 117
V548 Cyg 209
V729 Cyg 527
V1329 Cyg 547
Nova Cyg 1975 60
Nova Cyg 1978 60, 448, 450

Cyg X-1 (HD226868) 39, 40, 41, 57, 60, 61, 117, 314, 315, 323, 343-345, 355-356, 517
Cyg X-2 316, 317, 319, 321, 323
Cyg X-3 223

W Del 209

TW Dra 203, 204, 209, 213, 308
WW Dra 384
AI Dra 53, 60, 209, 213
BY Dra 60, 63

S Equ 209

RZ Eri 32, 384
AS Eri 203, 204, 209, 310

σ Gem 32, 35, 398
U Gem 434, 443, 444, 445
RW Gem 209
RX Gem 209, 261, 290, 308
RY Gem 203, 204, 213
YY Gem 90
AL Gem 209

X Gru 308

θ Her 60
u Her 209
Z Her 60, 63, 384, 392
TT Her 213
TX Her 91
UX Her 209
AD Her 209, 290
AH Her 434
AK Her 521
AM Her 318, 319, 453, 454, 459-460, 467, 468, 471
AW Her 384
DI Her 60
DQ Her 193, 443, 444
HZ Her (also Her X-1) 39, 43, 44, 45, 239, 316, 317, 323, 325, 339, 349-353, 453
MM Her 384
PW Her 384
V338 Her 209
Her X-1 (see HZ Her)

TT Hya 54, 60, 203, 204, 252, 288-292
VZ Hya 90
EX Hya 434, 439
GK Hya 384

VW Hyi 435, 439

RT Lac 32, 33, 384, 385, 392, 395, 398, 419-422, 427
AR Lac 32, 35, 384, 385, 398, 403, 409
CM Lac 90
HK Lac 32, 398

Y Leo 213
TX Leo 60
UV Leo 90
AM Leo 60, 525

T LMi 209

δ Lib 53, 152, 208, 209

β Lyr 32, 33, 34, 35, 60, 63, 252, 253, 258, 259, 263, 271-284
FL Lyr 54
MV Lyr 454, 460, 464, 465

RW Mon 209
TU Mon 209, 213
VV Mon 384
AR Mon 209, 384
AU Mon 252, 290
IM Mon 60
Nova Mon 1975 318

θ Mus 182
TU Mus 529

U Oph 91, 92
RV Oph 209
RZ Oph 261
WZ Oph 90
V451 Oph 90
V566 Oph 60
V1010 Oph 60, 527

α Ori 32, 550
δ Ori 117
CN Ori 434
DN Ori 310

AR Pav 252, 261, 541

AG Peg 535-542
AQ Peg 264, 266
AW Peg 203, 204, 209, 290
EE Peg 54, 90
II Peg (see HD224085)

β Per (Algol) 35, 53, 54, 60, 61, 67, 68, 93, 203, 204, 208, 209, 213, 215, 217-219, 221-222, 223-224, 225-227, 229-232, 577
θ Per 60
φ Per 199, 316
48 Per 60
X Per 235-236, 316, 361-365, 367-373
RT Per 209
RW Per 252
RY Per 203, 204, 213, 264, 266
ST Per 209
IQ Per 54
IZ Per 209
KS Per 573
KT Per 434
LX Per 434

ι Psc 55
Y Psc 209, 308
SZ Psc 32, 384, 385, 395, 403, 415-418
UV Psc 32, 384

UZ Pup 17
VV Pup 453, 454, 456-458, 471
XY Pup 203, 204, 209
XZ Pup 209
AU Pup 527
KQ Pup 60, 550

RZ Pyx 527
TY Pyx 384

U Sge 53, 152, 203, 209, 264, 265, 267, 301, 302
V Sge 17
UU Sge 572
WZ Sge 60, 62, 439, 441, 442, 444, 447-450
FX Sge 562

μ Sgr 60
υ Sgr 60, 272-285
ψ Sgr 55
V Sgr 573
XZ Sgr 209, 306, 308, 310
V356 Sgr 209, 252, 363
V505 Sgr 152, 209
V1148 Sgr 562, 563
V1434 Sgr 562
V1555 Sgr 562
V2359 Sgr 562

α Sco 32, 35, 550
μ' Sco 209
T Sco 562
V393 Sco 290
V453 Sco 116
V701 Sco 137, 527, 529
V861 Sco 321
V906 Sco 54
Sco X-1 317, 323
Sco X-2 323

RT Scl 137

RZ Sct 252, 264, 265, 267

W Ser 60, 62, 252-255, 258, 259, 261
CV Ser 116, 182
FH Ser 448

λ Tau 60, 209, 293, 296
28 Tau (Pleione) 60
RW Tau 209, 213, 233-235
V711 Tau (see HR1099)

X Tri 151, 209
RW Tau 443-444

W UMa 60, 501, 502, 525
RW UMa 384, 398
TX UMa 54, 60, 209
UX UMa 443
VV UMa 209
XY UMa 90, 384, 385, 423-426
AN UMa 453, 471
AW UMa 60

ε UMi 60
W UMi 151, 209
RS UMi 384

γ Vel 116, 182
Vel X-1 (see HD77581)

α Vir 35

Z Vul 209
RS Vul 203, 204, 209
BE Vul 209
ER Vul 384, 486-488

HR Numbers

HR 266 56
HR 1099 (V711 Tau) 32, 35, 46, 47, 48, 60, 61, 63, 395, 398, 403
HR 2130 56
HR 2142 293-297
HR 4665 46, 47, 48, 398
HR 5110 32, 60, 398, 403, 413, 414
HR 7084 293, 296, 297

HD Numbers

HD 108 63
HD 1337 (see AO Cas)
HD 5303 398
HD 17034 (see RY Per)
HD 19820 (see CC Cas)
HD 25487 (see RW Tau)
HD 30353 (see KS Per)
HD 34333 (see EO Aur)
HD 36486 (see δ Ori)
HD 39937 386
HD 47129 60
HD 50896 (see EZ CMa)
HD 51480 251
HD 57060 (see UW CMa)
HD 58713 (see RY Gem)
HD 60414-5 (see KQ Pup)
HD 72754 251
HD 77581 (Vela X-1) 39, 41, 42, 43, 44, 51, 117, 314, 315, 359
HD 90657 116, 177, 178, 180
HD 92740 182, 187, 190
HD 93033 (see TX UMa)
HD 93205 117
HD 94546 177, 178, 180, 182
HD 97152 182
HD 101379 386
HD 137569 575
HD 139319 (see TW Dra)
HD 149404 116
HD 152218 117
HD 152248 117
HD 152270 116, 182
HD 153345 (see AI Dra)
HD 153919 (4U 1700-37) 42, 117, 314, 315, 323
HD 155555 386, 398
HD 159176 117
HD 163181 (see V453 Sco)
HD 165052 117
HD 165590 55
HD 166734 116

HD 167771 117
HD 168206 (see CV Ser)
HD 174459 386
HD 186943 116, 182
HD 190967 (see V448 Cyg)
HD 191201 117
HD 192163 182, 183, 184
HD 192641 187-189
HD 193077 187-189
HD 193576 (see V444 Cyg)
HD 193928 182
HD 197406 182, 183, 184, 185, 190
HD 203345 55
HD 206267 117
HD 211853 (see CP Cep)
HD 216489 32
HD 224085 (II Peg) 32, 60, 63, 398
HD 226868 (see Cyg X-1)
HD 228766 117, 120, 182
HD 228854 117
HD 311884 117, 179, 180, 182, 185

Durchmusterung Numbers

BD - 2°2331 (see TU Mon)
BD + 17°3117 (see TT Her)
BD + 26°1981 (see Y Leo)
BD + 40°4220 117
BD + 61°1211 398

CoD - 42°14462 440

Other Designations

A 36 568, 569

AO 0620-003 (see Nova Mon 1975)

A 0535-26 335, 338, 339
A 0538-66 357
A 1118-61 339

B1985 (see KQ Pup)

BV 382 (see GK Cep)
BV 845 (see V747 Cen)

GX 339-4 314

He 2-36 568
He 2-131 568
He 2-138 568

LMC X-1 323, 357
LMC X-2 357
LMC X-4 117, 314, 315, 357

M5-V101 562
M14-Nova 1938 568
M30-V4 562
M33-Var A 159
M71-V3 562

NGC 3201-V65 562
NGC 6712-V9 562

NGC 246 568
NGC 1360 568, 572
NGC 1535 568
NGC 2346 568, 572
NGC 3132 568
NGC 4361 568
IC 418 568

SAO 015338 (see BD + 61°1211)

SMC X-1 117, 314, 315, 323

SS 433 319, 375-379, 577

4U 0115+63 316, 339
4U 0900-40 323
4U 1223-62 316
4U 1538-52 314, 315
4U 1626-67 318
4U 1700-37 (see HD 153919)
4U 2129+47 316

The manufacturer's authorised representative in the EU is Springer
Nature Customer Service Centre GmbH, Europaplatz 3, 69115 Heidelberg,
Germany. If you have any concerns regarding our products, please
contact ProductSafety@springernature.com

Printed and bound by CPI Group (UK) Ltd, Croydon, CR0 4YY
25/09/2026
02222839-0020